Microreaction Technology

Springer-Verlag Berlin Heidelberg GmbH

M. Matlosz · W. Ehrfeld · J. P. Baselt (Eds.)

Microreaction Technology

IMRET 5:
Proceedings of the Fifth International Conference
on Microreaction Technology

With 405 Figures and 41 Tables

 Springer

Editors:

Prof. Dr. Michael Matlosz
CNRS-ENSIC
Laboratoire des Sciences du Genie Chimique
1, rue Grandville
54001 Nancy, France

Prof. Dr. Wolfgang Ehrfeld
Kehlweg 22
55124 Mainz, Germany

Prof. Dr.-Ing. Jörg Peter Baselt
DECHEMA e. V.
Theodor-Heuss-Allee 25
60486 Frankfurt, Germany

ISBN 978-3-642-62706-4

CIP data applied for

Die Deutsche Bibliothek - CIP-Einheitsaufnahme
Microreaction technology : proceedings of the Fifth International Conference on Microreaction
Technology ; with 41 tables / IMRET 5 . M. Matlosz ... (ed.)-Berlin ; Heidelberg ; New York ;
Barcelona ; Hong Kong ; London ; Milan ; Paris ; Tokyo : Springer, 2002
     ISBN 978-3-642-62706-4   ISBN 978-3-642-56763-6 (eBook)
     DOI 10.1007/978-3-642-56763-6

http://www.springer.de

© Springer-Verlag Berlin Heidelberg 2001
Originally published by Springer-Verlag Berlin Heidelberg New York in 2001
Softcover reprint of the hardcover 1st edition 2001

Typesetting: Camera ready by authors
Cover-design: de'blik, Berlin
Printed on acid-free paper   SPIN: 10845567   62 / 3020 hu - 5 4 3 2 1 0 -

Preface

IMRET 5, the 5th International Conference on Microreaction Technology, has marked the coming of age of this recent discipline as a genuine source of original, innovative and creative concepts for the use of microtechnology in chemical and biological applications. With the onset of the new millennium, miniaturization and microstructuring of reactors and system components have gone beyond simple laboratory curiosities to become proven worktools for industrial development, and the strong, sustained participation of many active industrial and academic researchers at IMRET 5 has demonstrated the continuing vitality of the field.

Held from May 27 – 30, 2001 in Strasbourg, France, seat of the European Parliament and the Council of Europe, IMRET 5 was also an opportunity to emphasize the international character of microreaction technology, by welcoming more than 300 participants and a particularly wide range of scientific contributions from more than a dozen countries. As its predecessor conferences, IMRET 5 was organized conjointly by the Society for Chemical Engineering and Biotechnology (DECHEMA), the Institute for Microtechnology Mainz (IMM), the American Institute of Chemical Engineers (AIChE) and the Battelle Pacific Northwest National Laboratory (PNNL). For the venue of the IMRET series for the first time in France, the French National School for Advanced Study of the Chemical Industries (ENSIC–Nancy) participated as local organizer for the conference.

The conference featured more than 80 oral and poster communications covering the entire interdisciplinary field from design, production, modeling and characterization of microreactor devices to application of microstructured systems for production, energy and transportation, including many analytical and biological applications as well. In addition to the uses of microtechnology for the acquisition of chemical and biological information, a particular strong point of IMRET 5 was the investigation of the potential of microstructuring of reactors and system components for process intensification, an area of increasing importance for industrial processes, and in many domestic applications as well. Contributions to IMRET 5 also explored the perspectives of combining local, in situ, data acquisition with appropriate microstructuring of actuators and components within chemical and biological devices in order to enhance process performance and facilitate process control.

The present volume contains as far as it was successful in obtaining full papers from the presenters nearly completely all contributions to IMRET 5 and provides accordingly a substantial overview of the current state of the art in the field. The conference chairs and the international organizing and scientific committee wish to thank all IMRET 5 authors and participants for their contribution to the success of the conference. We hope that this new proceedings volume will be a useful and timely addition to the growing literature in the field of microreaction technology and a solid foundation for future developments in this rapidly expanding discipline.

Oktober 2001

M. Matlosz   W. Ehrfeld   J.P. Baselt

Table of Contents

Part 3
Mathematical Modelling and Systems Analysis for Microreaction Technology and Bioanalytical Devices

Lecture Session

Poster Session

Part 4
Characterization of Microstructured Unit Operation and Reaction Modules

Lecture Session

Poster Session

Part 5

Application of Microdevices for Production, Energy and Transportation Systems

Lecture Session

Poster Session

Part 7
Analytical and Biological Applications

Lecture Session

Poster Session

Author Index

Part 1

Keynote Lectures

Micro Fabrication for Process Intensification

Wolfgang Ehrfeld and Ursula Ehrfeld
wolfgang.ehrfeld@ehrfeld-mikrotechnik.de

Abstract: Microreaction technology is a novel approach for process intensification in chemical engineering. It allows to enhance mass and heat transfer processes, to adjust precisely initial and boundary conditions as well as residence time for continuous chemical reactions, and to realize nearly inherently safe plant concepts. A number of microfabication methods exist which cover prototyping and manufacturing of a few microchemical items as well as cost-effective mass production of integrated and modular microdevices for unit operations, reactions and control. A detailed analysis of such methods like LIGA technology, wet and dry micro-etching, mechanical micromachining and laser processing shows that a wide variety of materials is applicable to meet most requirements of chemical reactions. As a matter of fact, much effort has still to be spent to transfer the promising research results achieved to date in microreaction technology into commercial application.

1. Advanced design philosophy for chemical plants

During the past fifty years, the general technological progress has been dominated essentially by a unique strategy of success which constantly aims at comprehensive miniaturization and integration of functional elements in technical systems. The most outstanding development took place in microelectronics where meanwhile integrated circuits with hundreds of millions or even billions of transistors have become products of our daily live. More recently, micromechanical, microoptical, microfluidic, and many other microdevices have become the basis for a further multi billion dollar business, the market of microtechnology [1 - 3].

Researchers in chemical engineering are also intensively analyzing the possibilities which are potentially offered by the general strategy of miniaturization and integration to realize a radical change in design philosophy for modern chemical plants. This research and development work started at ICI in the late 1970s and the term process intensification was used to characterize this novel concept in chemical engineering. The main intention was to achieve much lower investment, operating, and maintenance costs for chemical plants without decreasing their production capacity by means of a dramatic reduction in plant size where they aimed at a reduction factor of 100 or even 1000 [4, 5].

This intention may rather look like a dream than a serious concept. However, the technological progress even in standard plant items has proven beyond doubt that this concept has a realistic basis. One may just consider, on the one hand, a standard stirred tank reactor with a cooling jacket having a volume of about 10 m^3

and, on the other hand, a potentially equivalent reactor for the same production capacity consisting of a static mixer and a compact heat exchanger having a volume of about 0.1 m^3 [6]. This simple comparison demonstrates the superiority of continuous operation over batch processing with regard to specific plant volume and its importance in process intensification. Many other potential examples exist like spinning disk reactors, vortex scrubbers, reactor-mixing systems and, of course, multifunctional reactors which integrate reaction and unit operations.

There is no doubt, that the ultimate development of process intensification leads to the novel field of microreaction technology [7 – 9]. It comprises completely different approaches in plant design like the concept of numbering-up instead of scale-up, extremely enhanced heat and mass transfer processes, strictly defined process conditions and, in particular, the possibility to utilize novel process routes. However, microfabrication methods have to be introduced into chemical engineering to profit from microreaction technology. In the following, a comprehensive analysis covering all these aspects will be given.

2. Effect of miniaturization on performance of chemical plant technology

2.1 Enhancement of heat and mass transfer processes

Diffusion, thermal conductivity, and viscosity are physically similar phenomena which involve the transport of a physical quantity through a gas or liquid. The driving forces for the corresponding transport fluxes of mass, energy and momentum are the gradients in concentration, temperature and velocity, respectively, where in all three cases the fluxes are in the same direction as the gradients. For given differences in these properties, a decrease in the characteristic dimensions results in an increase of these gradients and, correspondingly, in mass and heat transfer rates as well as in viscous losses. Accordingly, mixing and heat exchange systems with extremely high transfer rates per unit volume can be realized by miniaturization but ,on the other hand, the effect of viscous losses has to be taken into account.

With decreasing characteristic dimensions the amount of material in a system is reduced and, simultaneously, the surface area to volume ratio of the system increases. This results in a corresponding enlargement of the specific interface area, i.e. of the area per unit mass or unit volume, for transfer processes so that, in connection with the enhancement of the gradients, i.e. the driving forces for heat and mass transfer, extremely efficient mixers and heat exchangers can be realized by miniaturization.

It was meanwhile demonstrated in many cases that highly exothermal reactions can be performed under isothermal conditions using the channels of micro heat exchangers as reaction volumes [10]. Micromixers allow to achieve mixing times in the sub millisecond range. In addition, gas-liquid suspensions and liquid-liquid emulsions with extremely small bubble and droplet sizes, respectively, and high uniformity can be generated by means of such devices [11]. Accordingly, micro-mixers are promising tools to improve the performance of phase transfer and other exchange processes.

As a matter of fact, new problems arise, too. For instance, gravitational forces cannot be efficiently utilized to transport fluids at small characteristic dimensions since the effects of surface forces exceed by far those of mass or bulk forces. This is immediately evident when regarding the reflux in a distillation column or the settler in a mixer settler system. Consequently, other methods for phase separation are required for miniaturized process devices like microfiltration to break emulsions, utilization of hydrophobic and hydrophilic surfaces, application of high centrifugal forces, etc.

2.2 Consequences for selection of reaction routes, plant design and process control

The extreme enhancement in mass and heat transfer rates through miniaturization of process devices results in fundamentally novel design possibilities in view of selecting alternative reaction routes, plant design and process control. In contrast to macro devices like large stirring tanks, the starting conditions for a chemical reaction can be set precisely in respect of time and concentration because of the much faster mixing of educts in a micromixer. The reaction starts at a precisely defined position with a spatially uniform composition so that unfavorable reaction conditions due to incomplete mixing which eventually result in undesired side and secondary reactions and, consequently, losses in yield and selectivity, are minimized. Since micro reactors – except for high-throughput screening in combinatorial material research – are usually operated under continuous conditions, it is simple to adjust the optimum residence time by means of a suitable delay loop or channel which is also favorable in respect of yield and selectivity.

The high heat transfer rates achievable in micro heat exchangers and reactors allow to avoid unfavorable reaction conditions resulting from hot spots or thermal runaway effects. An optimum temperature or temperature profile for the reaction can be chosen in respect of spatial distribution and time. Thus, a fast flowing fluid element can be cooled down or heated up very rapidly in fractions of a millisecond and, because of the small thermal mass of microdevices, a periodic change of temperature of the reactor can be realized with a typical time constant of some seconds. All these examples offer possibilities to improve yield and selectivity.

The inherent advantage of precise adjustment of starting and boundary conditions for chemical reactions and unit operations in microdevices provides a novel basis for process control. Taking into account in addition the small hold-up of microreaction devices, it is evident that a nearly inherent safety for the operating part of a plant can be achieved. As a result, there is a unique chance to utilize alternative reaction routes for chemical synthesis which were not applied so far commercially for reasons of safety, difficulties in process control or because it is fundamentally impossible to realize such reaction routes using macroscopic devices. This is the case in particular for controlled reactions in the explosive regime

[12]. It is accessible by means of microreaction devices since, due to their small characteristic dimensions, they act like flame retention baffles. Moreover, the small dimensions allow to perform reactions at extremely high pressure which is of importance for chemical processes using supercritical solvents.

The safety problems connected with storage of large quantities of educts and products remain , of course, unchanged when a conventional plant is replaced by a microreaction plant with the same production capacity. Nevertheless, this problem may be reduced by replacing a large plant by several small plants for distributed production on site and on demand. In contrast to conventional plants with macroscopic process devices where scale-up usually results in a considerable reduction of specific investment costs, microreaction plants may rather profit from mass production of microdevices in reducing specific investment costs. Scale-up for achieving the desired production capacity can be done only at one site while a plant comprising a large number of chemical microdevices according to the numbering-up concept can be split for production at several sites.

There are a number of further advantages of the numbering-up concept. Research results can be faster transferred into production, plants can be constructed in a shorter time and the production capacity can be adjusted more flexibly to variations in demand. Since mass production of microdevices may result in relatively low costs per piece, novel cost saving maintenance and repair concepts based on disposable elements might be introduced.

3. Microfabrication of reaction and unit operation devices

3.1 General requirements
In order to profit from ultimate process intensification through microreaction technology, microfabrication methods have to be introduced into chemical engineering There are a number of requirements which have to be met under technological as well as financial aspects.

According to the extremely wide variety of reactions, educts, products, and process conditions, a sufficiently broad spectrum of materials is required to realize suitable microdevices for chemical processes. Metals and metal alloys, plastics, glass, ceramic materials, semiconductor materials like silicon and various auxiliary materials for sealing, surface treatment, etc. have been meanwhile successfully applied for realizing microreaction devices. Since the production of chemicals in a continuous process is inevitably connected with a transport flow of material three-dimensional micromanufacturing processes are required in order to realize sufficiently large cross-sections for channels and ducts as well as reaction volumes. In contrast to the situation some ten years ago, where essentially the LIGA technology met the requirements mentioned above, development and manufacturing of microreaction devices have become much easier today. Besides LIGA, further highly efficient technologies are meanwhile available for manufacturing high aspect ratio micro structures. Merely standard thin film methods usually applied in semiconductor technology are less suitable for the generation of three-dimensional microreaction devices but are widely applied for surface processing and protection as well as for manufacturing of sensor elements.

Besides such basic aspects concerning materials and shape of microreaction devices, costs play a major role in the selection of a microfabrication process. In this respect, the number of pieces and the precision which is really required as well as aspects like availability, manufacturing experience etc. have to be considered. Mathematical modeling of the device function may help essentially to save costs since it allows to work out more realistic specifications in regard of functional requirements. In addition, mathematical. modeling of the process sequence for microfabrication and assembly will be useful for cost saving. Such hard and soft aspects will be considered in some more detail in the following analysis of microfabrication methods for reaction devices.

3.2 Micro fabrication methods

3.2.1 LIGA Technology

Some ten years ago, the LIGA method was regarded to be the only technology to produce ultra precise micro structures with extreme aspect ratio from a wide variety of materials [13]. A promising application potential was demonstrated by a number of impressive test structures at a very early status of development during the 1980s and great expectations were placed on a fast transfer from research activities to mass fabrication of LIGA products. However, it always takes a long time until a commercial success of novel technologies is achieved and novel fields like micro reaction technology can be taken up.

LIGA technology is based on a combination of deep lithography, electroforming and moulding processes. In the first step of the manufacturing sequence, a pattern from a mask or by means of a serial beam writing process is transferred into a thick resist layer on an electrically conductive substrate. Ultra precise micro structures with extreme aspect ratio can be generated by deep X-ray lithography. However, with special epoxy resists like SU 8, which utilize intrinsic optical wave guide properties of irradiated cross-linked regions, favourable results are also achievable by means of UV lithography. In the second step, the three-dimensional relief like structure of the resist polymer generated by means of deep lithography is transferred into a complementary metallic structure by means of electroforming starting from the electrically conductive substrate. This metal structure may be the final product in some special cases.

In general, however, it is used in a third step as a master tool for a replication process like injection moulding, casting or embossing for mass fabrication of micro structures. A wide variety of mould materials can be applied for micro moulding, e.g. organic polymers, pre-ceramic polymers, ceramic and metallic powders with organic binders for subsequent sintering etc. so that most material requirements for chemical microdevices can be favourably met. Nevertheless, there are also some restrictions concerning the spectrum of materials since LIGA cannot be directly applied to generate microstructures of e.g. mono crystalline semiconductor materials and the choice of ceramic and metallic powder mould materials is rather limited.

Today, there are a number of LIGA products which evidently have promising markets in the fields of microoptics and integrated optics, molecular biotechnology, and micro actuators and promote technological development. More recently, LIGA components and systems have also been successfully applied for chemical engineering and micro reaction technology, respectively. A number of chemical companies and, of course, research institutes utilize meanwhile such devices like micromixers, micro heat exchangers, and micro bubble columns as well as modular systems with integrated functional elements for reaction, heat transfer, mixing, separation and fluid distribution for process development. LIGA devices are also seriously considered for the production of fine chemicals by chemical industry.

3.2.2 Wet and dry etching processes

Wet etching processes are widely used to produce micro structures by means of transferring resist patterns into various materials. However, for most materials only isotropic etching processes exist so that, because of lateral under-etching of the resist pattern, only shallow micro channels or other shallow structures can be generated at the surface of a bulk material. Three-dimensional structures can be manufactured when the pattern is etched completely through thin foils which have then to be stacked for the realization of deep micro channels with high aspect ratio. Isotropic etching has been applied several times for manufacturing micro-reaction devices. The technological expenditure is relatively low but there are a number of restrictions concerning accuracy, surface roughness and geometrical design.

Wet chemical anisotropic etching of monocrystalline silicon has been widely applied in microtechnology. This method is based on the dependence of etching velocity on crystal orientation so that only a few basic geometries can be realized. Besides silicon, only very few manufacturing experience exists with other monocrystalline, inevitably very expensive materials. Consequently, wet chemical anisotropic etching is in general not very attractive for manufacturing chemical microdevices because of strong restrictions in respect of shape and material. Nevertheless, the technological expenditure is low and material problems can also be solved by deposition of protection layers.

Besides anisotropic etching of mono-crystalline materials a further wet chemical etching process exists which uses a special type of photo-sensitive glass [14]. A wafer of such glass is irradiated through a mask with UV light and subsequently heated to a temperature between 800 and 900K. This results in a crystallization of the irradiated regions which can be dissolved much faster in hydrofluoric acid than the non-irradiated parts. Meanwhile, this method has been successfully applied to produce microreaction devices from glass. Restrictions of this method result from limitations in miniaturization, large surface roughness and, of course, that only one special, relatively expensive glass can be applied.

Much more precision and freedom in design of microstructures is offered by anisotropic plasma etching methods where again silicon is the most important and proven material [15]. Usually, a mask pattern is transferred into a thin layer consisting of a material resistant against plasma etching on a silicon wafer. Subsequently silicon is etched by means of a fluorine containing low pressure plasma which generates gaseous silicon compounds. The directed etching process is usually accompanied by a deposition process from the plasma where the walls oriented in parallel to the etching direction are covered with a plasma polymer resistant against the reactive plasma. By means of this combination of directed etching and side wall passivation channels and other structures with nearly vertical walls can be realized and, accordingly, extremely high aspect ratios are achievable for nearly any cross-sectional shape. This is the case in particular for the so-called advanced silicon etching (ASE) process. It is based on multiple etching and sidewall protection steps where the composition of the plasma is periodically changed using etching and plasma polymer generating gases. The etching velocity is typically in the order of 0.2 mm per hour.

The ASE dry-etching method has, of course, also specific limitations concerning e.g. material selection and surface smoothness or the brittleness of silicon which makes it nearly impossible to use it directly as a mould insert in micromoulding processes. Nevertheless, it is possible to transfer ASE silicon structures into complementary metal structures by electroforming. For a number of applications, ASE is evidently a favourable alternative to LIGA in manufacturing devices for microreaction technology.

3.2.3 Mechanical micro machining

In the past few years, an impressive progress was achieved in so-called mechanical micromachining utilizing technologies based on so-called ultra-precision machining. Complex three-dimensional microstructures have been generated with shape accuracies in the sub-micrometer range by means of milling, turning and grinding [16]. Three- and five-axis ultra-precision micro milling machines are meanwhile available as commercial products. Using diamond tools, an excellent surface quality of a few nanometers rms is achievable for nonferrous materials and progress has also been made in machining stainless steel by using ultra fine grain hard metal tools and novel technologies like vibration cutting. In addition, mechanical micromachining has been successfully applied with brittle materials.

It is evident that there are nearly no limitations concerning the generation of microstructures for chemical microdevices with complex geometries, extremely high aspect ratio and high precision from a wide variety of materials by means of mechanical micromachining. Restrictions may rather exist in manufacturing closely packed channels or corresponding other structures because of the finite size of the tools. Also manufacturing costs may become a problem in mass fabrication but in such a case mechanical micromachining may be helpful for manufacturing mould inserts for mass fabrication by means of micromoulding. Moreover, there are further mechanical methods for high volume production like punching and embossing which have meanwhile been successfully applied in fabricating e.g. micro heat exchangers.

An interesting alternative to standard mechanical micro milling, turning, drilling and grinding methods is micro electro discharge machining (EDM) which is virtually unlimited in view of the geometrical shape of the work piece [17]. Material is removed in a discharge between the electrically conductive work piece and an electrode by small sparks in a dielectric fluid like oil or deionised water. The forces acting on the work piece in EDM are extremely low which is an important advantage in micromachining. Disadvantages of micro EDM are a relatively large surface roughness, limitations in miniaturization because of the finite size of the electrodes and the spark gap in the electrical discharge and very long machining times so that this method is essentially used to manufacture mould inserts or prototypes.

The methods of mechanical micromachining and micro EDM have been extensively applied for the fabrication of components like micro heat exchangers, mixers and reaction channels as well as chemical microsystems with integrated heat exchange, reaction, mixing and distribution elements.

3.2.4 Micro machining by means of laser radiation

Micro fabrication by means of laser radiation covers a wide range of different methods [18]. On the one hand, these are processes where material is removed in an intense electromagnetic field by melting, evaporation, decomposition, photo ablation or a combination of these phenomena. On the other hand, generating processes exist where structures are built up from liquid resins, laminated layers, or powders using e.g. photo chemically induced cross-linking of organic compounds in stereo lithography or powder solidification by laser sintering. In addition, welding by means of laser radiation is of major importance for connection and assembly of micro devices.

There are no restrictions worth mentioning concerning materials in micromachining by laser radiation which is a real advantage for chemical microdevices. Limitations rather exist to achieve critical dimensions below 10μm and low surface roughness. Removal of material is also often connected with the generation of debris which reduces accuracy. Since laser based microfabrication processes, except lithography, are essentially serial and not highly parallel machining methods, their productivity is comparatively low. Nevertheless, they offer a huge potential in rapid prototyping.

Laser based micromachining processes have been applied to date only on a relatively small scale for manufacturing chemical microdevices. This will probably change relatively soon since rapid prototyping will become more and more important for developing novel microreaction devices.

4. Conclusions

The future progress in chemical engineering will be strongly determined by process intensification through microreaction technology. It offers fundamentally novel opportunities to save direct costs in the areas development, investment, operation and maintenance as well as to reduce indirect follow-up expenditures in connection with storage, transport and changes in demand or market trends. Nearly all major chemical, chemical engineering, and pharmaceutical companies are meanwhile interested or even active in analysing the potential of microreaction technology. Moreover, there are a number of powerful three-dimensional micro fabrication technologies which should meet nearly all requirements concerning geometries as well as materials of microreaction devices in prototyping and mass fabrication.

However, the implementation of a novel technology needs time. It is necessary to prove carefully the potential advantages, to develop a sufficiently broad scientific basis, to implement reliable and cost-effective fabrication of chemical microdevices on an industrial basis, to gain experience in design, construction and operation of microreaction plants, and to demonstrate finally a real commercial success. Meanwhile, a lot of effort has gone into this direction but there is still a multitude of tasks to solve until decision makers will be completely convinced of the commercial prospects of microreaction technology to accept inevitable financial risks of technological progress.

5. References

[1] W. Ehrfeld, U. Ehrfeld, S. Kiesewalter: Progress and Profit through Microtechnologies, Proc. VDE World Microtechnologies Congress, MICRO.tec 2000, Vol. 1, p. 9 – 17 (2000)

[2] H. Wicht: Mikrosystemtechnik: eine Marktanalyse, Europäische Hochschulschriften, Reihe 5, Band 2427, P. Lang Verlag, Frankfurt (1999)

[3] Bundesministerium für Bildung und Forschung: Förderkonzept Mikrosystemtechnik 2000+, Bonn, Januar 2000

[4] A. I. Stankiewicz, J. A. Moulijn: Process Intensification: Transforming Chemical Engineering, Chemical Engineering Progress, p. 22 – 33, January 2000

[5] A. Green, B. Johnson, A. John: Process Intensification Magnifies Profits, Chemical Engineering, p. 66 – 73, December 1999

[6] M. Wood, A. Green: A Methodological Approach to Process Intensification, IchemE Symposium Series No. 144, p. 405 - 416

[7] K. F. Jensen, I-Ming Hsing, R. Srinivasan, M. A. Schmidt, M. P. Harold, J. J. Lerou, and J. F. Ryley: Reaction Engineering for Microreactor Systems, Proc. 1st Int. Conf. Microreaction Technology, p. 2 – 9, Springer, 1998

12

[8]   W. Ehrfeld, V. Hessel, V. Haverkamp: Microreactors, Ullmann's Encyclopedia of Industrial Chemistry, Sixth Edition, Wiley-VCH, 1999

[9]   K:-P. Jäckel: Microreaction Technology – Vision and Reality, Plenary Lecture, ACHEMA 2000, Frankfurt

[10]   J. Mayer, M. Fichtner, D. Wolf, and K. Schubert: A Microstructured Reactor for the Catalytic Partial Oxidation of Methane to Syngas, Proc. 3rd Int. Conf. Microreaction Technology, p. 187 – 196, Springer, 2000

[11]   H. Löwe, W. Ehrfeld, V. Hessel, Th. Richter, and J. Schiewe: Micromixing Technology, Proc. 4th Int. Conf. Microreaction Technology, AIChE Spring National Meeting, March 2000, Atlanta, GA

[12]   U. Hagendorf, M. Jänicke, F. Schüth, K. Schubert, and M. Fichtner: A Pt/Al$_2$O$_3$ Coated Microstructured Reactor/Heat exchanger for the Controlled H$_2$/O$_2$ Reaction in the Explosion Regime, Proc. 2nd Int. Conf. Microreaction Technology, p. 81 – 87, AIChE Spring Meeting, March 1998, New Orleans, LA

[13]   W. Ehrfeld and D. Münchmeyer: Three-dimensional Microfabrication Using Synchrotron Radiation, Nucl. Inst. Meth. Phys. Res. A303, p. 523 – 531, 1991

[14]   A. Freitag, T. R. Dietrich, and R. Scholz: Glass as a Material for Microreaction Technology, Proc. 4th Int. Conf. Microreaction Technology, p. 48 – 54, AIChE Spring National Meeting, March 2000, Atlanta, GA

[15]   I. W. Rangelow and R. Kassing: Silicon Microreactors made by Reactive Ion Etching, Proc. 1st Int. Conf. Microreaction Technology, p. 169 – 174, Springer, 1998

[16]   M. Weck: Ultraprecision Machining of Microcomponents, Proc. Int. Seminar on Precision Engineering and Microtechnology, M. Weck ed., July 2000, Aachen

[17]   F. Michel, W. Ehrfeld, O. Koch and H.-P. Gruber: EDM for Microfabrication – Technology and Applications, , Proc. Int. Seminar on Precision Engineering and Microtechnology, M. Weck ed., July 2000, Aachen

[18]   E: Bremus, A. Gillner, D. Hellrung, H. Höcker, F. Legewie, R. Poprawe, M. Wehner, and M. Wild: Laserprocessing for Manufacturing Microfluidic Devices, Proc. 3rd Int. Conf. Micro reaction Technology, p. 187 – 196, Springer, 2000

Microstructures for SMART reactors : Precision performance in industrial production

Michael Matlosz, Sabine Rode, Jean–Marc Commenge
Laboratoire des Sciences du Génie Chimique
ENSIC–Nancy, CNRS–INPL
F–54001 Nancy, France

The increasing availability of low–cost microstructured components and devices over the last three decades has led to revolutionary advances in electrical, optical and mechanical systems, and the recent extension of microtechnology to chemical and biological applications is at the heart of economic development in an ever larger number of industrial sectors. For the process engineer, the advent of microstructured unit operation and reaction modules has provided new possibilities for industrial innovation that may ultimately lead to significant changes not only in the processes themselves but also in the way in which process technologies are conceived, designed and developed. On the basis of recent advances in the use of microstructured components for process miniaturization and process intensification, the present communication seeks to define future directions for process engineering in terms of "smart reactors", a new generation of highly integrated, adaptive and "intelligent" devices for precision performance in chemical production.

Smaller : process miniaturization

The strong impetus for the development of "microreactors" (i. e., microstructures for chemical and biological applications) results from the clear advantages of miniaturization in many application areas where only very small quantities of chemical and biological substances are available or desirable. Portable energy sources, such as fuel cells for personal computers, and small–scale analysis systems, such as those required for combinatorial chemistry or medical diagnosis, are typical examples of current industrial applications of microtechnology in this area. Distributed and delocalized production of toxic or hazardous chemical intermediates with miniplant technology is another area in which microstructured components play a major role [1], and applications abound in many other areas including artificial organs, high throughput screening and "lab on a chip". A recent review of the major developments in the field of microchemical systems can be found in [2]. In all of these application areas, microstructured devices and systems are advantageous in large part simply because they are smaller.

Better : process intensification

The movement of microfabrication methods from silicon–based technology into plastics, glasses, ceramics and metals has enabled the use of microstructured components in applications involving a large variety of aggressive and corrosive process fluids, and under highly demanding operating conditions. Furthermore, process engineers have realized that the advantages of microstructured components are not limited to simple miniaturization, and a number of highly promising applications involving moderate and, in some cases, even large quantities of matter and / or energy have begun to emerge in the rapidly growing field of process intensification [3]. For process intensification, microstructures are not simply employed to make devices smaller, but rather to make them better.

Compact *macro*–devices containing engineered microstructures with characteristic dimensions on the order of typical momentum–, heat– and mass–transfer boundary layers can provide significant gains in performance, even for existing technologies, in many industrial and domestic applications. There is little doubt, therefore, that microtechnology will continue to play a major role in process intensification, leading not only to more compact but also to more economically efficient and reliable process devices.

In chemical process applications, a number of encouraging results have already been reported on the pilot scale for improvement of existing industrial designs by simple addition of microstructured components such as microstructured static mixers [4] or microstructured heat exchangers [5] upstream of traditional chemical reactors.

The specific topological features of micromachined static mixers generate substantial decreases in mixing times by reducing diffusion lengths through intimate contact of fluid streams, while at the same time maintaining laminar flow conditions and avoiding turbulence. Efficient rapid mixing under laminar flow conditions in microstructured devices not only provides considerable reduction in energy consumption, but the precise control of local micromixing and fluid contacting that can be obtained with the devices can be extremely effective for maintaining narrow weight and size distributions in many polymerization, emulsion and precipitation processes.

A number of specifically designed microstructured demonstration units has been reported in the literature for targeted applications. A microstructured gas–liquid reactor with integrated heat exchange (Figure 1), for example, has been developed for direct fluorination of aromatics [6], a reaction pathway that is impossible to employ safely with conventional designs. A key feature in this regard is the ability of microstructured heat exchangers to provide a particularly short thermal diffusion length (rapid removal of reaction heat) while at the same time avoiding contact between the process and heat–transfer fluids. In these applications, not only is the small scale an advantage, but also the specific topological features of micromachined objects can be of interest, as illustrated by recent studies of nanostructured surface treatments for catalyst supports on the walls of accurately machined microchannels [7]. Numerous possibilities for performing reactions in highly unusual operating regimes (for example, with explosive reactant mixtures [8, 9] or with periodic modulation of inlet composition [10]) have also been proposed.

In all of these areas, the development of relevant demonstration devices for experimental study, combined with theoretical analyses to determine reliable design rules and evaluation criteria, is an urgent necessity and should be actively pursued.

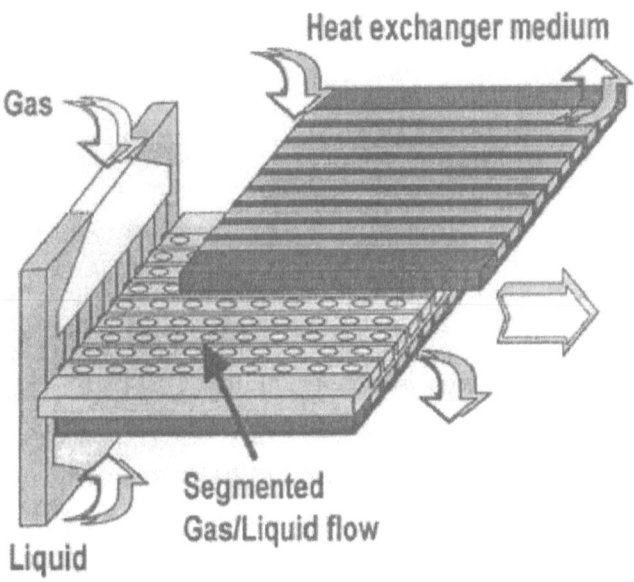

Figure 1. Process intensification. Microstructured gas–liquid reactor for direct fluorination of aromatics. (Source : IMM–Mainz)

The existence of these new technologies suggests strongly that traditional concepts in chemical process engineering will need to be re–examined carefully in the next few years, and that new process design concepts are likely to emerge. In some cases, the validity of traditional chemical engineering scaling laws may need to be re–evaluated [11], and the nature of scale–up itself may change.

The concept of "numbering up", for example, inspired by analogy with integrated circuit technology and discussed by many workers [12], seeks to replace traditional "scale–up" by a new approach involving "assembly" of a large number of elementary "micro"–modules to create a large–scale "macro"–device. As pointed out by Jensen [13], few efforts at real numbering up have been attempted, suggesting that the necessary assembly may constitute in itself a serious limitation. In this regard, research efforts concerning assembly of microreactor units into larger–scale devices (Figure 2) should offer insight into the nature of the difficulties likely to be encountered.

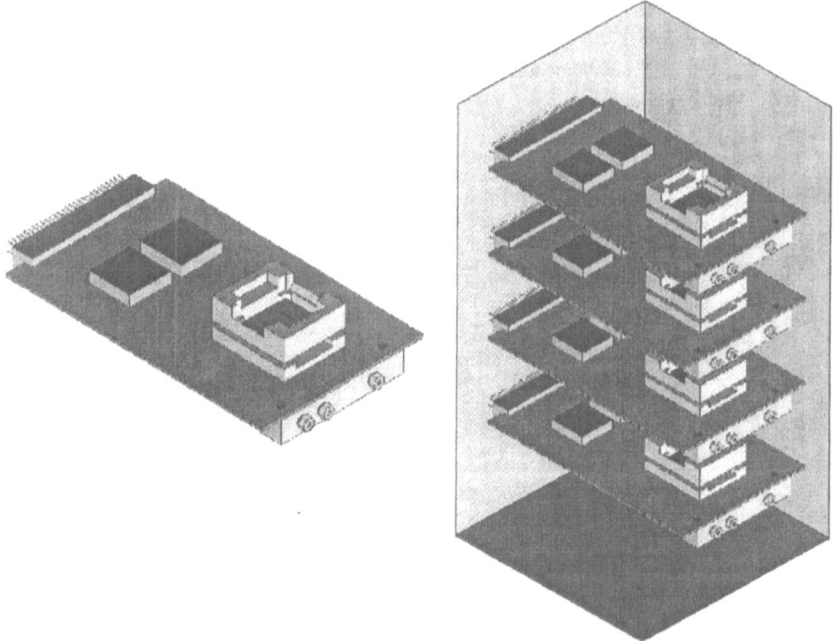

Figure 2. Numbering–up assembly. Macroreactor device constructed by assembly of microstructured building blocks. (Source : MTL––MIT)

In fact, although direct application of numbering up by a simple "LEGO" approach is unlikely to be directly operable, the concept of numbering up as a paradigm for process design has inspired much recent research. Rather than breaking down (or "deconstructing") a macrodevice into discrete microelements for analysis (traditional chemical engineering), a radically different approach involves building up (or "constructing") the macrodevice from elementary micro–scale building blocks for optimal performance. The appropriate design rules for such multi–scale devices, as well as the methods for "scale interconnection" that will be required for their construction, pose challenging theoretical questions that need to be addressed and investigated.

For the simplest case, scale–interconnection rules can be taken to be scale–independent. Macro–devices built up progressively from micro-scale components under scale–independent rules present a fractal structure, and the Devil's comb (Figure 3) proposed by Villermaux et al. [14] is a two–dimensional example of such a structure. When employed as a support for heterogeneous catalysis, the Devil's comb has been shown to exhibit particularly interesting properties of self–adaptability in the presence of catalyst poisoning that may have potential application [15].

By generalizing the approach to scale–interconnection rules that may vary from one scale to another, a new "constructal" theory for process optimization is under development that may lead to radically different process design methods and may even change the way processes are conceived and developed in the future [16].

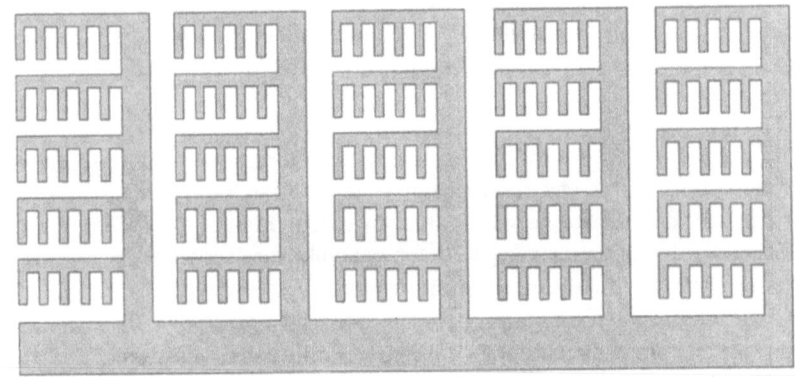

Figure 3. Multi–scale microstructured device. The Devil's comb, a two–dimensional fractal structure as catalyst support. (Source : LSGC–Nancy)

Smarter : process innovation

Innovative production methods are an essential factor in the pursuit of safe, sustainable development for the chemical and process industries, and engineered microstructures provide an extensive field of opportunity in this context for the establishment of new, creative design concepts for the chemical plants of the future.

Extrapolation of recent developments suggests that new applications of microstructured components in process engineering will involve not only miniaturization and intensification, but also advanced process control and systems design as well. This new generation of microstructured devices encourages true process *innovation*, opening the way to completely novel synthesis routes, unexplored operating regimes and dynamic operation. Microstructured reactors in the future will not only be "smaller" and "better", but also smarter as well.

Whereas current developments involving microstructured components for chemical production are essentially "passive" in nature, further conceptual advances will require a more "active" approach involving not only geometrical microstructures but also interactions between sensors, transmitters and actuators distributed in the structures. Application of these ideas suggests future directions for process development in chemical and biological reactors that go beyond "intense" or "compact" operation to include "intelligent" or "adaptive" aspects as well. Jacques Villermaux, in his address to the World Congress on Chemical Engineering in 1996 [17], proposed that increased efficiency, productivity and selectivity could be obtained through "intelligent operation and multi-scale control," an approach that he coined "Smart Chemical Engineering". "Smart" design would involve assemblies of structured, modular components and precise computer control based on information transfer between distributed arrays of local sensors and actuators.

Local, distributed process control is a key feature of "smart" reactors, and the multisectioned fixed bed electrochemical design described in [18] is an example of a demonstration device based on this approach (Figure 4). When combined with periodic operation, as suggested for example for "raster pulse electrolysis" [19], microstructured devices constitute totally new possibilities for innovative chemical synthesis. With this approach, one can envision a "programmable" reactor unit mounted directly onto a printed circuit board and providing optimal spatial and temporal profiles in reaction conditions that adjust automatically to the desired synthesis. Local process control is naturally not limited to electrochemically activated processes. Analogous systems involving optical fiber networks for photochemical reactions or distributed Peltier elements for thermally activated processes are also conceivable and deserve investigation.

19

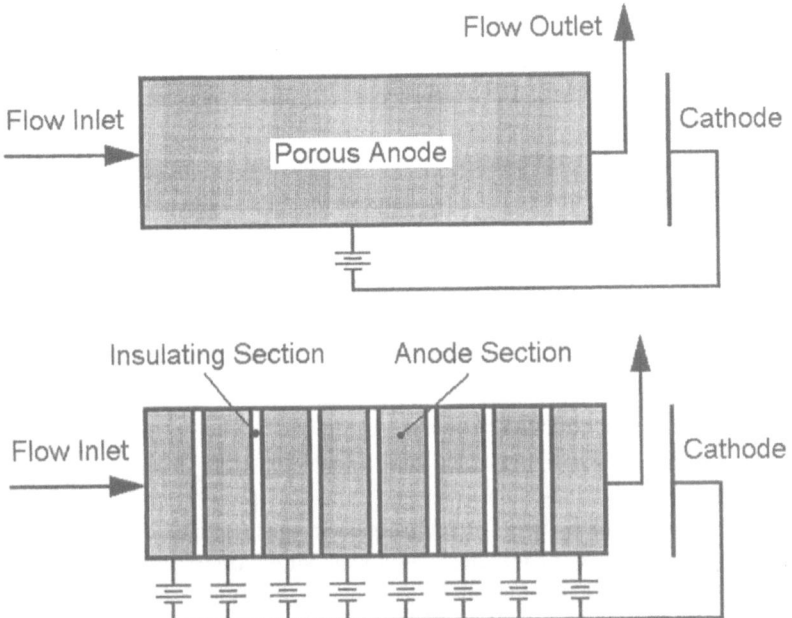

Figure 4. "Programmable" reactor with localized process control. Comparison of a traditional packed–bed electrochemical reactor and a microsectioned design.
(Source : LSGC–Nancy)

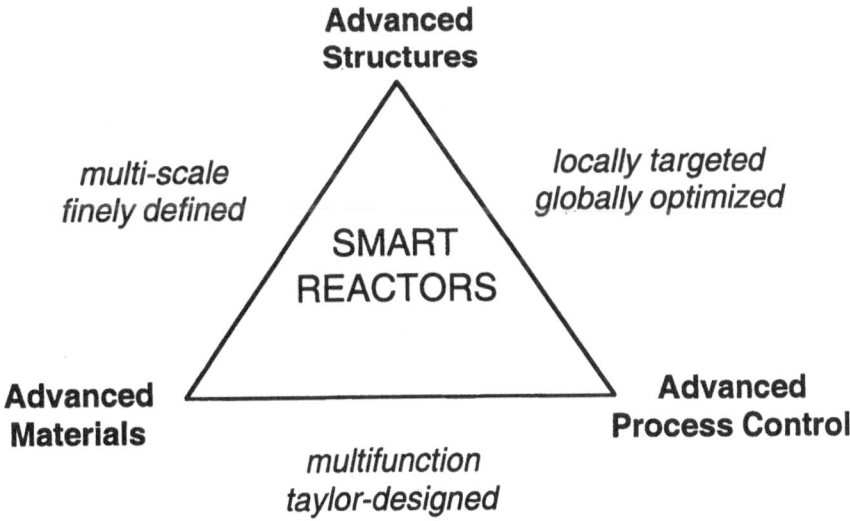

Figure 5. The essential features of SMART reactor technology.

SMART reactors : Structured, Multiscale devices for Advanced Reaction Technology

Recent reports on accelerated process development and concurrent engineering have emphasized the importance of multiscale analysis and multilevel approaches [20, 21]. SMART reactors, when viewed as a convenient acronym for "Structured Multiscale devices for Advanced Reaction Technology", build on these considerations as the basis for innovative design methods in industrial production. SMART designs are the result of increased interactions between structures (at the micro–, meso– and macro–levels) and materials (including materials for sensors, actuators and other functions), combined with localized process control (in space and in time). Figure 5 summarizes the essential features of these new, highly adaptive and efficient production systems.

Although relatively easy to imagine in principle, true industrial development and use of SMART reactor concepts must await further progress in theoretical and experimental research, combined with corresponding advances in the microfabrication methods required for their construction. Targeted laboratory and pilot–scale demonstrator devices are an urgent necessity in this regard, along with critical technico–economic analysis of potential practical applications. Despite the difficulties, prospects and potential markets for SMART reactors are considerable, and concerted research actions between industrial and academic institutions should lead to rapid advances and significant perspectives for their development in the near future.

[1] R. S. Benson and J. W. Ponton, "Process Miniaturisation. A Route to Total Environmental Acceptability ?" *Transactions of the Institution of Chemical Engineers, Part A*, **71**, 160–168 (1993).

[2] K. F. Jensen, "Microchemical Systems : Status, Challenges and Opportunities," *AIChE Journal*, **45**, 2051–2054 (October 1999).

[3] A. I. Stankiewicz and J. A. Moulijn, "Process Intensification : Transforming Chemical Engineering," *Chemical Engineering Progress*, **96**, 22–34 (January 2000).

[4] T. Bayer, D. Pysall and O. Wachsen, "Micromixing effects in continuous radical polymerization," *Proceedings of the Third International Conference on Microreaction Technology* (IMRET 3), Frankfurt/Main, Germany, April 19-21, Springer-Verlag, Berlin, 2000, pp. 165-170.

[5] H. Heinichen, I. Leipprand and T. Bayer, "Using a micro heat exchanger as diagnostic tool for the process optimisation of a gas phase reaction," *Proceedings of the World Micro-Technologies Congress*, Expo 2000, Hanover, Germany, September 25-27, VDE-Verlag, Berlin, 2000, vol. 2, pp. 493-497.

[6] V. Hessel, W. Ehrfeld, K. Golbig, V. Haverkamp, H. Löwe, M. Storz, Ch. Wille, A. E. Guber, K. Jähnisch and M. Baerns, "Gas/liquid microreactors for direct fluorination of aromatic compounds using elemental fluorine," *Proceedings of the Third International Conference on Microreaction Technology* (IMRET 3), Frankfurt/Main, Germany, April 19-21, Springer-Verlag, Berlin, 2000, pp. 526-540.

[7] G. Wießmeier and D. Hönicke, "Microfabricated components for heterogeneously catalysed reactions," *Journal of Micromechanics and Microengineering*, **6**, 285–289 (1996).

[8] R. Srinivasan, I.-M. Hsing, P. E. Berger, K. F. Jensen, S. L. Firebaugh, M. A. Schmidt, M. P. Harold, J. J. Lerou and J. F. Ryley, "Micromachined reactors for Catalytic Partial Oxidation Reactions," *AIChE Journal*, **43**, 3059–3069 (1997).

[9] U. Hagendorf, M. Janicke, F. Schüth, K. Schubert and M. Fichtner, "A Pt/Al$_2$O$_3$ coated microstructured reactor/heat exchanger for the controlled H$_2$O$_2$-reaction in the explosion regime", in *Process Miniaturization : Proceedings of the Second Int. Conf. on Microreaction Technology* (IMRET 2), Editor : American Institute of Chemical Engineers (AIChE), 81–87 (1998).

[10] A. Rouge, B. Spoetzl, K. Gebauer, R. Schenk and A. Renken", "Microchannel reactors for fast periodic operation : the catalytic dehydration of isopropanol," *Chemical Engineering Science*, **56**, 1419–1427 (2001).

[11] A. R. Oroskar, K. Van den Busche and G. Towler, "Scale Up vs Numbering Up – Can miniaturization change the rules in chemical processing,", *Proceedings of the World Micro-Technologies Congress*, Expo 2000, Hanover, Germany, September 25-27, VDE-Verlag, Berlin, 2000, vol. 1, pp. 385-392.

[12] W. Ehrfeld, V. Hessel and H. Löwe, *Microreactors : New technology for modern chemistry*, Weinheim, Wiley–VCH (2000).

[13] K.F. Jensen, "Microreaction engineering – is small better ?" *Chemical Engineering Science*, **56**, 293–303 (2001).

[14] J. Villermaux, D. Schweich and J. R. Authelin, "Le "Peigne du Diable", un modèle d'interface fractale bidimensionnelle," *Comptes Rendus de l'Académie des Sciences de Paris*, **304**, série II, n° 8, 307–310 (1987).

22

[15] P. Mougin, M. Pons and J. Villermaux, "Reaction and diffusion at an artificial fractal interface : evidence for a new diffusional regime," *Chemical Engineering Science*, **51**, 2293–2302 (1996).

[16] A. Béjan and D. Tondeur, "Equipartition, optimal allocation and the constructal approach to predicting organization in nature," *Revue Générale de Thermique*, **37**, 165–180 (1998).

[17] J. Villermaux, "New Horizons in Chemical Engineering," *Proceedings of the World Congress on Chemical Engineering*, San Diego, California (USA), pp. 16–23 (1996).

[18] C. Vallières and M. Matlosz, "A Multisectioned Porous Electrode for Synthesis of D–Arabinose," *Journal of the Electrochemical Society*, **146**, 2933–2939 (August 1999).

[19] M. Matlosz, "Electrochemical Engineering Analysis of Multisectioned Porous Electrodes," *Journal of the Electrochemical Society*, **142**, 1915–1922 (June 1995).

[20] O. Wörz, K. P. Jäckel, Th. Richter and A. Wolf, "Microreactors, a new efficient tool for optimum reactor design," *Chemical Engineering Science*, **56**, 1029-1033 (2001).

[21] J. J. Lerou and K. M. Ng, "Chemical Reaction Engineering : A Multiscale Approach to a Multiobjective Task," *Chemical Engineering Science*, **51**, 1595–1614 (1996).

Part 2

Design and Production of Microreactor Devices and Systems

Microfluidics and Microtechnology for Microreactor Systems

N.R. Tas, R.E. Oosterbroek, T.T. Veenstra, M. Elwenspoek and A. van den Berg
MESA+ Research Institute, University of Twente, P.O. Box 217, 7500 AE
Enschede, The Netherlands. Fax: +31-53-4893343.
E-mail: A.vandenberg@el.utwente.nl

1. INTRODUCTION

In this paper we outline the evolution of the silicon-based technology for micro-chemical systems in the MESA Research Institute. There is a tendency toward the handling of smaller volumes, which leads to the development of fully integrated micro-fluidic systems, based mainly on thin film technology. Several examples of microdevices and systems will be discussed.

2. STACKING OF SILICON AND GLASS WAFERS

The history of micro-liquid handling systems at Twente University goes back to the eighties, with the development of the piezoelectrically driven micropumps of Smits et al. [1] and van Lintel et al. [2]. Both pumps were based on stacking of glass-silicon-glass wafers, to create closed chambers and channels. An important step in system integration was the development of a micro-liquid dosing system [3], consisting of a pump and a flowsensors (fig. 1), fabricated simultaneously in glass-silicon-glass technology [3].

In order to support this bulk micromachining technology, our micromechanics group investigates etching of silicon, and both anodic and fusion bonding on a more fundamental level. In collaboration with the solid state chemistry group of the University of Nijmegen, physical-chemical theories of anisotropic and isotropic wet-chemical etching of silicon are developed. The results of this project include a simulation tool for anisotropic wet chemical etching [4], as well as an explanation for the formation and stabilization of pyramidal etch hillocks on Si{100} [5]. Another result of the more fundamental look at KOH etching is a design method to create thin {111} plates in <100> oriented silicon (fig. 2). Although the processing involves doublesided etching, precision alignment is only required at one side of the wafer [6]

Wafer Bonding Technology

To understand the conditions for fusion bonding, a fundamental study has been done into the bondability of wafers as a function of surface roughness, elastic and adhesive properties of the surfaces. The result is a model that predicts bondability with a single parameter, the dimensionless surface adhesion parameter [7]. With the knowledge of the relation between surface roughness and bondability, it is

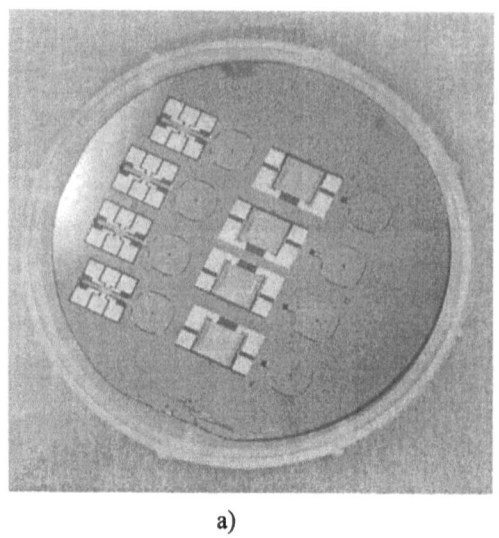

a)

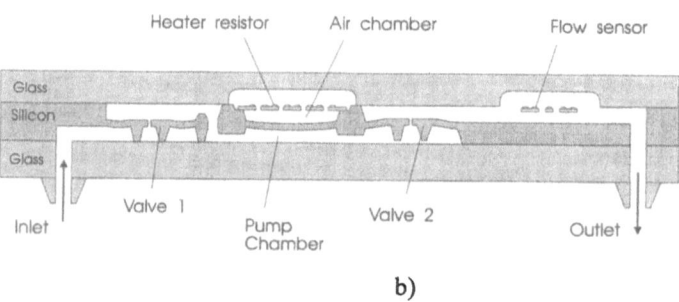

b)

Figure 1: a) Photograph of a wafer containing four micro-liquid dosing system [3]. b) Cross section of a dosing system, consisting of a peristaltic pump and a thermal flowsensor.

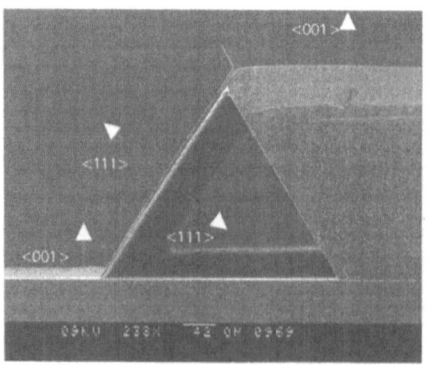 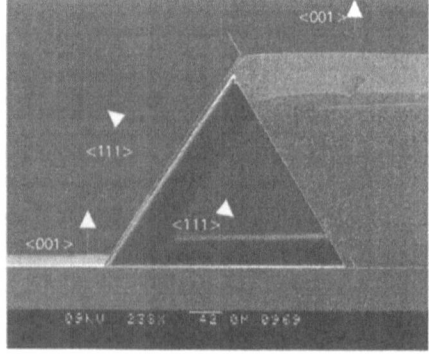

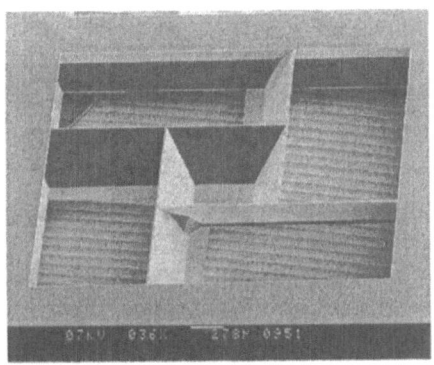

Figure 2: Thin {111} plate after fusion bonding to create a closed channel, magnification of the {111} plate, and 3D-structured rosette [6].

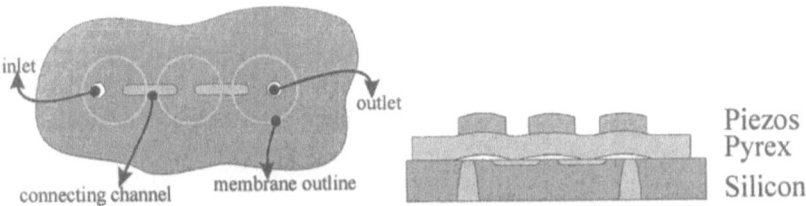

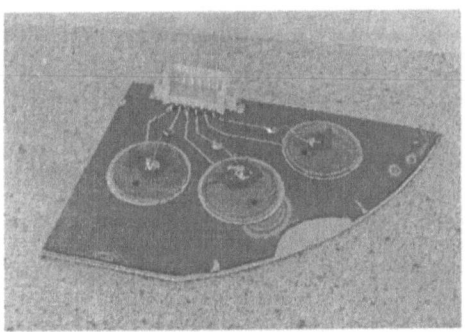

Figure 3: Peristaltic micropump. From left to right: top view and cross section showing the basic design, and photograph of a realized pump. In the cross section the pump membranes are shown in the activated upward position.

possible to obtain selective bonding: In regions with (tailored) increased surface roughness bonding can be prevented [8]. This technique can be used for example to prevent bonding of a bossed type valve seat to the counter surface [8].

Robust Peristaltic Pump

An improved peristaltic piezoelectrically driven micropump has been made using silicon-glass sandwich technology [9]. The improvement, compared to previously presented peristaltic micropumps, is that it is self-priming and it can pump any gas bubbles present in the liquid. The design and fabrication are extremely simple: the process is robust and in principle needs only one lithographic step. In order for a pump to be bubble tolerant, this compression ratio should be larger than 0.08. In the design of the peristaltic pump, the compression ratio is chosen to be 0.5. This results in a maximum pressure of 500 mBar. The design of the micropump is shown in fig. 3. It also shows a picture of a realized pump. The membranes are created from a 220 ± 10 μm thick pyrex wafer, which is anodically bonded to the silicon substrate wafer. In order to have low-dead volume pumping chambers, a selective bonding method has been used, in which the membranes are coated with chromium to prevent bonding locally [10]. The pump produces a maximum flow of 9 μl/min, at a stroke frequency of about 400 Hz [9].

3. MODULAR FLUID SYSTEMS (MFS)

The modular fluid system (MFS) concept has been developed in order to offer a modular approach to complex microfluidic system design. The MFS concept consists of a mixed circuit board (MCB) containing fluid channels as well as the

Figure 4: MFS demonstrator. It consists of two inlets, two flowsensors and two pumps, a mixing unit and a detection unit. The fluidic system is connected to the control electronics which are collected on the lower circuit board [11].

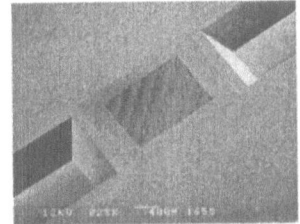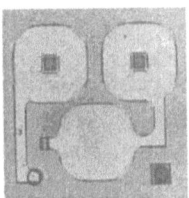

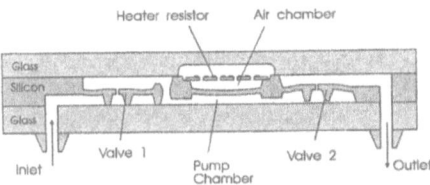

Figure 5: Close-ups of several silicon-based modules. From left to right: A thermal flowsensor, a filter module, a membrane pump and a cross section of such a pump [11].

electrical circuitry, in combination with silicon based modules of standard size [11]. The modules have standardized connections to the planar MCB both for fluids and for electrical signals. Figure 4 shows a demonstrator of a complete system, fig. 5 shows close-ups of several system components.

4. ELECTRO-OSMOTIC CONTROLLED MICROREACTORS

Synthesis of commercial chemical products like medicines takes a long and expensive development time. The most time-consuming step is finding the optimal mass ratio of the ingredients which results in the cleanest reaction product. Till now this screening step was worked-out by carrying-out many reactions by hand. or in a semi-automated way were titre plates were filled by robots. After the right set-points for the processes are found and some small prototype reaction products has been synthesized under laboratory circumstances, the experiments need to be transferred towards bigger production reactors. With microreactors this process can be strongly reduced. Since small fluid volumes are used, the reaction speed will strongly increase and thus reducing the total processing time. This means that more different reactions can be investigated per hour. This process can even be speeded-up since small systems take little space such that many systems can work in parallel. Together with computer-controlled injection and pumping systems, small, intelligent synthesis plants can be obtained which can work on for 24 hours a day. After the optimal settings have been found and prototype chemicals have been fabricated, small mass production can directly be started by using large amounts of identical microreactor chips in parallel such that no translation problem exists.

Synthesis lab-on-a-chip - present

For many microfluidic applications electro-osmotic flow (EOF) control has proven to be very effective. For the control only contact electrodes are needed which can easily be integrated in the micro-channels. By varying the voltage levels between the electrodes the flow-rate and flow direction can be controlled. Together with Avantium and the University of Hull we developed a reaction module optimized for Wittig reactions. Fig. 6 shows the module in which 3 fluids (3 entrance channels) can react to generate one reaction product. The micro reaction chip consists of $200 \times 100 \mu m$ channels fabricated by powderbasting [12] in two fusion bonded glass wafers. The integrated contact electrodes consist of sputtered platinum.

Figure 6: Microreactor chip with integrated electrodes for electro-osmotic flow control.

Synthesis lab-on-a-chip – future

The presented chip is a first step towards future miniaturized automated synthesis systems. The next step is to integrate more functionality in smaller-dimensioned chips. For the fluid control in micron or even sub-micron-sized channels EOF looks to be a promising pump mechanism. To get a more flexible design, small local EOF seems to be more suitable than EOF over the full channel length since lower voltages can be used to yield a high field strength and materials like silicon can be used which allows more fabrication flexibility. At MESA+ fluidic network systems are developed that use FlowFETs [13] to drive the fluid. These FlowFETs comprise two contact electrodes and an insulated in between to charge the channel wall such that the Zeta potential can actively be varied to adjust the flow direction and speed and higher flow-rates can be obtained at lower field strengths.

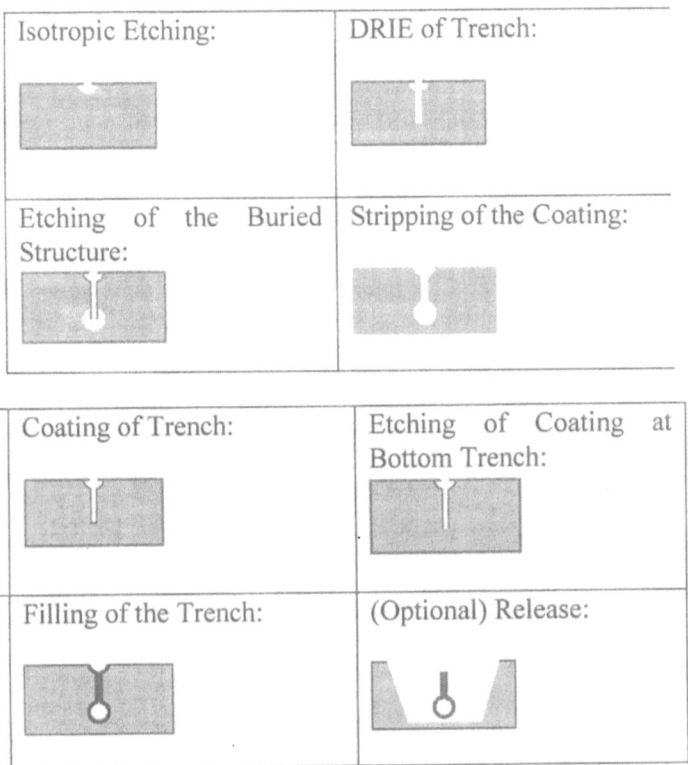

Isotropic Etching:	DRIE of Trench:
Etching of the Buried Structure:	Stripping of the Coating:

Coating of Trench:	Etching of Coating at Bottom Trench:
Filling of the Trench:	(Optional) Release:

Figure 7: Fabrication sequence of Buried Channels and Micro-Pipettes [14, 15].

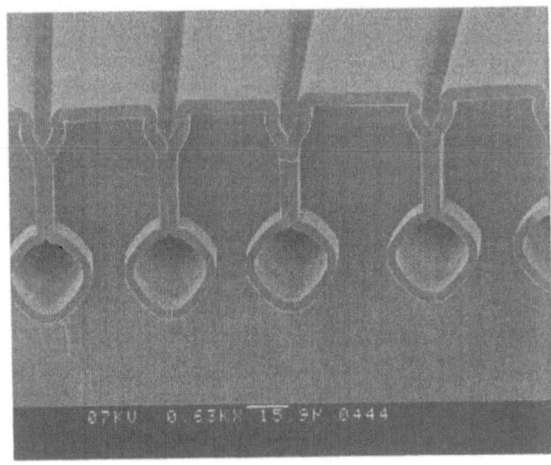

Figure 8: Buried Channels with a diameter of appr. 15 μm [14].

5. BURIED CHANNELS AND MICROPIPETTES

The Burried Channel Technology (BCT) is an alternative to channels made by bulk micromachining in combination with wafer bonding. The critical steps when using wafer bonding, alignment and void-less bonding, become problematic when channel diameters are scaled down to micrometer level. BCT is a monolithic, self-aligned fabrication strategy that avoids the critical bonding step. The basic fabrication sequence for BCT is shown in fig. 7.

Using the BCT in combination with a KOH bulk-etching step to remove part of the silicon substrate surrounding the buried channel, it is possible to make micro-pipettes with an open tip inside a flow channel [15]. Fig. 9 shows examples of fabricated micro-pipettes, which were made for handling of DNA-molecules that are connected to glass beads.

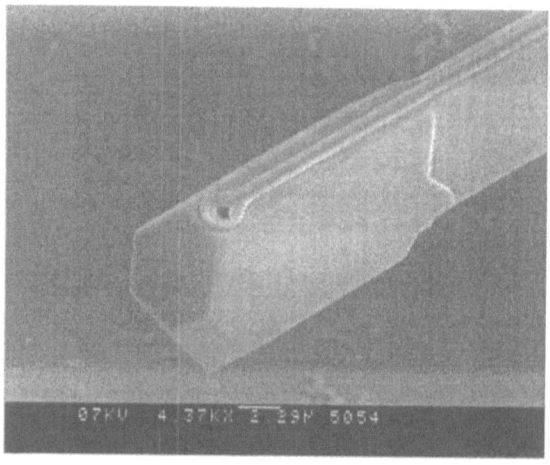

Figure 9: Micro-Pipettes fabricated using the Buried Channel Technology [15].

6. INTEGRATED FLUID SYSTEMS (IFS)

In general we aim toward the handling of smaller fluid volumes, down to the sub-nL level. Technologically this development is supported by a fabrication concept for Integrated Fluid Systems (IFS). This concept is based on deposition, patterning and selective removal of thin films. In analogy with the historical development of microelectronics, it is clear that the level of integration can be increased tremendously, at the expense of functional flexibility of the basic system components. Complex system functionality is created by integration of components with very basic functionality. Fig. 11 shows a thermal flow sensor designed for high sensitivity applications, with a flow channel made by sacrificial polysilicon etching [16]. The effective (hydrodynamic) channel diameter of such a device is typically in the order of 1 μm. Fig. 10 shows the basic steps in the sacrificial polysilicon technology (SPT).

7. MICROREACTOR IN A FLOW INJECTION ANALYSIS SYSTEM

In a Flow Injection Analysis (FIA) system we have developed for ammonia determination (the "MAFIAS" system), the ammonia is converted into a measurable colored product, indophenol, in a micro reactor. Since one of the reaction steps involved is highly temperature dependent, the temperature has to be controlled to stay within a very narrow temperature range (<0.5 C). In literature a few microfluidic heating chambers have been presented. Burns [17] presented a temperature controlled fluidic system for the pretreatment of DNA samples, whereas Miyake [18] shows us the design of a small reaction chamber, which is integrated into a small detecting system. The basic design of the reaction chamber for the MAFIAS is shown in

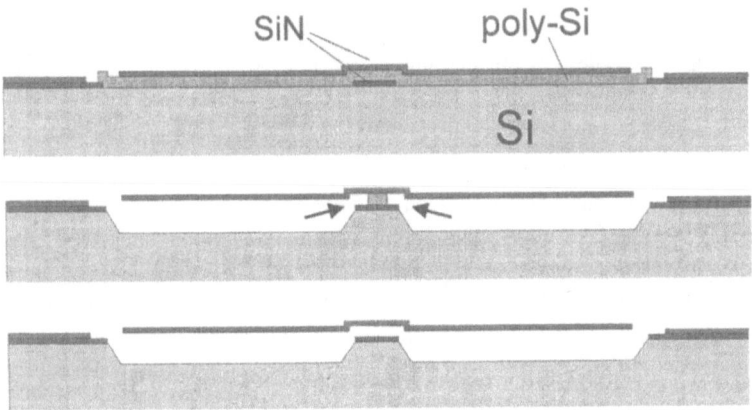

Figure 10: Principle of the sacrificial polysilicon technology (SPT). A polysilicon layer sandwiched between two silicon nitride layers is removed by KOH etching. At locations where the lower silicon nitride has been removed, the KOH also etches into the silicon substrate wafer, creating large, low resistance entrance and exit channels.

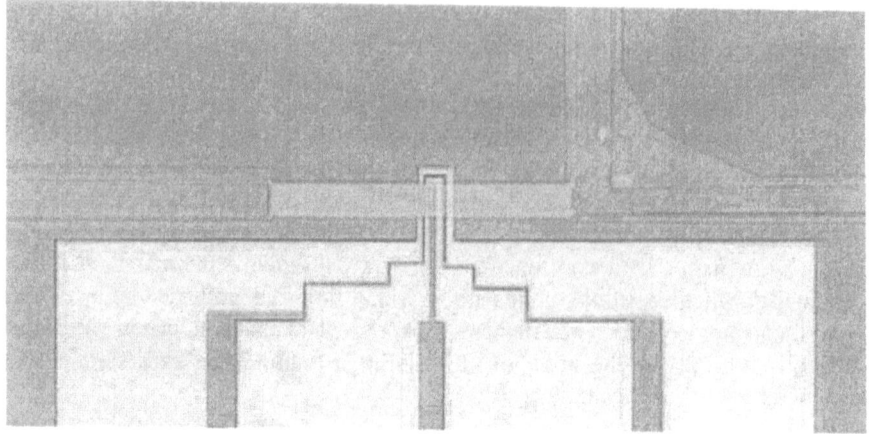

Figure 11: Microscope picture of a realized thermal flowsensor. It consists of a microchannel created by the process shown in fig. 10, in combination with two metal resistors, which both serve as heaters and temperature sensors at the same time [16].

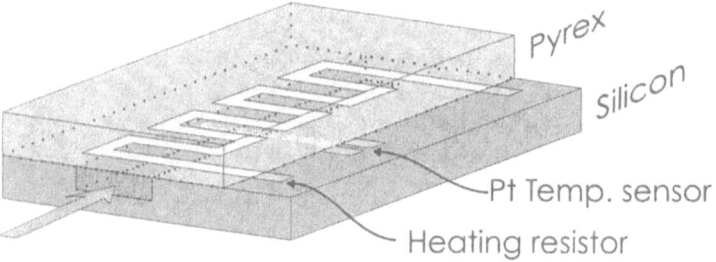

Figure 12: Principle of the design of the reaction chamber.

Figure 12. The reaction chamber will consist of a silicon base plate in which the flow channel is etched. The channel is closed with a Pyrex top plate. Thin film Platinum resistors are deposited on the bottom side of the Pyrex wafer. These resistors are used to (electrically) generate the power for heating of the fluid in the channel as well as for the measurement of the temperature of the fluid.

The silicon chip is covered with an insulating thermal oxide layer to prevent short-circuiting of the resistors through the silicon. The fluid has to be at 37 C for 30 seconds. This has been realized by choosing the channel length in such a way that the residence time of the liquid in the channel is 30 seconds. The temperature of the chip is controlled using a feedback loop. Therefore a temperature sensor is integrated on the chip. A platinum resistor in a four-point measurement configuration is used for this purpose. Temperature variations are calculated form the variations in the measured resistance via the temperature coefficient of resistivity. To estimate the required time and power for the heating of the liquid a simple one-dimensional model for the heat transfer has been made. The differential equation for the temperature distribution in a medium as a function of time and place is given by [19]:

$$\frac{\partial}{\partial t} U(x,t) = D \cdot \frac{\partial^2}{\partial x^2} U(x,t) \tag{1}$$

Where:
$U(x,t)$: the temperature at time t at position x from the heating source [K],
D: the thermal diffusion coefficient [m^2 sec]

A time constant τ [sec.] can be defined for heat diffusion over a distance d in a medium with thermal diffusion coefficient D (eq. 3) [20]:

$$\tau = \frac{d^2}{D} \tag{2}$$

This time-constant indicates the time needed for the temperature profile to travel over a distance d. After $t=\tau$ the initial temperature difference at a distance d will be reduced by 93% [21]. Table 1 gives the time constant τ for heating water for several channel depths.

Depth of the channel (d_h) [μm]	Time constant τ [msec]
100	17.9
200	71.4
400	285.7

Table 1 : Time constant τ for heating water for several channel depths

As is seen from the table, the model predicts that heating the fluid in micro channels is easily accomplished within a time of one second.

Samples were designed and fabricated in order to test the feasibility of the chosen reaction chamber principle. In figure 13 a cross-section of the reaction chamber is drawn. The long channel of the reaction chamber is KOH etched in a silicon wafer. The channel is closed by anodic bonding of a Pyrex wafer on top of the silicon wafer. The Platinum resistors that are used for the temperature measurement and heating of the chip are sputtered on the bottom side of the Pyrex chip. In order to prevent short-circuiting of the platinum resistors via the silicon, the silicon wafer has to be covered with a silicon oxide layer (wet oxidation). A silicon oxide layer (PECVD) also protects the resistors on the Pyrex wafer itself. This to ensure that a possibly conducting fluid will not short-circuit the resistors.

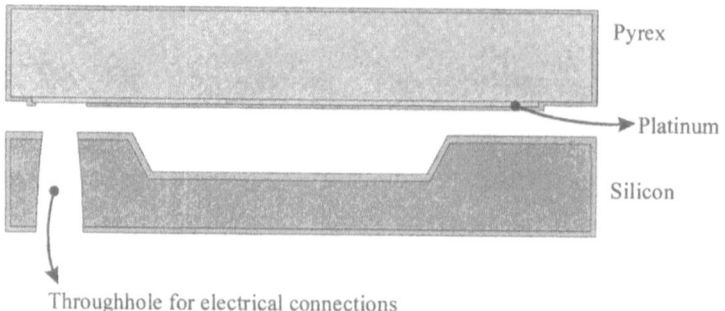

Throughhole for electrical connections

Figure 13: Cross-section of the chip design.

Access to the free-etched platinum contact pads is obtained by a powder blasted hole in the silicon wafer. The length of the channel of the reaction chamber corresponds to a sample residence time of 30 seconds for the chosen flow rate. With channel dimensions 200 200 m² and a flow rate of 10 l/min a length of 12.5 cm is desired.

Factors to be taken into account when designing the resistors for either the heater or the temperature sensor come from the control electronics and the anodic bonding process. In order to be compatible with the anodic bonding process the Platinum layer on the Pyrex can be only as thick as 50 nm. Next to this restriction it would be convenient for the control electronics if the heater doe not need more than 5 Volts of supply voltage. This results in a 400 m wide and 4.5 cm long resistor. The resistance then is around 240 Ohms (corresponding to 100mW of electrically generated heat at 5V driving voltage). The temperature sensors are designed 20 m wide and 4 cm length.

Figure 14: Photograph of the fabricated reaction chamber.

In fig. 14 one of the fabricated stand-alone reaction chamber is shown. A number of experiments were done in this reaction chamber. First, the resistor values of the temperature sensor were measured. From this sweep it is seen, that the temperature sensors behave very linear. It is also seen, that the calibration of the temperature sensors is necessary since the resistances of different temperature sensors vary easily 500 Ohms in absolute value.

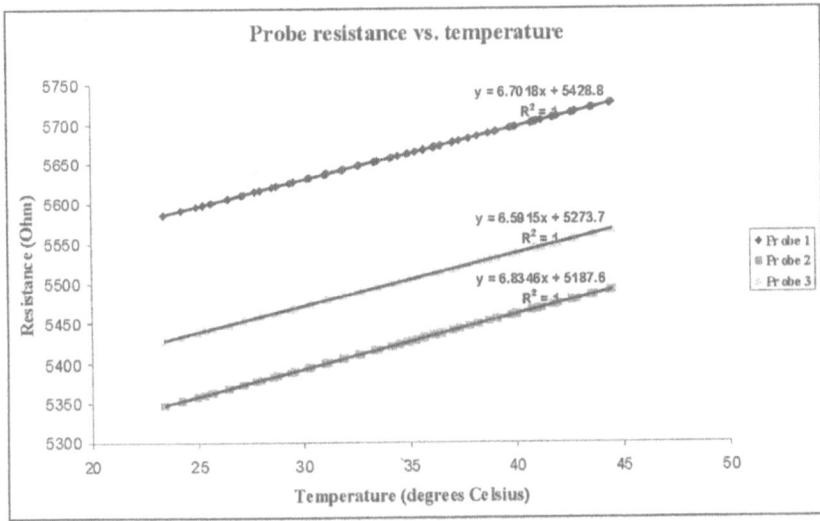

Figure 15: Measured resistance of three temperature sensors as function of temperature.

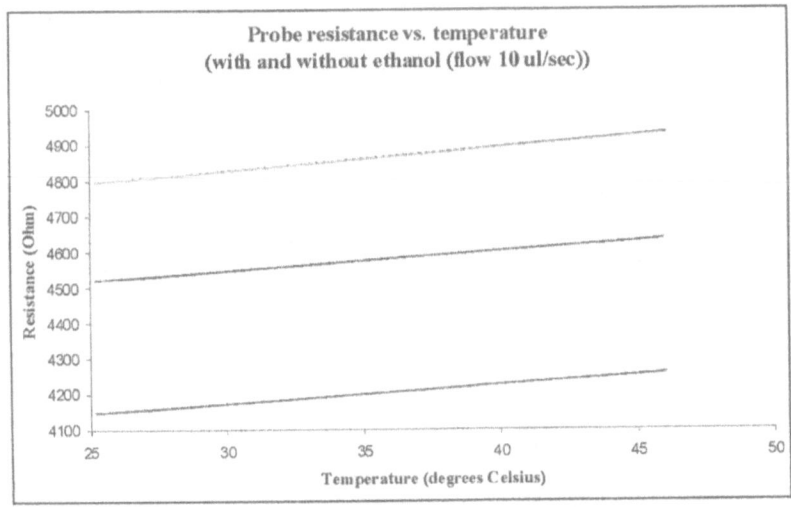

Figure 16: Test results of the reaction chamber filled with ethanol.

Figure 16 shows the results of an experiment in which liquid (ethanol) is heated inside the reaction chamber. In fig. 16 the control-loop is set to a certain temperature, after which the resistances of the temperature sensors are read. The lines seen in Figure 16 are the measurement results of the three temperature sensors located on one chip. Every line in fig. 16 consists of two lines, one for flowing ethanol (10 l/min) and one for stopped flow. Since the lines fall together it can be concluded that the reaction chamber with feedback loop keeps the temperature of the whole chip perfectly stable.

8. CONLUSIONS

The first silicon technology based fluidic systems were made by bulk micromachining in combination with wafer bonding to create closed channels and chambers. To increase the flexibility in system design, modularity was introduced, leading to a kind of surface mount technology. The critical issues when using wafer bonding, alignment and void less bonding, become problematic when channel diameters are scaled down to micrometer level. Solutions can be found in monolithic, self-aligned fabrication strategies like the BCT and SPT. For ongoing miniaturization of micro fluidic system we are looking at thin film monolithic fabrication strategies, in analogy with the past development of micro-electronics.

9. ACKNOWLEDGEMENTS

The authors acknowledge the helpful contributions of J.W. Berenschot, S. Schlautmann, R.M. Tiggelaar, R.B.M. Schasfoort, T.S.J. Lammerink, A.J. Nijdam, R.W. Tjerkstra, C.R. Rusu, R. Van 't Oever, and M.J. de Boer.

10. LITERATURE

1. Smits G.J., *A piezoelectric micropump with three valves working perestaltically*, Sensors and Actuators, Vol. 15, 1990, pp. 203-206. (NL-patent 8302860, 1985).
2. Van Lintel H.T.C., van de Pol F.C.M., and Bouwstra S., *A piezoelectric micropump based on micromachining of silicon*, Sensors and Actuators, vol. 15, pp. 153-167.
3. Lammerink T.S.J., Elwenspoek M., Fluitman J.H.J., *Integrated Micro-Liquid Dosing System*, Proc. IEEE MEMS Workshop 1993, Fort Lauderdale Fl., USA, Febr. 7-10 1993, pp. 254-259.
4. E. van Veenendaal, A.J. Nijdam, J. van Suchtelen, K. Sato, J.G.E. Gardeniers, W.J.P. van Enckevort, M. Elwenspoek, *Simulation of anisotropic wet chemical etching using a physical model*, Sensors and Actuators A, Vol. 84, 2000, pp. 324-329.

5. A.J. Nijdam, E. van Veenendaal, H.M. Cuppen, J. van Suchtelen, M.L. Reed, J.G.E. Gardeniers, W.J.P. van Enckevort, E. Vlieg, M. Elwenspoek, *Formation and stabilisation of pyramidal etch hillocks on silicon {100} in anisotropic etchants: experiments and Monte Carlo Simulations*, J. Appl. Phys., submitted.

6. Berenschot J.W., Oosterbroek R.E., Lammerink T.S.J., Elwenspoek M., *Micromachining of {111} plates in [001] oriented silicon*, Journal of micromechanics and microengineering, 1998 (nr: 8), (pp. 104-106). ISSN 0960-1317.

7. C. Gui, M. Elwenspoek, N.R. Tas, and J.G.E. Gardeniers, *The effect of surface roughness on direct wafer bonding*, J. Appl. Phys., Vol. 85, no. 10, 15 May 1999, pp. 7448-7454.

8. C. Gui, R.E. Oosterbroek, J.W. Berenschot, S. Schlautmann, T.S.J. Lammerink, A. van den Berg, M. Elwenspoek, *Selective fusionbonding by surface roughness control*, Proc. 5th Int. Symp. Semiconductor Wafer Bonding, Joint. Int. Meeting of The Electrochem. Soc., nr. 1017, 1999.

9. T.T. Veenstra, J.W. Berenschot, R.G.P. Sanders, J.G.E. Gardeniers, M. Elwenspoek and A van den Berg, *A simple selfpriming bubble tolerant peristaltic micropump*, Eurosensors XIV, 27-30 aug 2000, Kopenhagen, Denmark. In proceedings: pp371-372.

10. T. T. Veenstra, J. W. Berenschot, J. G. E. Gardeniers*, R. G. P. Sanders, M. C. Elwenspoek, and A. van den Berg, Use of Selective Anodic Bonding to Create Micropump Chambers with Virtually No Dead Volume, *Journal of The Electrochemical Society*, **148** (2) (2001) G68-G72

11. T.S.J.Lammerink, V.L. Spiering, M. Elwenspoek, J.H.J. Fluitman, A. van den Berg, *Modular Concept for Fluid Handling Systems*, Proc. IEEE MEMS Workshop 1996, San Diego, CA, USA, Febr. 11-15, 1996, pp. 389-394.

12. S. Schlautmann, H. Wensink, R. Schasfoort, *Powder-Blasting Technology as a Powerful Tool for Micro-Fabrication of CE-Chips with Integrated Conductivity Sensors*, Proc. MME 2000 Workshop, Uppsala, Sweden, Oct. 1-3 2000.

13. R.B.M. Schasfoort, S. Schlautmann, J. Hendrikse, and M. Elwenspoek, *Field-Effect Flow Control for Microfabricated Fluidic Networks*, Science, Vol. 286, 29 Oct. 1999, pp. 942-945.

14. M.J. de Boer, R.W. Tjerkstra, J.W. Berenschot, H.V. Jansen, G. J. Burger, J.G.E. Gardeniers, M. Elwenspoek, A. van den Berg, *Micromachining of Buried Micro Channels in Silicon*, J. MEMS, vol. 9, no. 1, March 2000, pp. 94-103.

15. C. Rusu, R. van 't Oever, M.J. de Boer, H.V. Jansen, J.W. Berenschot, M.L. Bennink, J.S.Kanger, B.G. de Groot, M. Elwenpoek, J. Greve, A. van den Berg, J. Brugger, *Fabrication of micromachined pipettes in a flow channel for sngle molecule handling of DNA*, Proc. IEEE MEMS Conf., Miyazaki, Japan, Jan. 23-27, 2000, pp. 429-434.

16. N.R.Tas, T.S.J. Lammerink, P.J. Leussink, J.W. Berenschot, H.-E. de Bree, M. Elwenspoek, *Towards Thermal Flowsensing With pL/s Resolution*, Proc. SPIE Conf. Micromachined Devices and Components VI (SPIE Vol. 4176), Santa Clara, USA, Sept. 18-19 2000, pp. 106-121.

17. M.A.Burns, B.N.Johnson, S.N.Brahmasandra, K.Handique, J.R.Webster, M.Krishnan, T.S.Sammarco, P.M.Man, D.Jones, D.Heldsinger, C.H.Mastrangelo, D.T.Burke, An integrated Nanoliter DNA Analysis Device, Science vol282, pp484-487, 1998.

18. R.Miyake, K.Tsuzuki, T.Takagi, K.Imai, A highly sensitive and small flow-type chemical analysis system with integrated absorptiometric micro-flowcell, proceedings MEMS, pp102-107, 1997

19. H.S. Carslaw and J.C. Jaeger, Conduction of heat in solids, Clarendon Press, Oxford, 1959

20. C.B.P. Finn, Thermal physics, 2nd edition, Chapmann & Hall, Londen, 1993

21. J.C.T. Eijkel, A. Prak, S. Cowen, D.H. Craston, A. Manz, Micromachined heated chemical reactor for pre-column derivatisation, Journal of Chromatography A 815, pp. 265, 1998.

Micromachined Straight-Through Silicon Microchannel Array for Monodispersed Microspheres

Isao Kobayashi,[1,2] Mitsutoshi Nakajima,[1,2]* Yuji Kikuchi,[1]
Kyoseok Chun[1] and Hiroyuki Fujita[3]

[1]National Food Research Institute, 2-1-2 Kannondai, Tsukuba, Ibaraki 305-8642, Japan, *Tel: +81-298-7997, Fax: +81-298-8122, e-mail: mnaka@nfri.affrc.go.jp
[2]Institute of Agricultural and Forest Engineering, University of Tsukuba, 1-1-1 Tennoudai, Tsukuba, Ibaraki 305-8572, Japan
[3]Institute of Industrial Science, The University of Tokyo, 7-22-1 Roppongi, Minato-ku, Tokyo 106-8558, Japan

An array of through-holes, referred to as straight-through silicon microchannel (MC) was developed as a novel emulsification device for preparing monodispersed microspheres (MS). Micromachining technology including deep reactive ion etching (RIE) yielded uniformly sized straight-through MC on a silicon substrate. The formation characteristics of MS were investigated using a newly fabricated straight-through MC with 17.3 µm equivalent diameter. Monodispersed oil-in-water (O/W) MS with an average diameter of 32.5 µm were successfully formed by forcing the dispersed phase into the continuous phase through an oblong cross-sectional shape of straight-through MC. Their elongated cross-sectional shape contributes significantly to the spontaneous formation of monodispersed MS without any turbulent mixing. Also, this straight-through MC allowed the stable formation of monodispersed MS in a wide range of the applied pressure of the dispersed phase.

1. Introduction

Microspheres (MS) are emulsion droplets or solid microparticles dispersed in a continuous phase; MS are greatly significant in industrial fields, such as foods, cosmetics, pharmaceuticals, and chemicals. The size and size distribution of MS critically affect the physical and qualitative stability of the MS and characterizes their properties [1]. Precisely size-controlled MS with a narrow size distribution are required to achieve better emulsion stability and to facilitate control of their properties. However, conventional mechanical emulsification devices, such as rotor-stator dispersing systems and high-pressure homogenizers, yield polydispersed MS over a wide range [1,2]. A novel emulsification that can prepare monodispersed MS is desired.

The remarkable development of micromachining technology over the last decade led to the precise fabrication of three-dimensional (3D) microstructures, termed microelectromechanical systems (MEMS) [3]. Potential chemical and biological applications of this technology have attracted interest, such as chemical and biochemical microreactors [4,5], capillary electrophoresis [6,7], DNA microchips [8-11]. Various principles of micromixers have also been developed on the basis of MEMS technology, performing laminar and fast mixing [12]. These microdevices are precisely replicated by combinations of photolithography, etching, and molding [11,13]. Better analytical and reactive performance of these microdevices can be achieved by hydrodynamics on a micron scale, which has the significant effect of interfacial tension, viscosity, and laminar flow.

Deep reactive ion etching (RIE) in MEMS fabrication techniques allows accurate etching of microholes with a high aspect ratio, in a silicon substrate. This etching technique should be applicable to the fabrication of an emulsification device consisting of uniformly sized through-holes. These through-holes are referred to as straight-through silicon microchannel (MC). A two-dimensional layout of this MC can yield a large number of channels on a plate, leading to MS formation with higher productivity. In addition, the microscope video system developed by Kikuchi *et al.* [14,15] enables real-time visual observation of the emulsification process [16].

In the present study, we developed a straight-through silicon MC plate as a novel emulsification device microfabricated by micromachining technology. The formation characteristics of MS were also investigated using a straight-through MC with newly fabricated oblong section.

2. Fabrication

2.1. Fabrication of straight-through silicon MC plate

Figure 1 schematically depicts the fabrication process of a straight-through silicon MC plate. A straight-through MC plate was microfabricated from a 4-inch silicon wafer with a 200 μm-thickness. An aluminum layer vapor-deposited on a silicon wafer works as a protective layer against deep reactive ion etching (RIE). A positive photoresist was spun on the substrate and was hardened by heating it. Designed channel patterns were transferred to the substrate via a mask and ultraviolet (UV) light as a radiation source. After this photolithography process, development, aluminum etching, and photoresist removal were performed. Deep silicon etching was performed by a deep RIE (inductive coupled plasma (ICP)-RIE, Plasma Thermo Inc.) until deep microholes had completely penetrated through the substrate. This process plays a critical role in fabricating uniformly sized a straight-through MC. Thermal oxidation was used to grow a hydrophilic silicon oxide layer to prevent wetting by the dispersed oil phase on the substrate, and then the substrate was cut in an individual chip.

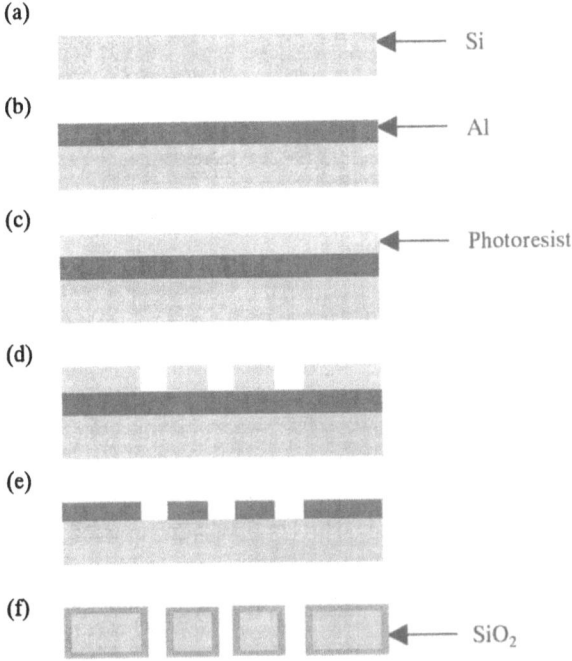

(a)   Si

(b)   Al

(c)   Photoresist

(d)

(e)

(f) SiO_2

Fig. 1 Schematic diagram of the fabrication process of a straight-thorugh silicon MC plate. (a) A 4-inch standard silicon wafer, (b) Aluminum vapor deposition, (c) Spin-coating of positive photoresist, (d) Photolithography and development, (e) Aluminum etching and photoresist removal, (f) Deep silicon etching by ICP-RIE and thermal oxidation of the substrate.

2.2. Straight-through silicon MC plate

Figure 2 displays a photograph of a micromachined straight-through MC plate. This plate measured 24 mm × 24 mm × 0.2 mm, with two 1-mm diameter feed-holes for a dispersed phase fabricated at the corners of the plate (Fig. 2a). A total of ten thousands channels is two-dimensionally placed around the center of this plate. Figure 2b shows a scanning electron microscopy (SEM) image of the micromachined straight-through MC. Each channel is placed at a 100 μm interval for the straight-through MC. This MC exhibits the considerable uniformity required as the emulsification device for monodispersed MS. The straight-through MC with an oblong section had an equivalent diameter of 17.3 μm. The equivalent diameter of one channel is defined as four times the cross-sectional area divided by the wetted perimeter of the channel. Hence, we succeeded in fabricating the straight-through MC plate as an emulsification device.

(a)

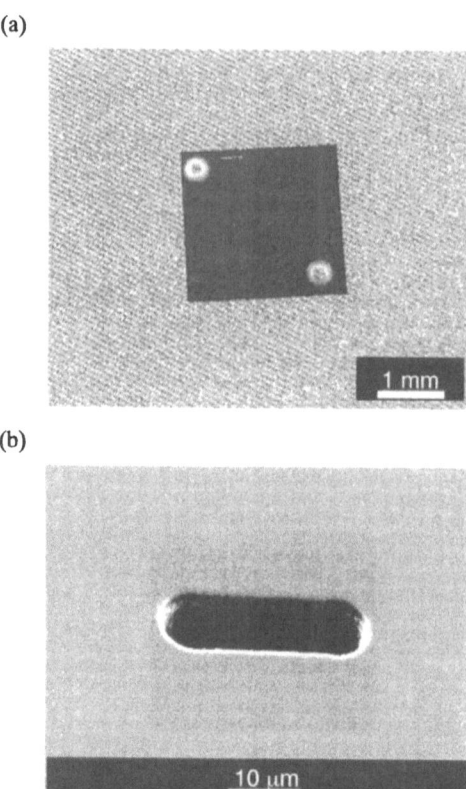

(b)

Fig. 2 A micromachined straight-through MC plate. (a) Photograph of a straight-through MC plate, (b) A SEM image of a straight-through MC with an oblong section.

3. Materials and Methods

3.1. Materials

We used soybean oil as the dispersed phase of the emulsification system and 0.3 wt% sodium dodecyl sulfate (SDS; hydrophilic lipophilic balance (HLB), 40) aqueous solution as the continuous phase.

3.2. Straight-through MC emulsification

Figure 3 depicts the experimental setup for straight-through MC emulsification, which consisted of a microscope video system [13-15], the straight-through MC module, the straight-through MC plate, a reservoir to feed the dispersed phase, and a syringe pump (Model 11, Harvard Apparatus Inc., Massachusetts, USA) to feed the continuous phase. The straight-through MC plate was tightly held between two rubber spacers in the module. The MS formation behavior was directly observed on a TV monitor (PVM-20M4J, Sony Co., Tokyo, Japan) through an inverted metallographic microscope (MS-511B, Seiwa Kougaku Seisakusho Ltd., Tokyo, Japan), a CCD camera (KP-C550, Hitachi, Tokyo, Japan) with a total magnification of 500. Video images taken during the experimental runs were analyzed using an image processing software (WinRoof, Mitani Co., Fukui, Japan).

The reservoir for the dispersed phase and the syringe pump for the continuous phase were connected to the inlets of the MC module, which was initially filled with the continuous phase. The pressurized dispersed phase was fed into the module, contacting the backside of the straight-through MC plate via the flow path for the dispersed phase. When the applied pressure reached the breakthrough pressure, the dispersed phase broke through the channels and MS formation commenced. The formed MS were recovered by the controlled continuous phase flow.

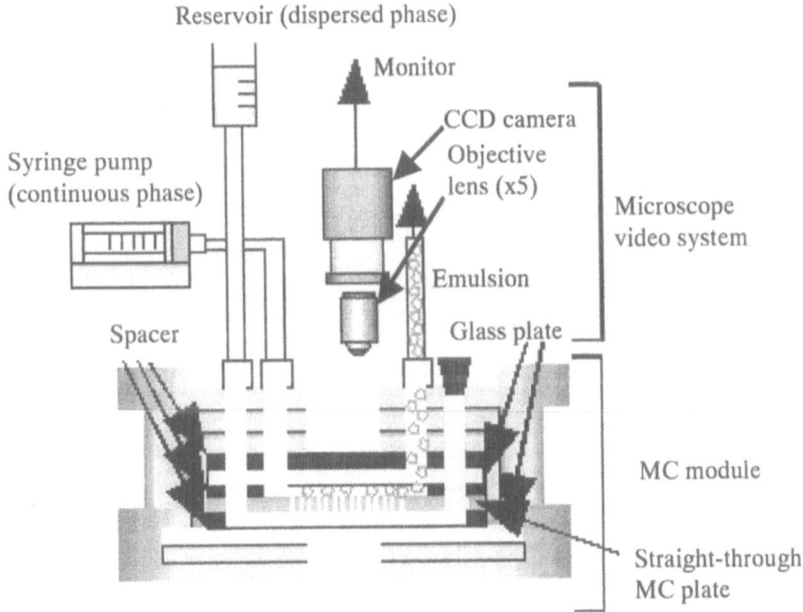

Fig. 3 Experimental setup for MS formation.

4. Results and Discussion

4.1. Formation of monodispersed O/W-MS using straight-through MC

Figure 4 depicts a microscopic image of the emulsification process using an oblong straight-through MC. The dispersed phase broke through the channels and O/W-MS formation commenced at a breakthrough pressure of 1.80 kPa. The formed O/W-MS had an average diameter of 32.5 µm and a coefficient of variation (*CV*, standard deviation × 100 / average droplet diameter) of 1.5%, verifying their excellent monodispersity. The monodispersity of the formed MS can also be confirmed by the image in Fig. 4. This demonstrates that the oblong straight-through MC made the formation of monodispersed MS possible. The average diameter of the MS formed to the MC equivalent diameter was 1.9, which was smaller than the ratio of 3 to 10 in MS formation by a circular glass capillary [17]. O/W-MS formation in the oblong straight-through MC was also performed without the continuous phase flow. We observed that monodispersed O/W-MS formed stably even without the continuous phase flow.

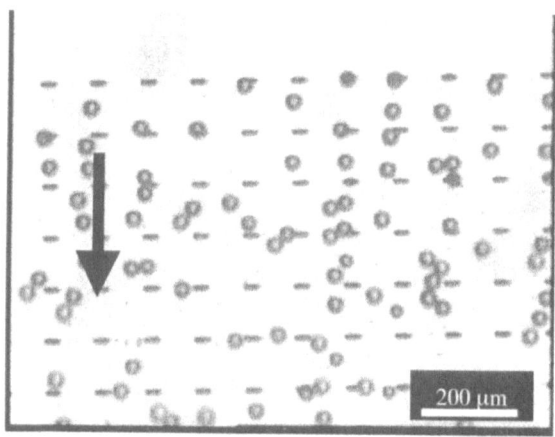

Fig. 4 A microscopic image of straight-through MC emulsification process.

4.2. Effect of pressure on straight-through MC emulsification

We investigated the effect of the applied pressure of the dispersed phase on the

MS size and size distribution for the oblong straight-through MC. Less than 10% of the channels worked at around the breakthrough pressure. The ratio of the channels that formed O/W-MS increased with the applied pressure over the breakthrough pressure, and more than 95% of the channels formed MS at 10.8 kPa. The size and size distribution of the formed O/W-MS changed little over the examined applied pressures, indicating that this elongated cross-sectional shape of straight-through MC yields stable formation of monodispersed MS over a wide range of applied pressures above the breakthrough pressure. The MS formation volume rate increased with the applied pressure, achieving up to 6.5 ml/h per MC plate. The MS formation rate depended on the site of the channels due to the significant pressure gradient on the micron scale. We observed that some channels stably formed uniformly sized MS at rates exceeding 10 droplets per second. Higher productivity of monodispersed MS would be realized by the optimization of the cross-sectional shape of a straight-through MC.

5. Conclusions

Rapidly developing MEMS technology brought the microfabrication of an ingenious emulsification device, called the straight-through MC plate, in which uniform through-holes are effectively positioned on a silicon microchip. This micromachined straight-through MC successfully prepared monodispersed O/W-MS with an average diameter of 32.5 μm and a coefficient variation below 2% by forcing the dispersed phase through the simply elongated cross-sectional shape of the channels even without any turbulent mixing. In addition, the stable formation of monodispersed MS was exhibited in a wide range of the applied pressures over the breakthrough pressure. The straight-through MS emulsification will provide a new insight to applicable nozzle-based dispersing systems. This emulsification device is also available for potentially functional MS that require monodispersity. Scaling up and numbering up of the straight-through MC plate lead to mass production of monodispersed MS.

Acknowledgement

This work was supported financially by the Program for Promotion of Basic Research Activities for Innovative Bioscience (MS project).

References

1. D. J. McClements, *Food Emulsions: Principles, Practice and Techniques* (CRC Press, New York, 1999).
2. H. Karbstein and H. Schubert, *Chemical Engineering and Processing*, **34**, 205 (1995).
3. S. E. Lyshevski, *Nano-and Micro-electromechanical Systems: Fundamentals of Nano-and Micro-Engineering*, (CRC Press, New York, 2000).
4. M. U. Kopp, A. J. de Mello and A. Manz, *Science*, **280**, 1046 (1998).
5. M. A. Burns *et al.*, *Science*, **282**, 484, (1998).
6. K. Chun, G. Hashiguti, H. Toshiyoshi, H. Fujita, Y. Kikuchi, J. Ishikawa, Y. Murakami and E. Tamiya, in Proceedings IEEE International MEMS 99 Conference (IEEE Piscataway, NJ, 1999), pp. 406-411.
7. J. Han and H. G. Craighead, *Science*, **288**, 1026 (2000).
8. C. S. Effenhauser, G. J. Bruin, A. Paulus and M. Ehrat, *Anal. Chem.*, **69**, 3451 (1997).
9. J. P. Kutter, *Trends in Anal. Chem.*, **19**, 352 (2000).
10. S. J. Haswell and V. Skelton, *Trends in Anal. Chem.*, **19**, 389 (2000).
11. T. McCreedy, *Trends in Anal. Chem.*, **19**, 396 (2000).
12. W. Ehrfeld, V. Hessel and H. Löwe, Microreactors (WILEY-VCH, Weinheim, 2000).
13. D. C. Duffy, J. C. McDonald, O. J. A. Schueller and G. M. Whitesides, *Anal. Chem.*, **70**, 4974 (1998).
14. Y. Kikuchi, K. Sato, and T. Kaneko, *Microvascular Res.*, **44**, 226 (1992).
15. Y. Kikuchi, K. Sato, and Y. Mizuguchi, *Microvascular Res.*, **47**, 126 (1994).
16. T. Kawakatsu, Y. Kikuchi, and M. Nakajima, *J. Am. Oil Chem. Soc.*, **74**, 317 (1997).
17. S. J. Peng and P. A. Williams, *Trans IchemE*, **76(A)**, 894 (1998).
18. A. W. Adamson and A. P. Gast, *Physical Chemistry of Surfaces* (John Wiley & Sons, New York, ed. 6, 1997).

Development of a cooled microreactor for platinum catalyzed ammonia oxidation

E.V. Rebrov, M.H.J.M. de Croon, J.C. Schouten

Laboratory of Chemical Reactor Engineering, Schuit Institute of Catalysis,
Eindhoven University of Technology,
P.O. Box 513, 5600 MB Eindhoven, The Netherlands
E-mail: J.C.Schouten@tue.nl

Abstract
This paper generalizes the results of our efforts aimed at research and development of a microstructured reactor/heat-exchanger for ammonia oxidation to nitrous oxide. The activity, selectivity, and the heat transfer characteristics of several microstructured reactors have been compared. Main factors determining the performance of a microreactor in N_2O production are considered. The use of a microreactor made of aluminum made it possible to estimate the intrinsic kinetic rate of the reaction. An ammonia oxidation kinetic model was developed based on both data published in the literature and our experimental observations. The model incorporates a reaction mechanism, which is combined with the flow characteristics of the reaction mixture and the heat evolution during the course of the reaction by means of the FLUENT® code. A regression analysis performed against experimental data resulted in a set of reaction parameters that describe quantitatively the reaction rate and selectivities to all end products at the steady-state conditions. Finally, the geometry of the inlet/outlet reactor chambers was optimized to obtain a desired flow distribution in the reaction and cooling channels. The results of numerical simulations with optimal design parameters showed that an isothermal reactor behavior could be approached even for conditions corresponding to an adiabatic temperature rise of about 1400°C.

1. Introduction

During the last decade there has been a growing interest in carrying out catalytic processes in reactors based on a microstructured catalyst bed [1-3]. Among such processes, a one-step oxidation of benzene by nitrous oxide [4] is an attractive way of phenol synthesis on zeolite catalysts. However, high production costs of nitrous oxide make the whole process economically unacceptable. At present, nitrous oxide is made either chemically by the thermal decomposition of ammonium nitrate or biochemically by nitrite reduction by heterotropic bacteria. In the latter case, anaerobic conditions and available organic carbon (e.g. acetate) are also required. In both cases the chief impurity of the product is molecular nitrogen. In comparison with these routes of producing N_2O, the catalytic ammonia oxidation seems to be more promising if nitrous oxide can be produced with high selectivity. The process can be conducted autothermally at relatively low temperatures and very short contact times due to the strong exothermicity and very high reaction rates of the underlying oxidation reactions. This allows for very high reactor throughputs and very small reactor sizes, which could make this process an interesting candidate for N_2O production. In recent work [3], we showed experimentally that metallic microstructured reactors can afford much higher reaction rates than conventional packed beds of catalyst pellets.

Furthermore, previously reported experiments on the Pt catalyzed ammonia oxidation in a microreactor demonstrated feasibility of chemical production on a microstructured catalyst bed [5]. In this way, the ammonia oxidation was performed at near isothermal conditions and the selectivity to N_2O was 15 % with e-beam deposited Pt as a catalyst and a 10 % NH_3 in O_2 reaction mixture. Thus, a considerable improvement was observed with respect to the performance of conventional fixed-bed reactors [6]. The latter are in general bound to a fatal loss of selectivity when increasing the ammonia inlet concentration above 4 vol %, allowing no further essential improvement. However, in the case of a very high inlet ammonia concentration (ca. 20 vol.%), even a microreactor may show significant axial and transverse temperature gradients, decreasing the selectivity to the desired product. In principle, cooled cross-flow microreactors could provide very high heat transfer rates, leading to markedly reduced hot spot temperatures, and improved conversion and selectivities.

It is obvious that a well-defined geometry of the microreactor can efficiently work only in combination with an optimal catalyst structure. Moreover, the catalyst loading is an important factor due to the structure sensitivity of the ammonia oxidation reaction. According to Ostermaier et al. [7, 8], there is an optimal catalyst particle size which favors N_2O production.

Here we present an experimental study as well as computer simulations of various prototypes of microstructured reactors with different catalyst supports in the ammonia oxidation reaction. The goal of this work was to optimise process conditions, microreactor geometry, and catalyst loading in order to obtain high selectivity towards nitrous oxide. Main factors determining performance of a microstructured reactor in the low-temperature ammonia oxidation on platinum catalyst are considered.

2. Lay-out of microreactors

All microstructured reactors were positioned in a specially designed reactor housing shown in Figure 1. The complete reactor module consists of three parts: the actual microreactor, the furnace, and the cooler. The reactor has a nickel housing with standard tube connections and can be heated up to 430 °C with an electrical furnace made of copper. The temperature was measured both inside the furnace (T1) and at an outer surface of the reactor (T2). Another thermocouple is positioned on the outer surface of the cooler (T3). To isolate the cooler section from the microreactor, a 2-mm thick ceramic ring is positioned between the microreactor and the cooler. The temperature of the cooler was maintained at about –20 °C in order to provide the fast removal of heat

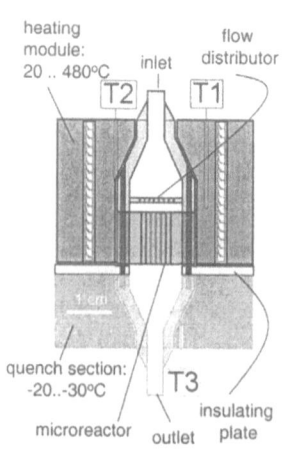

Figure 1. Microreactor test facility.

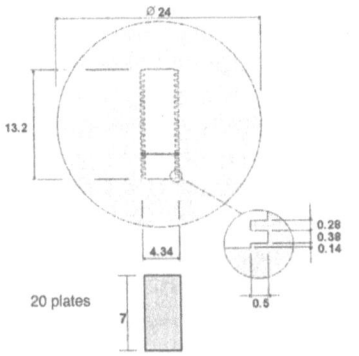

Figure 2. Schematic view of parallel plate microreactor A. Distances are given in millimetres.

produced in the reactor zone via four stainless steel screws to the cooler. The second use of the cooler is to quench the gas phase reactions and prevent further oxidation of the reactants and desired products downstream of the reactor.

Four microstructured reactors were designed and investigated. Three of them (A-C) contain a set of reaction channels of different size and geometry. The reactor D was a combined microreactor/heat exchanger which was designed and constructed based on the results of experimental studies carried out with reactors A-C and numerical simulation of temperature profiles obtained at higher inlet concentrations of ammonia. The following is a short description of these microstructured reactors.

A) *Parallel plate reactor.* The detailed view of this microreactor is shown in Figure 2. This microreactor consists of a stack of twenty removable plates. The plates were made of aluminum with an alumina layer on both sides produced by anodic oxidation. Platinum was immobilized by wet impregnation of a 1:1 solution of chloroplatinic and citric acids at room temperature for 6 hours. Calcination at 400 °C for 6 hours, followed by reduction in a hydrogen flow (30 ml/min) for 1 hour, was applied to make an active catalyst. To investigate the effect of Pt loading, three catalysts with different Pt loading of 0.05, 0.2, and 1.3 % wt/wt Al_2O_3 were prepared. The corresponding parallel plate microreactors are called A1, A2, and A3, respectively.

B) *Pt-monolith reactor.* This microreactor contains a set of 49 channels of 500 μm in diameter and a length of 9 mm produced by conventional machining in a metallic Pt cube with a volume of 10x10x9 mm^3. Each row contains 7 microchannels positioned at equal distance of 500 μm from each other. The distance between the rows was also 500 μm.

C) *Assembled $Pt/Al_2O_3/Al$ reactor.* This aluminum microreactor was designed and constructed in order to verify the feasibility of improving the heat transfer properties at the interface between the catalyst plates and the reactor housing. After anodic oxidation, the reactor is assembled from fourteen individual plates shown in Figure 3. Thus the reactor also contains 49 parallel channels. Pt impregnation was done after reactor assembly with a 1:1 solution of chloroplatinic and citric acids which was circulated through the reactor channels at room temperature for 6 hours. To obtain an

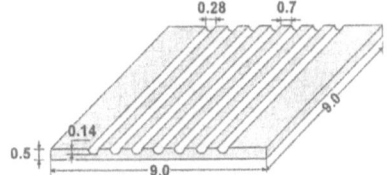

Figure 3. Schematic view of a single aluminum plate used in the microreactor C. Distances are given in millimetres.

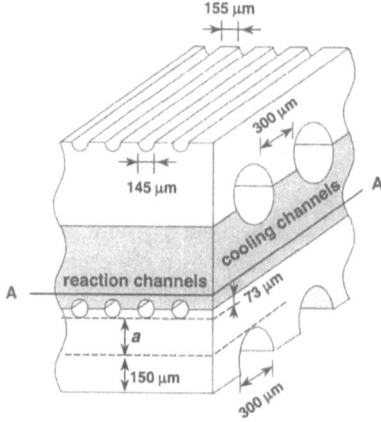

Figure 4. Assembling of microreactor D. The region of numerical simulation is shown in gray. Plane A-A indicates the position at which the temperature distribution is shown in Figure 8.

even Pt distribution throughout the total length of the channel, the flow direction was altered every 30 min.

D) *Microstructured reactor/heat-exchanger.* This reactor was designed and constructed with the aim to use high inlet ammonia concentrations while keeping the reactor temperatures within limited ranges to obtain still optimum conversion and selectivities. Each element of the microreactor/heat-exchanger is an aluminum plate containing two sets of microchannels (Figure 4). Each pair of adjacent elements forms a single periodic unit. The microreactor design was optimized based on CFD simulations of the temperature profiles in the reaction channels obtained at the ammonia inlet concentration as high as 20 vol.%.

The microstructured reactors A-C were tested in the NH_3 oxidation. The reactor feed streams consisted of 4-14 vol.% NH_3 in oxygen or oxygen/He mixture, with inlet flow rates in the range 2000-6000 cm^3/min (STP).

3. Comparison of performance of different microreactors

Prior to the kinetic study, the effects of the platinum loading, the reactor material, and the experimental conditions on selectivity were investigated to obtain the highest possible N_2O selectivity. As a starting point, the literature data for ammonia oxidation on a supported platinum catalyst were used. In these preliminary experiments, two different reactor materials and several catalysts with a different Pt loading were tested. Experiments were also performed to determine which NH_3/O_2 ratio and reaction temperature could give the highest N_2O selectivity.

Figure 5 compares the catalytic activities and selectivities of reactors A1-A3, B, and C under the standard operating conditions. These four reactors have identical alumina support, and differ by the Pt loading and the geometry. It is clear that reactor C is the most active, as it completes NH_3 conversion at about 325 °C, while reactor A1 is only starting to be active at this temperature and the activity curves of reactors A2 and A3 lie in between those of A1 and C. Maximum selectivity towards N_2O increases monotonously with increasing Pt loading up to 3.5 wt.%.

For proper evaluation, however, we should consider also the catalytic activity of a single Pt atom in these catalysts. Such comparison based on data of Figure 5 reveals that reactors A1 and A2 show similar performances, with the NH_3 consumption rate of about 20 s^{-1} at 300 °C. However, reactors A3 and C remain

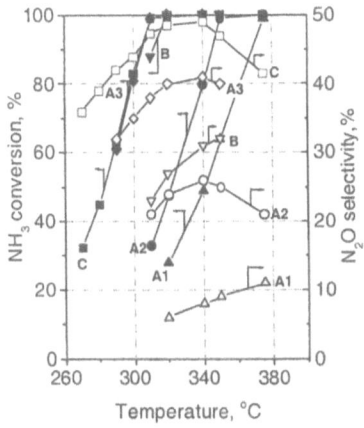

Figure 5. Activity and selectivity of microstructured reactors in the NH₃ oxidation. Reaction conditions: NH₃ 6 vol.%, O₂ 88 vol.%, balance - He. The flow velocity: 600, 1175, 4430, 580, and 4290 cm³/min (STP) for microreactors A1, A2, A3, B, and C respectively. Pt mass: 0.022, 0.086, 0.560, and 0.543 mg for microreactors A1, A2, A3, and C respectively.

considerably more active (TOF = 35 s^{-1}) and reactor C yields the highest N_2O selectivity also. In order to rationalize the advantage of the catalysts in reactors A3 and C, we could compare the data on Pt dispersion in these catalysts. It is well known that Pt catalyzed ammonia oxidation is a structure sensitive reaction [7, 8]. Therefore, the Pt cluster size plays an important role in the activity of these catalysts and also it could have an influence on the product distribution by a great extend. According to the catalyst dispersion data, Pt exists as isolated atoms in the reactors A1 and A2, and it is present in the form of clusters in the reactors A3 and C. The formation of Pt clusters is believed to be responsible for the better performance of reactors A3 and C in the ammonia oxidation.

The data obtained on the supported catalysts were compared with those of reactor B which is a Pt monolith microreactor. As can be seen from Figure 5, reactor B showed considerably lower nitrous oxide selectivity at the full conversion of ammonia as compared to reactors A3 and C. This can be explained based on the thermal behavior of those microreactor systems (Figure 6). The smallest thermal resistance is clearly associated with microreactor C. In this case the reactor temperature was always equal to or below the oven temperature. Because of the excellent intrinsic heat conductivity of aluminum (k_m = 240 Wm^{-1}K^{-1}), the only significant heat transfer resistance in this case was located at the interface between the external microreactor wall and the nickel housing (k_m ≈ 80 Wm^{-1}K^{-1}). The observed significantly greater thermal resistances (the furnace temperature was up to 15 °C above the reactor temperature) in case of the microreactors A and B, imply additional heat transfer resistances due to the reduced contact area between the catalyst slabs and the nickel housing (microreactor A) and due to the

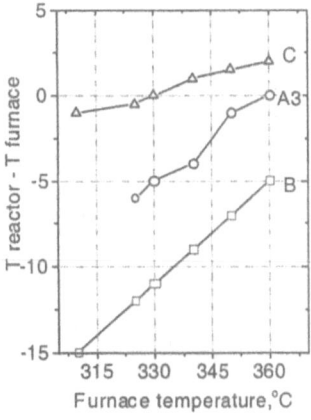

Figure 6. The differences between the furnace and microreactor temperatures for the microreactors A3, B, and C. The furnace load was kept at 90 W, the cooler was set at –20 °C. The inlet NH₃ concentration was changed to provide a desired microreactor temperature.

lower intrinsic heat conductivity of platinum ($k_m = 72$ Wm^{-1}K^{-1}, microreactor B).

Due to its better performance, the catalytic behavior of reactor C was further investigated by kinetic analysis, based on an extensive set of experimental runs which also included experiments at lower contact times, corresponding to feed flows in the range 4000-6000 cm^3/min (STP).

4. Mechanism of the ammonia oxidation

In the last decade, mechanisms of ammonia oxidative transformations were essentially revised, when selective reduction of nitrogen oxides by ammonia in an oxygen excess was studied in detail for environmental applications. Here ammonia oxidation to nitrogen oxides is an undesired side reaction, attracting thus a keen research attention. At present, all the reactions for most of the conditions are known fairly well at atmospheric or lower pressure. Despite of some uncertainties, a detailed mechanism based on a large number of elementary reactions could be drawn. For many of the reactions, also their rate constants are accurately known. For some reactions, however, large uncertainties in their rate constants exist. Current understanding of the ammonia oxidation reaction on Pt catalyst for surface temperatures below 400 °C is that NH$_3$ and oxygen after their adsorption on a surface form several active adspecies which react to form N$_2$, N$_2$O, and NO. N-containing adspecies are assumed to occupy single "on-top" adsorption sites [9], hereafter denoted by { }. All other species are assumed to occupy adsorption sites equivalent to that of oxygen, presumably hollow sites, denoted as ( ). The reaction mechanism along with the reaction rate parameters used in the present work is listed in Table 1.

A sensitivity analysis of the reaction rates has been performed to determine which reaction rate coefficients have the greatest influence on the formation of species of interest. The most obvious result from the sensitivity study is that parameters influence both the N$_2$ selectivity and NH$_3$ consumption rate in the same sense and have opposite influences on the N$_2$O selectivity. The reason for this can be easily understood when looking at the respective reaction steps: reaction step 5 leads towards N-adspecies followed by N$_2$ production in step 6.

Table 1. Reaction rate coefficients in form, $k = A \cdot \exp(-E/RT) \cdot \exp(\varepsilon \theta_i / RT)$

No	Reaction	A	E [kJ mol^{-1}]	ε [kJ mol^{-1}]
R1	NH$_3$ + { } → {NH$_3$}	2.0·10^8 s^{-1} atm^{-1}	0.0	0.0
R2	{NH$_3$} → NH$_3$ + { }	1.9·10^{13} s^{-1}	96.0	0.0
R3	O$_2$ + 2 ( ) → 2 (O)	4.3·10^6 s^{-1} atm^{-1}	0.0	0.0
R4	2 (O) → O$_2$ + 2 ( )	1.0·10^{13} s^{-1}	213.2	60.0
R5	{NH$_3$} + 3 (O) → {N} + 3 (OH)	3.0·10^{16} s^{-1}	141.0	0.0
R6	{N} + {N} → N$_2$ + 2 { }	8.0·10^{12} s^{-1}	124.0	0.0
R7	{N} + {NO} → N$_2$O + 2 { }	2.5·10^{10} s^{-1}	98.9	0.0
R8	{NO} + ( ) → {N} + (O)	1.0·10^{13} s^{-1}	118.0	0.0
R9	(OH) + (OH) → (O) + ( ) + H$_2$O	1.0·10^{13} s^{-1}	113.0	0.0
R10	{N} + (O) → {NO} + ( )	2.1·10^{13} s^{-1}	131.0	0.0
R11	H$_2$O + ( ) + (O) → (OH) + (OH)	2.0·10^8 s^{-1} atm^{-1}	60.5	0.0
R12	{NO} → NO + { }	1.5·10^{13} s^{-1}	143.0	0.0
R13	N$_2$O + ( ) → N$_2$ + (O)	2.5·10^8 s^{-1} atm^{-1}	72.2	0.0

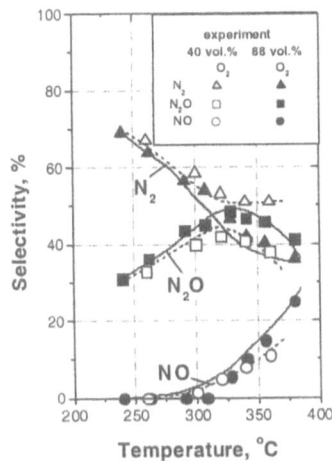

Figure 7. Selectivity as a function of temperature in microreactor C. Inlet conditions: NH₃ = 6 vol.%, balance –He. Contact time: 0.51 ms.

The increase of the rate of step 5 will increase the selectivity to N_2, at the same time decreasing the selectivity towards N_2O, because of increasing the rate of reaction step 8. This emphasizes the essential role of the competition for adsorbed oxygen in this reaction.

The reaction rate coefficients were found by regression based on experimental data obtained as a function of temperature, contact time, and inlet composition. At first, the regression analysis was performed in an isothermal plug-flow approximation. Then, parameters of steps 5-7, and 10, having the largest impact on the reaction rate and selectivities, were optimized in the Fluent® simulations. Figure 7 shows a comparison between simulation and experimental results from this study as a function of temperature at two different oxygen concentrations. The calculated selectivities lie all within a range of less than 5% of the experimental measurements, which is within experimental error bounds. In general, the simulation results show a very good agreement with experimental data over the whole range of investigated NH_3/O_2 ratios, contact times, and temperatures.

The experiments on the NH_3 oxidation show that the stoichiometric NH_3/O_2 ratio gives on average a two times less amount of N_2O than a mixture containing a large surplus of oxygen. This is quite remarkable because it would be desirable to avoid an excess of oxygen in order to prevent the further oxidation of N_2O. However, a cooler located downstream just after the reactor and maintained at –20 °C could effectively quench the gas phase side reactions which decrease the N_2O selectivity. As a result, in all reactors studied here, the large surplus of oxygen gave a positive effect on the N_2O selectivity, that could be explained on the basis of the proposed reaction mechanism. Furthermore, the oxygen abundant mixture, fed to a reactor at desired temperature, could consume more heat produced in the reaction improving the thermal behavior of the microreactor.

5. Microreactor optimization study

Despite of the fact that the application of an appropriate design and the proper choice of the reactor material could decrease drastically the temperature gradients inside the microreactor even at full conversion of 10 vol.% of ammonia, the further increase of ammonia concentration would increase the reactor temperature above the desired value decreasing selectivity to the required product – nitrous oxide. This problem was overcome by combination of a microreactor with a

56

microheat-exchanger in the "sandwich" cross flow design of the whole module (Figure 4).

In the optimization study, three-dimensional numerical simulation of the microreactor was performed using a commercial CFD computer code, the FLUENT® package. Calculations were performed with cross flow of the reaction mixture, containing 20 vol.% NH_3 in O_2 at an inlet temperature of 175 °C in one passage, and cold N_2 of 20 °C in the other passage. The reaction mixture was preheated to avoid an excessive nitrogen production upstream of the microchannels.

The detailed mechanism of ammonia oxidation was not used here and an adopted power law form of the rate expression was applied instead for temperature profile calculations:

$$r_{NH_3} = \frac{k_1 k_2 p_{NH_3}}{(1 + k_1 p_{NH_3}) p_{O_2}^{0.06}} \qquad (1)$$

Equation (1) includes two kinetic parameters with an Arrhenius temperature dependence, k_1 and k_2. The strong correlation between the activation energies E_1 and E_2 made it not possible to estimate the individual parameters. Therefore, the parameter k_1 was set constant (9.13 10^2 atm^{-1}) in the selected temperature range and the activation energy and the pre-exponent factor of constant k_2 (72.3 kJ/mol, 1.46 10^8 s^{-1} atm$^{0.06}$) were found by regression on data from ammonia oxidation runs, with ammonia conversion as the experimental response.

The effect of thickness of the aluminum layer on the temperature field and the microreactor performance was explored. Four different cases were studied, with a separate grid created for each configuration. In these microreactors the distance a between the reaction and cooling channels (Figure 4) was set at 125, 270, 470, and 670 μm. Furthermore, several non-uniform flow distributions in the cooling channels were investigated in order to reduce the large axial temperature gradients in the reaction channels which are caused by the varying reaction and heat production rates along the reaction channels. The idea was to locate the cooling channel with the maximum coolant flow close to the area of maximum heat production.

Table 2 demonstrates the mean-square deviations from the average catalyst temperature for the most interesting flow distributions listed in Table 3. A change in the distance between the reaction and cooling

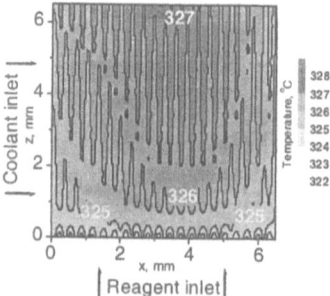

Table 2. Mean-square deviations from the average catalyst temperature obtained at different distance between the reaction and cooling channels and different flow distributions in the cooling channels.

Flow distri- bution	Distance between reaction and cooling channels, a (see figure 4)			
	125 μm	270 μm	470 μm	670 μm
A	5.26 °C	2.95 °C	1.91 °C	1.14 °C
B	5.03 °C	2.61 °C	1.24 °C	1.12 °C
C	6.90 °C	2.35 °C	1.02 °C	1.10 °C

Figure 8 Temperature distribution at the position A-A in the alumina layer. Distance a (see Figure 4) equals 470 μm.

Table 3. Flow distribution in the cooling channels. The data are in percent from the total flow (35 m/s).

Flow distribution	Cooling channel number								
	1	2	3	4	5	6	7	8	9
A	11.1	11.1	11.1	11.1	11.1	11.1	11.1	11.1	11.1
B	9.1	9.1	9.1	9.1	9.1	13.7	13.7	13.7	13.7
C	7.1	7.1	7.4	8.6	9.8	12.9	15.4	15.7	16.0

channels produces two effects which act in the opposite way. On the one hand, increasing the distance between two sets of microchannels increases the axial heat conduction through the wall of the cooling channels, thereby decreasing the temperature gradient between the 1st and 20th reaction channels. As a result, the difference in conversion between different channels is eliminated. However, this effect becomes smaller with further increasing distance a: enlarged heat losses to the environment create the temperature difference between the central and outside reaction channels. On the other hand, an increased heat conduction in the direction transverse to the cooling flow decreases the positive effect of flow distribution in the cooling channels. In this case, an axial temperature gradient is observed in the reaction channels. Therefore, in between 270 and 670 μm there is an optimal distance a at which the temperature differences in both axial and transverse dimensions are minimal. One can see in Table 3 and Figure 8 that the design with $a = 470$ μm combined with the optimal coolant distribution (case C) demonstrates that both axial and transverse temperature gradients across the whole plate do not exceed 4 °C.

6. Geometry design of a flow distributor

To obtain the required flow distribution and corresponding temperature distribution, a particular geometry of the inlet and outlet reactor chambers should be used. The pressure drop required for the desired flow distribution could not be obtained, if gas is fed and withdrawn from one side of the reactor. Therefore, at fixed position of the inlet pipe near the 9th cooling channel there is an optimal position for the outlet pipe at which the trans-chamber pressure drop could provide a desired flow distribution in the cooling microchannels. Several variations on the standard geometry of the outlet chamber were chosen for this study, with separate geometry and grid created for each configuration studied. Figure 9 shows the geometry and pressure distribution providing the flow profile which is closest to the optimal one (distribution C). The corresponding flow distribution is shown in Figure 10.

However, the temperature is not the only important parameter that must be carefully controlled. Also, the selectivity depends highly on the residence time. Therefore, the selectivity can be further enhanced by having a uniform residence time distribution in the reaction channels. In contrast to the coolant flow profile, the flow in the twenty reaction channels should be equal. It was found that by means of a mirror position of the inlet and outlet pipes, a uniform distribution is obtained in all reaction channels. Figure 11 shows the schematic view of microreactor/heat-exchanger D which was designed according to the results of the numerical simulations.

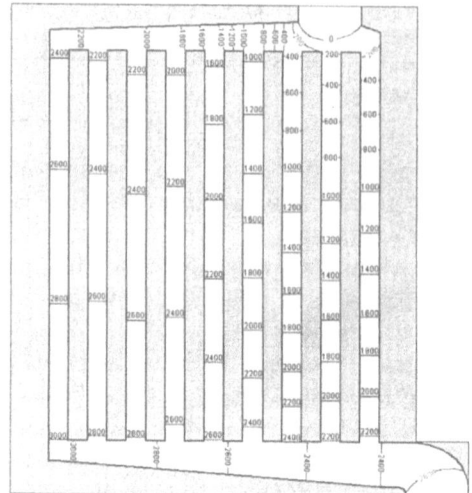

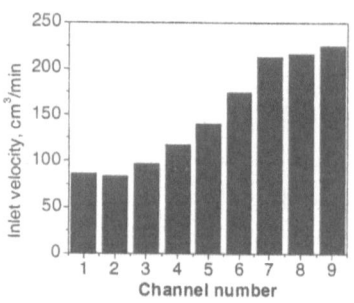

Figure 10. The optimal flow profile in the cooling channels. Channel # 1 is located near the reagent inlet.

Figure 9. The optimized geometry of the inlet and outlet flow chambers and the cooling microchannels. Numbers represent relative pressure in kPa.

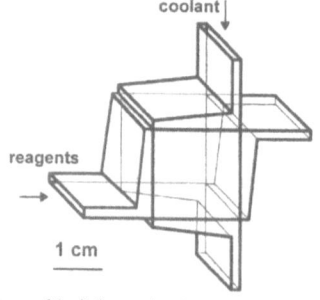

7. Conclusions

Main factors determining the performance of a microstructured reactor with a platinum catalyst in the low-temperature ammonia oxidation have been considered. Both the surface oxygen

Figure 11. Schematic view of reactor/heat-exchanger D with optimized configuration of the inlet and outlet chambers for the reagents/products and the coolant.

bonding strength determined by the platinum cluster size, and the intrinsic thermal conductivity of the reactor material appear to be extremely important. In case of a microreactor made of a highly conductive material (i.e. aluminum), there is an optimal distance between the reaction and cooling channels in a single microstructured plate. Simulations performed using the FLUENT® code on a cooled cross flow microstructured reactor showed that the inlet concentration of ammonia could be increased above the explosion limit up to 20 vol.% without decreasing the selectivity to nitrous oxide. Appropriate design of a single reactor plate makes it possible to decrease the temperature non-uniformity in the reactor to 4 °C even under conditions corresponding to an adiabatic increase of temperature of about 1400 °C. In combination with the optimized process parameters, this gives an N_2O selectivity as high as 52 %.

References

[1] B. Louis, P. Reuse, L. Kiwi-Minsker, A. Renken, Appl. Catal. A. 210 (2001) 103.

[2] A. Cybulski, J.A. Moulijn, in: Structured catalysts and reactors, Marcel Dekker, New York, 1998.

[3] E.V. Rebrov, G.B.F. Seijger, H.P.A. Calis, M.H.J.M. de Croon, C.M. van den Bleek, J.C. Schouten, Appl. Catalysis A: General 206 (2001) 125.

[4] V. Sobolev, G.I. Panov, A.S. Kharitonov, V.N. Romannikov, A.M. Volodin and K.G. Ione. J. Catal. 139 (1993) 435.

[5] A.J. Franz, D. Quiram, R. Srinivasan, I.-M. Hsing, S.L. Firebaugh, K.F. Jensen and M.A. Schmidt, "Process Minimization: 2nd Int. conf. on Microreaction Technology", Topical Conference Preprints, I. Rinard, R. Wegeng and W. Ehrfeld, AIChE, New York (1998) 33.

[6] T. Pignet and L.D. Schmidt, Chem. Eng. Sci., 29 (1974) 1123.

[7] J.J. Ostermaier, J.R. Katzer and W.H. Manogue, J. Catal., 33 (1974) 457.

[8] J.J. Ostermaier, J.R. Katzer and W.H. Manogue, J. Catal., 41 (1976) 277.

[9] J.M. Bradley, A. Hopkinson and D.A. King, J. Phys. Chem., 99 (1995) 17032.

Microchemical Systems for Direct Fluorination of Aromatics

Nuria de Mas, Rebecca J. Jackman, Martin A. Schmidt, Klavs F. Jensen
MicroChemical Systems Technology Center
Massachusetts Institute of Technology, Cambridge, MA 02139, USA

Abstract

We describe a silicon-based microchemical system that enables efficient and safe direct fluorination of aromatic compounds, a process difficult to implement on a conventional macroscopic scale. The microreactor has been designed, fabricated, and tested for the fluorination of toluene at room temperature using excess elemental fluorine. The reactor, fabricated by using silicon processing microfabrication and metal deposition techniques, provides controlled gas-liquid contacting, selective fluorination, and easy replication. Toluene was chosen as model chemistry because past studies of the fluorination of this compound can serve as a benchmark. Here we provide additional insights into the conversion and product distribution as a function of operating conditions: toluene concentration, molar ratio of fluorine to toluene, and temperature. Conversions of toluene on the order of 60% resulted in selectivities of monofluorinated toluenes of up to 30%. The substitution pattern of the ortho, meta, and para isomers, as determined from GC, was 4:1:2.

Introduction

Aromatic compounds containing fluorine ring substituents are commercially important as intermediates in the production of pharmaceuticals, pesticides, and as monomers for manufacturing high-performance aromatic polymers. The ideal method to selectively fluorinate an aromatic ring would be by directly reacting the aromatic with elemental fluorine. Under controlled conditions, this method would be a one-step process applicable to a large range of organic compounds. However, unlike chlorination or bromination, the direct fluorination of aromatic compounds is highly exothermic and thus very difficult to control on a large scale. In addition to the large heat of reaction, fluorine is poorly soluble in commonly used solvents resulting in reactions that proceed at the liquid-gas interface [1]. Consequently, localized hot spots are likely to form in large-scale systems, which can lead to unwanted side reactions, e.g., ring scission, polymerization, and charring. Since the late 1960's, several research groups have investigated the reaction of elemental fluorine with aromatic compounds [2-5]. They showed that this reaction can be performed with reasonable yields and as a true aromatic electrophilic substitution, provided the heat generated is removed efficiently. Heat management was achieved on a laboratory scale by diluting fluorine with nitrogen, operating at low temperatures, or processing at very low concentrations of organic substrates. Because these reaction conditions are clearly limiting for scale-up selectively fluorinated aromatics are commercially produced by indirect methods in which the heat of reaction is distributed over several synthetic steps. These methods are based on diazotization reactions or chlorine/fluorine exchange with alkali-metal fluorides. The main drawback of these methods is their low overall yield [6].

Microreactors could allow direct fluorination to be performed under more aggressive conditions and with higher yields than those achieved in bench top systems, thereby making this process potentially attractive for commercialization. Because of their small size, microreactors provide high area to volume ratios, high heat and mass transfer rates, and are inherently safe. Moreover, microreactors can be integrated with sensors and other process unit operations, including temperature sensors, cooling channels, and detection systems, creating capabilities exceeding those of conventional macroscale units [7]. The feasibility of controlling direct fluorination reactions on a small scale with reasonable yields has been recently demonstrated [8,9]. The devices used, however, were machined in metal blocks. Machining techniques are useful for making a small number of reactors with specific purposes, but ideally, one should be able to replicate microreactors easily and quickly in order to increase production rates by increasing the number of reactors.

Easy and low cost replication in addition to integration of sensors and actuators can be achieved by microfabrication strategies similar to those used in the microelectronics industry. Here we describe a microreactor system built in silicon and Pyrex and coated with thin nickel films for carrying out direct fluorination reactions. The performance of this microreactor has been characterized using the direct fluorination of toluene as a model chemistry. Conversion and product distribution has been investigated as a function of the operating conditions.

Design and Fabrication

The microreactor, schematically represented in Figure 1, consists of two channels formed in silicon by standard photolithographic techniques and anisotropic etching with potassium hydroxide. The channels are capped with Pyrex by anodic bonding. Since fluorine is corrosive to silicon and hydrogen fluoride (subproduct of the reaction) etches silicon oxide at ambient conditions, the regions of the silicon and Pyrex in contact with the reaction mixture are coated with nickel thin films. These films are patterned by photolithography and deposited by electron-beam evaporation. The channels have 400 μm wide, 280 μm deep triangular cross sections to facilitate the deposition of the nickel films. They also provide a large area to volume ratio (290 cm^{-1}) to enhance the contact between fluorine and the substrate solution. The two channels per device allow an increased productivity per device, while keeping the area to volume ratio high. Additional channels could be incorporated in future designs for higher productivities. Figure 2 shows a scanning electronic micrograph of a channel cross section. The liquid inlets are located at one end of the reactor channels, while the gas bubbles into the liquid streams through ports located downstream of the liquid ports.

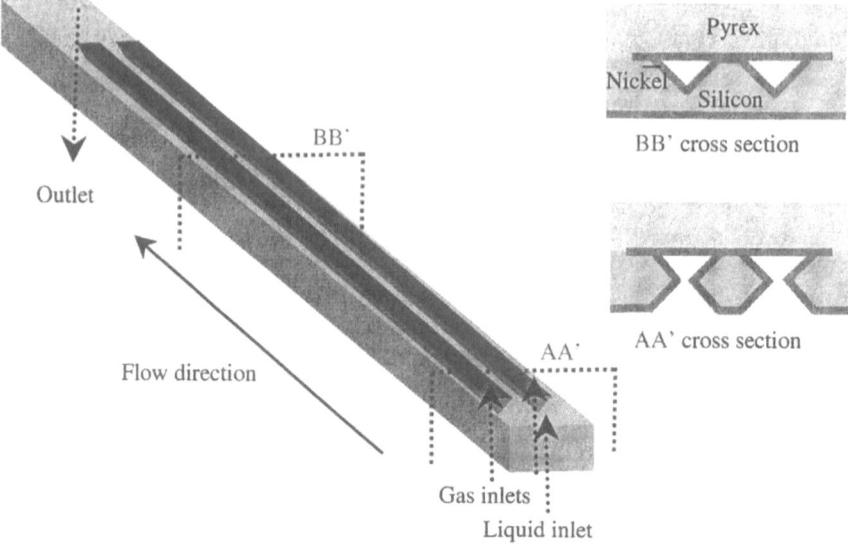

Figure 1 Microreactor configuration

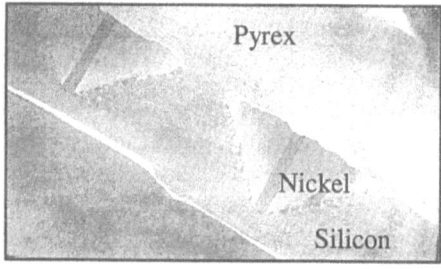

Figure 2 Scanning electronic micrograph of BB' cross section

The microreactor silicon chip is packaged by compression with a plexiglass plate against a Kalrez™ sheet (DuPont perfluoroelastomer) and a stainless steel base. Standard high-pressure fittings and fluidic lines, machined into the stainless steel, are used to deliver the reactants and collect the products. A syringe pump meters the liquid reagent, while a mass flow controller delivers the gas reagent. A thermocouple, inserted between the plexiglass and the stainless steel in contact with the silicon chip, monitors the chip temperature during the reaction. The reactor was designed to operate continuously at ambient conditions without requiring external cooling while fluorinating a 0.1M toluene solution.

Results

Fluid Flow Visualization

Gas/liquid flow contact patterns were characterized up to 10 ml/min of nitrogen (fluorine is pre-mixed with nitrogen prior to entering the reactor) and 100 μL/min of methanol (solvent used in the fluorination reactions) in a non-nickel coated reactor. Three types of flow regimes were clearly identified. The anisotropic etch used during channel fabrication moved the interchannel wall downstream the gas inlets, creating asymmetric pressure drop and channeling at low gas flow rates (below 4 ml/min). The transition from bubble flow (bubbles of gas are dispersed throughout the liquid) to annular flow (liquid flows as a film around the walls and the gas flows along the center of the channel) was difficult to observe in the 400 μm wide triangular channels due to the high linear velocities. Pulsing flow (alternating plugs of liquid and gas) occurred at high liquid rates.

Annular flow regime through the reactor was chosen to perform the fluorination reactions. In this regime the microreactor can be described essentially as a falling film reactor with a well-defined gas/liquid contact area. The contact area is an important design variable. Fluorinations are fast reactions, but due to the low solubility of fluorine in commonly used solvents, fluorination is virtually an heterogeneous process occurring at the gas/liquid interface. Hence, in this process the reaction rate is a direct function of the area available for contact between the gas and the liquid. The ratio of interfacial area to volume of reactor in the annular flow regime is on the order of 100 cm^{-1}.

Reaction Studies

Toluene was selected as a model chemistry to characterize the performance of the microreactor because it is easy to handle and monitor and there is literature conversion data for comparison [3,4,9]. Methanol was chosen as the solvent for its polar character and ability to form hypofluorites. A polar solvent serves several purposes: it is a good radical scavenger, provides a polar medium to encourage the aromatic electrophilic substitution mechanism, and acts as an acceptor for the developing F ion through hydrogen bonding. The ability of the solvent to form hypofluorites in the presence of fluorine should also be beneficial to promote the solubilization of fluorine [1,4].

We present two sets of experiments. The first set is the design case: a 0.1M toluene solution in methanol fluorinated at varying fluorine concentrations. In the second set, the substrate concentration was increased to 1.0M. The flow rates in all experiments were kept constant to maintain the system hydrodynamics. The reaction conditions are shown in Table 1.

The reactor effluent was collected in a sealed flask immersed in a salt-ice water bath upon stabilization of the system. The cold bath served two purposes: to

64

condense any toluene vapor from the gas effluent and to quench the reaction since unreacted fluorine could remain in the reaction mixture. The gas effluent was scrubbed with a potassium hydroxide solution and exhausted to the hood. The work-up steps included degassing of the sample with dry nitrogen, addition of sodium fluoride, and sample filtration. The products were then characterized by gas chromatography. A GC/MS was used for identification while GC/FID was used for quantitative analysis.

Table 1 Reaction conditions

Case A – Design case: 0.1M toluene/methanol (45 µl/min), 10 ml/min F_2/N_2

Experiment	A1	A2	A3	A4
% F_2/N_2	1.0	2.5	5.0	10.0
Ratio F_2 : toluene	1:1	2.5:1	5:1	10:1

Case B: 1.0M toluene/methanol (45 µl/min), 10 ml/min F_2/N_2

Experiment	B1	B2
% F_2/N_2	5.0	10.0
Ratio F_2 : toluene	0.5:1	1:1

Conversion of toluene (X), combined yield of monofluorotoluenes (Y), and selectivity of monofluorotoluenes (S), obtained by GC/FID, are plotted in Figure 3 for the design case. These parameters are defined as:

$$X = 1 - \frac{n_{tol,fin}}{n_{tol,in}}$$

$$Y = \frac{n_{mF,fin}}{n_{tol,in}}$$

$$S = \frac{n_{mF,fin}}{(n_{tol,in} - n_{tol,fin})}$$

where $n_{tol,in}$ is the molar amount of toluene before the reaction, $n_{tol,fin}$ is the molar amount of toluene after the reaction, and $n_{mF,fin}$ is the molar amount of ortho, meta, and parafluorotoluene isomers after the reaction. The substitution pattern, defined as the molar ratio of the ortho, meta, and para fluorotoluene isomers, was 4:1:2. The reaction data for the second set of experiments are shown in Table 2.

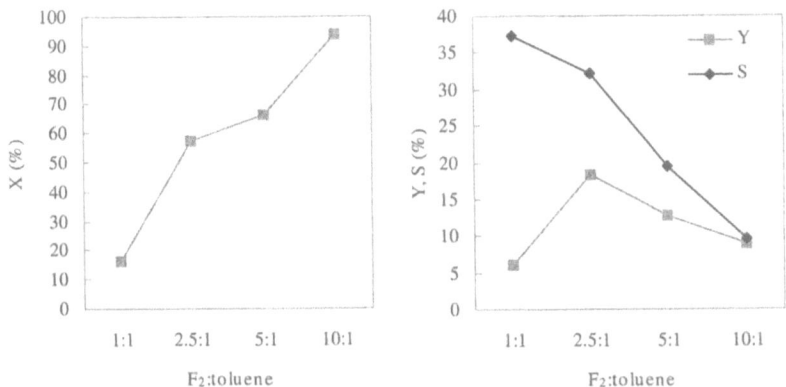

Figure 3 Reaction data for case A – design case: 0.1M toluene/methanol

Table 2 Reaction data for case B: 1.0M toluene/methanol

Experiment	B1	B2
Ratio F_2 : toluene	0.5:1	1:1
X (%)	35	52
Y (%)	7	9
S (%)	20	17

Discussion

Experiments showed that microreactors allow direct fluorination to be performed under more aggressive conditions (i.e. room temperature and high fluorine to toluene molar ratios) than bench-top scale systems. The trends in conversion and product distribution observed are consistent with previous fluorination studies at lower temperatures [3,9].

In the design experimental set, in which a 0.1M toluene solution was fluorinated, a mixture of fluorinated toluenes was identified by gas chromatography. The ortho, para, and metafluorotoluene isomers were the main products. Fluoromethylbenzene, 2,4-difluorotoluene, 2,6-difluorotoluene, trifluorotoluenes, and other unidentified high boiling compounds were also formed. It was also observed that methanol partially reacted with fluorine. Toluene conversion increased as the concentration of fluorine increased, while the selectivity decreased. When operating in excess of fluorine the increased reactivity is likely to lead to localized hot spots that enhance the formation of fluorine radicals and radical reaction byproducts. A maximum selectivity of 37% was obtained when operating at 1:1 fluorine to toluene molar ratio. If we account for the difluorotoluenes and trifluorotoluenes formed, the selectivity would increase to about 45%; these selectivities are comparable with the values obtained in previous studies at temperatures around -15°C [9]. The yield, equal to conversion times selectivity, reached a maximum of 18% for a level of conversion of 57% (2.5:1 fluorine to toluene).

The substitution pattern of the three ring-substituted monofluorotoluenes at room temperature, 4:1:2, is consistent with previous fluorination studies. In these studies it was observed that the proportion of ortho and para isomers increased as temperature was decreased and was dependent on the type of solvent used [3]; the fluorination of toluene in acetonitrile at -16°C resulted in a 5:1:3 substitution pattern [9]. The direct fluorination of aromatic compounds has been postulated to occur by the aromatic electrophilic mechanism when operating at low temperatures, low substrate conversions, and low fluorine concentrations [3]. The relative yields of the ortho, meta, and para isomers are consistent with this reaction mechanism, but a more systematic study of the substitution pattern of various ring-substituted benzenes will be necessary to accurately assess the predominant reaction mechanism.

In the second experimental set the concentration of toluene was increased to 1.0M. These experiments show similar trends as the design case but illustrate how poor temperature control of the reaction leads to poor yields and selectivities. All reactions, however, were carried out safely (i.e. without reactor failure). In this case, the maximum selectivity was of 20%, half of the dilute case. The maximum yield dropped to about 9%. Heat transfer calculations through the liquid layer show that appreciable temperature gradients may form in the device due to the increased concentration. Because of the low heat dissociation of fluorine (155 kJ/mol) with respect to the large heat of reaction (450 kJ/mol), efficient heat control is key to control the selectivity of the reaction. Since any local increase in temperature has the potential to promote the formation of fluorine radicals that can then initiate competitive radical reactions, external cooling or a more efficient way of dissipating heat is required to increase selectivities at high toluene concentrations. Operating in a flow regime that results in thinner liquid layers should improve heat dissipation and lead to increased selectivities at high levels of conversion. Alternatively, microstructures with larger interfacial areas available for reaction would permit higher levels of conversion at low fluorine to toluene molar ratios, thus high selectivities, as well as increase the area for heat dissipation by conduction.

The performance of the evaporated coatings is dependent on the operating conditions (concentrations of fluorine and toluene). The nickel-coated silicon microreactor has been routinely operated up to several hours over the range of reaction conditions presented in this paper. The films, however, appear to loose adhesion to the substrates as the operation time increases. For extended operating times (days) and high reagent concentrations, additional testing will be needed.

Conclusions

The feasibility of safely running direct fluorinations in a nickel-coated silicon chip at room temperature has been demonstrated. Conversion and product distribution have been determined as a function of toluene concentration and fluorine to toluene molar ratio. Selectivities on the order of 30% were obtained for conversion levels of 60% (0.1M solution). The substitution pattern of the ortho, meta, and parafluorotoluene isomers was consistent with previous work on the same reaction carried out at lower temperatures, possibly suggesting an electrophilic aromatic substitution mechanism. The ability to more efficiently control the heat transfer in the device via external cooling, as well as to increase the gas/liquid interfacial area available for reaction, should lead to increased selectivities and conversions over a wide range of substrate concentrations. Future work will include the direct fluorination of a second aromatic system in an improved microreactor design.

Acknowledgments

The authors thank Novartis Research Foundation for financial support. The assistance of Steven Vitale and the personnel of the Microsystems Technology Laboratories is gratefully acknowledged.

References

[1] Gambaretto, G. P.; Conte, L.; Napoli, M.; Legnaro, E.; Carlini, F. M. *Journal of Fluorine Chemistry* 1993, *60*, 19-25.
[2] Grakauskas, V. *Journal of Organic Chemistry* 1970, *35*, 723-728.
[3] Cacace, F.; Giacomello, P.; Wolf, A. P. *Journal of the American Chemical Society* 1980, *102*, 3511-3515.
[4] Conte, L.; Gambaretto, G. P.; Napoli, M.; Fraccaro, C.; Legnaro, E. *Journal of Fluorine Chemistry* 1995, *70*, 175-179.
[5] Misaki, S. *Journal of Fluorine Chemistry* 1981, *17*, 159-171.
[6] Hessel, V.; Ehrfeld, W.; Golbig, K.; Haverkamp, V.; Lowe, H.; Storz, M.; Wille, Ch. *Proceedings of 3rd International Conference on Microreaction Technology*, Frankfurt, 18-21 April 1999, Springer, Berlin, 2000, 526-540.
[7] Jensen, K. F. *Chemical Engineering Science* 2001, *56*, 293-303.
[8] Chambers, R. D.; Spink, R. C. H. *Chemical Communications* 1999, 883-884.
[9] Jähnisch, K.; Baerns, M.; Hessel, V.; Ehrfeld, W.; Haverkamp, V.; Lowe, H.; Wille, C.; Guber, A. *Journal of Fluorine Chemistry* 2000, *105*, 117-128.

DESIGN, FABRICATION AND EXPERIMENTAL STUDY OF NEW COMPACT MINI HEAT-EXCHANGERS

L.LUO[1], B.HOAREAU[1], U.D'ORTONA[1], D.TONDEUR[1], H.LE GALL[2], S.CORBEL[2]

1 : Laboratoire des Sciences du Génie Chimique du CNRS, Groupe ENSIC-INPL Nancy
2 : Département Chimie-Physique des Réactions, Groupe ENSIC-INPL Nancy

The mini heat-exchangers presented here are cubic, cross-flow type exchangers with several specific features, concerning the internal design, the material, and the method of fabrication.

The overall dimensions are of the order of several centimetres, while the internal channels have characteristic dimension in the millimetre range. The different "layers" or plates and their channels may be designed in such way as to produce an important local mixing of the fluid in three dimensions, in order to improve the fluid-side heat transfer. Different internal geometries have been designed.

The fabrication technique is "stereophotolithography" , in other words, 3D laser polymerisation. In this technique, an Argon laser beam, focused at any specific point at the surface of a bath of liquid monomer, produces a local polymerisation. The laser beam is controlled, via different galvanometric mirrors, to "draw" a layer of the object, according to a program that ensures its connectivity. A vertical displacement of the monomer tank allows the object to be constructed layer by layer, in three dimensions.

This technique allows the heat-exchanger to be manufactured as one single solid piece, including inlets and outlets, and passages for temperature or pressure sensors. It therefore avoids the assembling operations and the tightness problems encountered with assembled plates exchangers. On the other hand, some other mechanical problems arise, and the allowable temperature range is limited.

An interesting aspect is that of thermal conductivity. A poor thermal conductivity is not necessarily a detrimental factor, as analysed in detail in a companion communication. This is due to the small thickness of the exchanger plates, making the heat transfer resistance through the material negligible, but making also the thermal diffusion in the longitudinal flow direction small, a factor of importance in small, high performance exchangers.

The paper presents experimental results obtained on water to water exchange, both at steady-state and in transient regime, for two (possibly three) exchangers. The performances, in terms of specific power, of overall heat transfer coefficient, and of pressure-drop, are compared to that of exchangers of similar size, but of different designs and/or different materials, and to classical transfer correlations. Some possible applications, in portable, embarked, or domestic domains, and perspectives are discussed.

View of the internal structure (haff exchanger)

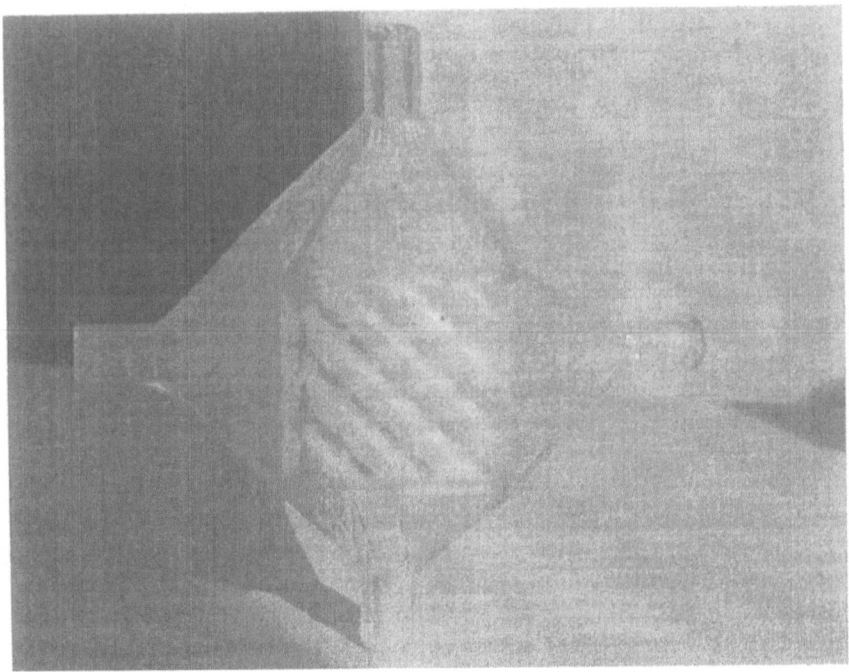

Overall view of an exchanger

Development of a Self-Heating Catalytic Microreactor

Katsuki Kusakabe[1], Daigo Miyagawa[1], Yunfeng Gu[2], Hideaki Maeda[3] and Shigeharu Morooka[1]

[1]Department of Applied Chemistry, [2]Venture Business Laboratory, Kyushu University, Fukuoka 812-8581, Japan
[3]Department of Inorganic Materials, National Institute of Advanced Industrial Science and Technology, Kyushu, Tosu 841-0052, Japan

Abstract

Microchannels were fabricated on the both sides of a (100) silicon wafer by wet chemical etching, after pattern transfer using a negative photoresist. The channel (upper width = 600 μm, lower width = 515 μm, depth = 60 μm, and length = 78 mm) on one side was used as a reactor. A heating element (Pt wire) was installed in the channel on the opposite surface of the reactor channel, and a thermocouple was installed in a channel adjacent to the reactor. A thin platinum layer was coated as a catalyst on the walls of the reactor channel by sputtering. In order to increase the surface of the catalyst, a γ-alumina support layer was formed in the reactor channel by a sol-gel process. The reactor channel, as well as the heating channel on the reverse side, was then sealed so as to be gas-tight with glass plates by an anodic bonding technique. A solution of H_2PtCl_6 was introduced into the reactor channel with the γ-alumina layer, and platinum was loaded on the support. Both platinum catalysts, prepared by sputtering and impregnation techniques, were activated in a flow of hydrogen at 773 K. The self-heating microreactor was then used for the hydrogenation of benzene as a model reaction. The reaction rate of the supported catalyst was one order of magnitude higher than that of the sputtered catalyst.

1. Introduction

Microreactors are ideal for use in reactions, the temperatures of which must be precisely and rapidly controlled. The formation of hot spots in catalysts can be prevented because heat is removed from the system efficiently. Table 1 shows the formation of catalysts and their reactions in microreactors. Catalysts are often formed by dry processes, such as modification of metal foils [3,4], sputtering [1], chemical vapor deposition [8,10], electron-beam vaporization [2] and aerosol deposition [6]. Such catalysts are effective for reactions at elevated temperatures where the reaction rates are sufficiently high. At moderate temperatures, however,

Table 1. Gas phase catalytic reactions in microreactors.

Properties of reaction and reactor	Catalyst and reaction	Ref.
Dehydrogenation of cyclohexane	Sputtered Pt	1
Oxidation of NH_3	Pt by electron beam evaporation	2
Oxidation of C_2H_4	Modification of silver foils	3
Oxidation of CH_4	Modification of Rh foils	4
$CH_4 + NH_3 + (3/2)O_2$	Pt catalyst with microchannel arrays	5
Oxidation of NH_3	Deposition of aerosol particles, deposition from solutions, PVD	6
Oxidation of methane	Anodized aluminum, impregnation of Pt	7
Oxidation, dehydrogenation	Anodized aluminum, wash-coating, CVD	8
Hydrogenation, Ru; oxidation, Ag	Anodized aluminum	9
Oxidation of H_2 with O_2	Pt impregnation on CVD alumina	10
Characterization of catalyst supports	Sol-gel, CVD, sintering of fine particles, anodized aluminum	11
Dehydration of alcohols	Wet coating of sulfated zirconia	12
CuO/ZnO, Pd/ZnO etc.	Coating of catalyst slurry of oxides	13
Coupling reaction	Sol-gel silica, wet impregnation of Pd	14
Catalyst supports	Packing of catalyst particles	15
Catalyst layers	Zeolite layers by hydrothermal synthesis	16

catalysts with a large surface area are needed. The anodic oxidation of aluminum produces porous alumina, which can be used as a catalyst support [7,8,9]. Fine catalyst particles are packed in a microchannel for this purpose as well [15].

The sol-gel process has been developed in order to produce a variety of finely divided inorganic materials, which can be used as catalyst supports [11]. The objective of the present study is to develop a self-heating reactor with a platinum catalyst layer. The catalyst was formed by sputtering, as well as by impregnation on a porous alumina support which was formed by the sol-gel process. The hydrogenation of benzene (a model reaction) was then performed to investigate the characteristics of the prepared microreactor.

2. Fabrication of a Microreactor

2.1 Formation of Microchannels

Figure 1 shows the microreactor which was prepared for use in the present study [17,18]. The microreactor was constructed on a (100) Si wafer of 10 mm x 40 mm. Both surfaces of the wafer were oxidized in a flow of air at 1273 K for 10 h, and one surface was spin-coated with a negative photoresist. The resultant resist was pre-baked at 413 K for 3 min, and a pattern for a preheater, a reactor and a thermocouple well was transferred using a microchannel photolithography technique. A channel pattern for the heating element was then transferred to the other surface of the wafer by the same procedure. After the resist was post-baked at 413 K for 10 min, the unexposed portion of the resist was removed. The SiO_2 layer on the unexposed area was then removed by immersing the wafer in an HF and NH_4F buffer solution at room temperature for 10 min. After the resist was completely removed, the microchannel pattern was formed by etching the wafer in a 30% KOH solution at 303 K for 80 min. The reactor channel was 600 μm in upper width, 515 μm in lower width, 60 μm in depth and 78 mm in length. In the upstream portion of the reactor channel, a preheater section, 200 μm in width, 60 μm in depth and 95 mm in length, was provided.

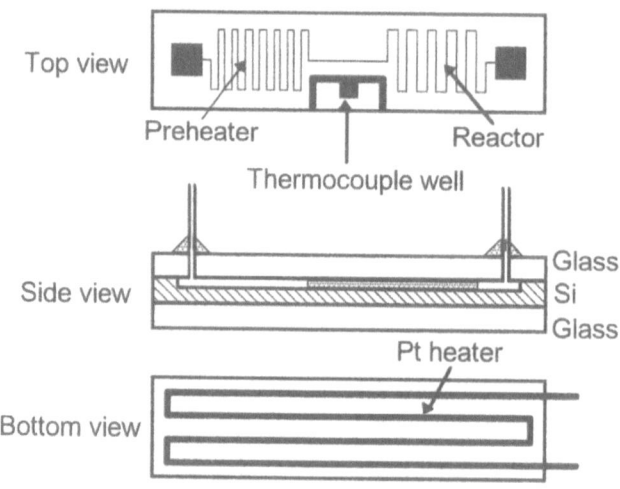

Fig. 1 A self-heating microreactor.

2.2 Formation of the Platinum Catalyst

The Pt catalyst was prepared by both dry and wet processes, as shown in **Figure 2**. The surface with the reactor channel   was spin-coated with the negative photoresist. A mask with the pattern of the reactor channel (no preheater section and no thermocouple well) was aligned to the wafer, and, by the pattern transfer procedure, all the area except for the reactor channel was covered with the resist. In the dry process, shown in **Figure 2(a)**, the entire wafer surface was coated with a platinum layer by sputtering, and the resist was then lifted off. Thus, the walls of the reactor channel were coated with a layer of platinum, which functioned as a catalyst. Details of the preparation method have been reported elsewhere [18].

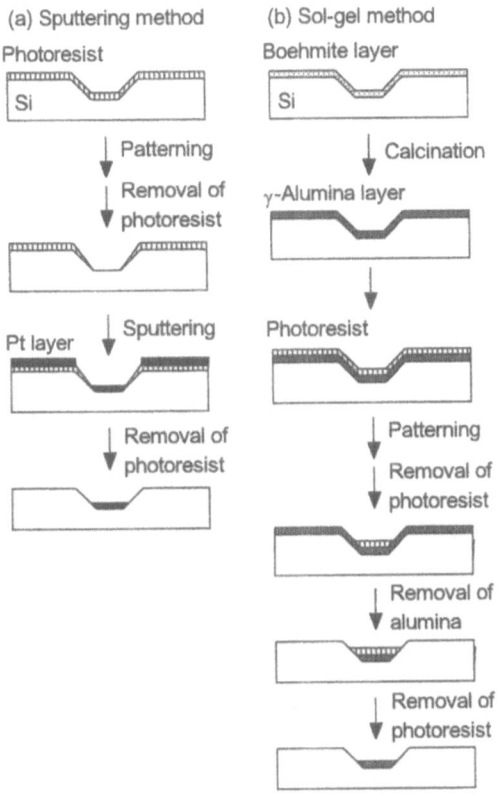

Fig. 2 Preparation of the Pt catalyst in the reactor channel.

The Pt thin layer catalyst prepared by the above procedure had an inherently low surface area. Thus, a microporous ceramic layer was formed by a sol-gel process [19], shown in **Figure 2(b)**, for use as the catalyst support. A boehmite sol (γ-AlOOH) was prepared using aluminum isopropoxide. The concentration of aluminum in the sol was 0.6 mol/L. The boehmite sol was then coated on the substrate surface with the reactor channels and dried in air. Using a heat-treatment at 873 K for 3 h, the boehmite was transformed to γ-alumina. The substrate surface was then coated by a photoresist, and, after the patterning, all the photoresist except for the reactor channel was removed. Thus the alumina layer in the reactor channel was protected with the photoresist. The unprotected γ-alumina layer was dissolved with an aqueous H_3PO_4 solution of 85% at 333 K for 30 min. The resist on the reactor channel was removed by firing in air at 773 K

2.3 Heating Element and Sealing

A platinum heating wire (100 μm in diameter and 20 cm in length) was placed in the channel, which was formed on the surface opposite to the reactor channel. Glass plates, 1-mm in thickness, were used to cover the reactor channel, and two 300-μm diameter holes were drilled at locations corresponding to the inlet and outlet of the reactor channel. A 300-μm diameter stainless steel tube was inserted in each hole and fixed in place with an inorganic adhesive. The wafer and the glass cover were then bonded by an anodic bonding technique at 673 K. The tightness of the gas seal was confirmed, since no bubbles were observed when the channel was pressurized and placed in water. The surface of the heating element was also covered with a glass plate by anodic bonding. A thermocouple, composed of 100-μm diameter element wires, was installed in the thermocouple channel.

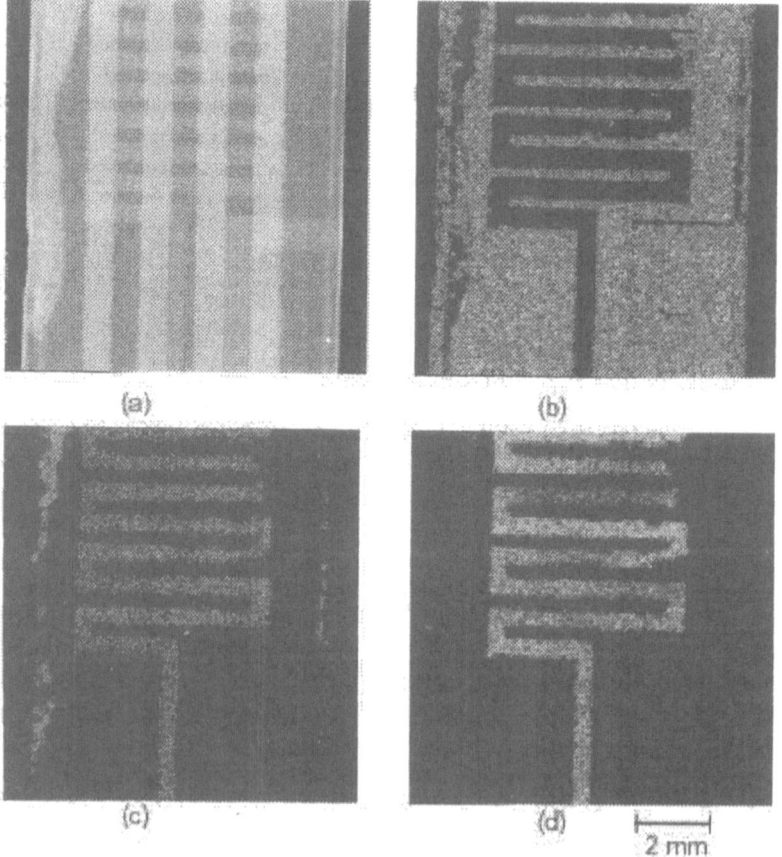

Fig.3 Elemental analyses of the microreactor.
(a) X-ray image; (b) distribution of Si; (c) distribution of Al; (d) distribution of Pt.

3. Evaluation of Catalysts

The γ-alumina layer, which was 2.5 μm in thickness, was contacted with a 0.1 mol/L H_2PtCl_6 solution. By calcination at 773 K for 3 h, a Pt/alumina catalyst layer was obtained. The Pt-loaded reactor channel in the gas-tightly sealed microreactor was characterized by X-ray microscopy (HORIBA Ltd., XGT-2700), which displayed two-dimensional distributions of the elements. **Figure 3(a)** shows the X-ray image of the microreactor. The horizontal lines indicate the reactor channel, and the vertical lines the heater channel. As shown in **Figures 3(c)** and **(d)**, aluminum and platinum, respectively, were homogeneously distributed in the reactor channel.

The hydrogenation of benzene was used as a model reaction. The catalyst was activated in hydrogen at 773 K prior to the reaction. The flow rates of benzene and hydrogen were controlled with mass flow controllers, and the maximum total flow rate was 1 mL/min. Concentrations at the outlet of the reactor channel were analyzed by means of a high-sensitivity gas chromatograph equipped with a thermal conductivity detector. No deactivation was observed under the reaction conditions used. **Figure 4** shows that the reaction rates obey the first-order kinetics. The reaction rate of the Pt/alumina catalyst was one order of magnitude higher than that of the sputtered Pt catalyst, as indicated in **Figure 5**.

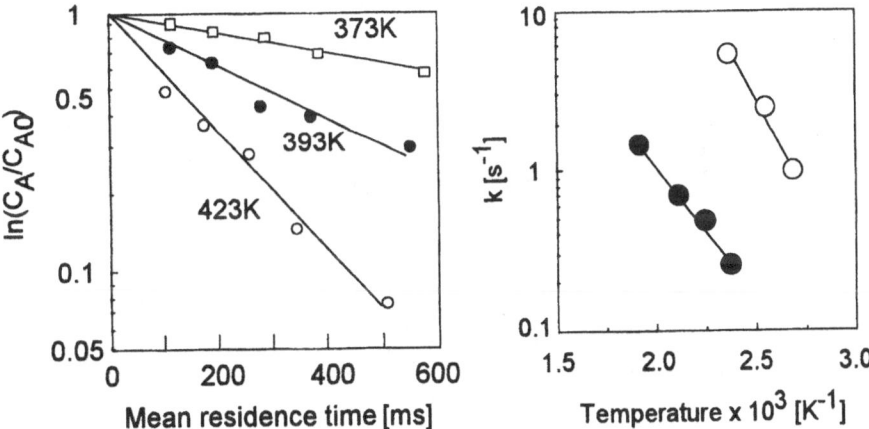

Fig.4 Effect of temperature on the hydrogenation reaction of benzene.

Fig.5 Reaction rates based on contact times. Open circles, sputtered Pt catalyst; closed circles, impregnated Pt catalyst.

4. Conclusions

A self-heating microreactor was constructed on a (100) silicon wafer by means of photolithography and wet etching. Platinum catalyst layers were then formed in the reactor channel by sputtering and impregnation techniques. The Pt catalyst formed by the impregnation showed a reaction rate, which was one order of magnitude higher than that achieved by sputtering.

Acknowledgments

This study was supported by the Japan Society for the Promotion of Science (JSPS) and the New Energy and Industrial Technology Development Organization (NEDO).

References

[1] T. Cui, J. Fang, A. Zheng, F. Jones and A. Reppond, *Sensors Actuators B*, **71**, 228 (2000).
[2] R. Srinivasan, I.M. Hsing, P.E. Berger, K.F. Jensen, S.L. Firebaugh, M.A. Schmidt, M.P. Harold, J.J. Lerou and J.F. Ryley, *AIChE J*, **43**, 3059 (1997).
[3] H. Kestenbaum, A. Lange de Oliveria, W. Schmidt, F. Schüth, W. Ehrfeld, K. Gebauer, H. Löwe and Th. Richter, Proc. IMRET 3, p.207 (2000).
[4] J. Mayer, M. Fichtner, D. Wolf and K. Schubert, Proc. IMRET 3, p.187 (2000).
[5] V. Hessel, W. Ehrfeld, K. Golbig, C. Hofmann, St. Jungwirth, H. Lowe, Th. Richter, M. Storz, A. Wolf, O. Wörz and J. Breysse, Proc. IMRET 3, p.151 (2000).
[6] A.J. Franz, S.A. Ajmera, S.L. Firebaugh, K.F. Jensen and M.A. Schmidt, Proc. IMRET 3, p.197 (2000).
[7] R. Wunsch, M. Fichtner and K. Schubert, IMRET 3, p.623 (2000).
[8] M.A. Liauw, M. Baerns, R. Broucek, O.V. Buyevskaya, J.-M. Commenge, J.-P. Corriou, L. Falk, K. Gebauer, H.J. Hefter, O.-U. Langer, H. Löwe, M. Matlosz, A. Renken, A. Rouge, R. Schenk, N. Steinfeldt and St. Walter, Proc. IMRET 3, p.224 (2000).
[9] A. Kursawe, E. Dietzsch, S. Kah, D. Hönicke, M. Fichtner, K. Schubert and G. Wießmeier, Proc. IMRET 3, p.213 (2000).
[10] M.T. Janicke, H. Kestenbaum, U. Hagendorf, F. Schüth, M. Fichtner and K. Schubert, *J. Catalysis*, **191**, 282 (2000).
[11] M. Fichtner, W. Benzinger, H. Haas-Santo, R. Wunsch and K. Schubert, Proc. IMRET 3, p.90 (2000).
[12] N.G. Wilson and T. McCreedy, *Chem. Commun.*, 733 (2000).
[13] P. Pfeifer, M. Fichtner, K. Schubert, M.A. Liauw and G. Emig, Proc. IMRET 3, p.372 (2000).
[14] G.M. Greenway, S.J. Haswell, D.O. Morgan, V. Skelton and P. Styring, *Sensor Actuators B*, **63**, 153 (2000).

[15] M.W. Losey, M.A. Schmidt and K.F. Jensen, Proc. IMRET 3, p.277 (2000).,
[16] Y.S.S. Wan, J.L.H. Chau and A. Gavriilidis, Microporous Mesoporous Mater., **42**, 157 (2001).
[17] T. Tsubota, D. Miyagawa, K. Kusakabe and S, Morooka, *Kagaku Kogaku Ronbunshu*, **26**, 895 (2000).
[18] K. Kusakabe, D. Miyagawa, Y. Gu, H. Maeda and S. Morooka, *J. Chem. Eng. Japan*, (2001) in press.
[19] B.E. Yoldas, *Ceramic Bull.*, **54**, 289 (1975).

Preparation of Microchannel Palladium Membranes by Electrolysis

Katsuki Kusakabe[1], Makoto Takahashi[1], Hideaki Maeda[2] and Shigeharu Morooka[1]

[1] Department of Applied Chemistry, Kyushu University, Fukuoka 812-8581, Japan
[2] Department of Inorganic Materials, National Institute of Advanced Industrial Science and Technology, Kyushu, Tosu 841-0052, Japan

Abstract

A negative resist was spin-coated on the surface of a 50-μm thick copper plate and solidified as a protective film. A palladium layer was then electrodeposited on the other uncovered surface. After removing the protection film, the resist was again spin-coated on the copper surface, and a pattern of microslits was transferred. After development, the microslits were anodically etched out. The final result was a palladium membrane, formed as an assemblage of thin layers in the microslits. The H_2/N_2 separation factor through the palladium membranes was dependent on the electrodeposition conditions employed. The flux of H_2 through the palladium membrane formed with a current density of 40 $mA \cdot cm^{-2}$ at an electrolysis temperature of 45°C was 0.06 $mol \cdot m^{-2} \cdot s^{-1}$ (H_2 pressure on the feed side = 0.1 MPa), and a H_2/N_2 ideal separation factor was 2000 at a permeation temperature of 300°C

1. Introduction

Palladium membranes are used to separate hydrogen from mixtures of gases [1], and are conventionally formed on porous supports by chemical vapor deposition, sputtering or electroless plating. However, these techniques have been developed to prepare palladium membranes for use in the large-scale production of hydrogen. The development of thin palladium membranes in the shape of microchannels can be effective for downsizing the gas purification unit and can be used in small-scale fuel cell systems. Porous supports are relatively thick, in order to support its strength, and have microscopically rough surfaces. Thus they are not suitable for the formation of pinhole-free, miniaturized palladium membranes. Franz et al. [2] prepared a thin palladium membrane on a silicon wafer patterned by photolithography. The H_2/N_2 separation factor was higher than 1800 at a permeation temperature of 500°C. In the present study, a microfabrication technique was developed for preparing palladium membranes in microslits which had been etched on a thin copper plate, and permeation properties of H_2 and N_2 through the membranes were determined [3].

2. Experimental

2.1. Fabrication of microchannel Pd membranes

Figure 1 shows the procedure for the preparation of microchannel palladium membranes [2]. A copper plate (1 cm x 1 cm, 50 μm in thickness) was used as the substrate. One side of the substrate was spin-coated with a negative photoresist, exposed, and then baked at 140°C for 3 min (step 1). The other side of the substrate was polished with fine alumina powders (particle size = 1.0 and 0.3 μm), and the plate was rinsed with distilled water. A palladium layer was formed by electrodeposition on the entire uncovered surface of the copper plate, which was used as the cathode. A platinum plate was used as the anode. The deposition of palladium was carried out at a cathode current density of 3-50 mA/cm^2 for 3-10 min at 25-50°C (step 2). The bath composition was as follows: $PdCl_2$ = 19.8 g/L; citric acid = 21.5 g/L; and $(NH_4)_2SO_4$ = 50.0 g/L. The pH was adjusted at 7 with NH_4OH.

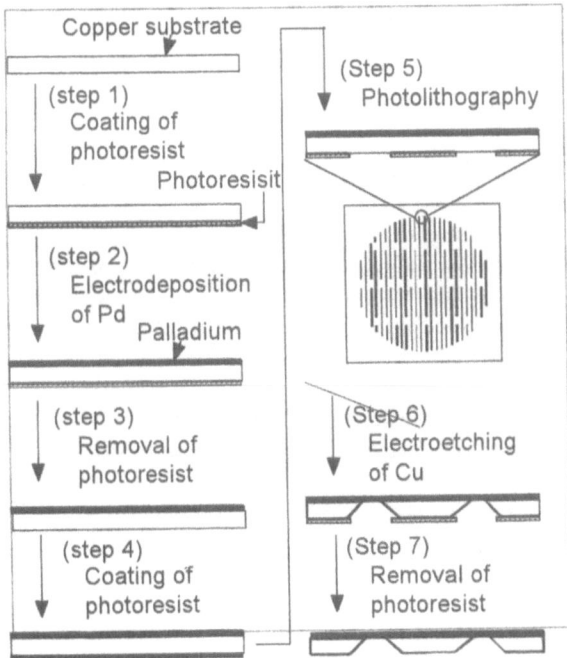

Fig.1 Preparation procedure of microchannel palladium membranes.

The resist on the copper surface was then removed (step 3), and the negative resist was again spin-coated on the copper surface and pre-baked at 140°C for 3 min (step 4). A mask with a pattern of microslits (80 μm in width) drawn within a circle of 5 mm in diameter was aligned to the substrate, and the pattern was transferred by photolithography via contact printing (step 5). The resist was exposed, postbaked at 140°C for 10 min, and the unexposed part of the resist was removed in a developer. The uncovered copper surface was then electroetched in a 3wt% H_2SO_4 solution at an anodic current density of 30 mA/cm^2 for 210 min (step 6). The resulting microslits were exposed the palladium layer on the other side. The residual photoresist was completely removed at the end of the procedure (step 7).

2.2 Gas permeation tests

The Pd deposited copper plate was cut in the form of a disk (diameter = 10 mm), for use in permeation tests, and was fixed with Cu gaskets (outer diameter = 10 mm, inner diameter = 5.5 mm) in a stainless steel holder. The area of the disk exposed to gas was 0.238 cm^2. Gas permeation tests were carried out using H_2 and N_2, which were fed into the feed side at permeation temperatures of 200-400°C. Argon was used as a sweep gas and was fed into the permeate side. The total pressure of the feed and permeate sides were maintained at 101.3 kPa. Flow rates were measured with soap-film flow meters, and gas compositions were determined using a gas chromatograph, equipped with a thermal conductivity detector (Shimadzu GC-8A).

Hydrogen flux, J_{H2}, through the palladium membrane is expressed as

$$J_{H2} \propto [(P_{feed})^n - (P_{permeate})^n] \tag{1}$$

When the rate-determining step is the diffusion of protons in the membrane, the flux is proportional to the difference in the square roots of the partial pressure of hydrogen on the feed and permeate sides ($n = 0.5$).   In the present study, however, the value of n was close to unity. Thus, the permeance, Π, is defined by following equation, assuming $n = 1$.

$$\Pi = (\text{mol of gas transferred per unit time}) \times$$
$$(\text{partial pressure difference})/(\text{membrane area}) \tag{2}$$

The ideal separation factor is defined by the ratio of permeances.   Details have been reported elsewhere [3,4].

3. Results and Discussion

3.1 Characterization of Cu substrate and Pd layer

Figure 2 shows the top surface of a glass mask and a copper substrate after etching. Although the width of the microslits on the glass mask was 80 μm, the width of the etched slits on the copper substrate was 100 μm at the center region and 120 μm at the peripheral region, because of the maldistribution of the current density on the substrate. The width of the microslits, which penetrated the copper substrate, was constant from the front surface to the rear surface. The total area of the microslits on the copper substrate was 0.054 cm^2.

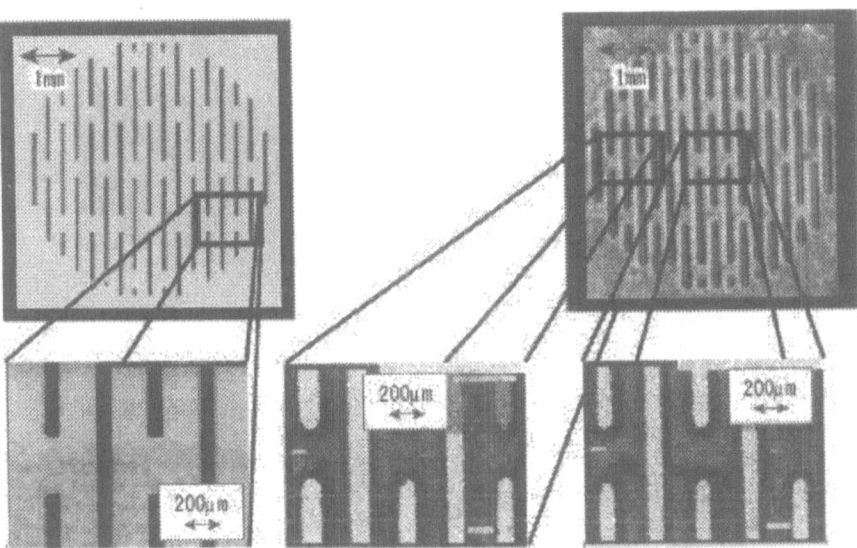

Fig.2 Top surfaces of the glass mask (left side) and the copper substrate after etching (right side).

Figure 3 shows the effect of current density on the surface morphology of the palladium deposited on the copper substrate. The Pd layers, which were deposited at current densities of 20-30 $mA\,cm^{-2}$, showed a metallic luster. The thickness of the Pd layer was approximately 4 μm. At current densities of 5 and 50 $mA\,cm^{-2}$, however, globular and cylindrical crystals, respectively, were observed. The metallic luster disappeared when large crystals were grown.

3.2 Hydrogen permeation

Electrolysis for Pd deposition was carried out at 25-50°C. The current density suitable for the hydrogen-selective membranes was in the range of 15-30 $mA\,cm^{-2}$ at electrolysis temperature of 25°C and 30-45 $mA\,cm^{-2}$ at 50°C. Figure 4 shows the effect of electrolysis temperature on the ideal separation factor of H_2 to N_2. The permeance to H_2 was approximately constant in the range of 25-50°C, but that

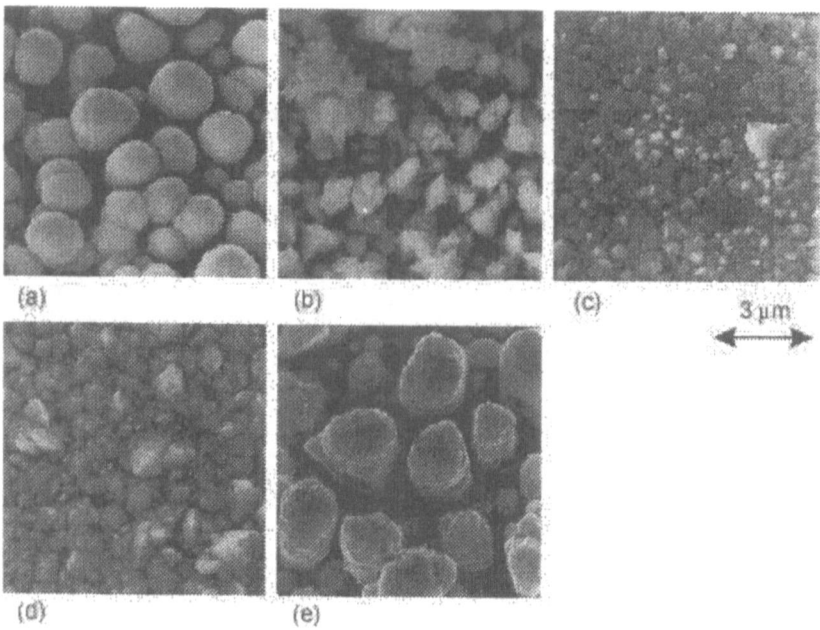

Fig.3 SEM images of the palladium deposition on the copper substrate.
(Current density; (a) 5 mA cm^{-2}, (b) 10 mA cm^{-2}, (c) 20 mA cm^{-2}, (d) 30 mA cm^{-2},
(e) 50 mA cm^{-2}; total electric charge per membrane area = 9000 mC cm^{-2};
electrolysis temperature = 25°C)

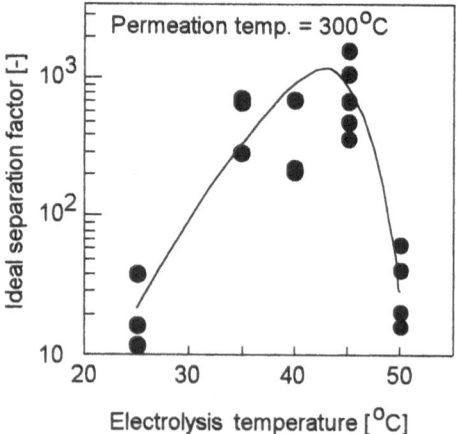

Fig.4 Effect of electrolysis temperature on ideal separation factor of H$_2$ to N$_2$.
(Current density; 15-30 mA cm^{-2} at 25°C, 30-35 mA cm^{-2} at 35°C, 25-40 mA cm^{-2}
at 40°C, 25-45 mA cm^{-2} at 45°C, 30-45 mA cm^{-2} at 50°C; total electric charge per
membrane area = 9000 mC cm^{-2}; permeation temperature = 300°C)

to N_2 showed a minimum in the range of 35-45°C, and, as a result, the ideal separation factor was increased by one order of magnitude in this temperature range.

Figure 5 shows the relationships between permeation properties of H_2 and N_2 and electrolysis time. During the electrolysis, the thickness of the Pd membrane increased, the defects, through which N_2 permeated, were plugged, and the H_2 permeance decreased to some extent. As a result, the ideal separation factor was at a maximum for an electrolysis period of 385 s.

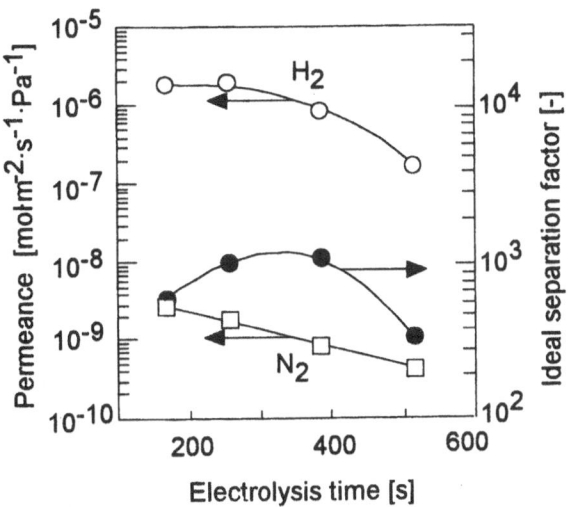

Fig.5 Relationship between permeation properties and electrolysis time.
(Current density = 35 mA·cm^{-2}; electrolysis temperature = 35°C; permeation temperature = 300°C)

Figure 6 shows the influence of permeation temperature on permeance and ideal separation factor. The H_2 permeance increased with increasing permeation temperature. However, the N_2 permeance, which approached the detection limit of the measuring system used, was temperature-independent. The H_2 flux and the ideal separation factor at 300°C were 0.06 mol·m^{-2}s^{-1} and 2000, respectively, at a pressure difference of 0.1 MPa. The ideal separation factor was essentially the same as the separation factor for H_2-N_2 mixtures and was approximately 6000 at a permeation temperature of 400°C. Yan et al. [4] prepared thin palladium membranes inside the porous wall of an alumina support tube using a metal-organic chemical vapor deposition method. The thickness of the Pd layer was approximately 4 μm. The H_2 flux through the Pd/alumina membranes was higher than 0.1 mol·m^{-2}s^{-1} at a pressure difference of 0.1 MPa. The values shown in **Figure 6** are based on the entire area of the disk. However, hydrogen molecules are dissociatively adsorbed on the surface of the entire Pd layer on the feed side, and are associatively desorbed on the permeate side only in the area of the palladium slits. Based on the open area of the microslits, the H_2 flux at 300°C is calculated to be 0.26 $mol \cdot m^{-2} \cdot s^{-1}$, which is high compared with the data reported for palladium membranes formed on porous supports [1].

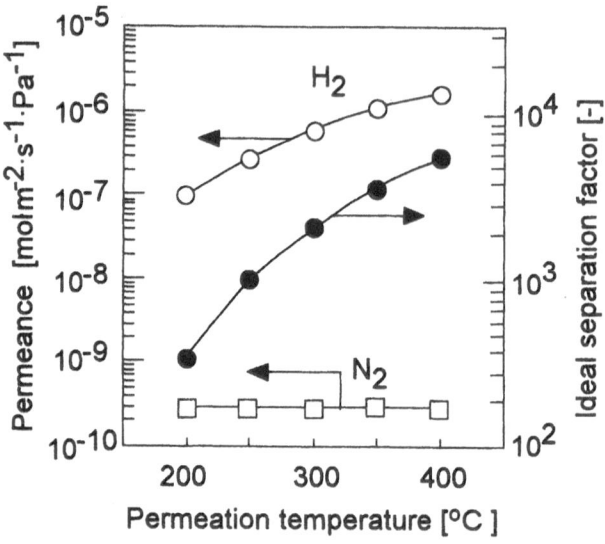

Fig.6 Effect of permeation temperature on permeation properties.
(Current density = 40 $mA \cdot cm^{-2}$; total electric charge per membrane area = 9000 $mC \cdot cm^{-2}$; electrolysis temperature = 45°C)

Conclusions

We have successfully formed thin palladium membranes in microslits of approximately 100 μm in width. The substrate was a 50-μm thick copper plate. Electrolysis conditions, such as current density and electrolysis temperature, strongly influenced the permeation properties through the prepared palladium membranes. The H_2 flux increased with increasing permeation temperature, while the N_2 flux was independent of the temperature. As a result, the H_2 flux through the palladium membrane formed with a current density of 40 $mA \cdot cm^{-2}$ at an electrolysis temperature of 45°C was 0.06 $mol \cdot m^{-2} \cdot s^{-1}$ (H_2 pressure on the feed side = 0.1 MPa), and the ideal separation factor of H_2 to N_2 was 2000 at 300°C.

References

[1] S. Yan, H. Maeda, K. Kusakabe and S. Morooka, "Thin palladium membrane formed in support pores by metal-organic chemical vapor deposition method and application to hydrogen separation," *Ind. Eng. Chem. Res.*, **33** (1994) 616.

[2] A.J. Franz, K.F. Jensen and M.A. Schmidt, "Palladium membrane microreactors," Proc. of IMRET 3, p.267 (2000).

[3] K. Kusakabe, M. Takahashi, H. Maeda and S. Morooka, "Preparation of thin palladium membranes by a novel method based on photolithography and electrolysis," *J. Chem. Eng. Jpn.*, (2001) in press.

[4] S.H. Jung, K. Kusakabe, S. Morooka and S.-D. Kim, "Effect of co-existing hydrocarbons on hydrogen permeation through a palladium membrane", *J. Membr. Sci.*, **170** (2000) 53.

Heating Concepts for Ceramic Microreactors

Regina Knitter, Ralf Lurk, Magnus Rohde, Stefan Stolz, Volker Winter
Forschungszentrum Karlsruhe GmbH, Institute for Materials Research
Postfach 3640, D-76021 Karlsruhe

Abstract. For high-temperature applications of a microreactor system the precise temperature control inside the reactor is important to optimize the yield of chemical reactions. If a microreactor system is operated inside a tube furnace, at best a homogeneous temperature distribution will be achieved inside the reactor. Heating elements that are integrated in the reactor have the potential to realize localized heating. Moreover, integrated heaters offer an efficient heating and a more convenient handling. Ohmic resistance heating for a modular ceramic microreactor system was investigated by direct connection and induction heating. The ceramic reactor system was modeled with FEM methods. Heat transfer and stress were calculated and compared with measurements taken.

Keywords ceramic microreactor, induction heating, screen printing, low-pressure injection molding, simulation, finite element model

1
Introduction

The use of microreactors at high temperatures or for reactions with aggressive reactants makes high demands on the reactor materials. For these applications the use of ceramic materials is of particular interest, as ceramics are superior to other materials due to their high thermal and chemical resistance. A modular ceramic microreactor system has been proposed for high-temperature gas-phase reactions [1], fabricated by means of a rapid prototyping process chain [2]. This reactor system, made of alumina, consists of a reactor housing and exchangeable inner components. In the model shown in figure 1 two gas flows are guided through two flow dividers separated by a partitioning into the reaction chamber that contains two catalyst carrier plates. In the outlet area two additional components are attached to the channels of the catalyst carrier plates.

For many chemical reactions thermal energy is needed to start the reaction or to maintain the reaction temperature, even in the case of exothermic reactions. The integration of heating elements into the reactor system instead of using an external heat source has advantages concerning the precise controlling and distribution of the temperature inside the reactor. Yet an internal heat source, especially inside a ceramic reactor, may also cause specific problems due to mechanical stress and thermal expansion, which have to be solved by material considerations and design.

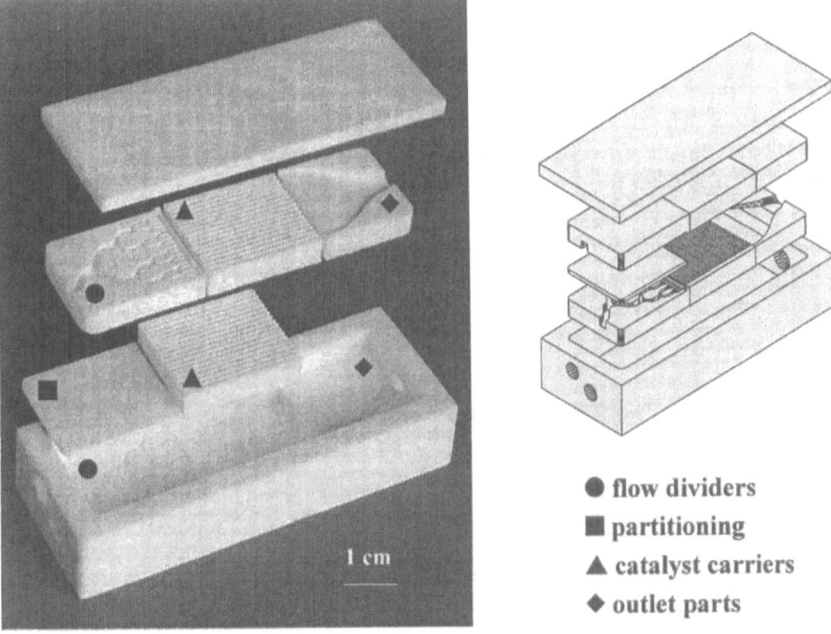

● flow dividers
■ partitioning
▲ catalyst carriers
◆ outlet parts

Fig. 1. Modular ceramic microreactor. (L, 68 mm; W, 25 mm)

A composite of Al_2O_3 / TiN is expected to be appropriate and to meet the requirements as a heater material for the alumina microreactor. As an important item, the coefficients of thermal expansion of both phases are rather similar. Therefore the two phases can be mixed at all ratios without generating intolerable internal stresses when these composites are used as heating devices [3]. Alumina, the insulating matrix, exhibits good thermomechanical properties and a superior resistance against corrosive media compared to most metals or alloys. However, due to the limited oxidation stability of the electrically conductive titanium nitride at temperatures above approx. 400°C, the composite has to be coated.

Depending on the volume fraction of TiN the electrical resistivity of the mixture can be adjusted within a range of several orders of magnitude. Due to the linear dependence of the resistivity on temperature [3], the temperature is easily controlled and measured simultaneously.

2
Experimental

Two different heating concepts were examined to operate the modular ceramic microreactor system at temperatures of about 1000°C. The resistance heating with directly connected bulk heaters and electromagnetic induction heating was investigated. The bulk heaters made of Al_2O_3 / TiN consist of a central heating zone with a relatively low conductivity and two outer sections with a higher conductivity to reduce the temperature at the connection areas. For the induction heating Al_2O_3 / TiN layers with a constant amount of TiN were chosen.

Two different fabrication methods were applied for the preparation of the ceramic heaters. The heating elements for direct connection were manufactured by low-pressure injection molding, similar to the fabrication of the reactor components [2]. Thick film heaters, however, were made by screen-printing on alumina carriers [4]. These devices may be located outside the reactor or integrated into the reactor and operated by electromagnetic induction. Both types of heaters were modeled and the thermomechanical stress and heat transfer were investigated by FEM simulation of the reactor system.

2.1
Low-Pressure Injection Molding of Heating elements

A rapid prototyping process chain was used for the fabrication of bulk heaters. In this process chain the components are designed as 3 dimensional CAD models and then built as polymer models by stereolithography, already taking into account the shrinkage of the ceramic parts during thermal treatment. The polymer models are replicated in silicon rubber molds that are directly used as tools in low-pressure injection molding.

The ceramic feedstocks, containing different amounts of TiN, were made by mixing paraffin, a dispersant, and 60-64 vol.% ceramic powders. The connection areas and the heating segment were separately molded with different feedstocks, joined in the green state and subsequently cofired in N_2 atmosphere at 1750°C for 2 hours. The connection areas and the heating segment contain 40 and 20 vol.% TiN, respectively. The extent of the heating segment was designed to cover the length of the reaction zone when the device is placed on the reactor.

To protect TiN from oxidation at higher temperatures the sintered parts were coated with an alumina layer by dip coating or screen printing and subsequently thermally treated at 1750°C. For electrical connection nickel wires were joined by active brazing.

2.2
Screen-Printing of Heating elements

For induction heating of the alumina reactor system homogeneous layers containing 50 vol.% TiN were fabricated in thick film technology by screen-printing. The pastes were prepared by mixing 35 vol.% ceramic powder with a solution of ethyl cellulose in terpineol and were directly printed on the back of the alumina catalyst carriers. After 10 minutes of leveling at room temperature, and the evaporation of the solvent at 100°C, the samples were sintered in N_2 at 1850°C for 10 minutes. As a coating for TiN an additional alumina layer was printed on top of the

Al_2O_3 / TiN layer and subsequently heat-treated. A layer thickness of 30-50 µm was obtained by each printing step.

2.3
Simulation

Using a finite element model (FEM) the influence of different heating concepts on the thermal behavior of the microreactor were studied. The modular features of the reactor were built into the numerical model. For each module a separate finite element mesh was generated. The exchange of thermal energy between the modules was modeled using thermal surface interactions. This procedure has the advantage that a design variation of a single module can be applied without the need to remesh and recalculate the whole FE model. Therefore, it can be used as a very flexible tool to study the thermal behavior of the reactor and furthermore to predict the effect of design variations on the heat transfer within the microreactor and the development of thermally induced mechanical stresses.

The finite element model of the ceramic reactor is shown in figure 2. For the generation of the FE mesh the preprocessing program FEMGEN / FEMVIEW (FEMSYS LTD.) was used, while the temperature distribution was calculated with the FE program ABAQUS. Since the FE mesh was generated separately for each reactor component the thermal coupling between the single parts was achieved by a given thermal contact conductance between adjacent surfaces. Heat losses through external surfaces were also considered in the model.

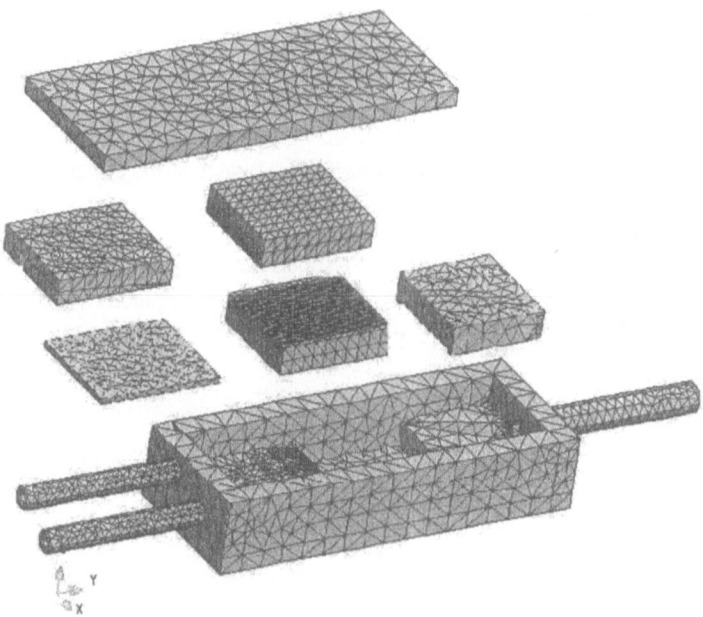

Fig. 2. FE meshing of the components of the ceramic reactor.

3
Results and Discussion

Heating tests of the fabricated heating devices were performed both, individually and together with the ceramic microreactor. A bulk heating element fabricated by injection molding is shown in figure 3. The rectangular heating zone contains two openings to increase the resistance. These devices were operated in air at temperatures of up to 1200°C, generating a heating power of approx. 120 W at 1000°C (Fig. 4). Due to the lower resistance of the connection areas, the temperature of these areas does not exceed approx. 400°C. For the heating test of the ceramic reactor two heaters were placed above and below the reactor housing and connected in series. Preliminary heating tests of the reactor showed, that it is necessary to insulate the housing in order to reduce the heat loss. The temperature was measured with a thermocouple inside the wall of the reactor housing near the catalyst carriers. The electrical voltage was raised stepwise, and thus the temperature was gradually increased to minimize thermomechanical stress inside the reactor. In the chosen arrangement a temperature of 1000°C ,was achieved with a power of about 250 W. The temperature was not affected by a constant air flow of 200 ml/min through the reactor.

For induction heating the samples were placed above the copper induction coil operated with a 3 kW HF-generator. In this case the temperature was measured either pyrometrically or with a thermocouple placed beneath the samples. Temperatures of 700°C were achieved with a single alumina sample coated with an Al_2O_3 / TiN layer. Heating tests of the microreactor were performed with two coated alumina catalyst carriers inserted into the reactor housing. In this case a maximum temperature of 620°C was measured with a thermocouple on the outside of the reactor housing.

The temperature rise in the reactor inside the channel structure of the catalyst module was calculated with the FE model for two heating concepts. Induction heating at the top and bottom surface of the catalyst carriers and also heating with a bulk heating element integrated into the top and bottom plate of the reactor housing were considered. Figure 5 shows the temperature rise as a function of time for the heating concepts. Considering, that the measured temperatures are affected by the heat losses and hence by the chosen insulation of the arrangement, the calcu-

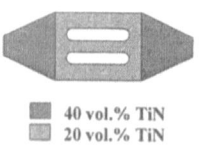
■ 40 vol.% TiN
■ 20 vol.% TiN

Fig. 3. Al_2O_3 / TiN heating element operated at about 1000°C. The connection areas and the heating section contain respectively 40 and 20 vol.% TiN.

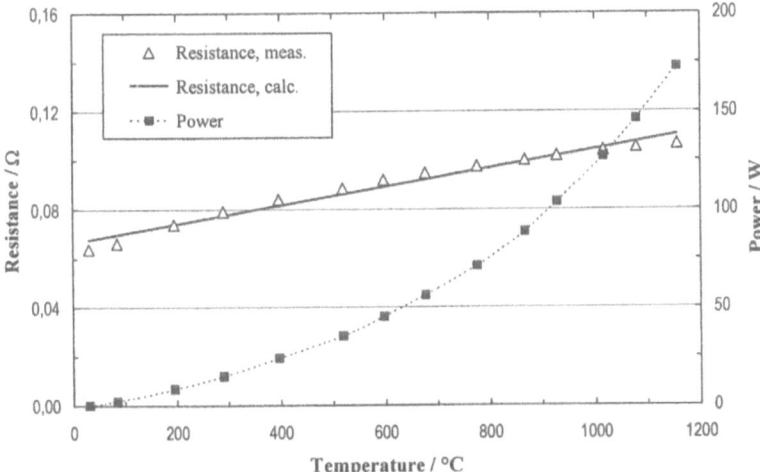

Fig. 4. Increase of resistance and power as a function of temperature for a single bulk heating element.

lated temperature data for the bulk heaters fairly agrees with the measured temperatures at the corresponding level of heating power. For the case of induction heating the comparison between the results of the FE calculation and the temperature measurements can not be made since the exact value of the RF heating power which is coupled into the catalyst carrier can only be roughly estimated.

Additional calculations were performed with the FE model to estimate the thermally induced mechanical stresses in the reactor. Based on the known tem-

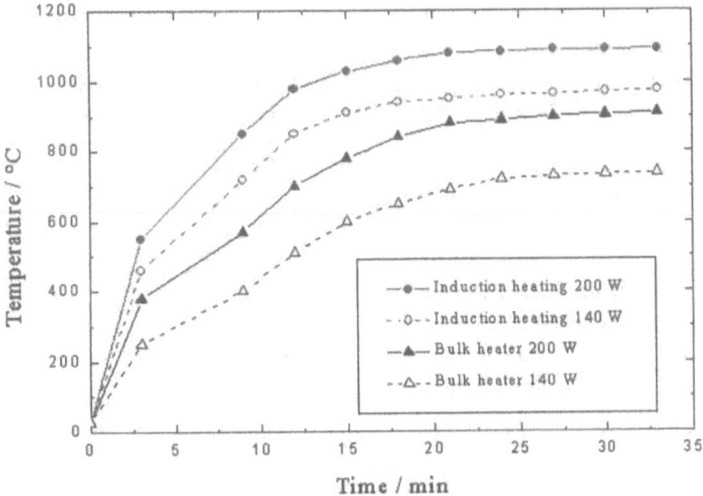

Fig. 5. Temperature rise as a function of time in the reacting zone i.e. within the channel structure of the catalyst carriers.

Fig. 6. Calculated stress distribution (von Mises equivalent stress) on the external surface of the reactor under thermal loading with the bulk heaters.

perature distribution from previous runs the von Mises equivalent stress was calculated using values of 360 GPa for elastic modulus, 0.24 for the Poissons ratio and $8 \ 10^{-6} \ K^{-1}$ for the thermal expansion coefficient [5] at 1000°C. The results of this calculation for the case of thermal loading with the bulk heaters at a maximum temperature of 900°C after 30 min are shown in figure 6. The average stress level obtained from the FE calculation is between 100 and 150 MPa. Only at the top and bottom surface small areas with a stress levels between 200 and 250 MPa can be observed which are due to the integrated bulk heaters. Since typical values of the tensile stress for alumina are between 250 and 280 MPa within this temperature range critical stress values which can lead to failure appeared only at the edges of the reactor. Estimations of the critical stress for failure based on fracture toughness values [5] in the temperature range between 500 and 1000°C and a typical crack length (10-40 μm) in alumina showed that this is higher than 300 MPa. For the case of induction heating the mechanical stresses due to thermal loading were strongly reduced compared to the bulk heating concept.

The bulk heaters offer the advantage of an easy handling and temperature control and they enable the ceramic reactor to be operated at a temperature of 1000°C. The size of these heaters may be adjusted to different purposes, but due to the thermomechanical stress caused by temperature gradients, a defined localized heating of different parts of the ceramic reactor housing cannot be realized. Furthermore, the use is limited, as any integration inside the reactor requires a gastight electrical connection through the reactor housing.

The induction heating enables the integration of the heat source in the reaction chamber without the need of an electrical connection. The inductor used here should be geometrically adapted to the heating application to ensure a homogeneous coupling into both catalyst carrier plates. Furthermore the electrical conductive layer may be printed on different elements, not only on the back of the catalyst carries but also on the elements of the inlet or outlet area. By varying the conductivity of the layer and the applied electromagnetic field, the temperature distribution will be adjusted in accordance with the focused application.

4
Conclusions

Ceramic heaters based on Al_2O_3 / TiN are suited for the operation of an alumina microreactor up to temperatures of at least 1000°C. In addition to bulk heating elements that can be placed as heating plates above and below the reactor, heating devices directly printed e.g. on the catalyst carrier plates were prepared and tested. Especially heating devices integrated inside the reaction chamber and operated by electromagnetic induction are appropriate to heat up the reaction zone efficiently. Thus the temperature distribution inside the reactor can be more precisely controlled compared with an external heat source. This feature will offer advantages for many chemical reactions.

The partial support of this work by the strategy fund of the Helmholtz Association of German Research Centers is gratefully acknowledged. The authors thank Dr. D. Göhring and P. Risthaus for their valuable contributions.

5
References

1. Knitter R, Göhring D, Bram M, Mechnich P, Broucek R (2000) Ceramic Microreactor for High-Temperature Reactions. IMRET 4, 4th Int. Conf. on Microreaction Technology, AIChE Spring Meeting, March 5-9, Atlanta, Georgia, Topical Conference Proceedings 455-460.
2. Knitter R, Bauer W, Göhring D, Haußelt J (2001) Manufacturing of Ceramic Microcomponents by a Rapid Prototyping Process Chain. Adv. Eng. Mater. 3 [1-2] 49-54.
3. Winter V, Knitter R (1997) Al_2O_3 / TiN as a Material for Micro Heaters. Micro Mat, 16.-18. April, Berlin, in: B. Michel, T. Winkler (eds.) Proceedings Micro Materials, ddp golden-bogen, Dresden, 1997, 1015-1017.
4. Stolz S, Bauer W, Ritzhaupt-Kleissl HJ, Haußelt J (2000) Ceramic Thick Films for Heating Applications. Electroceramics VII, 7th Int. Conf. on Electronic Ceramics and Their Applications, Portoroz, Slovenia, September 3-6, Abstract Book 222.
5. Murno R G (1997) Evaluated Material Properties for a sintered α-Alumina. J. Am. Ceram. Soc. 80 [8] 1919-1928.

Design and Fabrication of Zeolite-containing Microstructures

Y.S.S. Wan[1], J.L.H. Chau[2], A. Gavriilidis[1] and K.L. Yeung[2]

[1]Department of Chemical Engineering, University College London, Torrington Place, London WC1E 7JE, UK

[2]Department of Chemical Engineering, Hong Kong University of Science and Technology, Clear Water Bay, Kowloon, Hong Kong

Abstract

Zeolites can be employed as catalysts, membranes and structural materials in the design architecture of miniature chemical devices. Traditional semiconductor technology was employed in the fabrication of micromachined silicon structures. MFI-type zeolites were grown in-situ on silicon supports by hydrothermal synthesis. Another new fabrication procedure was developed based on microelectronic fabrication and zeolite thin film technologies. The fabricated miniature zeolite-based structures can find applications as catalytic microreactors, membrane sensors, electrochemical cells and membrane microseparators.

Introduction

Microtechnology researchers envisage that desktop miniature factories and micro-pharmacies can be developed through advances in the fabrication of micromixers, reactors and separators [1]. Microreactors have the advantages of better energy and material utilization, and result in more efficient chemical production and less pollution. At the same time, separation processes can benefit from the large surface area-to-volume ratio obtained in a microseparator. Chemical engineering separation processes, such as extraction and membrane separation have been successfully miniaturized [2].

In microreactors, an important task is to incorporate active catalyst within microchannels. Metal catalysts can be coated as a thin layer using sputter coating, thermal deposition or chemical vapour deposition [3]. An alternative method is to form a porous oxide layer by anodization, and impregnate it with solution of metal precursor [4]. However, deposition of zeolites requires a different strategy. Zeolite films can be grown directly on microstructured supports by *in-situ* synthesis. Free-standing silicalite-1 membrane has been grown onto silicon microchannels [5]. ZSM-5 zeolite has also been grown on machined stainless steel microchannels for selective catalytic reduction of NO with ammonia [6].

Zeolites are important catalysts in many chemical and petrochemical processes, and more recently, clean fine chemicals production [7]. The aluminum containing MFI zeolite, Al-ZSM5 is an important catalyst for many hydrogenation, disproportionation, isomerization and alkylation reactions [8].

TS-1 zeolite is known to be an efficient catalyst for selective oxidation of alcohols, epoxidation of alkenes and hydroxylation of aromatics [9,10]. Zeolites with their molecular sieving properties [11] and tunable pore structure [12] are also strong candidates for membrane separation in miniature devices. The zeolite pores can host different ions, atoms, molecules and clusters. Their structural, chemical, catalytic, separation, electronic and optical properties can be modified to suit different applications. Zeolite membranes can be utilized in membrane reactor configurations. They can exhibit better yield and higher selectivity than fixed bed reactors [13]. The membrane reactor's ability to control the addition of reactants and the selective removal of products can result in better material utilization, less waste and pollution, and safer operation.

The purpose of this paper is to investigate different designs and fabrication strategies for zeolite microreactors, membrane reactors and microseparators. MFI-type zeolites including silicalite-1 (Sil-1), aluminum ZSM-5 and titanium silicalite-1 (TS-1) were incorporated into the designs as structural material, catalyst or membrane by employing traditional semiconductor technology along with zeolite thin film technology.

Experimental
Microchannel fabrication

A T-shaped design pattern was employed for the microchannel reactor, which is similar to that reported by Srinivasan et al. [14]. This design consists of two inlets and a single outlet for connection to reactant sources and analyzer respectively. At the same time, it provides *in-situ* mixing of reactants after entering the microreactor.

The schematic diagram shown in Fig.1 displays the dimensions of the reactor pattern etched onto a chromium mask. Using standard microfabrication technology, the T-shaped pattern was etched onto the silicon wafer resulting in a trapezoidal channel cross-section with a depth of 250μm.

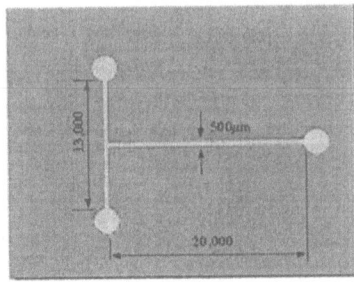

Figure 1. A magnified picture of the T-shaped pattern on a chromium glass mask used in lithography process with the dimensions in μm.

Zeolite synthesis and incorporation

Localized zeolite growth

In this method, zeolite growth is confined within the microchannel. This was achieved through surface modification using chemical functionalization and seeding to locally promote the growth of the zeolite layer. Details of the procedure are described elsewhere [15]. Mercapto-3-propyltrimethoxysilane was introduced in the microchannel, which was then seeded with colloidal zeolite using micropipette (Fig. 2A). The etched microreactor pattern was then placed on a specially designed Teflon holder such that the pattern was facing downwards (Fig. 2B). The sample along with the synthesis solution were transferred into a 125 ml Teflon vessel and sealed within a stainless steel sleeve. The autoclave was placed in a preheated oven (T = 398K) for 24 h. After the hydrothermal treatment, the silicon sample was rinsed with deionized distilled water and dried overnight. The zeolite synthesis solution contained both silica precursor and organic growth directing agent. For silicalite-1, the synthesis solution contained tetraethyl orthosilicate (98%, Aldrich) as the silica source, tetrapropyl-ammonium hydroxide (1M, Aldrich) as the template, and deionized water. Additional components for Al-ZSM5 were alumina sol (Vista) and sodium hydroxide (98%, BDH). For TS-1, the synthesis solution was similar to that of silicalite-1 with tetraethyl orthotitanate (95%, MERCK-Schuchardt) added as the Ti source.

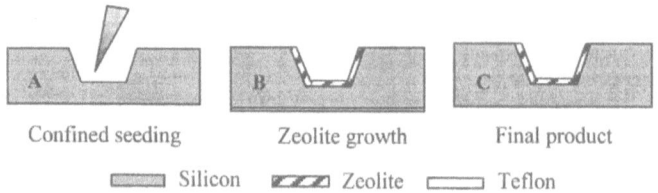

Confined seeding   Zeolite growth   Final product

▭ Silicon   ▨ Zeolite   ▭ Teflon

Figure 2. Process diagram for zeolite incorporation within microchannel.

Etching of zeolite-silicon composite wafer

In this method, zeolite is used as the structural material of the microreactor. The seeding procedure was repeated five times and zeolite growth took place for 48h. The zeolite film was silicalite-1 and the regrowth procedure is similar as above. After depositing photoresist (HPR-207) onto the composite wafer, the micropattern was transferred onto the wafer using standard photolithography (Fig. 3B). The design pattern was then developed following wet etching in BOE solution (1 HF: 6NH$_4$F, Olin) (Figure 3C). The photoresist was stripped using acetone (Lab-scan).

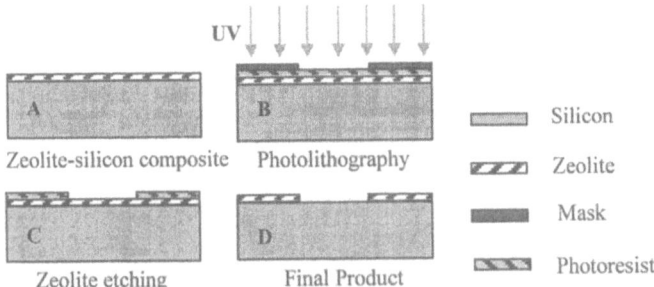

Figure 3. Process diagram for fabrication of zeolite-based micropatterns.

Free-standing zeolite membrane microseparator

The membrane microseparator includes a T-shaped channel on the front and a rectangular recess at the back. Figure 4 illustrates the fabrication procedure. The T-microchannel and rectangular recess were etched onto the front and the back side of the silicon wafer respectively using standard photolithography. The etching time was controlled until the separation between the microchannel and the recess was 50μm. The silicalite-1 zeolite film was then deposited onto the microchannel as mentioned before. The remaining 50μm thick silicon layer was then etched using tetramethylammonium hydroxide (25 % Moses Lake Ind.) to expose the free-standing zeolite membrane.

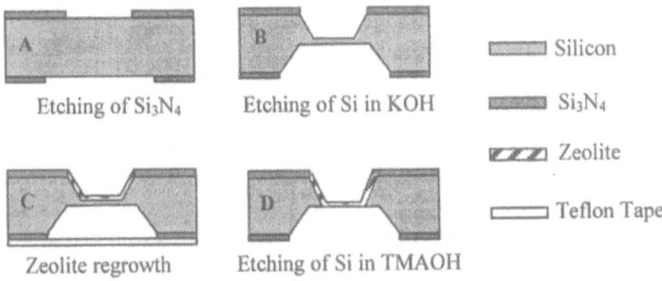

Figure 4. Process diagram for fabrication of zeolite membrane microseparator.

Results and Discussion

Localized zeolite growth

A prepared zeolite-based T-microreactor is shown in Fig. 5. The silicalite-1 film was well intergrown and uniform along the channel length (Fig. 5c). The zeolite crystals were oriented with their <101> crystallographic axis perpendicular to the silicon surface. Only a few zeolite crystals (one zeolite per 100μm^2) were deposited on the surface of the wafer (Fig. 5b). The channel cross-section shows that the zeolite film was 3μm. The film was smooth with a surface roughness (peak to valley) less than 0.3μm (Fig. 5d inset).

ZSM-5 was also deposited within the microchannel. The Si/Al ratio was 2.7 from XPS analysis and the zeolite film was 4.7μm thick with a zeolite loading of about 0.5mg inside the whole channel. ZSM-5 can be readily transformed into protonated HZSM-5 acid catalyst [12]. TS-1 zeolite was deposited in a similar way, a Si/Ti ratio of 17 was obtained using isomorphous substitution of titanium ions into the MFI zeolite framework.

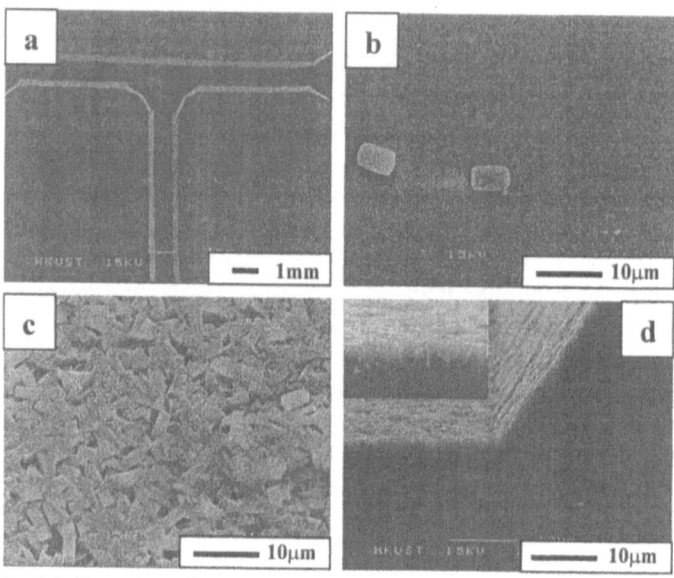

Figure 5. (a) Scanning electron micrograph of the zeolite-based miniature T-reactor. (b) Image of the wafer surface outside the T-reactor channel (c) Microstructure of the zeolite layer grown within the confines of the microchannel. (d) Cross-sectional view of the well-oriented (101) zeolite film within the reactor channel (The figure inset is at 2× higher magnification.)

Etching of zeolite-silicon composite wafer

Different designs of micropatterns were etched onto the zeolite-silicon composite. Complex microchannel geometries and networks, as well as zeolite arrays with high aspect ratio were successfully fabricated (features <10 μm). The zeolite micropatterns were stable even after repeated thermal cycling between 303 K and 873 K for prolonged period of time. Fig. 6 displays the test patterns etched onto a 10μm thick silicalite-1 (101) film. A series of microchannels are shown in Fig. 6a. The cross-section of the channels was rectangular with a width of 5μm. Fig. 6b displays an example of microfabricated zeolite grid. Each line of zeolites forming the 8×16μm rectangular grids was 5μm wide and 10μm high. Microfabricated catalysts can be pre-designed to optimize the reactor performance. Using this technique, complex features as small as 3µm can be precisely reproduced.

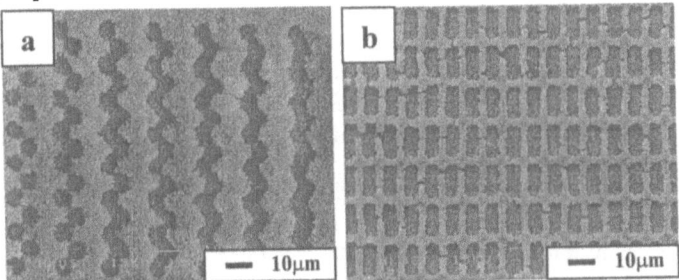

Figure 6. Examples of zeolite micropatterns, (a) microchannel and (b) microfabricated grid.

Fig. 7a displays a T-shaped microchannel employing zeolite film as the structural material of the microreactor. The microchannel is magnified as shown in Fig. 7b. The channel was etched all the way through the zeolite film to expose the silicon surface underneath forming the floor of the channel (Fig. 7c). Zeolite films ranging from less than a micron up to 100µm thick were prepared by varying the synthesis conditions. A modification in the T-shaped microreactor design is shown in Fig. 7d in which the serpentine structure is used to provide a longer channel.

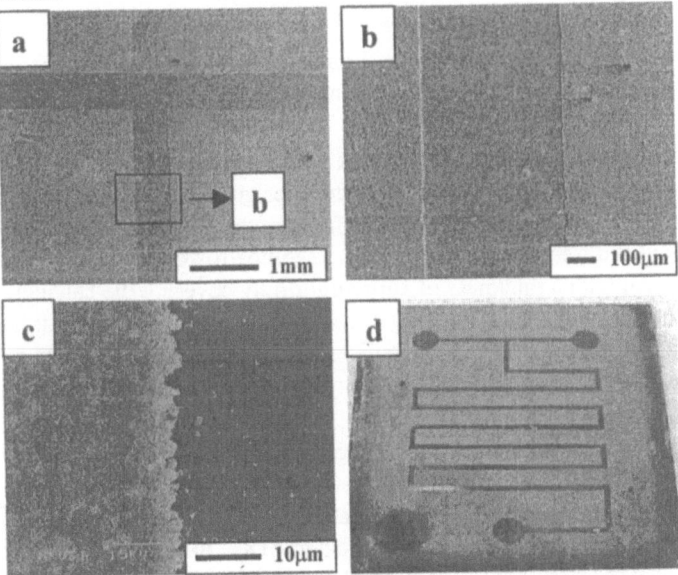

Figure 7. (a) Zeolite-based T-shaped microreactor etched onto a (101)-oriented zeolite silicon composite. (b) Top-view of the reactor microchannel etched through the zeolite layer. (c) Magnified view of the zeolite microchannel. (d) An example of serpentine-shaped zeolite-based microreactor.

Zeolites with their uniform nanometer-sized pores and molecular sieving properties are ideal materials for micromembranes. Miniature electrochemical cells and sensors, in which membrane barriers are an integral part of the design, can also be fabricated using this technique. A simple design for a miniature membrane contactor based on a Wicke-Kallenbach cell is shown in Fig. 8a. The zeolite wall between the two microchannels allows the selective exchange of chemical components between the two passing streams. Hydrophobic compounds can be transferred across the membrane from an aqueous to an organic solution [16]. Fig. 8b displays one possible design for incorporating zeolite membrane in an electrochemical cell and sensor. The I-shaped pattern design consists of four entry ports, two for electrodes and two for fluid inlets. Two exit ports are available to allow continuous flow of electrolyte solution. The two channels are 400μm wide separated by a 300μm thick zeolite membrane barrier. Using this method, zeolite membranes can be fabricated with thickness of 5μm.

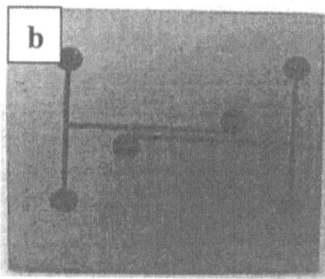

Figure 8. (a) Design of a silcialite-1 membrane microcontactor. (b)An I-shaped pattern fabricated onto zeolite silicon composite.

Free-standing zeolite membrane microseparator

The zeolite membrane microseparator design was fabricated as described earlier in the experimental section. The lighter colour on the microchannel (Fig. 9a) displays the free-standing membrane. Zeolite growth was confined within the microchannel with a thickness of 16μm uniformly covering the channel. A part of this film formed the free-standing membrane. Fig. 9b shows an image of the recess etched onto the back of the silicon wafer exposing the zeolite membrane. As defined by the channel width and length of the recess, the exposed membrane has dimensions of 0.7mm × 4.8mm.

For high throughputs, a much larger membrane area is required for the separation of chemicals. However, scale-up of zeolite membranes is difficult due to stress build-up during high temperature and pressure operating conditions, making them more susceptible to cracking. This limitation can be overcome through miniaturization. Membrane scale-up can be achieved through replication, while maintaining the individual area small to prevent stress-related failure. A larger surface area-to-volume ratio can also be attained in microsystems, which facilitates higher throughputs.

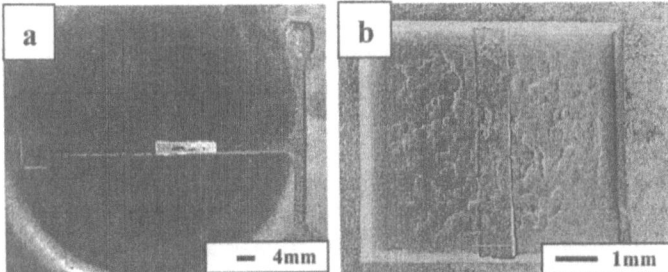

Figure 9. (a) Image of the T-microchannel coated with a layer of silicalite-1 membrane while the light rectangular area is the free-standing zeolite membrane. (b) Scanning electron micrograph showing the recess fabricated at the back of wafer with the exposed zeolite membrane.

Concluding Remarks

A new fabrication procedure was developed for incorporating zeolites catalysts, membranes and structural materials in the design architecture of miniature chemical devices based on microfabrication methods and zeolite thin film technologies. The chemical and catalytic properties of zeolites can be engineered for various application requirements.

Different strategies for incorporating zeolites in microchemical devices were discussed. Zeolite catalysts (ZSM-5 and titanium silicalite-1) were deposited within the confines of microchannels. High-resolution micropatterns of complex microchannel geometry and network, as well as particle arrays were successfully etched onto zeolite silicon composites. The zeolite micropatterns can be fabricated with size of 3 μm and can also withstand prolonged thermal treatment under harsh conditions (temperature oscillation between 303 to 873K). Using this technique, design models of zeolite microcontactor and microelectrochemical cell were fabricated.

A zeolite membrane microseparator was constructed using the new fabrication technique. Silicalite-1 film was deposited within the T-microchannel. The recess at the back of the silicon wafer was etched to expose a free-standing zeolite membrane. Studies are now underway to study the performance of zeolite-based microchemical devices for chemical reaction and separation.

References

1. Kawahara, N., T. Suto, T. Hirano, Y. Ishikawa, T. Kitahara, N. Ooyama and T. Ataka, "Microfactories; new applications of micromachine technology to the manufacture of small products", *Microsystems Technol.*, **3**, 37-41 (1997).
2. Quiram, D.J., I.M. Hsing, A.J. Franz, K.F. Jensen and M.A. Schmidt, "Design issues for membrane-based, gas phase microchemical systems", *Chem. Eng. Sci.*, **55**, 3065-3075 (2000).

102

3. Thomas, J.M., and W.J. Thomas, Principles and practice of heterogeneous catalysis, VCH Publishers, Weinheim, Germany, 1-60 (1997).

4. Wießmeier, G., and D. Hönicke, "Microfabricated components for heterogeneously catalysed reactions", *J. Micromech. Microeng.*, **6**, 285-289 (1996).

5. den Exter, M.J., H. van Bekkum, C.J.M. Rijn, F. Kapteijn, J.A. Moulijn, H. Schellevis and C. I. N. Beenakker, "Stability of oriented silicalite-1 films in view of zeolite membrane preparation," *Zeolites*, **19**, 13-20 (1997).

6. Rebrov, E.V., G.B.F. Seijger, H.P.A. Calis, M.H.J.M. de Croon, C.M. van den Bleek and J.C. Schouten, "The preparation of highly ordered single layer ZSM-5 coating on prefabricated stainless steel microchannels", *Appl. Catal. A.*, **206**, 125-143 (2001).

7. Maxwell, I. E. and W. H. J. Stork, "Hydrocarbon processing with zeolites," *Stud. in Surf. Sci. and Catal.*, **58**, 571-630 (1991).

8. Jacobs, P. A. and J. A. Martens, "Introduction to acid catalysis with zeolites in hydrocarbon reactions," *Stud. in Surf. Sci. and Catal.*, **58**, 445 (1991).

9. Maspero, F. and U. Romano, "Oxidation of alcohols with H_2O_2 catalyzed by titanium silicalite-1", *J. Catal.*, **146**, 476-482 (1994).

10. Clerici, M.G., G. Bellussi and U. Romano, "Synthesis of propylene oxide from propylene and hydrogen peroxide catalysed by titanium silicalite," *J. Catal.*, **129**, 159-167 (1991).

11. Jansen, J.C., D. Kashchiev and A. Erdem-Senatalar, "Preparation of coatings of molecular sieve crystals for catalysis and separation," *Stud. in Surf. Sci. and Catal.*, **85**, 215-250 (1994).

12. Karge, H.G., and J. Weitkamp (Eds.), Molecular sieves, Synthesis : Science and Technology, Vol. 1, Springer, Berlin, 1998.

13. Zaman, J. and A. Chakma, "Inorganic membrane reactors," *J. Membrane Sci.*, **92**, 1-28 (1994).

14. Srinivasan, R., I.M. Hsing, P.E. Berger, K.F. Jensen, S.L. Firebaugh, M.A. Schmidt, M.P. Harold, J.J. Lerou and J.F. Ryley, "Micromachined reactors for catalytic partial oxidation reactions," *AIChE J*, **43**, 3059-3069 (1997).

15. Wan, Y.S.S., J.L.H. Chau, A. Gavriilidis and K.L. Yeung, "Design and fabrication of zeolite-based microreactors and membrane microseparators", *Microporous Mesoporous Mat.*, **42**, 157-175 (2001).

16. Wu, S.Q., C. Bouchard and S. Kaliaguine, "Zeolite containing catalytic membranes as interphase contactors," *Res. Chem. Intermediates*, **24**, 273-280 (1998).

New process for manufacturing ceramic microfluidic devices for microreactor and bioanalytical applications

C. Provin, S. Monneret, H. Le Gall, H. Rigneault*, P. -F. Lenne*, H. Giovannini*

DCPR - ENSIC - 1, rue Grandville – BP 451 – 54001 Nancy Cedex – FRANCE
* Institut FRESNEL - ENSPM – Domaine universitaire de St-Jérôme, 13397 Marseille Cedex - FRANCE

1. Introduction

The advances of the past few years in microreactors have demonstrated that microchips have numerous significant advantages with respect to cost, safety, throughput, kinetics and scale-up [1-3]. The whole aspect of heat management, enabling mass and heat transfer to be extremely rapid, leads to a higher level of reaction control and reactant manipulation at any one point within the chip.

Chips generally referred to as chemical microreactors represent devices that can perform chemical synthesis in volumes less than one microliter. Then, the fluidic channels in microreactors are typically 50-300 µm in diameter. Such values are significantly high to prevent the microfabrication of complete monolithic chips with conventional lithography-based microfabrication processes. Moreover, electrokinetic pumping, which can be considered as the method of choice for transporting liquid samples in microchannels, can not be used with silicon substrates because of the high voltage needed for electroosmotic flow. At last, biomolecules often tend to create a bond to silicon surface groups, which prevent them to circulate freely, unless a surface coating (*e.g.*, silanisation) is deposited on the substrate.

Polymers do not present such drawbacks and then are applicable for electrophoresis flow of biomolecules inside microchannels. As a consequence, they are considered as promising materials for microsystems with regard to microfluidic applications. A review of polymer microfabrication methods for microfluidic analytical applications has been proposed recently [4].

Microstereolithography is one of the process of manufacturing polymer complex microparts [5-9]. It corresponds to a layered manufacturing methodology in which objects are built as a series of horizontal cross sections, each one being formed individually from the relevant polymeric raw materials, and bonded to preceding layers until it is completed.

The aim of this paper is to present a special issue of such a micro-sterolithography (µSL) process, which has been recently developed in order to manufacture freeform three-dimensional alumina-based composite micro-components. Some different types of complex microstructures are presented, which could be used for microfluidic applications. As an example, we also present a chip which has been designed to be driven by electrophoresis flow for bioanalytical application on the molecular level. At last, first examples of sintered alumina microparts are presented. The high potential of such ceramic structures comes mainly from the fact that they tolerate high temperature or chemically corrosive environments, but also because they are considered to be excellent candidates for catalyst supports.

2. Manufacture of composite micro components

Stereolithography has been developed for rapid prototyping applications. It is a process of creating a solid physical object directly from a three dimensional Computed-Aided-Design (CAD) file [10].

The basic principle of the process relies upon a space-resolved laser-induced polymerisation, leading to a liquid to solid transformation. Once the object to be built is designed, it is then numerically sliced to define the different cross sections of the object to be manufactured. Then, the numerical data defining each of the sections are sequentially sent to the machine, in order to built them. The metamorphic transition from CAD model to physical part takes place in a matter of hours.

After demonstrating the reliability of a mask-based micro-SL technique for the microfabrication of polymeric parts [11], we present the apparatus still including a dynamic mask but optimised for the manufacture of alumina-based composite or pure sintered alumina microcomponents, according to the different stages of the process presented on figure 1 [12,13].

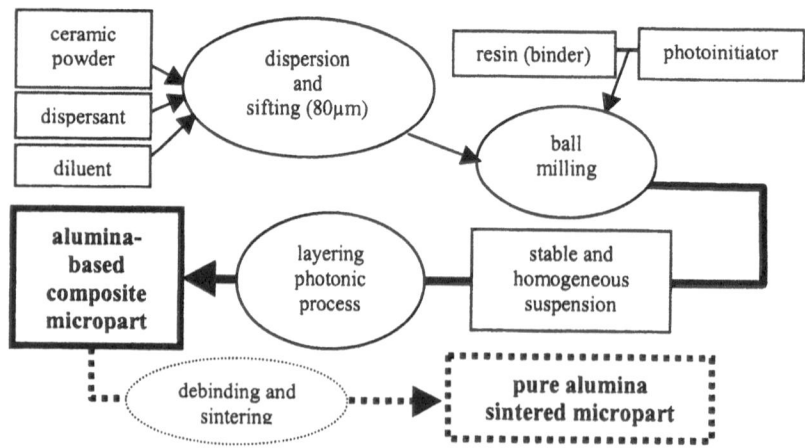

Figure 1 : Global process for the manufacture of ceramic microcomponents

The reactional medium then consists of ceramic particles dispersed in a suitable photocurable resin in which a photoinitiator is dissolved. Once polymerised, the organic phase constitutes a though matrix (binder) around ceramic particles and confers the cohesion to the composite body, which can be directly used as a component for microfluidic applications. But, when pure ceramic parts are needed, it is also possible first to subsequently remove the organic phase by an appropriate thermal treatment (debinding) then to sinter the ceramic powder. Yet debinding and sintering steps make necessary to minimize the organic concentration in the suspension in order to obtain homogeneous and dense ceramic parts. This implies the viscosity of the material to increase, which could pose problems during the manufacturing of the part by the µSL process. That is why most of the microparts which are presented in this paper are composite ones.

Figure 2 shows the experimental set-up of the build-up process. The light beam comes from a high pressure mercury lamp. An adapted optical system allows to image the dynamic binary mask on the surface of the phocurable resin, with a variable magnification in order to control the size of the object. The contrast ratio of the dynamic mask is sufficient to use it also as a shutter. The liquid resin hardens only where touched by the light pattern. Thin layers are sequentially solidified and stacked from bottom to up to create complicated three dimensional shapes, leading to true 3D microparts.

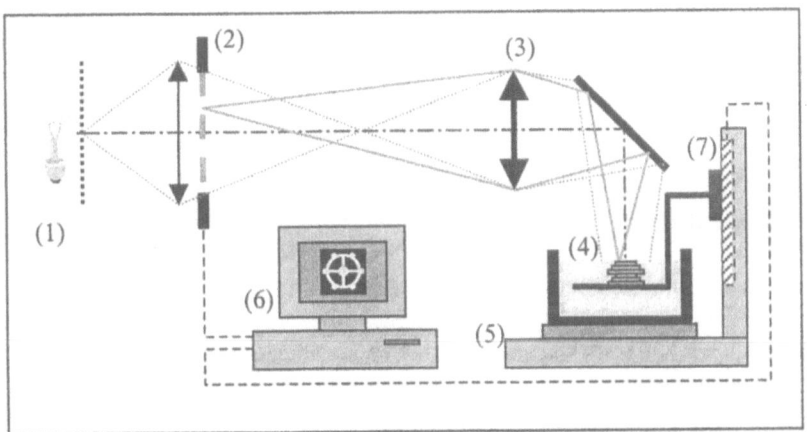

Figure 2 : *Experimental set-up of the actual process*
(1) high pressure Hg light source ; (2) LCD ; (3) imaging system ; (4) photoreactor ;
(5) temperature regulation;(6) computer ; (7) vertical moving stage.

The dynamic binary mask is composed of a liquid crystal display (LCD) spatial light modulator (SLM) which is inserted between two polarizers, enabling the liquid crystals to act as an intensity modulator. The XGA2 SLM was supplied by CRL Ltd. It is a 33mm diagonal poly-silicon TFT twisted nematic LCD, with a pixel pitch of 26x26 μm^2. The contrast ratio of the device is given to be better than 100:1.

The interface allows the SLM to be driven directly by the video adapter of a personal computer set to XGA resolution (1024 x 768 pixels) and running at a frame of 60 Hz., so that it gives a replication of all which is displayed on the computer monitor. In order to display not only the different binary masks but also fabrication and control parameters in real time, a restricted area has been reserved for the masking zone, so that the effective dynamic mask corresponds to a limited area of 640 x 480 pixels only.

The photocurable materials must have a lowest as possible viscosity, according to the process to deposit the fresh layers of resin [11]. No scraper can be used because such a device could damage the micro-objects during their manufacture. So, after the exposure step, the part is largely immersed in the reactive medium and then moved upwards, such that between the last cured layer and the surface there is a new layer of fresh resin. The relaxation time which allows to obtain a homogeneous thickness of the fresh layer of liquid depends mainly on the rheological properties of the resin. Thus, only low viscosity photocurable resins have to be used to decrease both the fabrication time and the layer thickness.

The curable system then consists of a photoinitiator (Irgacure 819, Ciba Specialty Chemicals) dissolved in a low viscosity photocurable monomer (HDDA, UCB Chemicals). The ceramic powder is an α-alumina powder with a mean diameter of 0.5 μm and a specific area of 10.2 $m^2.g^{-1}$ (P172SB, Péchiney). A dispersant is used to decrease the viscosity and to increase the stability of the suspension. The final mixture used in this paper is prepared with 55 wt. per cent of alumina, dispersant (1.5 wt. per cent with respect to alumina powder) and photoinitiator (5.6 wt. per cent with respect to the monomer).

3. Resolution of the process

Because of the photocuring step which is used to manufacture micro-objects, the global resolution of the process comes obviously from the quality of the imaging optical system, but also from the chemical and physical properties of the photocurable mixture.

3.1. Manufacture from pure polymer

In case of pure polymer, the lateral resolution of the process can be considered only as dependent on the imaging system quality. Single layer manufactured objects presenting a pixel size of 2x2 μm^2 have been obtained, showing the reliability of the process [11]. The photocuring of the resin only occurs where it has been illuminated, which defines the lateral object resolution as the imaged mask pixel size on the resin.

On the other hand, vertical resolution strongly depends on the illuminating conditions of the photosensitive material, but also on its absorption and reactivity, because of the absorption of light as it progresses into the liquid. A simple model would give its intensity, in case of a monochromatic light source, as :

$$I(z) = I_0 \exp(- z / D_p),$$

where I represents the intensity of the light inside the material versus its vertical propagation z from the free surface of the liquid, I_0 is the intensity of the incident light beam, and D_p is the penetration depth of the material. D_p is then a fundamental property of a resin.

Progress of photocuring is characterised by the development of stiffness in the liquid resin under the influence of incident light. Any region of liquid resin that is impinged upon by light of the right wavelength undergoes the photochemical polymerisation reactions known as curing. However, it is only after the region receives a certain critical amount of light energy E_c that it begins to gel, or acquire stiffness that indicates it is turning solid.

As a consequence, the values D_p and E_c are of a great importance because, together with the intensity and duration of incident light, they determine the gel depth D_g. It corresponds to the effective depth for which the resin turns solid under a particular exposure.

3.2. Manufacture from ceramic-filled polymer

In case of alumina/polymer composite, the adjunction of ceramic fillers in the photosensitive material strongly modifies the behaviour of the resin under photopolymerisation, because of scattering effects and absorption. The penetration depth of the mixture then strongly decreases, leading to D_p of only a few tens of micrometers. But the scattering of the light, due to the alumina particles, also reduces the lateral resolution, which becomes larger than one pixel size on the photosensitive surface.

As a consequence, it is very difficult to give a single value which would characterise the vertical resolution in the µSL process, because it strongly depends on the shape of the manufactured object, but also on the penetration depth of the raw material. Figure 3 gives a good idea of the ceramic suspension behaviour employed in this paper. Current work is under development to propose a reference micropart which could help to properly estimate the process with respect to its accuracy.

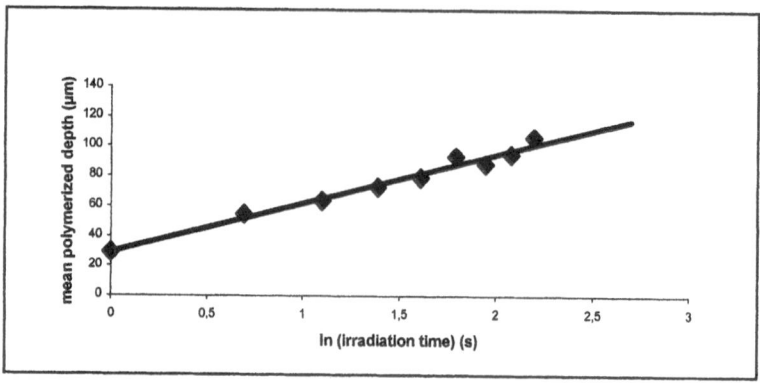

Figure 3 : Polymerized depth (Dg) versus the irradiation time.
HDDA monomer filled with 55% wt of alumina powder.

4. Examples of microfabricated structures for application in microfluidics

This part of the paper is devoted to demonstrate the possibilities of the µSL process when concerned with composite alumina-based microstructures.

Structures which could be used as microexchangers are first presented. Figure 4 shows a component containing two channels. The first one go through the structure, while the second one is circulating 7 times around it. The total size of the part is 7.8 x 3 x 3.3 mm³. Figure 5 shows a detailed view of a central zone of a crossed micro-exchanger structure, composed of two groups of 30 parallel channels, separated with walls of thickness 75 µm. The global size of the part is of 5.5 x 5.5 x 4.3 mm³.

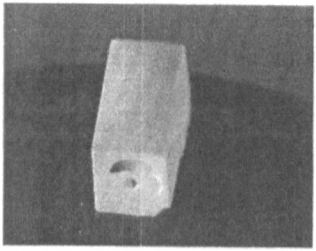

Figure 4 : Composite microstructure with two internal channels.
(110 layers of thickness 30 µm)

Figure 5 : Central microstructure of a composite crossed microexchanger.
(170 layers of thickness 25 µm)

Figure 6 shows a demonstrating part containing several thanks connected together by different kinds of channels, including internal ones. The global size of the part is of 5.3 x 4 x 2 mm³.

 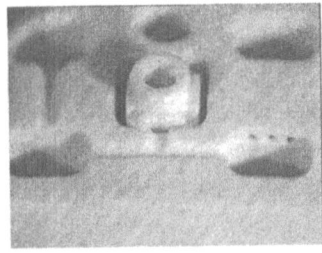

Figure 6 : Composite demonstrating part containing thanks and channels.
(70 layers of thickness 35 μm)

5. Example of a microfabricated structure for a bioanalytical application at the single molecule level

Detection and spectroscopy of single molecule has attracted considerable attention in the past few years at both cryogenic [14] or ambient [15] temperatures. Since the first successful detection of single fluorophore in a relevant biological environment has been reported [16], strong efforts have been brought to study single biomolecules (for a review see [17]). Enormous breakthrough are expected in DNA analysis [18], molecular dynamics [19] and in fundamental studies of physical, chemical and biological phenomena at the single molecule level.

Detection at the single molecule level allows the sample size to be reduced down to the micrometer scale (or even nanometer scale) : averaging over a large population of molecules is not anymore a prerequisite to characterize an assay.

As an example of application, high throughput screening (HTS), devoted to immunoassay and drug manufacturing, uses detection systems that monitor molecular binding at the single molecule level [20]. The holy grail is to achieve 'on line' detection of single molecule flowing through micro-capillaries etched on hard surfaces (biochips).

We have manufactured a custom microstructure in which flowing single molecules of Cyanine 5 (Cy5) can be detected. The monolithic microfluidic structure (figure 7, center), made thanks to our μSL apparatus, is 4 mm long and 2 mm large and permits the flow of dilute solution in a 30 μm deep reaction chamber. The chamber is set in front of a microscope objective lens devoted to detect luminescent molecule passing through (figure 7, left). Molecular motion is detected by the fluorescent correlation spectroscopy (FCS) technique [21,22] which analyses the fluctuations of light coming from molecules passing through a detection volume defined by a confocal microscope apparatus. The detection volume in our apparatus is of the order of 10 femto-liters (10 μm³) and allows single Cy5 molecules to be detected. From a practical point of view, FCS gives the Cy5 mean time of residence in the detection volume while diffusing or flowing through it.

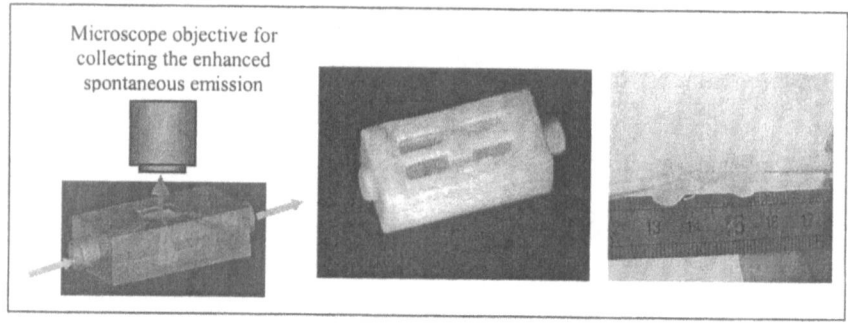

Figure 7 : Example of a microdevice developed for bioanalytical application.
From left to right : CAD design (without the additional external mirror),
ceramic microdevice alone, and with external capillaries.

Figure 8 shows the mean residence time evolution versus applied voltage for negatively charged Cy5 molecules flowing in a glass capillary of diameter 50 μm. At zero voltage, Cy5 diffuse freely and the mean residence time is directly related to the diffusion coefficient. At non zero voltages an electrophoretic flow superimposes on the Brownian motion, shortening the mean residence time. A similar work, under investigation, consists in applying a voltage between the reservoir chamber (containing the luminescent Cy5 negatively charged molecules) and the waste chamber of our μSL device to set an electrophoretic flow in the reaction chamber.

At last, we plan to coat a dielectric mirror on the bottom of the flat detection chamber (see figure 6, left) in order to take advantage of spontaneous emission control for Cy5 molecule flowing above the mirror [23]. In this case, we have shown [24] that the collected signal per molecule is enhanced three fold if the mirror is highly reflecting for the Cy5 luminescence light. This significant improvement in the signal to noise ratio would benefit to high speed test for HTS applications. Alternatively, low efficient luminescent species could be studied with this mirror coated reaction chamber.

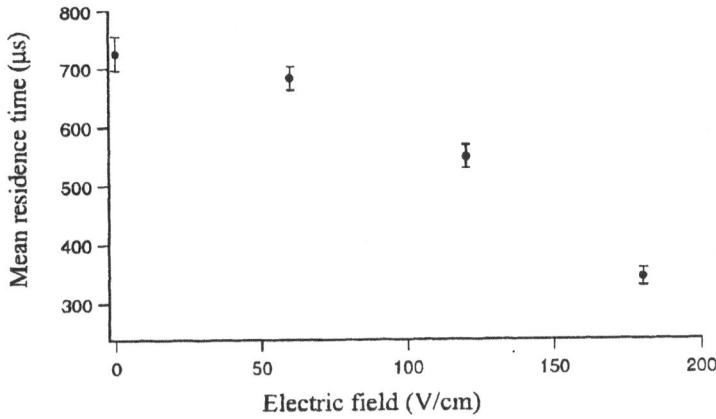

Figure 8: FCS mean residence time evolution versus applied voltage for negatively charged Cy5 molecules flowing in a glass capillary of diameter 50 µm

6. Perspectives: manufacture of sintered microparts

As shown in figure 4, the organic phase of the composite polymerised material can be subsequently removed by appropriate thermal treatment (debinding), and then sintered to give pure ceramic solid components. Indeed, our ultimate goal is to produce not only composite but also pure ceramic microparts, in order to ensure their interesting thermal, chemical and mechanical properties. This requires to find a way to increase the ceramic loading in the photosensitive resin used as a binder. As an exemple, the structure of the figure 7 (center) is a solid sintered micropart, obtained from a mixture prepared with 65% wt. of alumina.

7. Conclusions

A new process for manufacturing ceramic monolithic microchips with a spatial resolution of 30 µm in the three directions of space has been presented, which shows a high potential to develop new types of chemical microreactors or complex microfluidics chips like biosensors. It could also be realistic to use such a microstereolithography technique in order to test the manufacture of Micro Total Analytical Systems (often referred to as µTAS) in a next future.

8. References

1. H. Löwe, W. Ehrfeld, "State-of-the-art in microreaction technology: concepts, manufacturing and applications", Electrochem. Acta **44** (1999), 3679-3689.
2. S. H. DeWitt, "Microreactors for chemical synthesis", Current Opinion in Chemical Biology **3** (1999), 350-356.
3. T. McCreedy, "Fabrication techniques and materials commonly used for the production of microreactors and micro total analytical systems", Trends in analytical chemistry, **19** (2000), 396-401.
4. H. Becker, C. Gärtner, "Polymer microfabrication methods for microfluidic analytical applications", Electrophoresis, **21** (2000), 12-26.
5. C. Carroza, N. Croce, B. Magnani, and P. Dario, "A piezoelectric-driven strereolithography-fabricated micropump", J. Micromech. Microeng., Vol. 5, pp.175-179, (1995).
6. T. Nakamoto, K. Yamaguchi, P. Abraha, and K. Mishima, "Manufacturing of three-dimensional micro-parts by UV laser induced polymerization", J. Micromech. Microeng. Vol. 6, pp. 240-253, (1996).
7. X. Zhang, X.N. Jiang, C. Sun, "Microstereolithography of polymeric and ceramic microstructures", Sensors and Actuators A, Vol. 77, pp. 149-156, (1999).
8. A. Bertsch, S. Zissi, J.Y. Jézéquel, S. Corbel, and J.C. André, "Microstereolithography using a liquid crystal display as dynamic mask-generator", Microsyst. Techn. 3, pp. 42-47, (1997).
9. M. Farsari, F. Claret-Tournier, S. Huang, C.R. Chatwin, D.M. Budgett, P.M. Birch, R.C.D. Young, J.D. Richardson, "A novel high-accuracy microstereolithography method employing an adaptive electro-optic mask", J. Materials Proc. Technol., 107, 167-172, (2000).
10. P.F. Jacobs, Rapid Prototyping and Manufacturing: Fondamentals of Stereolithography, The Society of Manufacturing Engineers, Dearborn, MI, 1992.
11. S. Monneret, V. Loubère, S. Corbel, "Micro-stereolithography using a dynamic mask generator and a non-coherent visible light source", Proceedings of the SPIE **3680** (1999), 553-561.
12. M.L. Griffith, J.W. Halloran, "Freeform fabrication of ceramics via stereolithography", J. Am. Ceram. Soc., Vol. 79, N°10, pp. 2601-2608 (1996).
13. Hinczewski, S. Corbel, T. Chartier, "Stereolithography for the fabrication of ceramic three-dimensional parts", Rapid Prototyping Journal, Vol. 4, N°3, pp. 104-111, (1998).
14. W. E. Moerner, M. Orrit, Science **283**, 1670 (1999),
15. S. M. Nie, R. N. Zare, Annu. Rev. Biophys. Biomol. Struct. **26**, 567 (1997)
16. E. B. Shera, N. K. Seitzinger, L. M. Davis, R. A. Keller, S. A. Soper, Chem. Phys. Lett. **174**, 553 (1990)
17. S. Weiss, Science 283, 1676 (1999)
18. W. P. Ambrose, P. M. Goodwin, J. H. Jett, M. E. Johnson, J. C. Martin, B. L. Marrone, J. A. Schecker, C. W. Wilkerson, R. A. Keller, Phys. Chem. **97**, 1535 (1993)
19. T. Funatsu, Y. Harada, M. Tokunaga, K. Saito, T. Yanagida, Nature **374**, 555 (1995)
20. C. Bühler, K. Stöckli, M. Auer, in Methods and applications of fluorescence spectroscopy , Sringer Verlag, 2000.
21. M. Ehrenberg, R. Rigler, "Rotational Brownian motion and fluorescence intensity fluctuation", Chem. Phys. **4**, 390 (1974)
22. E.L. Elson, D. Magde, "Fluorescence correlation spectroscopy", Biopolymers **13**, 1 (1974)
23. H. Rigneault, S. Monneret, "Modal analysis of spontaneous emission in a planar microcavity," Physical Review A, **54**, 2356-2368 (1996)
24. C. Begon, H. Rigneault, P. Jonsson, J. G. Rarity, "Spontaneous emission control with planar dielectric structures : an asset for ultrasensitive fluorescence analysis", Single Molecules **1**, 207-214 (2000)

Macroscale production of microsystems

M. Schreiber, C.M. Hagg, U. Schubert, A.H. Whitehead

Funktionswerkstoffe F & E GmbH, Marktstrasse 3, A-7000 Eisenstadt, Austria

Objectives and Background

Much effort has been exerted by many researchers in the development of techniques for the production of microsystems and components. The current array of micromachining techniques has largely been developed from the semiconductor industry. For example etching techniques such as electron beam etching, wet chemical etching, focused ion beam machining and constructive techniques such as LIGA.

However, the equipment required to produce such systems is typically large, expensive and suited to a limited number of materials. Furthermore, construction of a single device may require many process steps. The objective of this work was to use proven technologies to form a production process capable of cheap mass production. A further objective was to demonstrate this process by producing microscale, three-dimensional heat-exchange structures.

Heat exchangers are generally comprised of a body, which is made from a highly thermally conducting material, in which a fluid is made to circulate. The fluid transfers heat from heated regions of the body to a radiator, where the heat may be dissipated easily. Such systems are widely used in, for example, cooling automotive engines and electrical transformers.

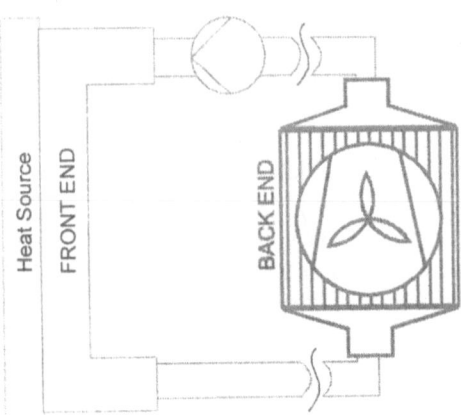

Fig. 1 - schematic representation of a heat exchange system for cooling of electrical components

In general the cooling needs of small electronic components are met with passive heat sinks (elements with a large thermal mass, high thermal conductivity and high surface area) or with forced air-cooling. Central processing units (CPUs) in computers presently require combinations of both passive heat sinks and forced air-cooling. However, the trend is towards more powerful CPUs which will require either more powerful fans, which are undesirable due to the increased noise and cost, or active heat-exchange devices. Cooling of CPUs is one possible application of the devices outlined in this work. A schematic representation of a heat-exchange system is shown in Fig. 1, where the component to be fabricated is denoted as "back end". The back end is used to transfer heat between an internally circulating liquid and the air (which is forced over the surface with a fan). This system has several potential advantages over current systems which just employ a forced-air cooling; including higher heat dissipation (as the back end may have a significantly higher surface area than the device being cooled), and lower noise levels (as the fan and back end assembly may be located remotely).

The "front end" is a second hollow, metallic structure that is connected directly to the device requiring cooling. A pump maintains fluid flow in the system (it would also be possible to vary the pump or fan speeds in a "smart" system to accommodate fluctuating heat output from the device being cooled).

Experimental

Electroplating is often used to produce a metallic layer on a structure. It is a process which may be readily automated. In addition many small items may be simultaneously plated in one bath, significantly increasing throughput. A further advantage of electroplating is that the rate and quantity of metal deposition may be very accurately and easily controlled, simply by adjusting the current and plating duration. Electroplating from aqueous solution is suitable for a number of materials, including copper, which has a very high thermal conductivity (4.0 W cm^{-1} K^{-1}) and good corrosion resistance.

If the underlying structure can be removed without significant expansion then a shaped, metallic body may be left. Classical galvanoforming techniques with removable templates has been used for many decades for producing large structures (e.g. [3]) and has been used recently to produce patterned, porous metallic structures on a meso- to macro- scale (e.g. [1, 2]).

In order to examine the flexibility of this approach it was decided to electroplate polymeric templates to make heat exchangers. It was necessary that the templates could be produced in large quantities at little expense and that their disposal would not create significant pollution. Polymeric templates with complex geometries may be made by a number of standard engineering techniques in large quantities cheaply. The polymers may then be removed thermally or by dissolution. In this instance dissolution was preferred as thermal removal often led to excessive expansion of the template, causing distortion or cracking of the overlying structure.

The polymeric template was electrically insulating and so a thin layer of electroless copper was applied prior to electroplating. The full electroplating process, including cleaning, rinsing, electroless plating and electroplating steps was carried out using commercial baths using an automatic dip-coater to transfer the work piece between baths.

Scanning electron micrographs of cross-sectional pieces were prepared by embedding the fractured samples in an epoxy resin and abrading the surface with progressively finer materials.

Pressure testing was conducted by sealing the back ends after filling with nitrogen at a pressure of 2 bars and monitoring the rate of pressure loss.

Results

Initially polymeric threads (ca. 30 μm diameter) were electroplated to determine whether the copper layer could follow fine structures. From Fig. 2, it may be seen that uniform, continuous layers were deposited (the copper appears light grey). It may also be seen from Fig. 2 that the metallic layer had a relatively smooth outer surface and was of even thickness (ca. 3 μm), as required.

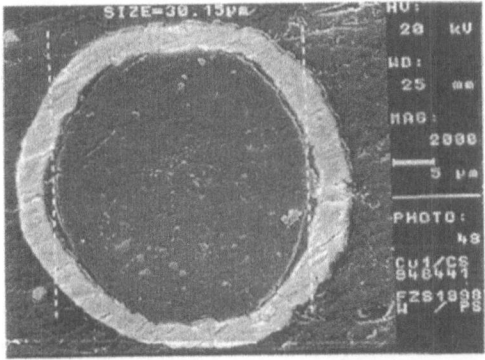

Fig. 2 - Scanning electron micrograph of a sectioned copper-plated polymeric fibre

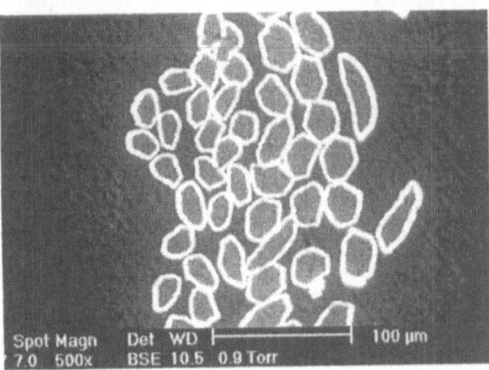

Fig. 3 - scanning electron micrograph of an assembly of copper-plated fibres

Fig. 3 shows an assembly of plated polymeric fibres. It may be seen that the electroplating bath had sufficient "throwing power" to plate each of the fibres uniformly and with no marked variation in copper thickness among the fibres.

Clearly for a heat exchanger the internal surfaces should be smooth so as to ensure laminar flow of the cooling fluid. Local turbulent flows around protrusion, although increasing heat transfer rates, would lead to increased rates of abrasive corrosion, shorting the device lifetime. In addition the plating thickness should be uniform so that no weak spots, which may be liable to failure, are formed.

A heat-exchanger back end was designed such that the internal area remained constant from any section perpendicular to the flow direction. This design was chosen to minimise pressure differences. A further feature of the back end was the inclusion of several through holes to increase strength in the hollow structure. Fig. 4 shows a polymeric template of the back end of the heat exchange system.

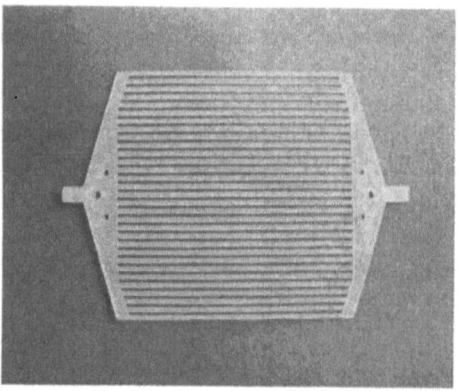

Fig. 4 - polymeric microsystem template

Fig. 5 shows a copper article prepared as above and with the sacrificial core removed, leaving the system with interconnected hollow tubes. It may be seen that the structure was not significantly deformed through removal of the polymer and that the shape of the copper closely followed that of the template.

One problem which can arise during copper plating is the formation of pinholes or other defects due to trapping of hydrogen gas bubbles, which can form as a side product of the plating process. However, from Fig. 5, the copper plating was relatively defect free. Pressure testing also confirmed that no pinholes were present.

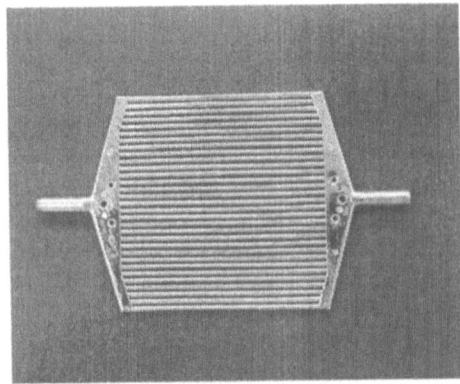

Fig. 5 - copper structure prepared from the template shown in Fig. 4

A back end was cut into sections to reveal the inner surfaces, Fig. 6. Fig. 6(a) reveals the round tube inlet and narrowing body section. Fig. 6(b) and (c) shows more clearly how the body section becomes narrower towards the cooling tubes. The cooling tubes had constant cross-sections throughout their lengths and were each of similar dimensions, Fig 6(d).

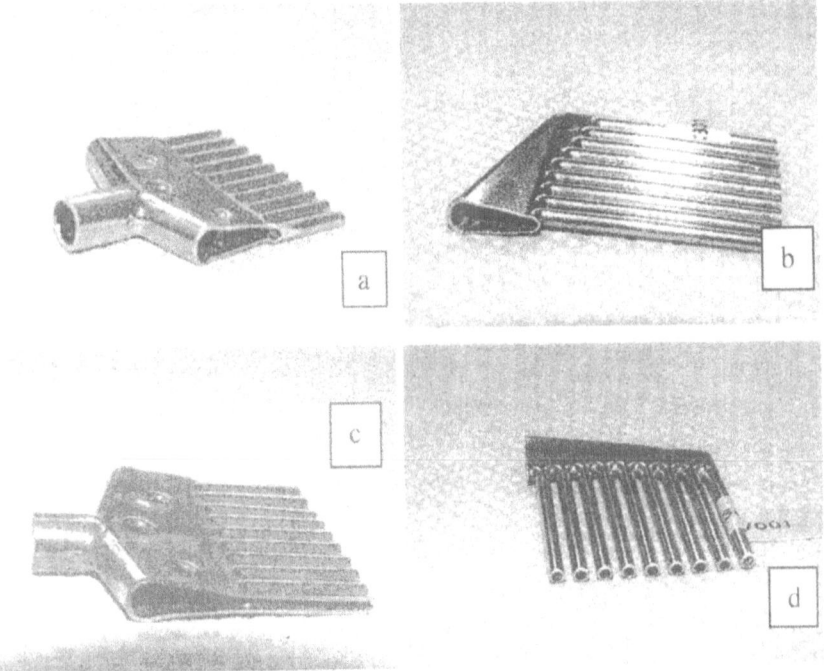

Fig. 6 - sections of a copper "back end" structure

From Fig. 6 it is also apparent that the plating thickness was uniform, as expected from the initial experiments.

Further work

Although copper is a suitable material for forming heat exchangers the hollow pipework structures formed by the methods described here could also be used for microreactors. Depending on the nature of the chemicals used in the microreactors the inner surface of the pipework structure may require a protective treatment. Further work in this area may focus on applying coatings directly onto the polymer template which will then be retained as a protective layer in the final structures. Preliminary work has already indicated that it may be possible to produce graphitic inner-surface coatings by this method.

References

1. A. H. Whitehead, J. M. Elliott, J. R. Owen and G. S. Attard, *Chem. Commun.*, 331 (1999).
2. P. N. Bartlett, P. R. Birkin and M. A. Ghanem, *Chem. Commun.*, **17**, 1671 (2000).
3. D. E. Batty and S. Whyte, in GB Patent **153231** (1920), *to The Associated Equipment Co.*

Part 3

**Mathematical Modelling and Systems
Analysis for Microreaction Technology and
Bioanalytical Devices**

Prediction of the potential of microreactors for heterogeneously catalysed gas phase reactions

Thomas Stief and Otto-Ulrich Langer

Karl Winnacker Institute of DECHEMA e.V., Theodor-Heuss-Allee 25, 60486 Frankfurt am Main, Germany

Abstract: *The practical applicability of microreaction technology is an unsolved problem. A program has been developed for potential users of microreaction technology to predict the feasibility of applying this technology for a specific synthesis in a microreactor.*

The program consists of two parts. In the first part basic process equations and parameters and a simplified model of the microreactor are used to obtain preliminary information on the applicability of the microreactor. The second part provides a more detailed consideration of the potential of applying a microreactor, using a detailed microreactor model (CPU intensive).

This study presents the program structure and features, and also test cases on technically relevant reactions.

Keywords: microreactor, modelling, technical use, potential

1   Introduction

Various studies have shown that it is possible to conduct heterogeneously catalysed gas phase reactions in microreactors. WIEßMEIER [1] investigated the selective gas phase hydration of cyclododecatriene, MAYER et al. [2] showed the possibility of catalytic partial oxidation of methane to syngas, and KURSAWE et al. [3] and KESTENBAUM et al. [4] the general possibility of epoxidation of ethylene to ethylene oxide over silver in microreactors. The course of the reaction is affected by the special conditions in the channels of the microreactor:

- small channel dimensions which lead to laminar flows in the channels
- high surface to volume ratio of the channels
- very good heat and mass transfer properties between the fluid in the channels and the wall material
- defined distribution of residence time.

However, the question whether and, if so, how these conditions in microreactors affect the characteristic performance data of the reactor, remains largely unanswered. In cases where evidence of the practical feasibility of chemical conversions in microreactors is given, hardly any information about their technical applicability is provided.

From a technical point of view it is necessary to verify whether a microreactor obtains the same or even better performance data for a given synthesis than a technical reactor under the same, or microreactor-specific, operating conditions.

2 Objective

The aim of this investigation is to equip a potential user of microreactor technology with a computer program which predicts the potential of microreaction technology for a specific synthesis. The program should, therefore, provide information on

- the basic design of the microreactor
- fundmental process engineering parameters and
- the anticipated reactor performance.

The programme is structured in two parts: a consultation and a verification part. The **consultation part** provides basic information on

- the potential of using the microreactor for a concrete synthesis
- the influence of an approximate isothermal behaviour in the reactor and
- the varied residence time distribution.

The consultation part bases its predictions on simple process engineering equations, parameters and a simplified model of the microreactor (isothermal; plug-flow).

The **verification part** provides a detailed mathematical model, permitting more exact data on the behaviour of the microreactor and the achievable performance data. It is then possible to take into account processes in the microreactor which the simplified model of the consulting part is not able to consider (non-isothermal behaviour, axial dispersion, periodic process control).

3 Status of development

3.1 Consultation part

The consultation part has been developed to the extent that it is possible to obtain information on the basic design and the influence of temperature and residence time distribution. After specifying the performance parameters of a technical reference reactor, the necessary physical characteristics and the reaction kinetics, the following are determined:

- the required microreactor geometry (particularly number of channels, necessary catalytic surface, reactor volume, etc.)

- heat transfer efficiency (heat transfer coefficient medium ↔ reactor material)
- mass transfer efficiency (mass transfer coefficient(s) medium ↔ catalytically active surface) and
- pressure drop.

These parameters are determined in dependence on the channel dimension (the hydraulically equivalent diameter of the channels) which can be varied for any given areas.

Parallel to this the consultation part provides the possibility of investigating the influence of the typical, axial temperature profile in the microreactor (highly isothermal) compared with the technical reactor. For this the achievable performance data of the technical reference reactor and the microreactor are determined in dependence on operating temperature. The variation range of the operating temperature has to be chosen by the program user.

In order to assess the influence of residence time distribution on the performance data of the microreactor these data are determined for the microreactor and the technical reference reactor and also for a CSTR and a tubular plug-flow reactor (the two limits for back-mixing). According to the TAYLOR-ARIS approach, for superimposed laminar flow and molecular diffusion in the microreactor the axial, effective dispersion depends on the channel dimensions. For this reason the above calculation for the microreactor is carried out in dependence on the hydraulic diameter of the channels.

3.2 Verification part

The verification part consists of a program framework which the user has to supplement with the data of the synthesis of interest. This procedure is necessary as, given the number of syntheses, reactor geometries and process control strategies, it is not possible to set up a universally valid model.

Such program frameworks can only be set up for a limited number of "basic models", to date for an uncooled heterogeneously catalysed gas phase reaction and for two thermally coupled, heterogeneously catalysed gas phase reactions. The latter can also be applied to the description of a cooled reaction.

4 Results

4.1 Consultation part

The consultation part was tested on two technically relevant reactions:

1. synthesis of phtalic anhydride (PA) from o-xylene and
2. synthesis of ethylene oxide (EO) by direct catalytic oxidation of ethylene on silver catalysts.

124

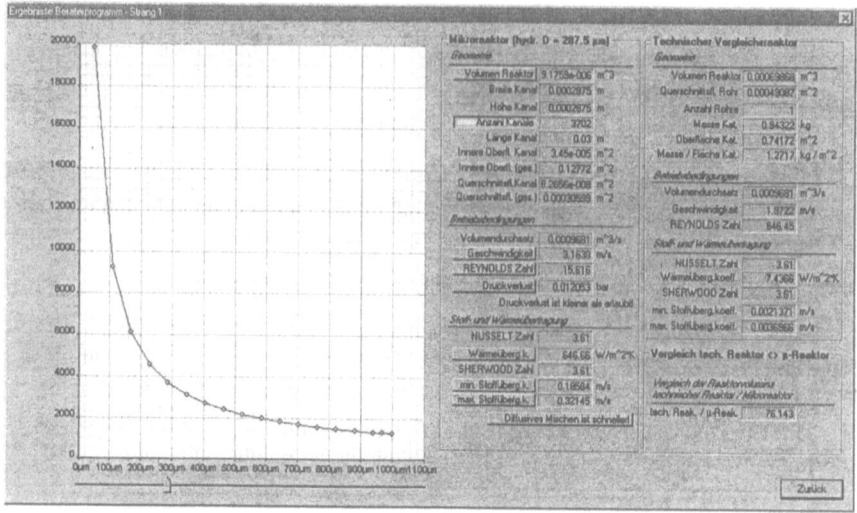

Fig. 1: Dependence of number of channels on hydraulically equivalent diameter
(PA synthesis)

The aim of these investigations was to see whether it is feasible to replace a technical reaction pipe 3 m long and with an inner diameter of 3 cm by a microreactor with suitable dimensions if the channel dimensions in the microreactor are set at 300 μm.

In spite of the analogous reaction schemes, very different results are obtained. Whereas for PA synthesis the reaction pipe of the technical reactor can be replaced by a microreactor with 3,700 channels (see Fig. 1), for EO synthesis a microreactor with more than 600,000 channels is required (see marking in Fig. 2).

Consequently according to the consultation program, in contrast to PA synthesis it does not make sense to carry out EO synthesis in a microreactor on an industrial scale.

Besides the basic data of the microreactor (i.e. number of channels, necessary microreactor volume) the corresponding values for the mass and heat transfer from the medium to the reactor material or to the catalytically active surface of the channels are determined. These data are significant, on the one hand for the quick reactions limited by mass transfer, on the other hand for reactions with high reaction heat. Corresponding parameters are determined in dependence on the channel dimension in the consulting part (see Fig. 2). At this point the good heat and mass transfer properties of the microreactor are clearly evident: even in the case presented here with channel dimensions of 300 μm, a very good heat transfer coefficient of 550 W/m^{2}K is recorded. This value increases considerably with increasingly smaller channel dimensions (NB for the heat transfer of a gas to a solid surface).

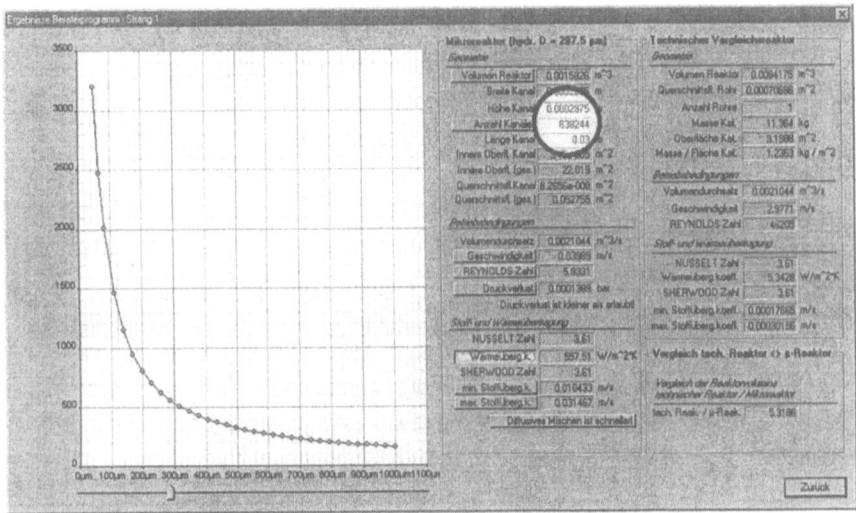

Fig. 2: Dependence of the heat transfer coefficient on hydraulically equivalent
diameter (EO synthesis)

Compared with the technical reference reactor, a microreactor with this good
heat transfer efficiency coupled with a large proportion of good heat conducting
wall material works to a great extent isothermally [5,7]. In order to quantify the
influence of this changed temperature control the performance data of each reactor
is established in dependence on operating temperature (see Fig. 3).

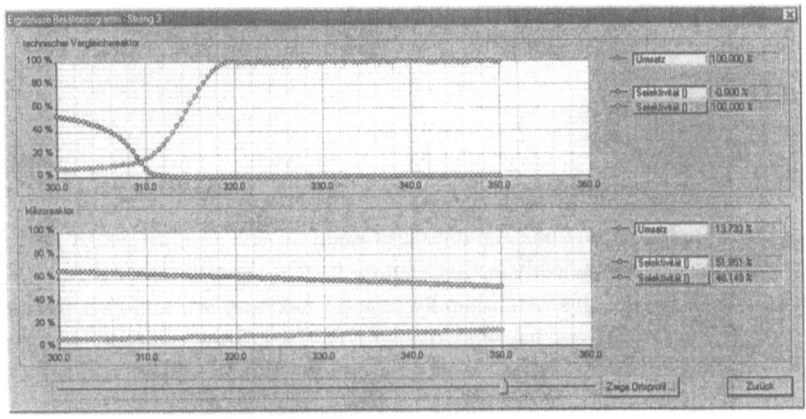

Fig. 3: Dependence of conversion and selectivity on the operating temperature of
the technical reference reactor and the microreactor

In this example (EO synthesis) it can clearly be seen how the technical reactor
"runs away" at operating temperatures above 315°C. This means the educt is
"burnt" completely, yielding total oxidation products (conversion = 100%;
selectivity = 0%).

Such a "runaway" effect does not occur with the microreactor. For this reason it is possible to run the microreactor even at significantly higher operating temperatures, which lead to higher conversions.

4.2 Verification part

One basic model, set up and tested in the verification part, is that of a heterogeneously catalysed gas phase reaction in an uncooled microreactor (i.e. without cooling channels).

This model assumes that the total microreactor can be described by a "representative channel" and the surrounding wall material. This is possible as the individual channels in the microreactor behave similarly – the individual channels are fed with identical educt supplies and are approximately isothermal to each other [7].

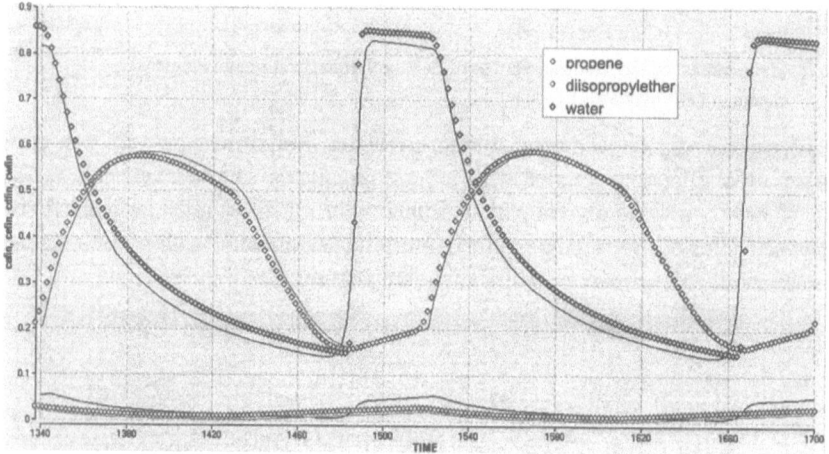

Fig. 4: Comparison of the simulation results with the results of the project partner EPFL

The model includes relations for the description of the mass balances in the channels and of the catalytically active centres in the catalyst material that are occupied by the components. Additionally, energy balances are included for the channels, the catalytically active wall material at the channel surface, and the remaining catalytically inactive wall material. However, first isothermal investigations were carried out, rendering the energy balances irrelevant.

The basic model was tested on the periodic processing of the dehydration of isopropanol to propene. The simulation results were verified on the basis of the simulation calculations of the project partner EPFL [6] and were found to be in good agreement with them (see also Fig. 4).

This reaction is characterized by a "STOP" effect, which means that the reaction is hindered by the educts themselves. In the case of a constant educt composition, only a limited amount of product can be obtained. This can be seen in Fig. 5 in the time domain up to 300 s. Only when the supply of educt is stopped and thus the inhibiting effect decreases to zero, can the educt adsorbed on the catalyst react under production of free catalytic centres and the reaction rate rises rapidly – until the adsorbed educt is consumed. This can be seen in Fig. 5 in the time domain of 350 s to 500 s.

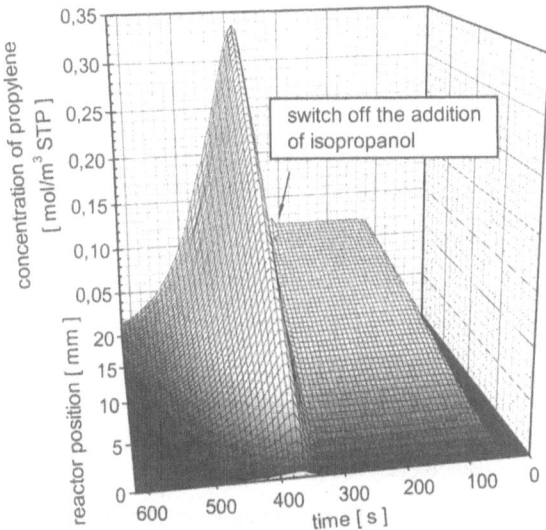

Fig. 5: Dependence of concentration of product propylene on the time and the position in the reactor with a variation of the educt isopropanol

It is immediately clear that with such a reaction mechanism a periodic variation of educt composition can lead to an increase in product yield. With such an operating mode, called periodic process control, the domain where the adsorbed educt is eliminated by reaction is run through periodically and the mean product yield can lie above the stationary yield.

The model was used for a closer analysis of the periodic operation of dehydration of isopropanol. For this the mean propene yield was determined in dependence on the cycle time of isopropanol modulations and the split. The results are shown in Fig. 6.

It was ascertained that with a larger split, that is a greater switch-off/switch-on ratio of the modulations, greater periodic effects occur – albeit both in a positive and a negative sense. It can be observed that in the region of a cycle time of 100 s distinct rises in yield are produced as compared with the stationary yield (detectable for cycle times → 0). With periodic processing a decline in yield to

128

below the stationary yield results in the case of cycle times greater than 500 s. With such long cycle times the educt, which has been adsorbed on the catalyst surface, is completely consumed and the reaction almost comes to a standstill over a long period of time, which is expressed in the depreciation in yield.

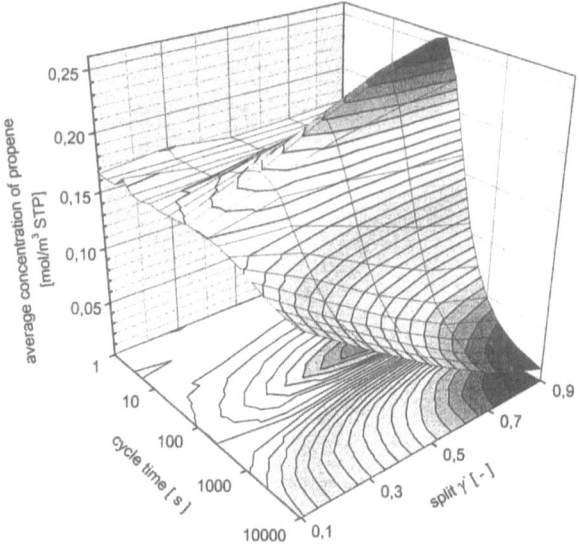

Fig. 6: Simulation results (mean concentration of propene) when dehydrating isopropanol to propene using periodic process control

Periodic processing of the dehydration of isopropanol to propene in the microreactor has a distinct optimum. With a cycle time of 100 s and a high switch-off/switch-on ratio (in this study up to 0.9 investigated) a maximum yield of propene was obtained. At this point the yield increases from 38% to 60%.

These investigations were carried out for various mean residence times of the microreactor in the region of 0.175 to 0.7 s. Table 1 gives a summary of the corresponding results.

Table 1: Optimal periodic processing in dependence on mean residence time

Mean residence time	Yield with stationary processing	Yield with periodic processing	Optimum periodic processing with
$\tau = 0.175$ s	38.6%	60.4%	$\Delta t = 100$ s $/ \gamma^- = 0.9$
$\tau = 0.350$ s	60.5%	81.8%	$\Delta t = 158$ s $/ \gamma^- = 0.9$
$\tau = 0.525$ s	75.4%	89.6%	$\Delta t = 158$ s $/ \gamma^- = 0.7$
$\tau = 0.700$ s	83.2%	92.1%	$\Delta t = 158$ s $/ \gamma^- = 0.7$

It is clearly shown that an increase in propene yield is achievable by using periodic process control, independent of mean residence time. Nevertheless, with higher mean residence times the increase decreases as, in this case, stationary processing already provides correspondingly high yields and a further increase is, of course, scarcely possible.

It is evident that the point of optimal periodic processing is dependent on the mean residence time in the reactor. The great leap in optimum cycle time and optimum split is a consequence of the intercept length used for the investigations.

5 Summary

A program has been developed that enables potential users of microreactor technology to predict the feasibility of applying this technology for specific syntheses. The program consists of two parts. The consultation part supplies fundamental information on applicability, based on basic process engineering parameters and relations, and on a simplified model of the microreactor. The verification part provides the user with a model framework which is used to create a detailed model of the microreactor and can give a more precise prediction of the microreactor's performance characteristics.

When tested, the consultation part produced distinct differences with regard to the feasibility of two reactions with analogous reaction schemes in the microreactor. This has to be attributed, therefore, to the high sensitivity of the microreactor behaviour to the physico-chemical parameters.

The investigation of periodic process control of a reaction with a STOP effect using algorithms supplied by the verification part produced interesting results with regard to the dependence of target quantities (here yield) on cycle time and split. The result is a distinct optimum.

6 Acknowledgement

This research was supported by the Federal German Ministry for Research and Technology (BMBF, contract number 03 C 0282 B 5), which is gratefully acknowledged.

7 Literature Cited

[1] Wießmeier, G.: Monolithische Mikrostruktur-Reaktoren mit Mikroströmungskanälen und regelmäßigen Mesoporensystemen für selektive, heterogen katalysierte Gasphasenreaktionen. Shaker Verlag, Aachen 1997: Dissertation Universität Karlsruhe (TH) 1996

[2] Mayer, J.; et al.: A Microstructured Reactor for the Catalytic Partial Oxidation of Methane to Syngas. In: Microreaction Technology: Industrial Prospects. 3rd International Conference on Microreaction Technology, Frankfurt, Germany, Springer Verlag: Berlin, Heidelberg, New York 2000, 187-196

[3] Kursawe, A.; Hönicke, D.: Epoxidation of Ethene with pure Oxygen as a Model Reaction for Evaluating the Performance of Microchannel Reactors. In: 4th International Conference on Microreaction Technology, Atlanta, GA, Topical Conference Preprints, The American Institute of Chemical Engineers 2000, 153-166

[4] Kestenbaum, H.; et al.: Synthesis of ethylene oxide in a microreaction system. In: Microreaction Technology: Industrial Prospects. 3rd International Conference on Microreaction Technology, Frankfurt, Germany, Springer Verlag: Berlin, Heidelberg, New York 2000, 207-212

[5] Stief, T.; Langer, O.-U.: Modellierung und experimentelle Verifizierung zyklischer Prozeßführungen in Mikroreaktoren. Abschlußbericht zum Teilprojekt des Verbundprojektes MIKREAK (16SV671 to 16SV677)

[6] Rouge, A.; Renken, A.: The dehydration of isopropanol in a microchannel reactor under periodic operation: Performance enhancement. In: Froment, G.F.; Waugh, K.C. (eds.) / Reaction kinetics and the development and operation of catalytic processes, Elsevier Science, Amsterdam 2001, in press

[7] Stief, T.: Numerische Untersuchungen des dynamischen Verhaltens von Mikrostrukturen im Hinblick auf die periodische Reaktionsführung. eingereicht als Dissertation Universität Dortmund 2001

Microchannel reactors for kinetic measurement: Influence of diffusion and dispersion on experimental accuracy

J. M. Commenge, L. Falk, J. P. Corriou, M. Matlosz
Laboratoire des Sciences du Génie Chimique
CNRS - INPL, Groupe ENSIC - Nancy
1 rue Grandville, BP 451, F-54001 Nancy (France)

Introduction

Microstructured reactors offer new possibilities not only for industrial-scale production of fine chemicals but also for kinetic measurements of heterogeneously catalysed reactions. In addition to excellent heat management for accurate temperature control and the ability to work with small volumes of highly reactive or hazardous products, microreactors should also exhibit enhanced mass transfer from the bulk fluid to the catalyst-coated walls due to the small characteristic dimensions of the reactor channels. By eliminating diffusional limitations, microreactors should enable reliable measurement of intrinsic reaction rate constants, even for highly reactive systems, and thereby contribute significantly to shortening delays from laboratory development to industrial scale–up.

The goal of the present study is to evaluate quantitatively, by application of previously published numerical methods, the role of radial concentration profiles and axial dispersion on the determination of kinetic constants from measurements in microchannel reactors. The study indicates that a simple dimensionless group, the heterogeneous Damköhler number representing the ratio of a characteristic heterogeneous reaction rate to a corresponding radial diffusion rate, can be used effectively to determine whether channel dimensions for a given kinetic system are sufficiently small.

The results of the study should be of particular interest for estimating the channel dimensions necessary for reliable use of microreactor devices in chemical research.

Two dimensional model

Governing equations

To quantify diffusional limitations in the microchannel reactors typically employed for kinetic measurement in practice, numerical simulations have been performed to determine axial and radial concentration profiles under laminar flow conditions in tubular reactors with heterogeneous first order reaction at the walls. A detailed review of studies of this type can be found in Cheremisinoff (1986), and the results presented here are based on a model similar to that of Dang and Steinberg (1980).

In an axisymmetric laminar-flow tubular reactor, the equation governing axial and radial species transport can be written in dimensionless form as follows:

$$\frac{\partial^2 C^+}{\partial r^{+2}} + \frac{1}{r^+}\frac{\partial C^+}{\partial r^+} - Pe_D(1 - r^{+2})\frac{\partial C^+}{\partial z^+} + \frac{\partial^2 C^+}{\partial z^{+2}} = 0. \tag{1}$$

The boundary conditions used in this study consider:
(i) a uniform dimensionless species concentration at the inlet of the tubular reactor

$$C^+ = 1 \quad at \quad z^+ = 0, \tag{2}$$

(ii) an axisymmetric concentration profile along the tube axis

$$\frac{\partial C^+}{\partial r^+} = 0 \quad at \quad r^+ = 0, \tag{3}$$

and (iii) a condition at the catalyst coated wall for the heterogeneous first-order reaction

$$\frac{\partial C^+}{\partial r^+} + \alpha C^+ = 0 \quad at \quad r^+ = 1. \tag{4}$$

The dimensionless coordinates, dimensionless concentration and characteristic dimensionless groups used in the equations and boundary conditions are defined as

$$r^+ = \frac{r}{R}, \quad z^+ = \frac{z}{R}, \quad C^+ = \frac{C}{C_0}, \quad Pe_D = \frac{2u_m R}{D} \quad and \quad \alpha = \frac{k_s R}{D} \tag{5}$$

where r^+ is the dimensionless radial coordinate, z^+ the dimensionless axial coordinate, R the tube radius [m], C^+ the dimensionless concentration, C the molar concentration [mol/m^3], C_0 the molar concentration at the tube inlet [mol/m^3], u_m the mean fluid velocity [m/s], D the molecular diffusion coefficient [m^2/s] and k_s the heterogeneous rate constant [m/s]. The solutions only depend on two dimensionless groups. The first dimensionless group is the radial Peclet number for diffusion, denoted Pe_D, which represents the ratio of convective transport in the axial flow direction to radial diffusion from the channel axis towards the reactive channel walls. The second group, α, represents the ratio of heterogeneous reaction rate at the channel walls to radial diffusion from the channel axis toward the walls. The dimensionless group α is frequently referred to as the heterogeneous Damköhler number, usually denoted Da. Nevertheless, to avoid confusion with the axial dispersion coefficient D_A, introduced later, the Damköhler number will be systematically denoted α in the present paper.

Numerical solutions

Numerical solutions of concentration profiles resulting from equations (1), (2), (3) and (4) have been presented by several authors. The solution to these equations in this work follows the approach of Dang and Steinberg (1980), where a full development of the method can be found. The solution is obtained by assuming that radial and axial coordinates may be separated, so that the solution can be written in the form

$$C^+(r^+,z^+) = \sum_{i=1}^{\infty} A_i \psi_i(r^+) \exp(-\lambda_i^2 z^+) \qquad (6)$$

By substituting this development into equations (1) to (4), a set of equations governing the eigenfunctions $\Psi_i(r^+)$ associated with the eigenvalues λ_i can be written. The eigenfunctions $\Psi_i(r^+)$ are then developed on a specified basis of cosine functions.

Numerical determination of the eigenvalues λ_i and the eigenfunctions $\Psi_i(r^+)$ can be performed conveniently with the software code Matlab®. The calculated results are validated by substitution into the governing equations. The number of eigenfunctions is chosen such that the inlet boundary condition is satisfied to within 1 %. The concentration fields obtained are identical to those obtained by Dang and Steinberg (1980) and Solomon and Hudson (1967) for the case without axial diffusion. Further details of the model and of the numerical methods used for the calculations presented here can be found in Commenge (2001).

Characteristic results

The numerical results enable calculation of the concentration field as a function of r^+ and z^+ for a species that undergoes first-order heterogeneous reaction at the channel wall. The radial concentration profile at each axial position can then be deduced in order to quantify the mass transfer limitations. This paragraph presents typical solutions obtained by this method and examines the influence of the parameters involved.

Figure 1 presents dimensionless axial and radial concentration profiles for $\alpha = 1$ and $Pe_D = 10$, corresponding to a microchannel radius R of 100 µm, a molecular diffusivity D of 2.10^{-5} m²/s, a mean fluid velocity u_m of 1 m/s and a rate constant of the heterogeneous reaction k_s of 0.2 m/s.

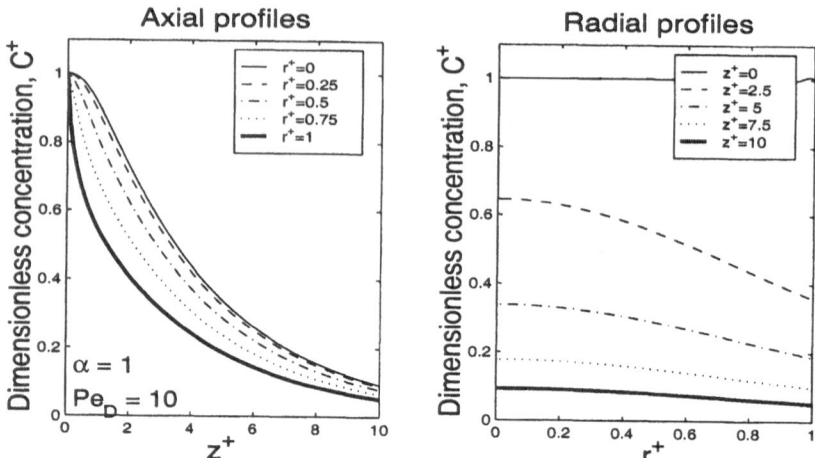

Figure 1: Axial and radial concentration profiles for a single
heterogeneous reaction with $\alpha = 1$ and $Pe_D = 10$.

For rapid heterogeneous reaction, radial diffusion limits the mass transfer from
the bulk fluid to the walls and significant radial concentration gradients can
appear. For $z^+ \geq 2.5$, for example, the concentration at the wall is almost 45 % lower
than the concentration at the channel axis.

Mass transfer between the bulk fluid and the wall for a given reaction can be
improved either by reducing the characteristic dimension of the channel R, or by
increasing the molecular diffusivity D. For a given system, increasing the
diffusivity requires changing the operating temperature or pressure, thus
modifying the kinetics. As a consequence, reducing the channel radius R appears
more adapted for rate constant measurement.

Indeed, reducing the characteristic radius R modifies simultaneously Pe_D and
α. Figure 3 presents the concentration profiles obtained when the radius is divided
by a factor 10, that is $R = 10$ μm for the example above, which leads to $Pe_D = 1$
and $\alpha = 0.1$. The figure clearly shows the enhanced mass transfer between the bulk
fluid and the catalyst coated walls. Radial gradients are practically absent and can
even be neglected. The measured average concentration is very close to that in
contact with the catalyst, allowing reliable measurement for kinetics on the basis
of the difference in average concentration between the inlet and outlet of the
reactor.

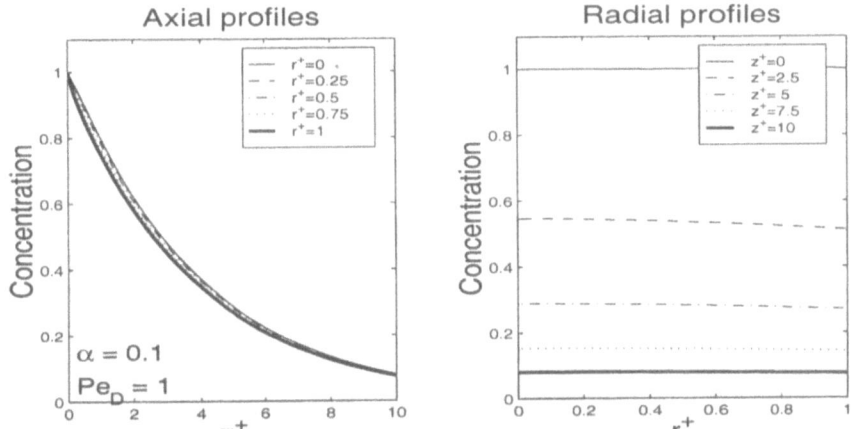

Figure 2: Axial and radial concentration profiles for a single heterogeneous reaction with $\alpha = 0.1$ and $Pe_D = 1$.

Decreasing the radius results in a decrease in radial diffusional limitations, thus in improved concentration uniformity. The study of different cases, with varying values of Pe_D and α, shows that the presence of radial gradients can be considered practically independent of the Peclet number Pe_D in most cases and thus, depend only on the parameter α. The average limiting value between the chemical regime where radial uniformity is verified and the diffusional regime is near $\alpha = 0.1$.

Comparison with pseudo-homogeneous models

The use of a simple plug-flow model for the rate constant determination of a pseudo-homogeneous reaction gives

$$k_{ph} = -\frac{L}{u_m}\ln\left(\frac{C(z=L)}{C(z=0)}\right) \tag{7}$$

where the concentrations C(z=L) and C(z=0) are assumed constant across the radius of the microreactor tube, and

$$k_{ph} = k_s \frac{2}{R} \tag{8}$$

Radial variations in concentration can represent a significant source of inaccuracy in kinetic measurement. The mean concentration over the cross section of the channel, defined as the cup-mixing concentration is equal to:

$$C_{cup} = \frac{\int_0^{2\pi} \int_0^R u(r)C(r,z)r\,dr\,d\theta}{\int_0^{2\pi} \int_0^R u(r)r\,dr\,d\theta} \qquad (9)$$

This cup-mixing concentration, which is the experimentally measured concentration, is larger than the concentration at the wall, that is in contact with the catalyst. In such a case, the concentration difference is likely to cause inaccuracies in the kinetic measurements. In fact, even in the absence of chemical reaction, the laminar velocity profiles in microchannel systems lead to significant axial dispersion that can have a detrimental influence on radial concentration uniformity. For systems without reaction, Taylor (1953) and Aris (1956) proposed a simple one–dimensional equivalent dispersion model, based on the value of Pe_D, that can be used to evaluate accurately the deviation from plug–flow behavior due to the laminar velocity profile.

The pseudo-homogeneous Taylor and Aris model considers that the radial concentration is always uniform, which makes it similar to a plug-flow reactor with axial dispersion. The axial dispersion is introduced by using the Taylor and Aris law, that defines the dispersion Peclet number Pe_{D_A} as a function of the diffusion Peclet number Pe_D

$$\frac{1}{Pe_{D_A}} = \frac{1}{Pe_D} + \frac{Pe_D}{192} \quad with \quad Pe_{D_A} = \frac{2u_m R}{D_A} \qquad (10)$$

where D_A is the effective axial dispersion coefficient that accounts for the influences of both molecular diffusion and the laminar velocity profile. The equation governing the axial mass transfer of a species that undergoes a first-order pseudo-homogeneous reaction is

$$\frac{d^2 C^+}{dz^{+2}} - \frac{Pe_{D_A}}{2}\frac{dC^+}{dz^+} - \beta_{ph}C^+ = 0 \quad with \quad \beta_{ph} = \frac{k'_{ph}R^2}{D_A} \qquad (11)$$

where β_{ph} is a dimensionless parameter representing the pseudo-homogeneous reaction, i.e. the homogeneous Damköhler number, D_A the axial dispersion coefficient [m/s] and k'_{ph} the rate constant of the pseudo-homogenous reaction [s^{-1}] in the plug-flow reactor with axial dispersion. The stable solution of this equation for a unit dimensionless concentration at the channel inlet is

$$C^+(z^+) = \exp\left(\frac{Pe_{D_A}}{4}\left[1 - \sqrt{1 + \frac{16\beta_{ph}}{Pe_{D_A}^2}}\right]z^+\right) \qquad (12)$$

To compare the two-dimensional model and the pseudo homogeneous model, the rate constants k_s and k'_{ph} are related so that equivalent results are obtained when radial uniformity is verified. This is achieved by comparing the pseudo-homogeneous and heterogeneous reaction rates by length unit. Since the two-dimensional model takes into the molecular diffusion whereas the plug-flow reactor with axial dispersion only includes a dispersion coefficient, the relation is different from equation (8). The following equation is obtained

$$k'_{ph} = k_s \frac{2}{R} \frac{Pe_{D_A}}{Pe_D} \tag{13}$$

With this relation, the agreement between both models could be successfully checked when no radial gradients appear.

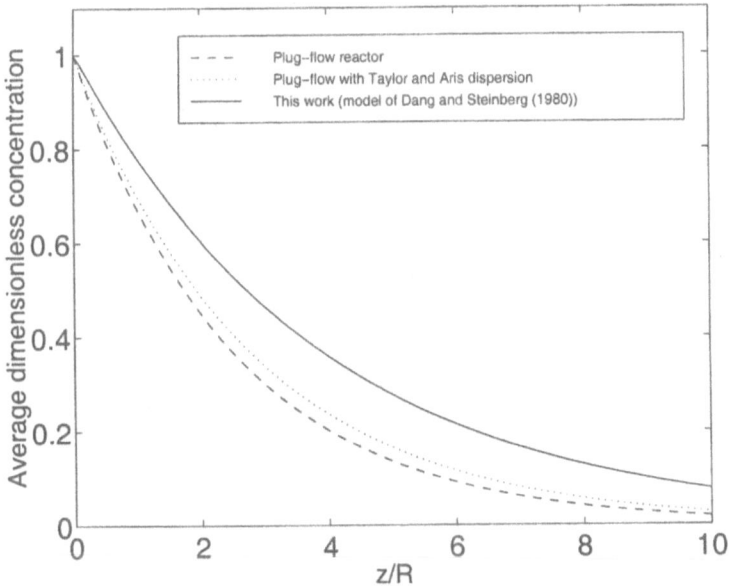

Figure 3: Comparison of axial profiles in cup–mixing average reactant concentration with three models (full line: model of Dang and Steinberg (1980), dashed line: plug-flow reactor, dotted line: plug-flow with Taylor and Aris dispersion). The figure is presented for $\alpha = 1$ and $Pe_D = 10$, corresponding to $R = 500$ μm, $D = 2.10^{-5}$ m²/s, $u_m = 0.2$ m/s and a $k_s = 0.04$ m/s.

Figure 3 compares the axial profile in cup-mixing average concentrations both to simple plug flow and to the pseudo-homogeneous model with Taylor and Aris dispersion. The figure shows clearly that application of the Taylor and Aris dispersion correction alone is not sufficient for an accurate representation of axial concentration profiles for channel structures with significant heterogeneous reaction. The deviation from simple plug–flow behaviour in such systems is primarily due not to Taylor–Aris dispersion but rather to radial concentration gradients created by the high rate of reactant consumption at the channel walls. Determination of kinetic rate constants directly from experimental measurement with the plug–flow assumption requires therefore that the role of radial concentration gradients be taken into account when choosing channel dimensions.

Interest for reliable rate constant measurement

Comparison of simulations calculated over a wide range of operating parameters suggests that the influence of Taylor–Aris dispersion on deviations from plug–flow behavior is frequently much weaker than that of the radial diffusion limitations created by reaction at the channel walls. Under such conditions, values of the dimensionless heterogeneous Damköhler number α lower than 0.1 generally guarantee that radial diffusion will be sufficiently rapid for significant error in rate constant determination to be avoided. Figure 4 presents an example of the variation in error for determination of a first–order rate constant as a function of the ratio α for several cases.

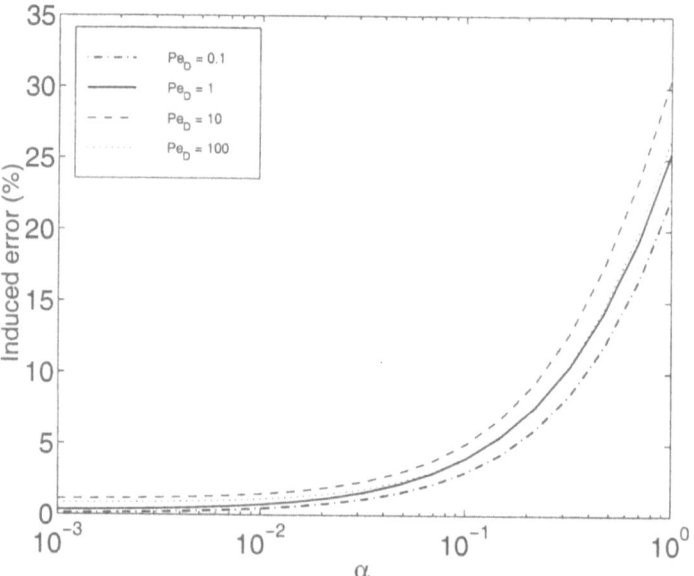

Figure 4: Variation in the error induced on the measured rate constant k_s determined with the plug–flow assumption as a function of the Damköhler number α, for various values of Pe_D, and a microreactor channel length $L = 10\ R$.

The figure shows that for a measurement system with a reactant whose diffusivity is 2.10^{-5} m^2/s and a heterogeneous reaction rate constant of 0.04 m/s, microchannels with a radius $R = 25$ μm, implying $\alpha = 0.05$, would be operating in a transition zone and rate constant determination with the plug–flow assumption would generate less than 3 % error, whereas for the same system a microchannel radius of 250 μm would lead to significant diffusional limitations and a measurement error of practically 20 %. This figure also illustrates the weak influence of the Peclet number Pe$_D$ on the accuracy of the kinetic measurement. Indeed, the error induced strongly depends on the heterogeneous Damköhler number α with only negligible variations induced by Pe_D.

In general, application of the α–criterion suggests that the channel radius R required to avoid diffusional limitations in a given system should be inversely proportional to the rate constant k_s to be measured. Accurate rate–constant determination for a system exhibiting a ten–fold increase in k_s should require, therefore, a ten–fold decrease in channel radius. As a result, use of microchannel systems appears particularly advantageous for rapid heterogeneous kinetics for which diffusional limitations in larger devices could lead to significant measurement error.

Conclusions

This study examines the influence of radial mass transfer on the accuracy of heterogeneous kinetic measurements in microchannel reactors. To quantify this influence, convective diffusion for laminar flow in a tubular reactor has been numerically solved with a first-order reaction at the channel wall. The solutions obtained depend only on two dimensionless groups: the Peclet number Pe_D and the heterogeneous Damköhler number, here denoted α. The two-dimensional solutions are radially averaged and compared to those obtained for a plug-flow reactor with simple axial dispersion and an equivalent first-order pseudo-homogeneous reaction.

The comparison allows quantification of the error induced on the heterogeneous rate constant measurement by the plug-flow assumption. The error is shown to be strongly dependent on the parameter α, whereas Pe_D exhibits only a weak influence. Values of α lower than 0.1 limit the error in rate constant measurement to less than 3 %. The study confirms the intrinsic advantages of microstructured reactors for the study of rapid heterogeneous reactions.

Notations

A$_i$	eigenfunction coefficients	[-]
C	concentration	[mol/m^3]
C$^+$	dimensionless concentration	[-]
C$_0$	inlet concentration	[mol/m^3]
D	molecular diffusion coefficient	[m^2/s]

140

D_A	Taylor and Aris dispersion coefficient	[m^2/s]
k_s	heterogeneous rate constant	[m/s]
k_{ph}	pseudo-homogeneous rate constant	[s^{-1}]
k'_{ph}	pseudo-homogeneous rate constant	[s^{-1}]
L	channel length	[m]
R	channel radius	[m]
r^+	radial dimensionless coordinate	[-]
r	radial coordinate	[m]
u_m	mean velocity	[m/s]
z	axial coordinate	[m]
z^+	axial dimensionless coordinate	[-]
λ_i	eigenvalue	[-]
Ψ_i	eigenfunction	[mol/m^3]
α	heterogeneous Damköhler number	[-]
β_{ph}	homogeneous Damköhler number	[-]
Pe_D	diffusion radial Peclet number	[-]
Pe_{D_A}	dispersion radial Peclet number	[-]

References

Aris, R., "On the dispersion of a solute in a fluid flowing through a tube", Proc. Roy. Soc., vol. 235, pp. 67-77, 1956.

Cheremisinoff, N. P., "Handbook of heat and mass transfer", Gulf publishing company, 1986.

Commenge, J. M., "Modélisation de microréacteurs en Génie Chimique", Ph. D. Thesis, Institut National Polytechnique de Lorraine, Nancy, France, 2001.

Dang, V. D. and Steinberg, M., "Convective diffusion with homogeneous and heterogeneous reactions in a tube", J. Phys. Chem., vol. 84, pp. 214-219, 1980.

Solomon, R. L. and Hudson, J. L., "Heterogeneous and homogeneous reactions in a tubular reactor", AIChE J., vol. 13, n° 3, pp. 545-550, 1967.

Taylor, G. I., " Dispersion of soluble matter in solvent flowing slowly through a tube", Proc. Roy. Soc., vol. A 219, pp. 186-203, 1953.

A vertically-averaged formulation of wall catalytic reactions in microchannel flows: single isothermal & non-isothermal reactions

D. Gobby[1], I. Eames[2] and A. Gavriilidis[1]

[1]Department of Chemical Engineering, University College London
Torrington Place, London WC1E 7JE

[2]Department of Mechanical Engineering, University College London
Torrington Place, London WC1E 7JE

Abstract

A single reaction in a parallel plate catalytic wall microreactor under isothermal and non-isothermal conditions is studied within an analytical framework. Microreactors are characterised by small values of Peclet number to aspect ratio of the duct (Pe_m/R) and this is exploited in the mathematical treatment of the problem. When axial diffusion can be neglected, the solution of the governing equations reduces to determining a single eigenvalue for the reacting species. The vertically-averaged species concentration is then determined from first order differential equations. The analytical solutions developed are in excellent accord with full numerical solutions.

1 Introduction

Microreactors are a relatively new type of reactor which are undergoing significant development. The main benefits are high heat and mass transfer rates in comparison to macroscale apparatus, and this is attributed to their small dimensions, with typical channel widths 100-1000 μm, which results in high surface area to volume ratio (Wegeng et al. 1996, Knight et al. 1998, Hessel et al. 2000). The low reaction volume inherent in microscale reactors enables exothermic reactions to be carried out safely at higher temperatures and pressures (Veser, 2001). Furthermore, uniform velocity fields within the microreactors, short reaction pathways, residence times and quench rates all provide the potential for greater product selectivity through the avoidance of side reactions (Hsing et al. 2000). Such benefits have already been demonstrated for various reaction systems such as oxidations (Franz et al. 1998) and hydrogenations (Weißmeier and Hönicke 1998). Modelling and simulation approaches of microreaction systems have been typically numerical (Hsing et al. 2000) or CFD based (Bibby et al. 1998).

Prototype catalytic microreactors are costly to construct and can present manufacturing difficulties. In order to enable rapid prototyping of these devices and optimising the performance, it is useful to have simple reactor models, which include the appropriate mass and energy balances.

There is a large amount of literature on the solution of advection-diffusion partial differential equations (PDE's) which are analyzed often by means of orthogonal eigenfunction series expansions. Such complex expressions do not easily give insight into the dominant parameters. A note must be made of the works by Denbigh (1951) and Cleland & Wilhelm (1956) where, in the limits of slow and fast radial diffusion, analytical results for second order homogeneous systems have been obtained.

Simultaneous solution of the mass and energy balances (non-isothermal systems) has been mainly numerical. Heck & Katzer (1976), Hayes & Kolackzowski (1994), and Groppi et al. (1995) have all considered lumped and distributed numerical models for highly exothermic reactions.

In this work, a one-dimensional model is presented in §2 for isothermal systems by averaging the advection-diffusion equation describing the evolution of concentration of chemical components. The new model permits the key physical parameters, which control reactor performance, to be identified via closed form expressions. To illustrate how the method of solution may be applied to microchannel flows a parallel plate geometry is considered with a single first order reaction. The analysis is extended to a non-isothermal single first order reaction in §3, and main conclusions are drawn in §4.

2 Vertically-averaged formulation for isothermal flows

2.1 Defining equations

The mathematical model is based on laminar incompressible microscale flow between two parallel plates, where a heterogeneous first order catalytic reaction occurs only on the lower plate (A → B). Modelling assumptions applied are constant fluid properties, no volume change and dilute reacting solution. When $Pe_m \gg 1$, axial diffusion can be neglected and the mass balance for species A is

$$\frac{Pe_m}{R} u(\eta) \frac{\partial \theta_a}{\partial \zeta} = \frac{\partial^2 \theta_a}{\partial \eta^2}, \qquad (1)$$

where $u(\eta)$ is the velocity profile, θ_a is the dimensionless concentration, ζ and η are the dimensionless axial and vertical coordinate respectively, $Pe_m = Uh/D$ and $R = L/h$ (aspect ratio). The boundary conditions for the system are,

$$\theta_a\left(\eta, 0\right) = 1, \quad \frac{\partial \theta_a\left(1, \zeta\right)}{\partial \eta} = 0, \quad \frac{\partial \theta_a\left(0, \zeta\right)}{\partial \eta} = Da\theta_a\left(0, \zeta\right), \qquad (2)$$

where Da, defined as $Da = kh/D$, is the Damköhler number which characterises the ratio of diffusive to reactive timescales.

2.2 Vertically-averaged solution

The above model is simplified by considering how the reactant concentration averaged across the channel, $\overline{\theta_a}$, where

$$\overline{\theta_a}(\zeta) = \int_0^1 \theta_a(\eta, \zeta)\mathrm{d}\eta, \qquad (3)$$

varies with downstream position. A separable solution of the form $\theta_a(\zeta, \eta) = \overline{\theta_a}(\zeta)f_a(\eta)$ is assumed (where f_a is the vertical function). The solution method employed is essentially equivalent to the eigenvalue method proposed and applied by previous researchers (Walker 1961).

For microreactors a key variable that controls reactor performance is Pe_m/R, typically $O(0.01 - 1)$. Under these conditions only the first term in the eigenvalue expansion is important (Walker 1961), and it is this term we seek in the formulation proposed.

If the separable form is substituted into (1) and separated then

$$\frac{\mathrm{d}\overline{\theta_a}}{\mathrm{d}\zeta} = -\lambda_a\overline{\theta_a}, \qquad (4)$$

and

$$-\frac{\lambda_a Pe_m}{R}u(\eta)f_a = \frac{\mathrm{d}^2 f_a}{\mathrm{d}\eta^2}, \qquad (5)$$

where λ_a is a concentration decay constant. Integrating (4) and applying the entrance condition $\overline{\theta_a}(0) = 1$, gives

$$\overline{\theta_a}(\zeta) = \exp(-\lambda_a\zeta). \qquad (6)$$

Reactor performance is weakly dependent on the flow profile when $Pe_m/R < 1$ (Gobby et al. 2001), hence (5) can be solved for a plug velocity profile. Equation (5) reduces to a constant coefficient second order ODE and imposing flux conditions on reactor walls yields the following solution for the decay rate

$$\alpha \tan \alpha = Da, \quad \text{where} \quad \alpha = \sqrt{\lambda_a Pe_m/R}. \qquad (7)$$

In the limits of slow and fast reactions corresponding respectively to $Da \ll 1$ and $Da \gg 1$, the decay rate of A can be determined from

$$\frac{\lambda_a Pe_m}{R} = \begin{cases} 3\left(\frac{Da}{2}+1\right) - \sqrt{\left(\frac{9Da^2}{4}+3Da+9\right)}, & Da < 2.0, \\ \left(\frac{\pi Da}{2(Da+1)}\right)^2, & Da > 2.0. \end{cases} \qquad (8)$$

In the limiting case of $Da \gg 1$, the decay constant tends to $\lambda_a \to \left(\frac{\pi}{2}\right)^2 \frac{R}{Pe_m}$ and the conversion of A $\left(X = 1 - \overline{\theta_a}\right)$ tends to $1 - \exp\left(-\left(\frac{\pi}{2}\right)^2 \frac{R}{Pe_m}\right)$. Thus, even in the limit $Da \to \infty$ (or more specifically for $Da > O(10)$), the conversion of A is not complete as the reactor begins to operate in the mass transfer limited regime (some packets of fluid cannot reach catalyst) whilst convective transport still removes reactant from the system. This can also be seen from (8) where the decay group reaches a threshold with increasing Da. However when $Pe_m/R \to 0$, complete conversion can be achieved.

2.3 Entrance length considerations

The vertically-averaged solutions indicate that the concentration profile is ultimately determined by the ratio $\lambda_a Pe_m/R$ and Da. However the vertically-averaged solution method is not able to account for entrance effects which arise from the fact that the concentration of reactant is uniform at the channel entrance cross section (i.e. a Dirichlet boundary condition). To determine the ability of the analytical solutions to describe the reactant concentration, a full numerical code was written to solve the governing equation (1) and determine the influence of entrance effects on the downstream development of the concentration profile.

The time taken for reactant to diffuse from the upper wall to the lower wall is $O(h^2/D)$, and in this time a parcel of fluid has been advected a distance $O(Uh^2/D)$. In order that entrance effects are negligible, this distance must be smaller than the length of the channel L, which necessarily requires $Uh^2/D \ll L$ or $Pe_m/R \ll 1$. After solving (1) the vertical function f_a is calculated from $f_a(\eta, \zeta) = \theta_a(\eta, \zeta)/\overline{\theta_a}(\zeta)$. This function provides an appreciation of how transverse concentration profiles change and is plotted in Fig. 1 at downstream positions $\zeta = 0.05, 0.1, 0.15$ when $Pe_m/R = 1.0$ and $Da = 5.0$; the vertically-averaged solution is plotted as a solid line. The figure shows that convergence to the vertically-averaged solution is good, even through the condition Pe_m/R being small is not strictly true.

3 Extension to non-isothermal flows

In order to further illustrate this technique, we proceed to examine non-isothermal flows, where heat is produced on the catalyst by an exothermic

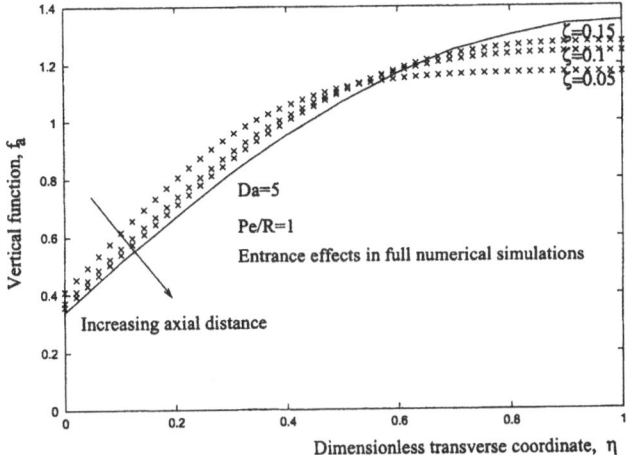

Figure 1: Vertical function profiles of reactant A at three positions downstream from the inlet for $Pe/R = 1$ and $Da = 5.0$. The analytical vertical function is shown as a solid line.

reaction, and no heat exchange with the surroundings takes place. The dimensionless mass equation is given by (1) where the mass flux condition on the catalyst is now

$$\frac{\partial \theta_a(0,\zeta)}{\partial \eta} = Da \exp\left(\gamma\left(\frac{y}{y+1}\right)\right)\theta_a(0,\zeta), \qquad (9)$$

where $y = T/(T - T_0)$, $\gamma = E_a/RT_0$ and the Damköhler number is now defined as $Da = A\exp(-\gamma)\,h/D$. This equation is solved in conjunction with the energy equation, which takes on a similar form

$$\frac{Pe_e}{R}u(\eta)\frac{\partial y}{\partial \zeta} = \frac{\partial^2 y}{\partial \eta^2}, \qquad (10)$$

where $Pe_e = U\rho c_p h/k$, with the inlet condition

$$y(\eta,0) = 1. \qquad (11)$$

The boundary conditions for the energy balance are

$$\frac{\partial y(1,\zeta)}{\partial \eta} = 0, \quad \frac{\partial y(0,\zeta)}{\partial \eta} = -\frac{\beta Da}{Le}\exp\left(\gamma\left(\frac{y}{y+1}\right)\right)\theta_a(0,\zeta), \qquad (12)$$

where $Le = k/\rho c_p D$ and $\beta = (-\Delta H)c_0/\rho c_p T_0$. The above system of equations constitute a highly non-linear problem. We explore briefly how the

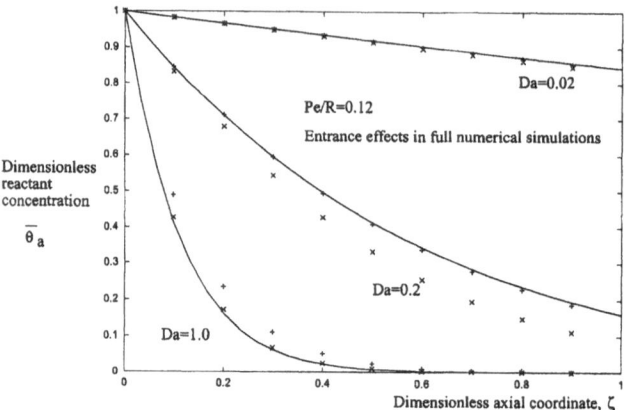

Figure 2: Dimensionless axial reactant profiles for different Damköhler and β numbers; + and × correspond to full numerical solutions for β 0.1 and 0.3 respectively, solid and dashed lines correspond to analytical solutions for β 0.1 and 0.3 respectively.

limiting case of a weakly exothermic reaction, corresponding to $\beta \ll 1$, may be obtained analytically. Using a similar averaging method as in §2.2, the vertically-averaged analytical solution for weakly exothermic flow is

$$\overline{\theta_a}(\zeta) = \frac{\left(1 + \frac{\gamma\beta Pe_m}{2LePe_e}\right) \exp\left(-\frac{Da\zeta RA}{Pe_m}\right)}{1 + \frac{\gamma\beta}{2LePe_e} \exp\left(-\frac{Da\zeta RA}{Pe_m}\right)}, \tag{13}$$

where $A = 1 + (\gamma\beta Pe_m/(2LePe_e))$. Figure 2 shows these average profiles of reactant for different β and Damköhler numbers. As expected the higher the β and Damköhler number the higher the reactant conversion. A comparison is made in Fig. 2 with full numerical simulations, which are displayed as discrete points. It can be seen that there is reasonable agreement for $Da < 0.2$ and $\beta < 0.1$. The discrepancies observed are partly due to the formulation of the analytical solution, which involves a binomial expansion in terms of β and Damköhler number of the exponential terms in (12).

For the higher Damköhler number of 1 in Fig. 2 it can be seen that although the profiles are not well matched the axial position corresponding to near complete conversion (ca. $\zeta = 0.7$) is successfully predicted by both solutions.

4   Conclusion

In catalytic wall microreactors, where transverse diffusion is fast and axial diffusion can be neglected, the advection-diffusion equation for a first order reaction can be solved analytically since the problem reduces to the solution of a single eigenvalue problem. The vertically-averaged formulation is illustrated for the case of a parallel plate isothermal catalytic microreactor and the analytical results are tested against full numerical simulations of the governing equations. In addition, the vertically-averaged scheme is also tested against numerical simulations for non-isothermal flows. The primary conclusion from this work is that under the conditions considered ($1 \ll Pe \ll R$) the transverse transport of reactants is fast and the advection-diffusion equations may be replaced by a one-dimensional vertically-averaged formulation. This method can also be extended to second-order kinetics, multiple reactions and ducts of arbitrary cross section (Gobby et al. 2001). Reduction of the dimension of the problem, makes this technique a useful tool for rapid evaluation of microreactor performance without residing to computationally demanding methods such as computational fluid dynamics which require grid generation and significant computing time.

Nomenclature

Symbol	Assignment
A	pre-exponential factor
c_0	inlet concentration
c_p	specific heat capacity
D	diffusivity
Da	Damköhler number
E_a	activation energy
f_a	vertical function
h	channel height
ΔH	heat of reaction
k	thermal conductivity or reaction rate constant
L	reactor length
Le	Lewis number
Pe_e	energy Peclet number
Pe_m	mass Peclet number
R	aspect ratio (L/h) or ideal gas constant
T	temperature
T_o	inlet temperature
u	fluid velocity
U	bulk fluid velocity
y	dimensionless temperature
α	decay group
β	heat of reaction group
η	dimensionless transverse coordinate
γ	activation energy group (set at 5.0)
λ_a	decay constant
ρ	density
θ_a	dimensionless reactant concentration
ζ	dimensionless axial coordinate

References

Bibby I. P., Harper M. J. & Shaw J. (1998). Design and optimization of micro-fluidic reactors through CFD and analytical modelling. *Proc. 2nd Int. Conf. of Microreaction Technology (New Orleans)*, 335-339.

Cleland F. A. & Wilhelm R. H. (1956). Diffusion and reaction in viscous-flow tubular reactor. *A.I.Ch.E. J.*, 2, 489-497.

Denbigh K. G. (1951). The kinetics of continuous reaction processes: application to polymerization. *J. Appl. Chem.*, 1, 227-236.

Franz A. J., Quiram D., Srinivasan R., Hsing I.-M., Firebaugh S. L., Jensen K. F. & Schmidt M. A. (1998). New operating regimes and applications feasible with microreactors. *Proc. 2nd Int. Conf. of Microreaction Technology (New Orleans)*, 33-38.

Gobby D., Eames I. & Gavriilidis A (2001). A vertically-averaged formulation of wall catalytic reactions in microchannel flows: single and multiple isothermal reactions, to be submitted for publication.

Groppi G., Belloli A., Tronconi E. & Forzatti P. (1995). A comparison of lumped and distributed models of monolith catalytic combustors. *Chem. Engng. Sci.*, 50, 2705-2715.

Hayes R. E. & Kolackzowski S. T. (1994). Mass and heat transfer effects in catalytic monolith reactors. *Chem. Engng. Sci.*, 49, 3587-3599.

Heck R. H. & Katzer J. R. (1976). Mathematical modeling of monolithic catalysts. *A.I.Ch.E.J.*, 22, 477-484.

Hessel V., Ehrfeld W., Golbig K., Haverkamp V., Löwe H., Storz M., Wille C., Guber A., Jähnisch K. & Baerns M. (2000). Gas/liquid microreactors for direct fluorination of aromatic compounds using elemental fluorine. *Proc. 3rd Int. Conf. of Microreaction Technology (Frankfurt)*, 526-540.

Hsing I.-M., Srinivasan R., Harold M. P., Jensen K. F. & Schmidt M. A. (2000). Simulation of micromachined chemical reactors for heterogeneous partial oxidation reactions. *Chem. Engng. Sci.*, 55, 3-13.

Knight J. B., Vishwanath A., Brody J. P. & Austin R. H. (1998). Hydrodynamic focusing on a silicon chip: mixing nanoliters in microseconds. *Phys. Rev. Lett.*, 80, 3863.

Veser G. (2001). Experimental and theoretical investigation of H_2 oxidation in a high temperature catalytic microreactor. *Chem. Engng. Sci.*, 56, 1265-1273.

Wegeng R. S., Call C. J. & Drost M. K. (1996). Chemical system miniaturization. *PNNL*, 1-13.

Weißmeier G. & Hönicke D. (1998). Strategy for the development of micro channel reactions for heterogeneously catalysed reactions. *Proc. 2nd Int. Conf. of Microreaction Technology (New Orleans)*, 24-32.

Walker R. E. (1961). Chemical reaction and diffusion in a catalytic tubular reactor. *Phys. Fluids*, 4, 1211-1216.

Part 4

**Characterization of Microstructured Unit
Operation and Reaction Modules**

Intensification in Microstructured Unit Operations
Performance Comparison Between Mega and Micro Scale

Anil R. Oroskar[1], Kurt M. VandenBussche and Suheil F. Abdo

UOP LLC, 25 East Algonquin Rd
Des Plaines, IL 60017, USA

Process intensification refers to technologies and strategies that enable the physical sizes of conventional unit operations to be significantly reduced. The concept was pioneered by ICI in the late 1970s, when the primary goal was to reduce the capital cost of a production system. The motivation behind this approach was the recognition that the main plant items involved in the process (i.e. reactors, heat exchangers, separators etc.) only contribute to around 20% of the cost of a given plant. The balance is incurred by installation costs that involve pipe-work, structural support, civil engineering and so on. A major reduction of equipment size, coupled preferably with a degree of "telescoping" of equipment function - for example reactor/heat exchangers or combined condenser/distillation/re-boilers - could generate very significant cost savings by eliminating support structure, expensive column foundations and long pipe runs. Process intensification has the potential to deliver major benefits to the process industry, and many other sectors, by accelerating the response to market changes, facilitating scale-up and providing the basis for rapid development of new products and processes.

There are several drivers for process intensification through miniaturization. These are: safety, product properties, improved yields, distributed or just in time production, operational flexibility, and, of course, high return on capital. In-situ production of HCN as catalyst in a chemical process, numbering up of fine chemical production and distributed production of hydrogen for fuel cells are a few examples of process intensification/miniaturization practiced commercially. Even though advantages of a "miniaturized" process appear clear, the most significant drawback is the increased cost of production [1]. To overcome the financial impact of "scale-down" one has to look for enhancement in one or more of the fundamental transport processes given in Table 1.

Fundamental aspects of Micro/Mega Scale performance

Unit Operation	Transport Processes	Illustrative Examples

Mixing
    Gas-Gas Mixing
    Gas-Liquid Mixing
    LL Mixing
	Heat Transfer
	Mass Transfer
	Momentum Transfer
		1. Hydrogen/Oxygen Mixing
		2. Liquid Dispersions/ Emulsions

Separation	Heat Transfer Mass Transfer	1. Adsorption 2. LL Extraction
Heat Exchangers	Heat Transfer Momentum Transfer	1. Plate type heat exchangers
Pumping Compression	Momentum Transfer	1. Micro-Pumps
Reaction	Heat Transfer Mass Transfer	1. Gas Phase Heterogeneous 2. Gas Phase Homogeneous 3. Liquid Phase

Table 1. Overview of fundamental processes in Unit operations.

---

[1] Corresponding author

The objective of this paper is to go through the transfer processes by means of illustrative examples and discuss differences between a mega scale and a micro scale operation. Where appropriate, an actual measure of possible process intensification will be given. In most cases, a reaction system will be used to show the potential of the improved transfer process. The transfer of momentum will not be addressed. The literature abounds with information on micro-devices that accomplish this [31], but to the best of our knowledge, no efficiency comparisons between mega and microscale have been published. In addition, some time will be spent on the catalysis area, focussing on synthesis techniques and their potential to bring this technology to the next level.

1. Gas Phase Mass Transfer

The mixing of gases in mega scale plants is usually not problematic, since hydrodynamic turbulence and the good diffusion characteristics of these fluids enable complete mixing over appreciable distances within seconds. Zech et al. [12] studied mixing of gases by micro-multilamination in detail and confirmed mixing lengths under 1 mm for a typical micro-mixer.

Generally, one would consider the use of micro-systems for gas/gas mixing excessive. Less expensive large scale equipment handles this task equally well. However, in some cases, the 'seconds' time scale might not be short enough. Examples of this are found where very fast, often homogeneous, competing reactions occur and imperfect mixing affects the desired selectivity or where the mixture goes through an explosion envelope as a function of composition to end up as a non combustible mixture.

An example of the latter is the direct synthesis of hydrogen peroxide from hydrogen and oxygen. This technology has received significant attention over the last 10 years [2-4], to replace the cumbersome and inefficient anthraquinone route to hydrogen peroxide. All of this work has focused on a range of H_2 contents below 3.5 %, staying away from the explosion limits. One would expect H_2/O_2 ratios in the explosive range or even at stoichiometric ratios to be preferred from reaction chemistry, yield and economics point of view. However, the process is inoperable at this ratio because of safety considerations. This problem has two facets, both of which can potentially be overcome by using microscale equipment. On the one hand, the actual process of mixing, starting from hydrogen and oxygen will always take the mixture through the explosion envelope, whatever the resulting ratio of hydrogen to oxygen. In order to avoid reaction, it is important to conduct the mixing of the gases in a characteristic time lower than the time needed for reaction. Figure 1 shows a first pass approximation to the ratio of mixing to reaction time as a function of the characteristic distance in the (micro-) mixer. It is clear that safe mixing of the gases will occur at characteristic distances below 100 micron. The second angle to this problem occurs when the resulting mixture is also explosive. In such a case, the mixer and the subsequent lines should act as flame arrestors. Applying the Semenov and Frank-Kamenetsky theories [5] to calculate the characteristic distances under relevant conditions for this synthesis again yields numbers in the 100 micron range.

The safe mixing and reaction of hydrogen and oxygen in micro-systems has been shown by Veser et al. [6]. Similar considerations can be made for the combination of oxygen and ethylene, demonstrated by Kestenbaum et al. [7] and Hoenicke at al. [12].

2. Liquid-Liquid Mass Transfer

Mixing of liquids is more difficult, due to the lower diffusivities that exist in this environment. On the large scale, mixing is typically conducted in the turbulent regime. If the laminar regime is chosen, other means must be employed to reduce the characteristic distance over which the molecules must diffuse to mix. A good overview of the options in that case is given by Loewe et al. [8]. They have also adapted a mixing characterisation technique, originally developed by Villermaux [9], to the micro-environment and can thus quantify mixing quality using UV-VIS spectroscopy. Figure 2 summarises some of Loewe's results.

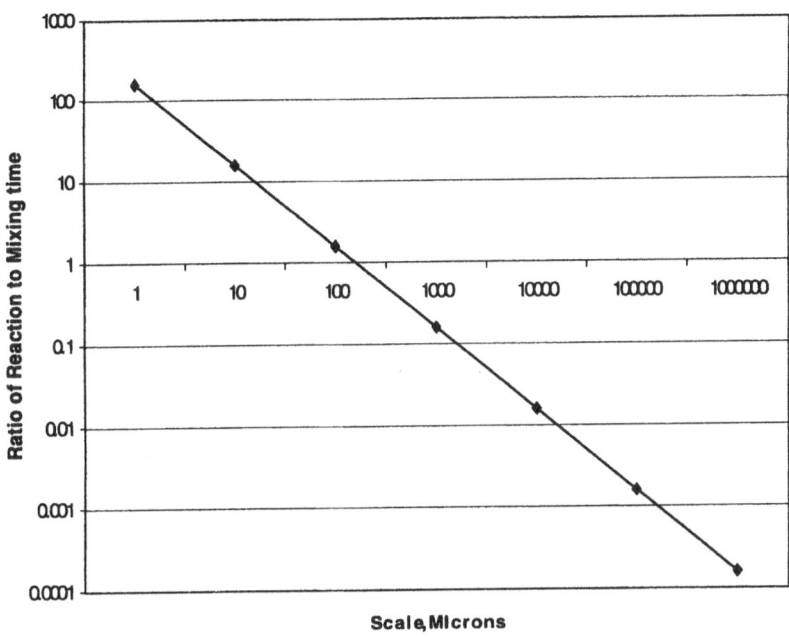

Figure 1. Ratio of characteristic mixing vs reaction time as a function of characteristic channel size.

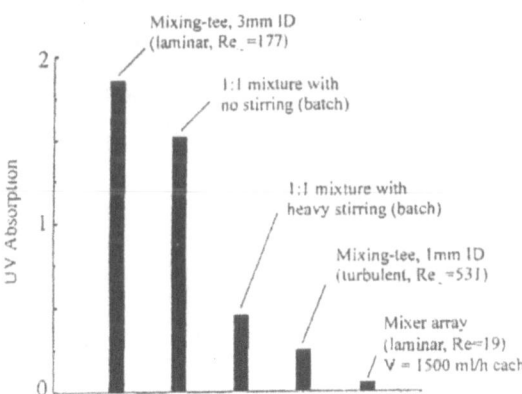

Figure 2. Comparison of mixing quality of a 40x300 micron mixing array to conventional technology. [from Loewe et al., [8]]

A feature of liquid-liquid mixing, when non miscible liquids are contacted, is that emulsions can be generated. In the chemical industry (on the mega as well as the microscale) fine emulsions have many useful applications in, e.g., extraction processes or phase transfer catalysis. Additionally, they are of interest for the pharmaceutical and cosmetic industry for preparation of creams and ointments. Micromixers based on the principle of multilamination have been found to be particularly suitable for the generation of emulsions with narrow size distributions [8]. Haverkamp et al showed the use of micro-mixers for the production of fine emulsions with well defined droplet diameters for dermal applications [10]. Bayer et al. [11] have recently reported a study of silicon oil and water emulsion in micro-mixers and compared the results to that obtained in a stirred tank. Their results are plotted in Figure 3 and indicate that the micro-mixers produce emulsions more efficiently than a stirred tank. Droplet size and their distribution are comparable for both systems.

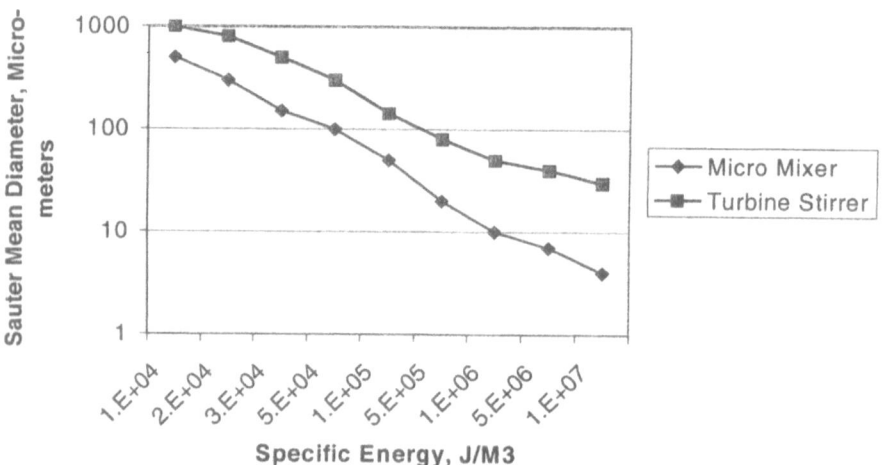

Figure 3. Energy efficiency comparison between micro mixer and conventional technology. Data by Bayer et al. [10]

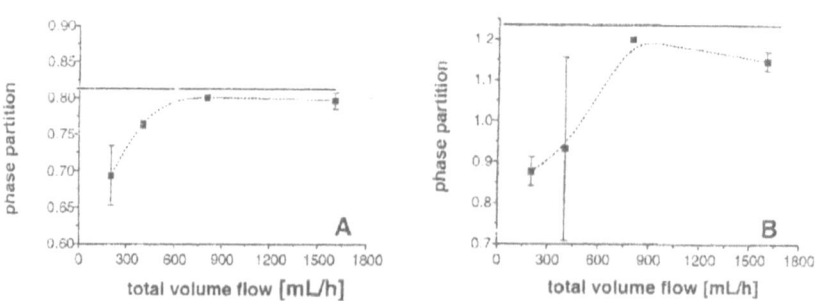

Figure 4. Phase Partition as a function of volume flow using a micromixer, with 2*15 microchannels. A= water (acetone)/toluene; B= water (succinic acid)/n-butanol. Solid horizontal lines indicate thermodynamic equilibrium. Data by Loewe et al. [8].

The work of Loewe et al. [8] on micromixers also suggests that they are highly efficient for liquid-liquid extraction processes. These authors carried out engineering studies on water, toluene, n-butanol system with acetone and succinic acid as extractants. They found that for both systems thermodynamic equilibrium was achieved in one mixer settler stage. As shown in Figure 4, the extraction efficiency strongly depends upon the flow rate. Loewe explained the shape of the phase partition curves for both model systems by the increase in specific surface area at higher flow rates.

3. Gas/Liquid Mass Transfer

Extremely widely practiced in industry, this type of mass transfer is at the basis of the absorption and distillation unit operations, two or three phase reaction systems and others. The gas-liquid interfacial area is, of course, of prime importance and this is where the process intensification opportunities for microtechnology lie. Table 2 [13], shows the potential of miniaturisation. An application in a reaction environment, for a two phase fluorination system, is given in [14].

Type of conventional reactor	Specific interface $[m^2/m^3]$	Type of microreactor	Specific interface $[m^2/m^3]$
Packed column Countercurrent flow Co-flow	10-350 10-1700	Micro bubble column (1100 μm x 170 μm) Isopropanol (observation)	5,100
Bubble columns	50-600	Micro bubble column (300 μm x 100 μm)	9,800
Spray columns	10-100	Micro bubble column (50 μm x 50 μm)	14,800
Mechanically stirred bubble columns	100-2000	Falling film microreactor (300 μm x 100 μm)	27,000
Impinging jets	90-2050		

Table 2. Specific interfacial areas of selected conventional and miniaturised reactor types. Data from [13].

4. Mass transfer in Gas/Solid Systems

Unit operations to consider in this area are adsorption, heterogeneous gas phase systems as well three phase reaction systems where the gas phase is continuous (sprays). Adsorption as a miniaturised unit operation has not yet extensively been discussed in the literature, but it is clear that the combination of superb hydrodynamics and residence time distribution, as well as precise temperature control (see below), adsorption processes can be conducted in a superior fashion. It opens perspectives for miniaturised TSA (temperature swing adsorption) technologies, where gases are separated from each other by differences in their adsorption isotherms, in a semi-continuous fashion.

Extreme control of the hydrodynamics and the residence time distribution has been shown to lead to higher yields in reaction systems on various occasions. The work by Hoenicke et al. [15, 16] on partial hydrogenation of various aromatics is classic in this area. A summary of these results is given in Figure 5.

158

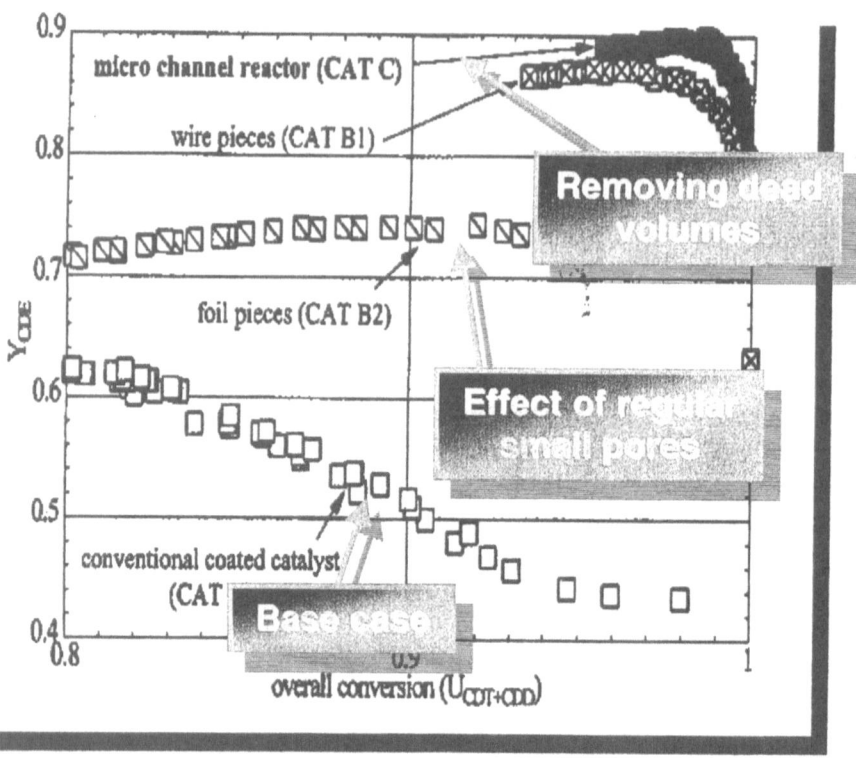

Figure 5. Effect of the various contributions of the RTD on the overall yield in the selective hydrogenation of CDT. Data from [16]

6. Heat Transfer

Figure 6 shows the surface area of various type of heat exchangers as a function of their cost of manufacture. Where as micro-systems are typically expensive, they do open up a region of specific interfacial area that was previously not attainable. Numerous papers have been published on this topic. Among the more recent ones, Brandner et al [17] studied cross flow heat exchangers with different micro structures. They found micro-column structure to be highest in heat transfer efficiency with overall heat transfer coefficients of 54000 W/m^2K. A similar conclusion was reached by Hardt et al. [20] who used computational fluid dynamics to study the effect of various fin configurations, as well as the effect of heat exchanger material conductivity on the overall performance of the microstructured equipment.

The ability to precisely control the temperature by means of high surface area and high heat transfer coefficients can be very valuable in highly endo or exothermic systems. Examples that come to mind for endothermic systems are dehydrogenation reactions, where isothermal conditions lead to smaller beds or higher conversions and the steam reforming system [30], where sufficient heat transfer can avoid that the catalyst deactivates by coking in the Boudouard region. In exothermic systems, poor temperature control often results in reduced selectivity (as in selective oxidation systems, [30]) or catalyst deactivation by reduction of the active surface area (as in methanol synthesis).

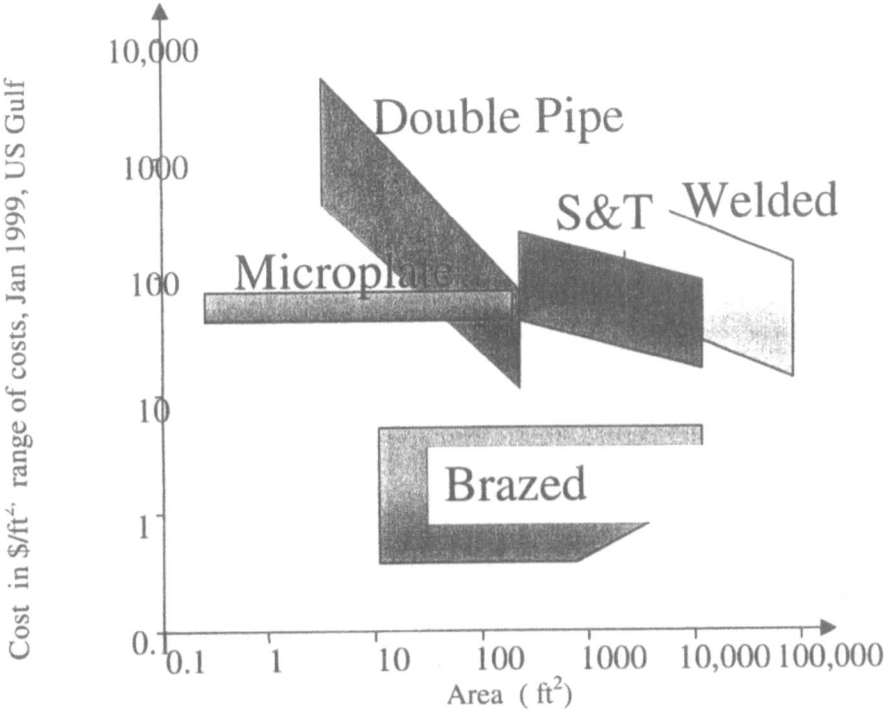

Figure 6. Comparison of various types of heat exchanger for specific area and cost.

7. Heterogeneous catalysis

In the previous paragraphs, reaction systems have been used illustrate the potential of micro-systems. Most of the improvements however, were based on better residence time distributions and improved mass transfer. An aspect that has not received sufficient attention so far is the use of the intrinsic activity of a catalytic phase on the nano-scale.

Figure 7 shows the intrinsic potential of the catalyst for a typical dehydrogenation process. The activity of current day catalysts is often significantly reduced by the way they are loaded and utilised in the large scale reactor. These reactors often can not provide sufficient heat transfer, making the catalyst bed run too hot or cold. In addition, pressure drop or fluidisation restrictions may reduce the space velocity to below what the catalyst can intrinsically handle. Furthermore, catalysts are typically manufactured in pellets of appreciable size, again reducing their effectiveness. In some extreme cases, e.g. for noble metals, the catalyst is poisoned on purpose to suit the reactor environment. In the case of micro-systems, some, if not all of these limitations can be avoided. Flow distributions, heat and mass transfer can be carefully controlled, the catalyst can be coated on the micro-channel wall etc....All of these factors may allow to use the catalyst at its intrinsic activity, which can lead to a process intensification of several orders of magnitude.

A good example of the latter is the work by Lanny Schmidt [21, 22], where thin foam monoliths, coated with noble metals, are run at space velocities as high as 10^8 per hour, replacing technologies conventionally running at 3.10^4/hr at most.

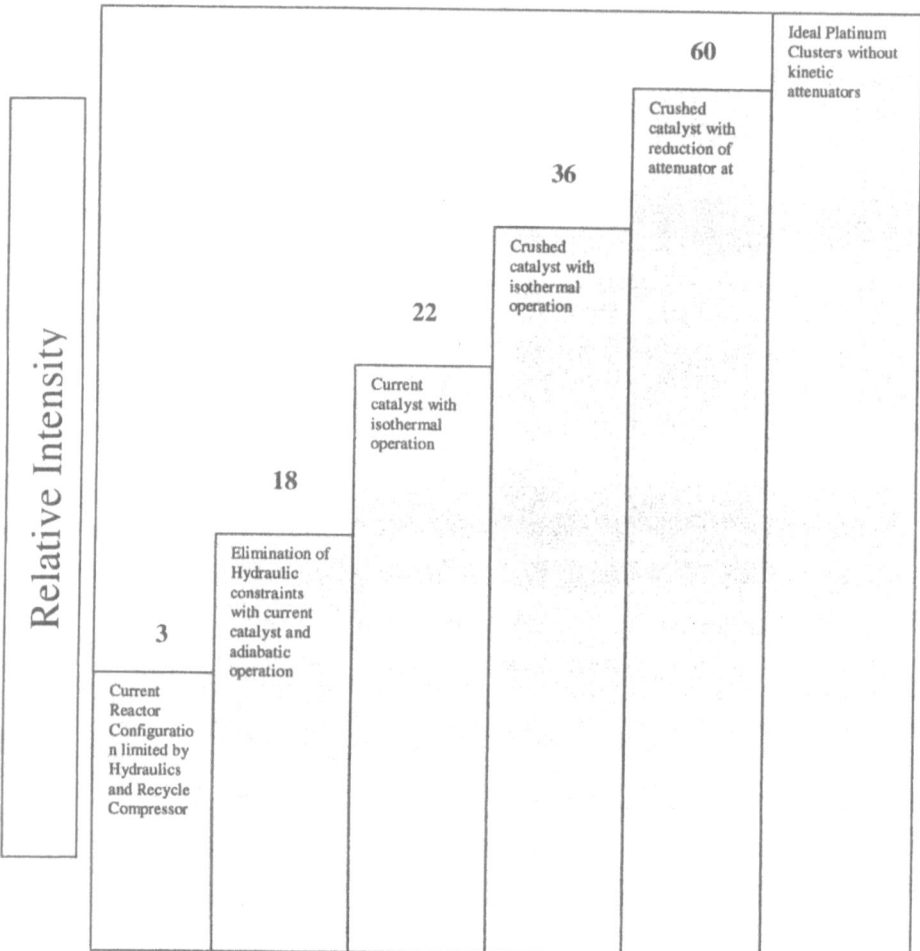

Process Geometry

Increased Temperature can significantly raise the
degree of intensification

Figure 7. Intrinsic potential of the catalyst active phase for a typical dehydrogenation system. Data from [1].

Along the same lines is the nano-tailordesign of catalytic systems [23]. The ability to design and fabricate elements at the submicron level is a rapidly growing area of science and technology. These tools can be mated to computational chemistry techniques, to custom design the catalyst of the future [24]. In a sense, catalysis in microstructures represents an improved method of delivery of the catalytic sites to the reaction medium. A wide variety of approaches to prepare nano-scale particles have been discussed in the recent scientific and technical literature. Some of the major approaches currently utilized, or proposed, for the manufacture of nano materials include a variety of chemical vapor deposition approaches, electron beam or photolithography, diffusion bonding to create very thin layers and anodic oxidation, sometimes employed to create porosity. Of course, the development of many of these approaches has been spurred on by the needs of the electronic industry. With manufacture of sub-micron structures having become common, references to nano-materials have also become common place. However, this scale is still significantly larger than that of catalytic materials and especially noble metal catalysts which has been based on molecular-scale chemical synthesis and manipulation of catalytically active components to maximize active site availability and improve metal utilization. Examination of current literature on nano-scale particles reveals that particles sizes derived by a variety of nano-scale approaches are typical in the range of tens of nanometers. Metal particles in catalysts, on the other hand, are often in the subnanometer size range. Zeolite-based noble metal catalysts, for example, are typically prepared by ion exchange procedures that render the metal ions (Pd or Pt) in an atomically dispersed state when freshly prepared. A successful exploitation of microstructure reactors will come from the effective use of suitable combinations of nano-technology approaches with appropriate techniques derived from molecular-scale methodologies. This is truly the gap which must be bridged in opening up the exciting potential of microreactor technologies.

A list of methods which are potentially useful in development of catalytic microreactors must then include [25],

> Lithography: electron beam and photo
> Chemical Vapor Deposition (CVD)
> Plasma Spraying [27]
> Anodic Oxidation to create porosity
> Pore forming additives and burnout agents to create porosity
> Diffusion Bonding and etching to create catalytic or anchoring layers in microstructures
> Solution coating of activated substrates
>> Spincoating techniques
>> similar to methods to form highly dispersed particles in the very small zeolite pores or encapsulation in zeolites
>> adjustment of solution composition such as type and concentration of sites and pH can be employed to tailor the size distribution and location of the nanoparticles
> Use of organometallics introduced from solutions or by sublimation to form supported clusters of controlled size and morphology
> Use of microemulsions and sols

Other methods while not widely discussed at the moment will offer new routes to design in microcatalytic systems. Examples of such possibilities are the in-situ growth of specific geometries or extended structures, e.g., zeolitic frameworks and approaches relying on leaching and dissolution of metals, such as raney nickel, after anchoring to the microstructures.

Use of selected combinations of the approaches listed above should allow us to build 'atomically precise' catalytic structures by combining chemical reactions with cutting edge material science. Thus a typical approach would require treatment of followed by deposition of active components. Alternatively, a base layer such as an alumina coat can be deposited and bonded to the microstructures in order to anchor catalytic metals. Addition of catalytically active metals may be accomplished employing salt solutions or organometallic complexes. Adjustment of metal distribution can happen in various ways. For example, high affinity of some solution species for a given substrate can be exploited to control the deposition profile and geometric placement of active components in the microstructures. In some selected instances, materials of construction containing certain active metals such as nickel aluminides may also be applied.

Varying the thickness of the catalytic coating can be effected to optimize mass and heat transfer and minimize undesirable side reactions. In this manner we may be able to discover catalytic behavior (improved activity and selectivity) unknown in large scale systems due to the inability to operate in such regimes. This is well demonstrated by the work of Becker et al.[26] for the oxidation of propylene to acrolein. When the thickness of catalyst coating was varied between 150 and 400 microns, the reaction rate constant dropped to 77% of its starting value along with a 4% drop in yield.

A subset of CVD techniques, plasma spraying approaches have been discussed recently by the group from Novosibirsk [27] as a method to deposit protective coating and catalytically active components on substrates. They point out the potential for this approach in designing catalytic materials of varying composition. The porosity of the deposited catalytic material can be controlled by varying spraying distances, deposition rate as well as the chemical composition of the plasma-forming gas. Control of the particle size of the sprayed material as well as thermal treatment of the sample can also be employed to set the final properties. With such methods, one can easily create very interesting catalytic composites of varying compositions by multiple-stage spraying and varying the sequences of sprayed materials.

Anodic oxidation of aluminum is also a well-established experimental technique for the formation of aluminas with controlled porosity and catalytic properties [28]). Pore size, length and wall thickness can be created by this approach. Further modification of pores thus created has been shown to significantly improve their catalytic behavior thereby pointing to yet another route to impact catalyst performance [29].

8. Conclusions

This paper attempts to assess the potential of microtechnology to enable process intensification in the chemical industry. Direct comparison of intrinsic properties clearly shows that mass and heat transfer can and has radically been improved. In many cases, reaction systems have been used to clearly illustrate the potential. However, the link with nanocatalysis has not fully been explored at this time, despite the early indications that there is significant potential. One can easily imagine the exciting possibilities offered by combining the molecular-scale approaches of catalysis with the new possibilities offered by nanotechnologies to design new types of catalytic systems in microreactors. This will not be limited to better utilization of catalytic materials but will also open up the possibilities for systematic placement of catalytic functions in unique spatial arrangements which are unattainable with conventional systems.

References

1. Oroskar et al., " Process Intensification by Miniaturisation." Proceedings of Microtec 2000, Hannover.
2. Gosser et al., "Method for the production of hydrogen peroxide", US 5135731.
3. Gosser L., 'Catalytic preparation of hydrogen peroxide from hydrogen and oxygen". US 506893.
4. Chuang et al., "Process for the direct synthesis of hydrogen peroxide", US5082647.
5. Warnatz and Maas, "Technische Verbrennung', Springer Verlag, Berlin, 1998.
6. Veser G. et al., 'A modular microreactor design for high temperature catalytic oxidations', in Proceedings of IMRET-3, Springer-Verlag, Berlin, 2000.
7. Kestenbaum et al., 'Synthesis of Ethylene Oxide in a catalytic micro-reactor system. Proceedings of IMRET-3, Springer-Verlag, Berlin, 2000.
8. Loewe et al., 'Micromixing Technology', Proceedings of IMRET-4, p31-47, 2000.
9. Villermaux et al., "Use of parallel competing reactions to characterise micro-mixing efficiency", AIChE Symposium Series, 88, 289, 1991.
10. Schiewe et al. "Micromixer based fomration of emulsions and creams for pharmaceutical applications. " Proceeding of IMRET-4, 2000.
11. Bayer T. et al., "Emulsification of silicon oil in water, comparison between a micromixer and a conventional stirred tank.", Proceedings of IMRET-4, p167-173, 2000.
12. Zech T. et al., 'Superior performance of static micromixers', Proceedings of IMRET-4, p390-399.
13. Ehrfeld et al., "Extending the knowledge base in microfabrication towards chemical engineering and fluid dynamic simulation". Proceedings of IMRET-4, p3-20, 2000.

14. Hessel et al., 'Micro-reactors for the liquid phase fluorination of toluene using fluorine", Proceedings of IMRET-3, Springer-Verlag, Berlin, 2000.
15. Hoenicke et al., 'The formation of cycloalkenes in the partial gas phase hydrogenation of c,t,t 1,5,9-cyclododecatriene, 1,5-cyclooctadiene and benzene in micro-channel reactors. Proceedings of IMRET-4, 89-99, 2000.
16. Wiessmeier and Honicke, Proceedings of IMRET-2, 1997.
17. Kursawe A. and Hoenicke D., Epoxidation of Ethene with pure oxygen as a model reaction for evaluating the performance of micro-reactors. Proceedings of IMRET-4, 2000.
18. Brandner et al., "Improving the efficiency of micro-heat-exchangers and reactors", Proceedings of IMRET-4, 244-249, 2000.
19. Burns J. and Ramshaw C., "A micro-reactor for the nitration of benzene and toluene". Proceedings of IMRET-4, p133-140, 2000.
20. Hardt et al., "Strategies for size reduction of heat exchangers by heat transfer enhancement effects." Accepted for publication in Chem. Eng. Reviews.
21. Schmidt L. et al., 'Catalytic partial oxidation reactions and reactors" Chem. Eng. Sci., 49, 3981-3994, 1994.
22. Dietz A. and Schmidt L. , "Effect of pressure on three catalytic partial oxidation reactions at millisecond contact times". Catal. Letters, 33, 15-29, 1995.
23. Baer D. et al., "Formation and stabilisation of Nano-sized Clusters on TiO_2 surfaces", Proceedings of IMRET-4, 2000.
24. Mavrikakis et al., "Making gold less noble", Catal. Letters, 64, 101-106, 2000.
25. G. A. Somorjai, J. Phys. Chem. 97, 9973
26. D. Becker, M. Kotter, L. Riekert, Chem.-Ing.-Tech. 61, 748, 1989.
27. Z.R Ismagilov, O. Yu. Podyacheva, O. P. Solonenko, V. V. Pushkarev, V. I. Kuz'min, V. A. Ushakov, N. A. Rudina, Catalysis Today, 51, 411-417, 1999
28. D. Hoenicke, App. Catal. 5, 199-206, 1983
29. G. Patermarakis and N. Nicolopoulos, J. Cat., 187, 311-320, 1999
30. Arakawa et al., 'Novel temperature controlled reactor technology increases productivity' Proc. AIChE Annual Meeting, 1997, Los Angeles, paper 265c
31. Hessel, Volker, Personal Communication, October 1998.

A New Microstructure Device for Fast Temperature Cycling for Chemical Reactions

J. Brandner[1*], M. Fichtner[1], K. Schubert[1], M. A. Liauw[2], G. Emig[2]

[1] Forschungszentrum Karlsruhe GmbH, P.O. Box 3640, D - 76021 Karlsruhe, Germany

[2] Lehrstuhl für Technische Chemie I, Universität Erlangen – Nürnberg, Egerlandstraße 3, D - 91058 Erlangen, Germany

1 Abstract

A microstructure reactor for fast periodic temperature changes will be presented. The developed reactor allows to reach temperature differences of several 10 K in the time range of several seconds. The device is made of stainless steel and electrically heated with high power resistor cartridges. It is cooled with deionized water as cooling fluid. The test facility is based on a microcomputer control system, which allows the half cycle time, i.e. the time between the lowest and the highest temperature of one cycle, to be adjusted from minutes to seconds. Depending on the half cycle time temperature changes of up to 200 K are achievable.

In first experiments conducted with a nitrogen flow in the reaction passage to simulate the flow of a reaction gas mixture, temperature changes of the reactor device of about 190 K could be achieved with a half cycle time of 30 seconds. At shorter half cycle times of 2 seconds and below, a temperature change of 100 K was still possible.

Keywords: Periodic Temperature Changes, Fast Temperature Cycling, Unsteady State, Catalytic Reactions

2 Introduction

Usually, chemical reactors are designed to run a desired reaction in stationary conditions („steady state"). This implies several constant environmental parameters like constant composition of the reactants at the inlet of the reactor, constant pressure inside of the reaction device or homogenous temperature of the whole device. These parameter sets can be changed to obtain an optimum steady state, where the yield or selectivity of the reaction is maximized.

Another possibility of running chemical reactions is to create an artificial „unsteady state" by changing some of the parameters continuously. The advantages and disadvantages of running chemical reactions not in steady state conditions have been discussed since the early 70s (e.g. [1], [2]). An unsteady state may result in better adsorption/desorption behaviour of reactants, reducing or prohibiting a side reaction or increasing the selectivity in heterogeneously catalyzed reactions. Especially a continuous change of the composition of the reactants at the reactor inlet and a continuous cycling of the temperature of the reactor itself was discussed, because those parameters seem to have a large impact. But while the researches on concentration cycling was done by many groups, the temperature cycling was meant to be not possible [1]. This was because of the use of conventional devices like fixed bed reactors or continuous flow stirred reactors (CFSR) with relatively high thermal masses, i.e. masses that have to be heated and cooled to obtain a temperature cycling, and limited heat transfer capabilities [1], [3].

In microstructure systems, the heat transfer capabilities can be several orders of magnitude higher than in devices made by conventional techniques. Beside this, the thermal mass of microstructure devices can be small compared to that of conventional reactor devices [4].

First experiments to obtain temperature changes in microstructure reactors have been done in the late 90s. It was tried to combine heated and cooled zones of a microstructure reactor to generate a quench directly behind a hot zone inside a microstructure device [5], [6], [7]. With those devices it is possible to change the temperature of the reaction mixture periodically. This is of special interest for homogeneous reactions. The devices used have been made of silicon and have limited heating and cooling power. The throughput of the devices is normally in the range of a few ml per minute.

It was discussed in some literature that fast temperature changes of the catalyst inside of a reactor device could possibly lead to significant increases of the rates of certain heterogeneously catalyzed reactions [8]. But no device was available yet in which the temperature of the whole reactor device could be changed in the second or subsecond time range and realize a *Fast Temperature Cycling (FTC)* of a chemical reactor. With the microstructure device described in the following, both methods of temperature cycling, periodic changes of the reaction mixture temperature and periodic changes of the reactor temperature are possible.

3 Fast Temperature Cycling

3.1 Boundary Conditions for Fast Temperature Cycling

In general, fast heating of reactor devices is easy by, for example, laser heating, microwave heating or resistance heating. Also, fast cooling can be achieved by microstructure quenching. The task is to build a device for the combination of fast heating and fast cooling with defined half cycle times.

To obtain fast changes in temperature, the device has to fulfill several boundary conditions.

The reactor has to be appreciably small, with low thermal masses and large heat transfer area. The reason can be seen from equation (1)

$$(1) \qquad Q_R = m_R \, c_p \, \frac{\Delta T}{\tau} \Leftrightarrow \tau = \frac{m_R}{Q_R} c_p \, \Delta T$$

In (1), Q_R is the thermal power which is needed to heat up or cool down the thermal mass m_R (i.e. the mass of the reactor, the heater cartridges, the coolant hold-up and the gas volume flow) with a given specific heat capacity c_p by ΔT in the half cycle time τ. Here, ΔT means a mean temperature difference of the device.

From (1) it is clear, that, with a given ΔT, a short half cycle time can be achieved by reducing the thermal mass m_R, reducing the specific heat capacity c_P or increasing the thermal power Q_R. The specific heat capacity c_P is depending on the material the reactor is made of. Assuming a given material to guarantee the stability of the reactor system and the possibility to integrate some catalytically active materials, the only parameters to optimize are m_R and Q_R.

This reason makes it difficult to build a conventional reactor system with cycle times in the second or subsecond range because the thermal mass is too high and the surface achievable for heat transfer is considerably small. This limits the transferable heat per time unit.

Microstructure devices fit the boundary conditions very well: A reactor for Fast Temperature Cycling made with microstructure technology provides a very small thermal mass combined with a large heat transfer area, quite high heat transfer rates and inherent safety [9].

3.2 Heating and Cooling

Heating and cooling of the microstructure system can be done using fluids or electrical power as it is shown in several papers [4], [9], [10].
In principle, there are three possibilities to generate temperature cycles:
- constant heating and periodic cooling
- periodic heating and constant cooling
- periodic heating and periodic cooling

A decision for one of the three combinations is influenced by boundary conditions which are due to the aspired operation regime of the reactor.

A limiting factor is the maximum temperature the system should reach. A temperature of several hundred °C is difficult to obtain with thermo fluids. Moreover, a flow control system for a fluid with high temperature is a source of danger and rather cost intensive. Therefore, it is much easier to use electrically heated systems. This allows to reach temperature ranges up to several hundred °C.

Similar considerations have to be made for the coolant side. Even if the reactor is small, the thermal power to be transferred is considerably high with short cycle times. Electrically driven cooling systems like Peltier elements have limited cooling power and an upper temperature limit which may be too low for the desired temperature range.

Controlling a cold fluid flow is more suitable. The transfer of thermal power is only limited by the size of the heat transfer area, the heat transfer coefficient of the device, the specific heat capacity and the volume flow of the cooling fluid and the temperature difference which is accessible with the fluid. Even evaporation can be used to consume thermal power, which leads to extremely high heat transfer rates.

Constant fluidic cooling and periodical electric heating would lead to higher half cycle times, because the fluid hold-up inside the device increases the thermal mass of the reactor system. A remarkable part of the applied electrical power would be used to heat this hold-up. Beside this, the lifetime of the heater cartridges would be reduced by fast cycling of the electrical power. A constant electric heating combined with periodic cooling by fluids works fine. The electrical power is easy to control [10], as well as the heat transfer to the cooling fluid. The fluid hold-up can be reduced by application of a high-pressure gas pulse of several milliseconds. All together, a combination of electrically driven constant heating and fluid driven periodical cooling should be a good compromise for Fast Temperature Cycling.

3.3 Reactor Design

Around a stainless steel block to insert high power electrical heater cartridges, six microstructured stainless steel foils for the cooling passage and four microstructured stainless steel foils to build the reaction passage are arranged in alternating order. Those numbers have been chosen as a compromise between a low pressure drop at a gas throughput of several 100 Nml/min and a minimum temperature difference between the innermost and the outermost reaction passage. Two thin stainless steel plates are used as top and bottom of the stack built by these eleven elements. A schematic drawing of the reactor device is shown in figure 1. A photo of the described device is shown in figure 3.

The heater block has dimensions of 44.0 x 33.0 x 6.8 mm^3. Six holes to insert heater cartridges with dimensions of 6.35 mm diameter and 33.0 mm length are drilled into the material. The heater cartridges used have a maximum electrical power of 1050 W each, so the maximum overall electrical power is 6300 W.

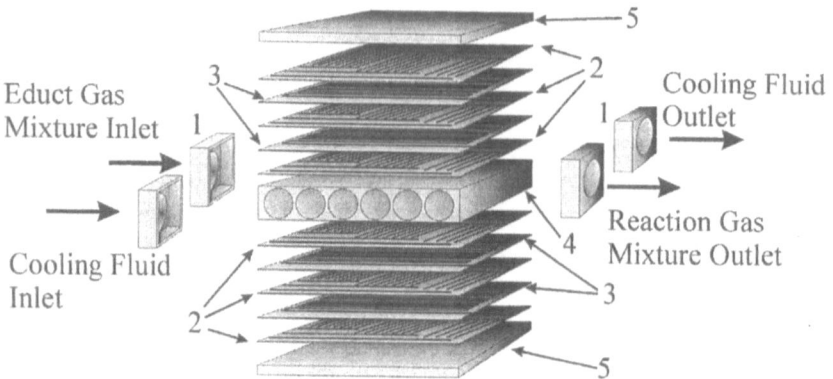

Fig. 1: Schematic explosion drawing of a microstructure reactor device for Fast Temperature Cycling. 1: Adapter, 2: Microstructured foils for cooling, 3: Microstructured foils for reaction fluid, 4: Heater block for electrical heater cartridges and thermocouples, 5: baseplate.

To insert small thermocouples, four grooves of 550 μm width and height and 33 mm length are machined into the heater block, two close to the fluid inlets and two close to the fluid outlets. The thermocouples inserted into this grooves are used to measure the temperature right beneath the innermost two foils of the cooling passage.

Each of the foils of the reaction and the cooling passage is 44.0 x 33.0 mm² in area and 0.2 mm thick. The microstructure in the reaction passage consists of three meandering microchannels which are 380 μm wide and 150 μm deep. The fins between the channels are 150 μm wide. Each channel has a total length of 710 mm. A schematic drawing of the structure is shown in figure 2a.

In the cooling passage, 81 microchannels of 350 μm width, 150 μm depth and 22.5 mm length are arranged under a specific angle. The fins between those channels are 150 μm thick. The fluid is ducted to the microchannels by eleven larger channels (600 μm wide) with staggered length (between 2.1 mm and 40.0 mm). The microstructure was designed to achieve a compromise between large heat transfer area and low pressure drop. A schematic drawing of the structure is shown in figure 2b.

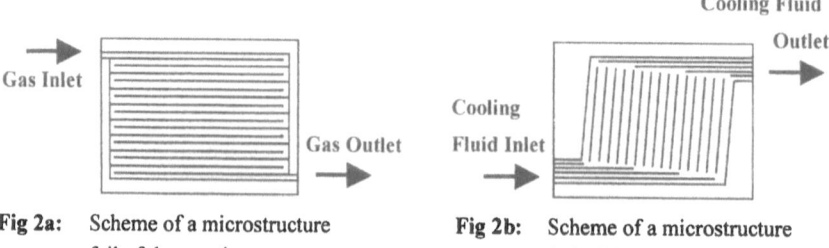

Fig 2a: Scheme of a microstructure foil of the reaction passage.

Fig 2b: Scheme of a microstructure foil of the cooling passage

All parts of the device are stacked and then connected either by diffusion bonding or electron beam welding. The reaction passage and the cooling passage are welded to fittings with small stainless steel adapters.

The design described before is optimized to reduce the thermal mass of the reactor and to maximize the heat transfer area. The heat transfer area is calculated to be 1230 mm^2 per foil of the cooling passage. A stainless steel reactor of the described type has a thermal mass of only 117 g but an electrical heating power of up to 6300 W, as mentioned before.

Fig. 3: Photo of a microstructure reactor device for Fast Temperature Cycling made of Stainless Steel and some of the heater cartridges. The arrow is pointed to a opening for a thermocouple to measure the temperature of the reactor device.

The reaction channels can be covered with a porous layer by, i.e., sol – gel – technique. If the complete reactor is made of aluminum, the reaction passage can also be anodically oxidized to generate porous material. This carrier layer can be wet impregnated with a catalyst precursors to obtain a catalytically active layer inside the reaction channels [11].

4 Experimental setup and first results

4.1 Experimental setup

As described before, a combination of continuous electric heating and periodic cooling by a fluid was applied to obtain Fast Temperature Cycling. The subsystem to obtain this is computer controlled. The computer system is also used for recording temperature and pressure data from the FTC reactor and the fluid in- and outlets. In figure 4, a scheme of the flowchart of the test facility subsystem for Fast Temperature Cycling is shown.

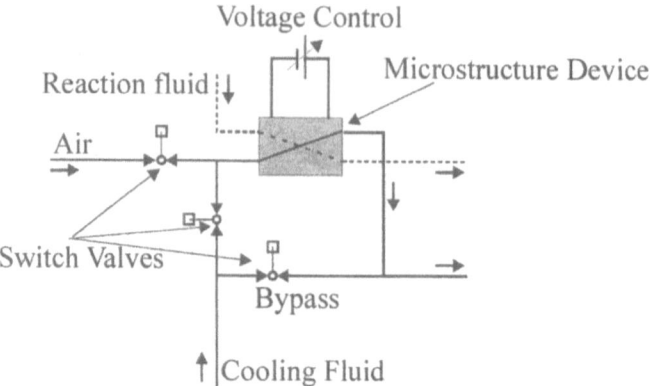

Fig. 4: Schematic drawing of the subsystem for Fast Temperature Cycling

At the beginning of each test a defined gas flow was applied by a mass flow controller. A N_2 flow of 100 Nml/min at 2 bar was used to simulate a reaction gas flow. This gas volume flow has no measurable effect on the reactor temperature distribution.

As coolant, deionized water was used in a flow range between 15 kg/h and 45 kg/h, leading to a pressure drop of around 10 bar in the cooling passage. The water flow was switched to bypass the FTC reactor using the switch valves. The desired voltage for the heater cartridges was applied and kept constant. After the cycle time was set, the water flow was automatically switched from flowing through the FTC reactor for a half cycle time to the bypass for the next half cycle time.

The measured half cycle time of heating might be slower than the desired one. The reason is a hold-up of coolant inside of the microstructures. Although the hold-up is rather small, it was a problem for the second to subsecond cycle time range. Hence, short (in the range of several 10 milliseconds) high-pressure pulses of air have been applied to clear the cooling passage from remaining coolant at the start of the heating half cycle. In the described experimental setup, each time the water was switched to the bypass, an air pulse of 200 ms length and a pressure of 12 bar was applied to the cooling passage.

The switching of the electropneumatic valves is time controlled but not temperature controlled. When a valve has been switched, time counting starts. After the programmed half cycle time is reached, the valve is switched to the other position, and counting starts again. The switch order of the valves used was:

Heating half cycle: first open the bypass, then close the cooling passage, then apply the air pulse
Cooling half cycle: first open the cooling passage, then close the bypass

This switch order should led to a behaviour of the measured temperature signals which is similar to some sawtooth behaviour.

The timing of the valve switching had to be done very carefully to obtain steep temperature changes. It may be possible that artificial delay times have to be inserted to obtain the desired temperature behaviour.

Through the test runs the temperatures inside the FTC reactor (see figure 3), the pressures and temperatures at the inlet and the outlet of the reaction gas passage and of the cooling passage have been measured and saved to a datasheet.

4.2 Experimental results

The tests conducted with the described device have been performed with special regard to obtain a high temperature difference at a low half cycle time. To obtain this, the power transfer ratio is of major interest. In a preliminary experiment, the power transfer ratio was tested. An electrical heating power of 1675 W was applied. At a water mass flow of 45 kg/h, the transferable thermal power was measured to 1550 W, which leads to a power transfer ratio of 0.93.

First temperature cycling tests have been done with half cycle times of 30 seconds (30 seconds heating time, 30 seconds cooling time). A voltage of 60 V (electrical power of around 450 W) and a water mass flow of 15 kg/h was applied, a temperature difference of about 190 K was reached. The temperature of the FTC reactor has been measured inside of the central heater block, right beneath the innermost cooling passage foil and close to the fluid inlet. The temperature of the N_2 flow of 100 Nml/min has been measured about 1 mm behind the outlet of the reactor, right in the middle of the gas flow.

In figure 5 a and b, results of such a test are shown. In the first half cycle at about 100 s of the timescale, the water flow was switched to the bypass. The FTC reactor was heating up fast to about 250°C after 30 seconds. At this time, the water flow was switched to the cooling passage, leading to a steep decrease of the temperature down to around 70°C after 30 seconds. Here, the next half cycle started, the water flow was switched to the bypass again and the air pulse was applied. This led to a steep increase of the reactor temperature. The measured temperature signals showed a sawtooth-similar behaviour.

The temperature measured in the outlet stream of the reaction passage followed the FTC reactor temperature but did not reach the minimum and maximum values due to some boundary effects of the reactor device. IR-thermographic measurements showed an almost homogeneous temperature distribution (temperature differences are in the range of 10 K) over the device surface. The surface area followed the temperature changes inside of the device immediately. Outer parts of the device such as adapters and fittings have a time delay due to the larger diffusion length for the heat to and from these regions. Hence, to reach the maximum or minimum temperature in this regions, the half cycle time simply was too short.

172

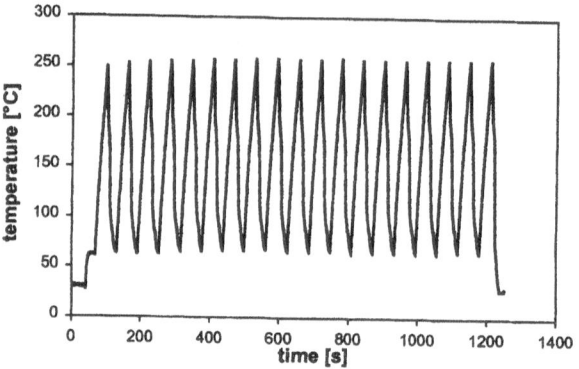

Fig. 5a: Temperature inside of the reactor device.

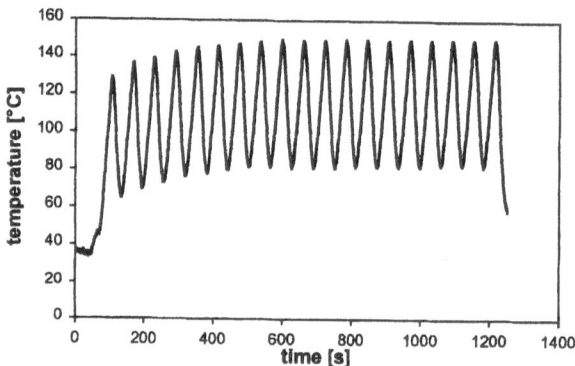

Fig. 5b: Temperature of the N_2 flow at the outlet of the device. For both figures 5a and b, the half cycle time was 30 seconds. The gas flow was 100 Nml/min, the water mass flow was 15 kg/h at a pressure drop of around 2.5 bar. With a voltage of 60 V an electrical heating power of around 450 W was applied.

To obtain shorter half cycle times, the electrical heating power and the water mass flow was increased. With a voltage of 70 V (580 W electrical power) and a water mass flow of about 30 kg/h, a temperature difference of around 100 K at a half cycle time of 10 seconds was achievable.

Even shorter half cycle times in the range of two seconds and below have been reached. With an electrical power of around 1700 W and a water mass flow of 45 kg/h it is still possible to reach a temperature difference of 100 K at a half cycle time of two seconds ± 100 ms. In figure 6, the temperature of the reactor device and the nitrogen flow at the reactor outlet is shown for two cycles.

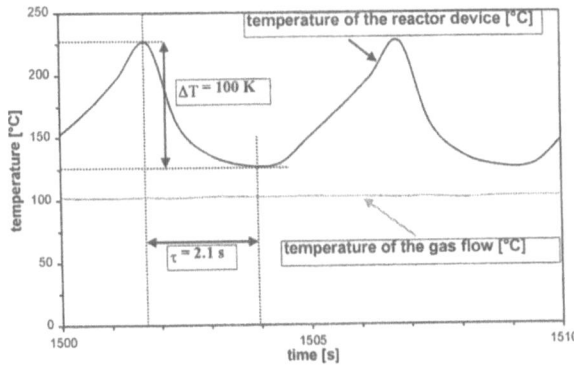

Fig. 6: Temperatures of the reactor device and the N_2 flow at the outlet of the device. The half cycle time was slightly above 2 seconds. The gas flow was 100 Nml/min, the water mass flow was 45 kg/h at a pressure drop of around 10 bar. At a voltage of 125 V an electrical power of around 1700 W was applied. A temperature jump of 100 K could be achieved.

It is interesting that the gas flow temperature at the outlet of the reactor was found to be nearly constant with short half cycle times. As described before, this is due to the heating and cooling delay the outer regions of the device have which act like thermal insulators. At short cycle times, those areas gain an averaged temperature, and the gas flow takes on the temperature of this area. They cannot follow the fast temperature changes of the other parts of the reactor device. Those areas are acting as an averaging low pass filter for the temperature, which might be beneficial for running reaction experiments.

5 Summary and Outlook

For Fast Temperature Cycling of chemical reactions, a microstructure device has been developed and thermally tested. The reactor is continuously heated with high power resistor heater cartridges and periodically cooled with deionized water. With each cycle, the cooling passage was cleared from its water hold-up by a short high - pressure pulse of air. A nitrogen flow of 100 Nml/min was applied to the reaction passage to simulate a reaction gas flow.

To obtain the short half cycle times, a microcomputer based control system was implemented into the test facility. The minimum half cycle time attained with the described system was about 2 seconds. It was shown that with this specially designed microstructure device and the FTC subsystem a temperature jump of 100 K with a half cycle time of around 2 seconds is possible.

It is planned to integrate a catalyst into the microstructure device. Then, a heterogeneously catalyzed reaction will be investigated under steady state and FTC conditions.

174

Beside this, Fast Temperature Cycling might be useful in materials research to obtain data about materials stability at fast temperature changes. Applications in analytical chemistry as well as biomedical applications may benefit from FTC.

With an optimized reactor and a faster control system it appears to be possible to reach half cycle times in the range of 0.1 seconds with temperature differences around 100 K. Shorter half cycle times should be possible at lower temperature differences. This could perhaps help to clear some kinetic aspects of catalyzed gas phase reactions [8] or lead to significant increases in the rates of certain reactions.

6   References

[1]   J. E. Bailey: Periodic Operation of Chemical Reactors: A Review;
        Chem. Eng. Comm. 1973, Vol. 1, pp. 111-124

[2]   J.E. Bailey, F.J.M. Horn, R.C. Lin: Cyclic Operation of Reaction Systems: Effects of Heat
        and Mass Transfer Resistance; AIChE Journal 1971, Vol. 17 No.4, pp. 818 - 825

[3]   P. Silveston, R.R. Hudgins, A. Renken: Periodic Operation of Catalytic Reactors:
        introduction and overview; Catalysis Today, 1995, Vol. 25, pp. 91-112

[4]   M. Fichtner, J. Brandner, G. Linder, U. Schygulla, A. Wenka, K. Schubert: Microstructure
        Devices for Applications in Thermal and Chemical Process Engineering;
        Proc. Int. Conf. on Heat Transfer and Transport Phenomena in Microscale,
        October 15-20, 2000, Banff, Canada, pp. 41 - 53

[5]   Ch. Alépée, R. Maurer, L. Paratte, L. Vulpescu, Ph. Renaud and A. Renken: Fast Heating
        and Cooling for High Temperature Chemical Microreactors;
        Proc. 3rd Int. Conf. on Microreaction Technology, 1999, pp. 514 – 525

[6]   J.M. Köhler, U. Dillner, A. Mokansky, S. Poser, T. Schulz: Microchannel Reactors for Fast
        Thermocycling; Proc. 2nd Int. Conf. on Microreaction Technology, 1998, pp. 241 – 247

[7]   S. Poser, R. Ehricht, T. Schulz, S. Uebel, U. Dillner, J.M. Köhler:
        Rapid PCR in flow-through Si chip thermocyclers;
        Proc. 3rd Int. Conf. on Microreaction Technology, 1999, pp. 410 – 419

[8]   F.J.R van Neer, A.J. Kodde, H. den Uil, A. Bliek: Understanding of Resonance Phenomena
        on a Catalyst under Forced Concentration and Temperature Oscillations;
        Can. J. of Chem. Eng., Vol. 74, October 1996, pp. 664 – 673

[9]   K. Schubert, W. Bier, J. Brandner, M. Fichtner, C. Franz and G. Linder:
        Realization and Testing of Microstructure Reactors, Micro Heat Exchangers and Micro-
        mixers for Applications in Chemical Engineering;
        Proc. 2nd Int. Conf. on Microreaction Technology, 1998, pp. 88 – 95

[10]   J. Brandner, M. Fichtner, K. Schubert: Electrically Heated Microstructure Heat Exchangers
        and Reactors; Proc. 3rd Int. Conf. on Microreaction Technology, 1999, pp. 607 – 616

[11]   M. Fichtner, W. Benzinger, K. Haas – Santo, R. Wunsch, K. Schubert: Functional Coatings
        for Microstructure Reactors and Heat Exchangers; Proc. 3rd Int. Conf. on Microreaction
        Technology, 1999, pp. 90 – 101

Photochemical Reactions and Online Product Detection in Microfabricated Reactors

Hang Lu, Martin A. Schmidt, and Klavs F. Jensen

MicroChemical Systems Technology Center, Massachusetts Institute of Technology, Cambridge, MA, USA 02139

Abstract

This work presents the application of microfabricated reactors to photochemical reactions. Two fabrication schemes of microreactors were realized for the integration of the reaction and the detection modules: coupling individually packaged chips, and monolithic integration of the two functions. In the latter scheme, we have successfully bonded two quartz wafers and a DRIE-patterned silicon wafer by a Teflon™-like perfluoropolymer (CYTOP™) at a low temperature (160°C). We studied the pinacol formation reaction of benzophenone in isopropanol as a model reaction to evaluate the performance of the microreactors. The reaction was continuously monitored by online UV spectroscopy. The data illustrated that controlling the flow rate of the reactant, or equivalently the residence time and irradiation time, is a viable method for controlling the extent of reaction. For photochemical reactions, miniaturization enables process intensification – the reaction is initiated on-chip, but is completed in an off-chip storage device with no further complications.

Introduction

Compared to other types of reactions, photochemistry is usually cleaner and more efficient because the key reactant is a source of photons [1]. The use of photochemistry is, however, limited by concerns about scalability, efficiency, and safe operation of the processes. We demonstrate here an improvement in the applicability and efficiency of photochemical reactions by using microfabricated reactors.

Large-scale photochemical reactions are usually performed with macro scale lamps immersed in the reaction vessel. In most cases, it takes considerable effort to transform a successful lab-scale reaction to its industrial counterpart [1]. Issues include the scalability of light sources, heat and mass transfer in the process, and safety concerns (explosions caused by excess heat) [1, 2]. Many of the photochemical reactions proceed via a radical mechanism. If the radicals, which are formed near the light sources, do not diffuse quickly to react further with other species, they are likely to recombine, generating heat instead. Radical recombination reduces the quantum efficiency of the overall process [1]. By miniaturization, however, the diffusive transport of these species effectively reduces the concentration near the light source, and increases the probability that they will collide with other molecules to produce the desired products. A more important issue, unique to photochemical reactions, is light absorption. Most photochemical reaction mixtures have components (either the solvent or solutes) that strongly absorb the incident light. These components reduce the efficiency of many large-scale photochemical reaction units. Miniaturizing photochemical reactors takes the advantage of the high surface-to-volume ratio and the small length scale of microfabricated devices. At the micro scale, the solution in the reactor is only hundreds of microns deep, allowing light to penetrate through most of the reactor depth. Motivated by these advantages, we have designed, fabricated, and characterized a microreactor for photochemistry using microfabrication technologies. For fine chemical synthesis and analysis, the micro-scale devices have faster mass and heat transfer properties, are dominated by surface forces, have high surface-to-volume ratio, and exhibit laminar flow. These properties are all attributed to the small length scale [3]. Performing photochemical reactions at this scale would benefit from these advantages. In designing our devices, we also considered material compatibility and process availability.

Device Design and Fabrication

In the design of the microreactor, the needs of photochemical reactions are considered: (1) the cover of the device must be transparent to the incident light; (2) the reactor must have a large surface area to receive maximum amount of light possible; and (3) the device must avoid crystallization and clogging. Two prototype reactors were produced: one for demonstrating a sequential detection of benzopinacol formation reaction (with incident light of 366 nm in wavelength), a second one for lower wavelength reactions and online detection. The two prototype reactors use two different fabrication schemes that are outlined below.

The first prototype reactor (in setup I) was designed to have a serpentine shape to yield a long channel with a large surface area (Figure 1). The channel was 500 μm in width, and approximately 250 μm deep with vertical sidewalls. This design was realized by timed Deep Reactive Ion Etching (DRIE) of a silicon substrate [4]. Isotropic etching of either silicon or glass substrates would have produced undesirable shapes at the turns of the reaction channel, which might induce crystallization.

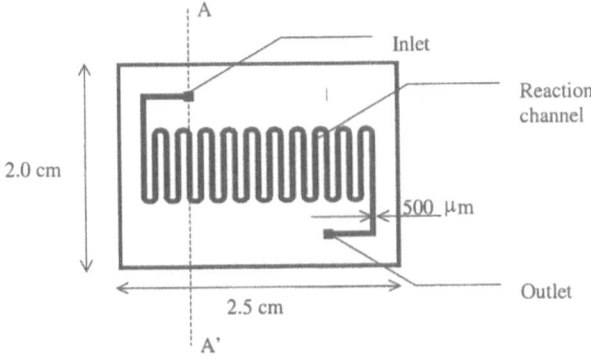

Figure 1. Design layout of the photochemical reactor chip

Figure 2 shows the fabrication sequence of this device.

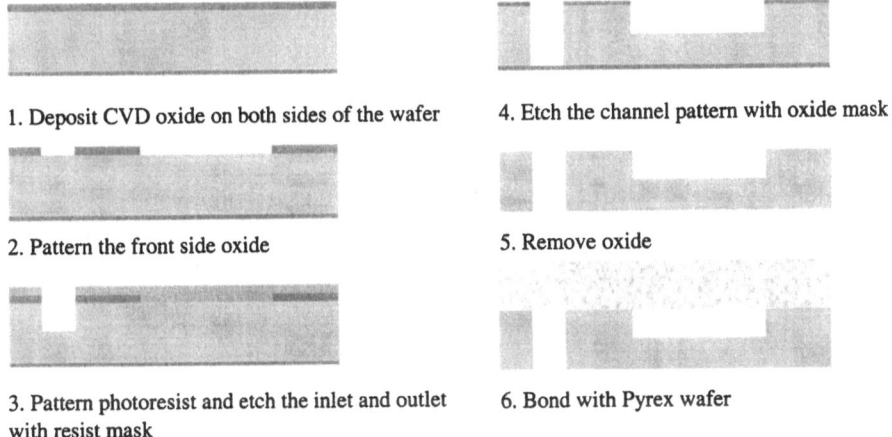

1. Deposit CVD oxide on both sides of the wafer

2. Pattern the front side oxide

3. Pattern photoresist and etch the inlet and outlet with resist mask

4. Etch the channel pattern with oxide mask

5. Remove oxide

6. Bond with Pyrex wafer

Figure 2. Fabrication sequence of the silicon-Pyrex device (cross section AA')

The detection unit accompanying this reaction unit was fabricated with quartz substrates and a photo-definable epoxy (SU-8) as described by Jackman *et al.* [5], and is transparent to wavelengths as slow as 200 nm. The particular device used in the present work is 50 µm deep, and has a straight channel configuration.

The need to incorporate a substrate for lower wavelength detection and reaction requires the use of quartz wafers in place of Pyrex wafers. However, the bonding of quartz and silicon wafers poses a different fabrication problem. Anodic bonding between silicon and Pyrex wafers relies on the presence of sodium ions, not present in quartz. Furthermore, direct fusion bonding cannot be used because it is a high temperature process; the thermal mismatch of quartz and silicon creates

178

large stress at the interface after annealing. To solve the bonding problem, we used CYTOP™, a cyclized perfluoro polymer (CPFP) developed as a low dielectric-constant material by Asahi Glass Co., Ltd. (Japan). This material is chemically similar to polytetrafluoro ethylene (PTFE), and therefore chemically inert [6]. CYTOP™ can be easily coated on substrates, and used as a bonding material [7].

We designed the second prototype to integrate the reaction and detection units. It was also designed to accommodate reactions that require light of low wavelengths (e.g. excitations for many reactions are at 254 nm). To design a detection unit that has an optical path perpendicular to the reactor, part of the reactor has to be transparent from top to bottom. A sandwich structure of a silicon wafer between two quartz wafers was then designed.

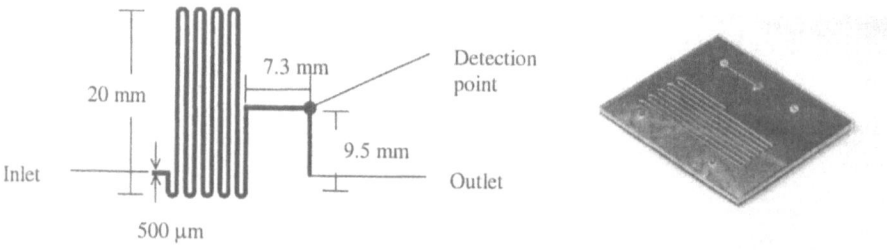

Figure 3. Design of the integrated reaction-detection unit

The reaction channel formed in silicon is 500 μm wide, and 500 μm deep (wafer thickness). Corresponding inlet / outlet holes (1 mm in diameter) were drilled in the bottom quartz wafer. CYTOP™ (Sigma-Aldrich, USA) was used as a bonding material [6, 7]. The fabrication process is shown in Figure 4.

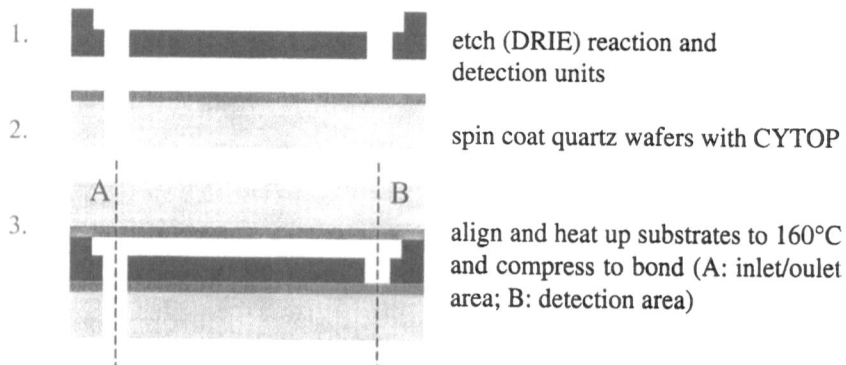

1. etch (DRIE) reaction and detection units

2. spin coat quartz wafers with CYTOP

3. align and heat up substrates to 160°C and compress to bond (A: inlet/oulet area; B: detection area)

Figure 4. Fabrication sequence of the integrated device

In our devices, CYTOP™ provided a fluidic seal, and was inert to chemicals used in all experiments.

The reactors (or detector chip) were packaged between a plexiglass plate and a steel or Teflon mount with opening to house the optical fibers and/or the miniature UV lamp (BF325-VU1, 365 nm, JKL component, CA).   In setup I (the two reactor setup), the chip-mount modules were connected by a 10-cm long HPLC tubing (PEEK) with standard fitting.

Model Chemistry

To test the design ideas of the microfabricated reactors, the pinacol formation reaction of benzophenone in isopropanol reaction was used as a model reaction.
Overall Reaction

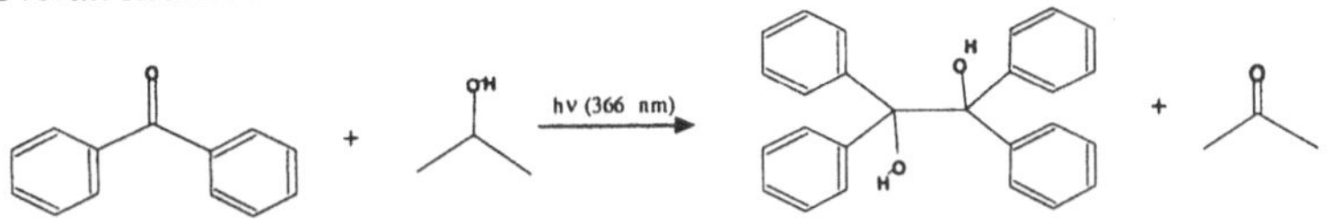

$$Ph_2CO + h\nu \rightarrow Ph_2CO^* \rightarrow Ph_2CO^{**} \tag{2}$$
$$Ph_2CO^{**} + (CH_3)_2CHOH \rightarrow Ph_2COH\bullet + (CH_3)_2C\bullet OH \tag{3}$$
$$Ph_2CO + (CH_3)_2C\bullet OH \rightarrow Ph_2COH\bullet + (CH_3)_2CO \tag{4}$$
$$Ph_2COH\bullet + Ph_2COH\bullet \rightarrow (Ph_2COH)_2 \tag{5}$$

Scheme 1. Mechanism of pinacol formation reaction of benzophenone in isopropanol

Reactant solutions were deoxygenated by bubbling nitrogen for at least 10 min before experiments.   To calculate the absorbance of a reaction mixture, the transmitted light through pure isopropanol was used as reference, and the following definition was used: $A = -\log((I-I_b)/(I_0-I_b))$, where I is the intensity of the reaction mixture, I_0 that of the isopropanol, and I_b the background reading. The same procedure was used to obtain the absorbance of the standard solutions with no reaction, achieved by turning off the miniature UV lamp. Samples were also collected for further analysis with HPLC.   A C-18 Econosphere™ column (5 μm packing, 250x4.6mm) from AllTech Associates, Inc. (Deerfield, IL) was used. The analyses were typically performed more than 48 hours after collection to ensure complete "dark" reactions (i.e. reaction steps following the initial light exposure, but not involving photons). Solutions of benzophenone (0.1M – 0.4M) in isopropanol were used to produce standards in UV response in the HPLC.

The model reaction is known to follow a radical reaction pathway (Scheme 1), in which benzopinacol, benzophenone, and the mixed-pinacol along with many radical species are generated [8].   Once the high-energy excited state of benzophenone is formed, the subsequent reactions proceed without additional photons. It has also been observed that a highly absorbent intermediate is formed in the reaction, but is short-lived and oxygen-sensitive [9]. After the irradiation, another molecule of benzophenone is to be reacted, and it takes a long period of time (>24 hr) for the completion of reactions [8]. We found that online UV spectroscopic analysis was indicative of the reaction course, and could be verified with off-line analysis when the reactions were complete.

In the monolithically integrated device, since the absorbance path length (500μm) was 10 times that of the first detection device made in SU-8 (~50 μm), the concentration of the benzophenone solution had to be reduced accordingly. All other conditions were kept the same as described above.

Results and Discussion

Since the 0.5 M benzophenone solution is a concentrated solution, Beer's law does not apply, and the commonly reported extinction coefficient cannot be used. The UV absorption was estimated experimentally before reactor design. The detector microchip has a path length of 0.05mm, which allows 60% of the light to reach the detector (measured experimentally); on the other hand, the reactor chip has a depth of ~0.25 mm, allowing the absorption of most of the incident light. If 0.5 M solution were used in macro scale reaction systems, it would be an optically thick solution where most of the light would be absorbed within a few hundreds of microns to the lamp. We have thus attained a more efficient absorption scheme by reducing the physical thickness of the reaction fluid in the microreactors.

Figure 5 shows UV absorption of reaction mixture at different flow rates superimposed on UV responses of standard solutions of the reactant, benzophenone. These curves follow identical shapes of the ones obtained with conventional UV-spectroscopic equipment found in literature [8]. The UV absorbance of the final pinacol product is small compared to that of the starting material, benzophenone, and therefore the UV absorbance of the benzopinacol is assumed not to contribute [8, 9]. The radical intermediates also absorb UV light at specific wavelengths between 300 and 400 nm. These radicals contribute to the slight shift of the maximum of the absorbance curves compared to the standards (Figure 5).

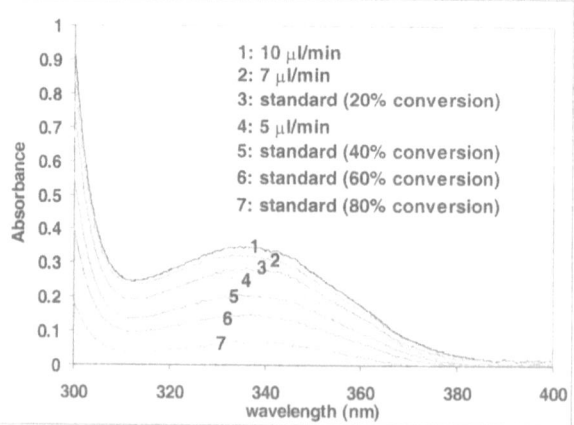

Figure 5. UV absorption of reaction mixture at different flow rates, thus different residence time.

Nonetheless, because the concentration of these intermediate species is small compared to the starting material and the product, the absorbance measurements are indicative of the reaction course, and agree well with the HPLC analysis of the final product mixture as described in the next paragraph.

The absorbance of reaction mixture as shown in Figure 5 can be used to estimate the on-chip conversion of benzophenone. The absorbance of the reactive solution was then compared to that of the standards. In this way, the rate of disappearance of the benzophenone on-chip was estimated for each run. Specifically, the values of absorbance at 333nm of standard solutions were used to plot a standard curve, and that of the reaction mixture was compared to the standard curve to yield an approximate concentration of benzophenone. The results from the two-chip setup demonstrate that the on-chip conversion of benzophenone due to UV irradiation is a function of the flow rate (Figure 6), or equivalently the residence time on chip. It should be noted that there is a finite volume between the two mounts, and therefore the delay in the detection varies inversely with the flow rate.

For the overall reaction, the longer the residence time, the greater the conversion as expected. An interesting observation from the result is that there is almost no measured on-chip conversion of the reactant for flow rates that are above 10 µl/min. A plausible explanation follows from the analysis of the various time scales. Because the residence time is so small, the amount of absorbed light by the reactant only creates a small amount of high-energy species near the light source. These species do not have long enough residence time to diffuse very far into the solution or to react, i.e. the time scale for flow (residence time) is smaller than both the diffusion time scale and the reaction time scale. With reduced flow rates (large residence times), the conversion improves because the amount of light absorbed increases, and there is sufficient time for the excited species to diffuse and react with benzophenone.

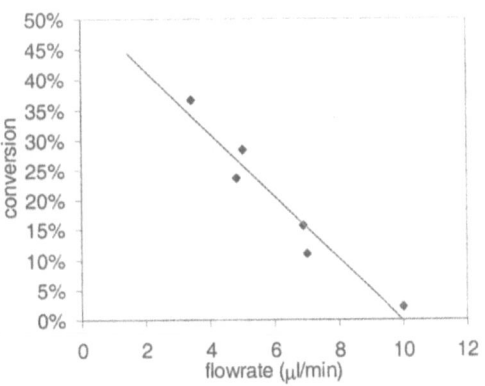

Figure 6. Conversion of benzophenone estimated from online UV spectral data as a function of flow rate of the starting material.

Benzopinacol has a low solubility in isopropanol. In bench scale reactions, benzopinacol separates out from the reaction mixture, forming white crystals in the reaction flask [10]. In most of the experiments in this work, no crystals formed while on the microreactor chip. When the samples were collected, however, crystals precipitated out of the solution in the storage vials after long period of "dark" reactions. The microreactor implementation is advantageous for reactions that only need initiation with UV light and propagate with radicals. Shrinking the length scale not only changes the characteristic time of transport relative to reaction and improves the efficiency, but also avoids undesirable crystallization (therefore eliminating the need for cleaning).

The conversion estimated from the online spectroscopic measurement is confirmed by the off-line analysis (after at least 48 hours). Figure 7 shows the conversion as a function of flow rate determined by HPLC separation (with UV detection at multiple wavelengths between 270 and 360 nm). The trend agrees with that obtained from online UV absorbance measurement, indicating that the conversion is a function of the feed flow rate. The difference is that at the end of the reaction as measured by HPLC analysis, the dark reactions made the conversion much higher than the corresponding online measurement. Nevertheless, it is evident that the online UV absorbance measurement is a practical indicator of the extent of reaction, and therefore can be used to monitor the progress of the reaction *in situ*.

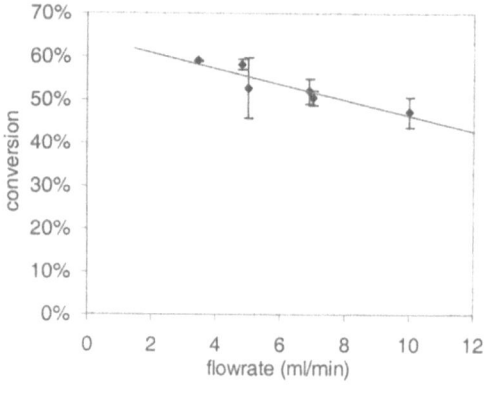

Figure 7. Off-chip HPLC analysis, showing the relationship between residence time and conversion.

In order to eliminate the delay in detection in prototype I, we have implemented a monolithic integrated device. Setup II detects the reaction mixture with only a few seconds of delay on chip (see Figure 3 for overall chip design). There was no interference from the miniature UV lamp for reaction on the UV spectra (data not shown).

The reactions were again monitored by online UV spectrometry. In these experiments, the UV spectra of the reactive solutions exhibited higher absorbance than unreacted solutions as shown in Figure 8, indicating the formation of the highly absorbent intermediate. We believe that there are two factors contributing to the different absorption spectra in the analogous experiments with the two-chip setup. First, reactions proceeded in the tubing between the reaction chip and the detection chip and their mounts, reducing the amount of the intermediate. Secondly, small amounts of oxygen might have permeated into the system between the reaction chip and the detection chip, reacting with the oxygen-sensitive intermediate. In general, for other reactions where no interfering intermediates are formed, the monolithic integration scheme would be more desirable because it eliminates the delay before detection. In this case, however, the presence of a strongly absorbing short-lived intermediate interferes with conversion measurements in the monolithically integrated device.

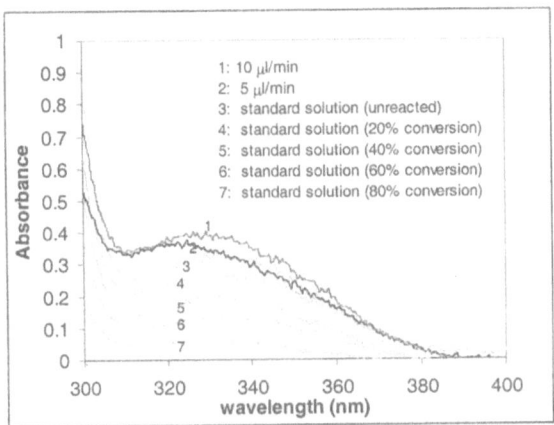

Figure 8. Reaction and online detection in the monolithically integrated device: the curves are noisier compared to prototype I due to the large path length in the detection scheme (larger absorbance and hence lower signal).

Conclusions

In the present work, we have demonstrated the practicality of photochemical reactions with miniaturized light source on microfabricated chips. We have also shown that it is possible to coupled reactions to an online UV spectral analysis of the reaction mixture. The potential applications of this technology are likely in organic synthesis and photo-initiated polymerization reactions of small scale. The advantages of the small-scale reactors are that the efficiency of photon transfer and reactions can be increased. In particular, with optically thick solutions, microreactors offer the opportunity of "thinning" the solutions to make the most use of the incident light. In this case, crystallization can be avoided in the reactor with a continuous flow system by controlling residence time, and consequently the extent of reaction on-chip, making downstream processes simpler. The two-chip integration is simpler in design, and because of the inherent delay to detection, it was suitable for the purpose of monitoring the benzopinacol formation reaction on chip. However, the monolithically integrated device provided a true online immediate analysis of the reaction mixture. Parallel operations of the miniaturized reaction devices presented in this work may open a door to process intensification, and the online detection provides an opportunity for fast process optimization.

Acknowledgments

The authors thank DARPA under the Microflumes Project (F30602-97-2-0100), and National Science Foundation graduate fellowship program for financial support. The technical assistance from the staff of the Microsystems Technology Laboratories, and Prof. T. F. Jamison and S. Patel in Department of Chemistry at MIT are gratefully acknowledged.

References

[1]   R. Roberts, R. P. Ouellette, M. M. Muradaz, R. F. Cozzens, P. N. Cheremisinoff, *Applications of Photochemistry* (Technomic Publishing Co., Inc., Lancaster, Pennsylvania, 1984).

[2]   H. Bottcher, J. Bendig, M. A. Fox, G. Hopt, H. Timpe, *Technical Applications of Photochemistry* (Deutscher Verlag fur Grundstoffindustrie, Leipzig, 1991).

[3]   K. F. Jensen, *Chemical Engineering Science* **56**, 293-303 (2001).

[4]   A. A. Ayon, R. Braff, C. C. Lin, H. H. Sawin, M. A. Schmidt, *Journal of the Electrochemical Society* **146**, 339-349 (1999).

[5]   R. J. Jackman, T. M. Floyd, R. Ghodssi, M. A. Schmidt, K. F. Jensen, *J. Micromech. Microeng.* (in press).

[6]   L. Asahi Glass Co., "Cytop Technical Report" (1997).

[7]   A. Han, K. W. Oh, S. Bhansali, H. T. Henderson, C. H. Ahn, "A Low Temperature Biochemically Compatible Bonding Technique Using Fluoropolymers for Biochemical Microfluidic Systems", Proceedings of IEEE MEMS Conference, Miyazaki, Japan (2000).

[8]   J. N. Pitts *et al.*, *J. Am. Chem. Soc.* **81**, 1068-1077 (1959).

[9]   J. Chilton, L. Giering, C. Steel, *J. Am. Chem. Soc.* **98**, 1865-1869 (1976).

[10]   B. S. Furniss, A. J. Hannaford, P. W. G. Smith, A. R. Tatchell, *Vogel's Textbook of Practical Organic Chemistry* (Addison Wesley Longman Ltd., Longman Group UK Ltd., 1989).

CHAOTIC MIXING IN ELECTROKINETICALLY AND PRESSURE DRIVEN MICRO FLOWS

Yi-Kuen Lee[a], Joanne Deval[a], Patrick Tabeling[b] and Chih-Ming Ho[a]
[a]Mechanical and Aerospace Engineering Department
University of California at Los Angeles, Los Angeles, CA90095
[b]Laboratoire de Physique Statistique de l'ENS, 24 rue Lhomond, 75231 Paris (France)

ABSTRACT

We present two micro-devices, fabricated by using MEMS technology, in which mixing of fluid and particles takes place. The systems are designed to induce folding and stretching of material lines, leading to chaotic-like mixing. In a first case, we use unsteady pressure perturbations superimposed to a mean stream, and in the second case, time-dependent dielectrophoretic forces to induce folding and stretching. The first device shows chaotic-like mixing is achieved in an efficient way, leading to rapidly homogenizing concentration fields. Folding and stretching effects inducing mixing are shown for the second system.

The systems are simple in their conception and may favorably be integrated within complex bio-MEMS or µTAS systems.

1) INTRODUCTION

Mixing is an important issue in bio-MEMS, and in micro Total Analysis Systems (µTAS). In integrated systems dedicated to biochemical analysis, all sorts of reactions take place ; if mixing is not achieved, reactants are not fully brought in contact with each other, and the system may not operate properly. One therefore requires mixing in these systems. The development of complex bio-devices may thus depend on the ability to implement simple efficient mixers along the channels conveying the bioreactants. For macro systems, mixing can be handled by driving propellers or moving magnetic beads within the fluids. In such systems, the inertia forces are large enough to induce turbulence, and in turn mixing can be achieved in an efficient way. In MEMS, inertia forces are extremely weak and as a consequence, turbulence does not develop. Flows in MEMS are laminar, which is advantageous for devising control strategies, but inconvenient for forming homogeneous mixtures.

In the last few years, an approach undertaken by several groups has been to reduce the channel cross-sections and increase their lengths, so as to bring fluids parcels closer, over longer period of time. Since the diffusive time varies as the square of the channel width, reducing the channel cross-section leads to vigorously enhancing diffusion processes, and in turn, mixing can be rapidly achieved. The design of such systems is however constraining;. In practice, the approach may lead to unacceptably long channels and prohibitive pressure drop.

In this context, it is worth observing that fifteen years ago, it has been demonstrated that chaos can be used to mix fluid in creeping flows. The idea stems from general mathematical properties of chaos and the reader may refer to textbooks such as Ottino's book [1] for a presentation of the topics. A fundamental result, relevant to the mixing problem, is that chaotic regimes are associated to stretching and folding of material lines. In these regimes, the blobs of markers, advected by the fluid, tend to adopt a complex, circonvoluted lamellar structure, which facilitates the final stage of mixing by molecular diffusion. This process is called chaotic mixing. Another useful theoretical result is that chaos arises in non-linear dynamical systems, provided that at least three degrees of freedom are present. For steady two-dimensional flows, the fluid particles are governed by a non-linear dynamical system, but with only two degrees of freedom. Adding a time dependent external perturbation to the flow provides a third degree of freedom and as a consequence, chaotic regimes, leading to chaotic mixing, may arise.

Following these ideas, attempts have been made to fabricate chaotic mixers in microfluidic devices. Devices, mimicking a source/sink system [2] and unsteady channels [3], have been proposed ; more recently, a mixer, consisting of a spiraling channel, has been fabricated and tested [4]] (in this case, three dimensionality is used to provide the additional degree of freedom). Encouraging progresses in enhancing mixing have been reported. In the first part of the paper, we demonstrate the path towards chaotic mixing and the existence of chaos in a pressure driven chaotic mixer. In the second part, we fabricate a DEP force driven micro-mixer, and reveal that folding and stretching are achieved, favoring mixing.

2) MECHANICAL CHAOTIC MIXER

The mechanically driven system is schematically represented on Fig 1. It consists of a main channel, along with a steady flow is driven. A time-dependent cross-flow is applied transversally to the main stream by means of two adjacent channels ; the configuration is similar to a previous work [5]. The cross flow varies sinusoidally in time, with a time-averaged mass flow kept equal to zero.

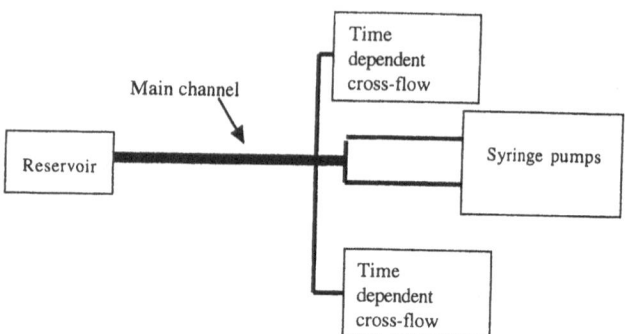

Fig 1 : Schematics of the mechanical mixer

The way the system works is schematically shown on Fig 2. We consider a material line, initially released in the main channel, along the symmetry axis. Without perturbation, the line is advected along the main stream, and remains straight. We caricature the temporal dependence of the perturbation by replacing the sinusoidal wave-form by a square wave-form, with two positions - on and off. As the transverse flow is switched on, the material line is bent as it crosses the intersection. After the transverse flow is switched off, a folding takes place, since the flow is faster at the center of the main channel than on the sides. This elementary process is a step towards a chaotic regime, favorable to mixing. To get a complete chaotic regime (i.e involving an infinite number of stetchings and foldings), one should replicate the structure *ad infinitum*, which is almost achievable by using MEMS technology.

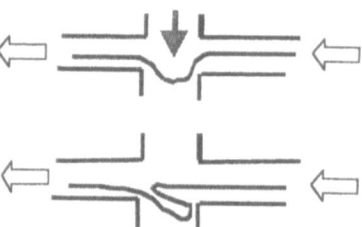

Figure 2 : Schematic way the systems works. One gets an elementary process of stretching/folding of the material lines, which is favorable to mixing.

To be more quantitative, we performed two-dimensional kinematical simulation, in which we assumed Poiseuilles flow both in the main and the adjacent channels; in the intersection region, owing to the fact that we work with creeping flows, we considered the two contributions essentially add up. Several regimes are obtained in the simulation. At small amplitude, the material line undergoes weak oscillations as it passes through the intersection region. As the perturbation amplitude raises, the oscillation amplitude increases. Chaotic like regimes appear at large perturbation amplitudes, when the wave periodically hits a corner of the intersection region. When this happens, the material line undergoes one or several foldings, and tends to adopt an extremely convoluted periodic pattern downstream, highly favorable for mixing. This is shown on Fig 3, which represents the evolution of an interface, initially flat, separating, in the upstream region collections of black and gray particles. After the pressure perturbation is applied, one sees the interface becomes extremely intricate, yielding a situation favorable for the mixing downstream.

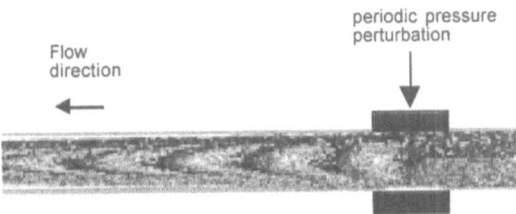

Fig 3: Production of an intricate interface in the downstream region. In the upstream region, the interface is flat.

As mentionned before, the system we discuss, replicated along the main channel, can operate as an efficient mixer since the interface tends to adopt increasingly convoluted shapes. From the practical viewpoint, we will see below that one unit is enough to achieve acceptable mixing provided the operating parameters are suitably chosen.

We further examine whether chaos does truly occur in this mechanical mixer. There are several ways to characterize chaotic mixing, and one instructive piece of information can be obtained by calculating Lyapunov exponents [6][7]. These exponents characterize the rate at which pairs of fluid particles diverge and thus provide an estimate on the stretching rate of the material lines. In the geometry under study, we find the Liapunov exponent rise up to 0.1 in the chaotic like regime. It seems changing the geometry can substantially increase this exponent. For instance, we obtained that, by adding an additional side channel, an exponent equal to 0.4 can be obtained.

For realizing chaotic mixing in microscale, we fabricate a mechanical micromixer, by using standard MEMS technology. The channels, etched into Silicon wafer using DRIE technique, are anodically bonded to Pyrex plates. The scales on Fig 5 and 6 indicate the channel sizes. There is possibility to inject two distinct fluids in the main channel, in the upstream region. For the first fluid we use sucrose solutions, or glycerol and for the second one the same solution, but labeled with a fluorescent tracer. Both fluids are injected with the same flow-rate, so that the interface between them lies right in the middle of the channel, in the upstream section. The adjacent channels are connected to a syringe pump operated in an

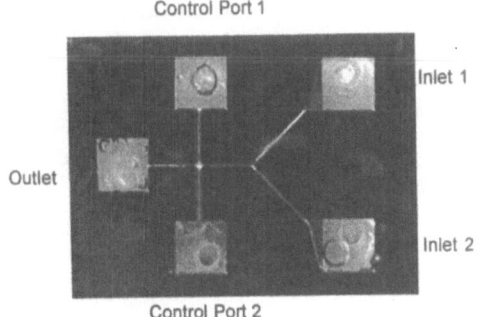

Fig.4 : Top view of the mechanical micromixer.

an oscillatory regime. The interface is visualized by fluorescence microscopy.

188

We could obtain the regimes found in the numerical simulation. In the chaotic regimes, the morphology of the interface can be extremely complicated. An example is given in Fig 5. The fact that in the experiment, the velocity profile is not uniform across the channels, but parabolic, reinforces the tendency to develop intricate interface patterns, in comparison to a plug flow. Here, contrarily to confined systems, the interface is folded only a finite number of times as it travels across the intersection.

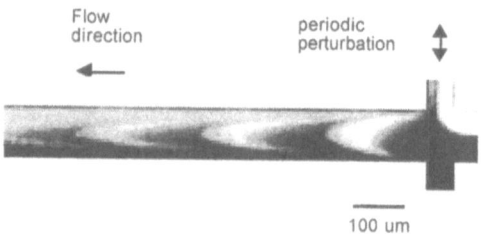

Fig 5 : Snapshot showing a highly convoluted interface

In the present experiment, and with the dye and fluids used, mixing is obtained with only one basic unit. This is shown on Fig 6, which represents the interface before, and after it passes the intersection, for particular values of the amplitude and frequency of the perturbation. One sees mixing is reasonably achieved, as confirmed by further concentration measurements, which clearly show the profiles flatten out downstream

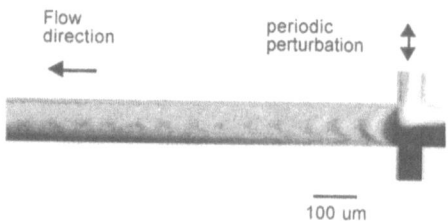

Fig 6 : Chaotic mixing by the mechanical micro-mixer

3) ELECTROKINETICAL CHAOTIC MIXER

For micro devices with charged particles or polarizable particles, electrokinetic force driven mixing is another approach. A micromixer has been fabricated, dedicated to the mixing of dielectric particles in a chamber; here we use dielectrophoretic forces induced by AC electric fields, periodically switched on and off. Dielectrophoresis is the translational motion of neutral matter caused by polarization effects in a non-uniform electric field. Neutral particles (such as polystyrene particles, cells, bacteria etc) of radius a subject to a spatially non uniform AC electrical field E experiences a DEP force given by [8] :

Force magnitude:

$$F = 2\pi a^3 \varepsilon \; \Sigma [K(\omega)] \; E^2$$

Clausius-Mossotti factor:

$$K(\omega) = \frac{\tilde{\varepsilon}_p - \tilde{\varepsilon}}{\tilde{\varepsilon} + 2\tilde{\varepsilon}}$$

Complex permittivities:

$$\tilde{\varepsilon}_{m,p} = \varepsilon_{m,p} - j\frac{\sigma_{m,p}}{\omega}$$

ε_m and ε_p being the permittivity and conductivity of the medium/particle. $\sum [K]$ stands for the real part of the Clausius-Mossotti factor. The force can be either attractive towards to the highest field gradients (positive DEP) or repulsive from them (negative DEP). For a given applied electrical field, different particle populations can therefore exhibit different behavior. Such phenomena have been studied in a number of papers [9], and have provided methods for particles manipulation [10], [11].

As for the mechanical device, we used kinematical simulations to search for the path towards chaos. The flow configurations we considered are represented in Fig 7. The flow is driven in a channel, along which a cavity is placed. Periodically, an electrical field is switched on and off. In Fig 7 we track a material line initially lying along the channel

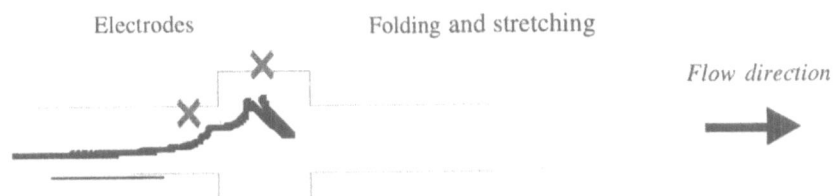

Fig. 7 : *Kinematic simulation of the electrokinetical mixer, showing that a folding process takes place as an electric field, delivered by two electrodes, is periodically switched on and off. The black line is formed by 1000 particles, previously released along a straight line in the inlet section*

symmetry axis, in the upstream region and display its shape after it penetrates the cavity region. Several regimes are obtained. At small amplitude of the electric field, the material line undergoes weak oscillations at it travels along the chamber. At large amplitudes, the effect of the electric field dominates and all the particles get trapped around the electrodes. In between, folding and stretching of the material line takes place, leading to a situation favorable for mixing. We are in the process of identifying the proper perturbations for achieving non-zero Lyapunov exponents.

A DEP force based micro mixer has been fabricated using MEMS technology (see Fig 8 and 9) The chamber dimensions

are 200x200x25 μm. The electric field is created by a 1MHz, 10V AC voltage applied between selected pairs of micromachined electrodes. Inlet and outlet holes are first anisotropically etched with KOH from the backside. Aluminum electrodes and external pads are patterned on the topside and the electrodes coated with insulation layer (SiO$_2$). Su-8 photoresist is spin coated, and then patterned to form the channel walls and the chamber. Cover glass is further bonded according to a technique currently under patenting. The electrical connections are done on the topside whereas fluidic connections are done on the backside, allowing convenient microscope visualization. In our device, we take advantage of the different DEP domains for polystyrene particles in aqueous suspensions. At a frequency of 100kHZ, particles are attracted to the edges of the electrodes, align with the field lines, and form pearl chains. At a frequency of 10 MHz, particles are repelled from the edges of the electrodes towards the field gradient minima..

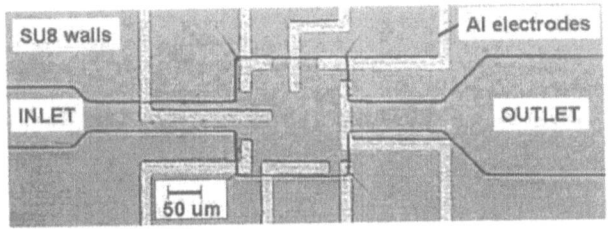

Fig.8 : *Top view of the electrokinetical mixer.*

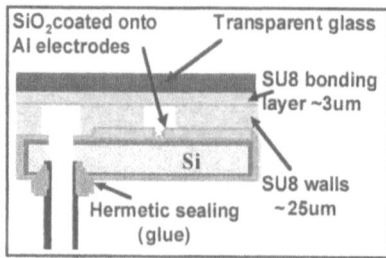

Fig. 9 : Cross-section of electrokinetikal mixer

The pictures of Fig 10 show the evolution of an interface as it travels through the chamber. In order to facilitate visualization, it has been underlined in white. As the electric field is permanently set to zero, the interface remains flat as it travels across the chamber. As it is periodically switched on and off, stretching and folding can take place, yielding a favorable situation for mixing. Inspection of the concentration profiles indicates they flatten out downstream.

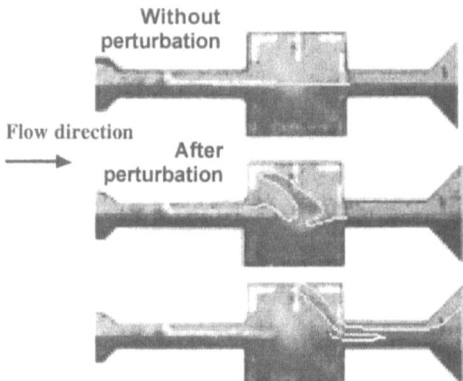

Fig 10: Stretching and folding as electrokinetic perturbation is applied. The yellow line indicates the evolution of the interface between particles solution (lower part) and DI water (upper part).

4) CONCLUSION

In a 2-D micromixer driven by unsteady hydrodynamic pressure, we have demonstrated the existence of chaotic behavior from the qualitative analysis of the material lines, and the non-zero Lyapunov exponents. The observed flow patterns in MEMS based devices correspond to the simulated patterns and lead to efficient mixing. In a flow with polarizable polystyrene particles, we use DEP force and can produce foldings and stretchings as predicted by simulation. Further investigation is under way to optimize the way how chaotic behavior can be exploited in this type of micromixers.

ACKNOWLEDGEMENTS

The authors would like to thank Prof W. Ditto at Georgia Institute of Technology, H. Chen and S. Huang at UCLA, Dr. D. Choi at JPL and C. Merkwirth at Drittes Physikalisches Institut, Universität Göttingen for their assistance. This work is supported by Defense Advanced Research Projects Agency, Microsystems Technology Office through contract from SPAWAR. The content of the paper is currently under patenting

REFERENCES

1   J.M. Ottino, *The Kinematics of Mixing:Stretching, Chaos,and Transport*, Cambridge University Press, New York, 1989.

2 - J. Evans, D. Liepmann, D., and A.P. Pisano, 1997, "Planar Laminar Mixer," Proceeding of the IEEE 10th Annual Workshop of Micro Electro Mechanical Systems (MEMS '97), Nagoya, Japan, Jan, pp.96-101.

3   Liu, R. H., et al., "Passive mixing in a three-dimensional serpentine microchannel," J. MEMS, pp. 190-7 ,2000.

4   M. Volpert, C. D. Meinhart, , I. Mezic, and M. Dahelh, "An Actively Controlled Micromixer," *Proceeding of MEMS, ASME IMECE*, Nashville, Tennessee, Nov., pp. 483-487, (1999).

5   Lee, Y. K., Shih, C., Tabling, P., Ho, C. M., 1999, "Characterization of Mixing Process in a Microchannel Flow," presented at the 52-th Annual Meeting of American Physical Society, Division of Fluid Dynamics, Nov., New Orleans.

6.   J. P. Eckmann and S. O. Kamphorst, "Liapunov Exponents from Time Series," *Physical Review A*, Vol. 34, No. 6, pp.4971-79, (1986).

7.   A. Wolf, J. B. Swift, H. L. Swinney, and J. A. Vastano, "Determining Lyapunov Exponents from a Time Series," *Physica D*, Vol. 16D, No. 3, pp. 285-317, (1985)).

8.   H.A. Pohl, "Dielectrophoresis: the Behavior of Neutral Matter in Nonuniform Electric Fields", *Cambridge University Press*, (1978).

9 - Ramos, H. Morgan, N.G. Green, and A. Castellanos, "The Role of Electrohydrodynamic Forces in the Dielectrophoretic Manipulation and Separation of Particles", *J. Electrostatics, 47*, (1999), pp. 71 – 81.

10 T. Schnelle, T. Müller, G. Gradl, S.G. Shirley, G. Fuhr,"Dielectrophoretic Manipulation of Suspended Submicron Particles", *Electrophoresis, 21* ,(2000), pp. 66 - 73.

11 M.P. Hughes, "Ac Electrokinetics: Applications for Nanotechnology", *Nanotechnology, 11,* (2000), pp. 124 - 132.

Filamentous Catalytic Beds for the Design of a Membrane Microreactor: Propane Dehydrogenation as a Case Study

O. Wolfrath, L. Kiwi-Minsker, A. Renken
Swiss Federal Institute of Technology, LGRC-EPFL,
CH-1015 Lausanne, Switzerland

Abstract

A novel design of microstructured reactor system in a micro-channel arrangement between closely packed catalytic filaments is reported. The system comprises filaments of 3-15 μm in diameter and shows a laminar flow with a short radial diffusion time. This leads to low pressure drop and to a narrow residence time distribution (RTD) during reactor operation.

The latter device in combination with a membrane permeable to hydrogen is used in non-oxidative propane dehydrogenation. The catalytic filaments of Pt/Sn on alumina were active/selective and sustained periodic regeneration. This innovative micro-reactor system allowed a precise control of the propane dehydrogenation due to a narrower RTD. Propane conversions exceeding equilibrium and selectivities towards propene up to 97 % were attained.

Keywords: catalyst, filament, microstructured, packing, membrane, propane, dehydrogenation

1. Introduction

Microstructured reactors consist usually of many parallel channels with diameters in the micrometer range. Compared to "macro-reactors", this design offers several advantages like a high surface to volume ratio, short response times, defined flow characteristics and short radial diffusion times. Microreactors are particularly suitable for fast endo- and exothermic reactions and for the design of autothermal systems. Furthermore, the narrow residence time distribution in microchannel reactors [1, 2] allows to optimise the contact time and to suppress formation of by-products in complex reaction networks.

One of the main problems in using microstructured reactors for heterogeneously catalysed reactions is the introduction of a catalytic active phase. The easiest way would be to fill microchannels by catalyst powder, as proposed by Tonkovich and collaborators for hydrogen generation [3-5]. The drawback of this method is the high pressure drop. In addition, each channel must be packed identically to avoid maldistribution, which is known to lead to a broad residence time distribution in the reactor system.

Therefore, research is focused on the development of thin catalytic layers deposited on the reactors walls within the microstructure. Hönicke and coworkers [6, 7] chose aluminium as construction material for microchannel reactors. The specific surface of the micro channels was increased by anodic oxidation of the aluminium surface resulting in a thin porous layer of α-Al_2O_3. The obtained oxide layer had a very regular pore structure oriented perpendicularly to the flow direction. This porous layer served as support for the catalytically active components. The use of aluminium is restricted to temperatures up to 450°C. Therefore, different methods of deposition of active catalytic layers on other materials are under development. Sol-gel methods are commonly proposed to obtain a porous support layer on the wall of the microchannels [8]. The catalytically active phase can be deposited on the porous layer by precipitation or impregnation.

In the present paper a novel concept of a microreactor system is proposed for the catalytic non-oxidative dehydrogenation of propane. It consists of a two zone tubular reactor of few millimeters in diameter filled with catalytically active filaments placed in parallel to the tube walls. The two zones are separated by a Pd-membrane, allowing to eliminate hydrogen from the reaction mixture, thus shifting the reaction equilibrium to higher conversions.

2. Reactor concept

The novel concept of a microreactor is based on a structured catalytic bed arranged with parallel filaments of few micrometers in diameter (3-10 μm). The arrangement gives flow hydrodynamics similar to multi-channel microreactors, known to have a narrow residence time distribution (RTD). The channels for gas flow between the filaments (see Fig.1) have an equivalent hydraulic diameter in the range of few microns ensuring laminar flow and short diffusion times in the radial direction.

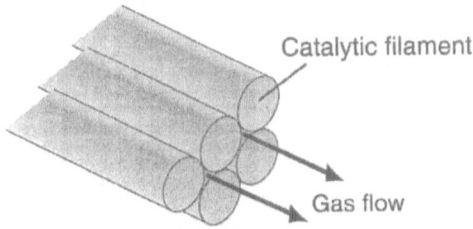

Catalytic filament

Gas flow

Figure 1: Axial flow of gases between catalytic filaments

The microstructured catalyst was used in a membrane reactor specially developed for the continuous production of propene from propane via non-oxydative dehydrogenation [9]. The main constrains of this reaction are as follows: the dehydrogenation is highly endothermic (129 kJ/mol at 823 K and 0.14 MPa), the propane conversion is limited by thermodynamic equilibrium (22% under the same conditions) and coke formation reduces quickly the catalyst activity. The proposed concept should overcome these constrains. The reactor design is schematically presented in the figure 2.

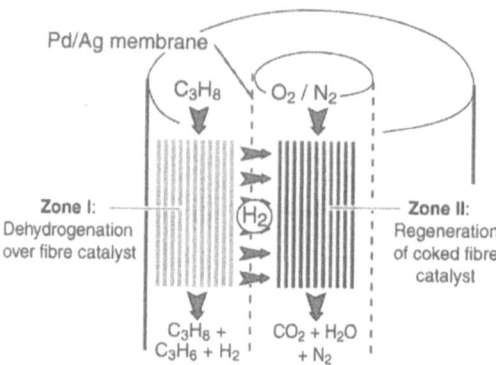

Figure 2: Scheme of the membrane reactor with two zones packed with microstructured filamentous catalyst

A Pd/Ag membrane permeable to hydrogen separates two concentric zones of the tubular reactor. On one side of the membrane (zone I), the dehydrogenation takes place with simultaneous coke formation on the catalyst surface and diffusion of hydrogen to zone II. On the other side of the membrane (zone II), hydrogen is oxidised by air, thus generating heat for the endothermic dehydrogenation in zone I. Moreover, due to the permanent oxidation of hydrogen, a high radial concentration gradient of hydrogen is obtained. This allows an efficient removal of hydrogen from the reaction zone and consequent shift of the reaction equilibrium. Simultaneously with the hydrogen oxidation, the deactivated catalyst is regenerated by burning off coke from the catalyst surface. The feeds of air and propane are switched periodically between the two zones to produce continuously propene.

3. Experimental

Aluminoborosilicate glass fibres in woven form (Vetrotex France SA) with a specific surface area of SSA=2 m^2g^{-1} were used as starting material for the catalyst preparation. First, the fabrics were treated at 90°C in 1.0 N aqueous solutions of HCl to leach out the non-silica components of the glass, then the material was rinsed in distilled water and dried in air at 50°C overnight [10]. By this procedure the specific surface area is increased up to 290 m^2g^{-1} indicating that porous filaments were obtained. The filament's surfaces were then covered by γ-alumina via deposition/precipitation of aluminium hydroxide from aqueous solution of a suitable salt followed by drying and calcinations in air at 650 °C during 3 h. The resulting support material – alumina/silica filaments (ASF) – is stable up to 800°C, and has a SSA in the range of 100-230 m^2g^{-1}.

The active metals (Pt and Sn) were deposited via two-steps impregnation from aqueous ammonia solutions (pH=10). $SnCl_2$ and hexachloroplatinic acid (H_2PtCl_6) (purum, Fluka Chemie AG, Buchs, Switzerland) were used as precursors.

Impregnation was followed by drying at 50°C overnight and calcination at 450°C in air for 1 h. The concentrations of the solutions were adjusted to attain a catalyst formulation of 0.5%Pt-1%Sn on ASF. This composition was reported to be selective towards propene and to have acceptable long-term stability [11].

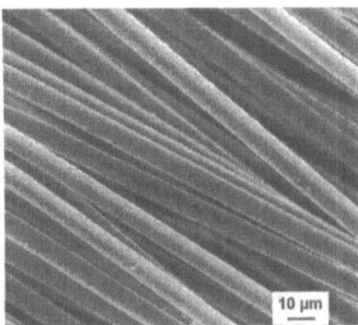

Figure 3: Catalytic filaments contained in a thread

The catalytic filaments were introduced into the tubular reactor in the form of threads. Each thread with a diameter of about 0.5 mm consists of a bundle of ~100 filaments, (figure 3), with a diameter of ~7 µm. The catalytic threads were placed in parallel into the tube to form a cylindrical catalytic bed of several centimetres length. The catalytic bed arranged in this manner has about 300 threads per cm^2 within the tube cross-section. The porosity of the filamentous packed bed is ε=0.8. The specific surface per volume is in the order of 108 m^2/m^3 and thus, about 50 times higher compared to washcoated tubes of the same inner diameter [12].

The membrane used in this study consists of a Pd/Ag(23wt%) alloy in the form of a tube of 6 mm internal diameter with a wall thickness of 70 µm (Johnson Matthey and Brandenberger SA, Zürich, Switzerland). The reactor has a length of L=140 mm. It is inserted in a quartz tube of 8.6 mm inner diameter (figure 2). The tube and shell volumes are equal to 4.0 ml.

The reaction was carried out at a temperature of T=823 K and a pressure of 0.14 MPa. Before the reaction, the catalyst was heated under nitrogen flow at 10 K/min up to the reaction temperature. After 15 minutes of temperature stabilization, nitrogen was replaced by a flow of pure propane and the reaction products were monitored. Conversion of propane and selectivities towards different products formation were calculated according to the following equations:

$$X_{C_3H_8} = \frac{y_{C_3H_6} + \frac{2}{3}y_{C_2H_6} + \frac{2}{3}y_{C_2H_4} + \frac{1}{3}y_{CH_4}}{y_{C_3H_8} + y_{C_3H_6} + \frac{2}{3}y_{C_2H_6} + \frac{2}{3}y_{C_2H_4} + \frac{1}{3}y_{CH_4}} \quad (1)$$

$$S_i = \frac{\frac{c_i}{3}y_i}{y_{C_3H_6} + \frac{2}{3}y_{C_2H_6} + \frac{2}{3}y_{C_2H_4} + \frac{1}{3}y_{CH_4}} \quad (2)$$

where y_i is the molar fraction of compound i (propene, ethane, ethene, methane), and c_i is the number of carbon atoms in this compound.

Residence time distribution in the filamentous catalytic bed was determined. A response on a step function (10%Ar in N_2) was measured via a quadrupole mass spectrometer (TSU 260D, Balzers, Balzers, Liechtenstein) in a larger tubular reactor: L = 230 mm, ID = 15 mm, OD = 18 mm.

4. Results and discussion
4.1 Hydrodynamics

Hydrodynamic of gas flow through the microstructured catalytic bed was studied and compared to different conventional packings. The residence time distribution (RTD) was measured in a tube packed with the filamentous catalyst and with particles of silica and γ-alumina of different shapes and sizes. Experimental results are presented in Figure 4. Under identical experimental conditions, randomly packed beds showed significantly broader RTD compared to the structured filamentous packing.

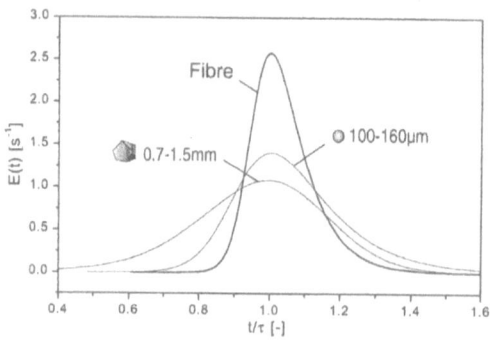

Figure 4:   RTD for the randomly packed beds in comparison with the structured filamentous bed: response of the switch 30 Nml/min 10%Ar/N_2 to N_2, 298 K, 0.1 MPa, internal tube diameter 15 mm, length 230 mm

The pressure drop during the passage of gas through the tube packed with spheres was compared to the pressure drop through the structured filamenteous packing. The hydraulic diameters of both beds are in the same order of magnitude. As can be seen from Fig. 5, the pressure drop in the randomly packed bed was found to be 5 times higher as compared to the pressure drop in the filamanteous packing.

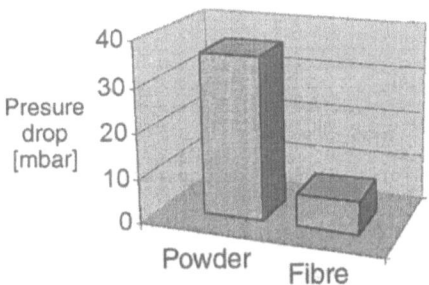

Figure 5: Pressure drop in a microstructured catalyst in comparison to an isotropic bed (100-160 μm spheres), 298 K, 1 bar, 120 ml(STP)/min N_2

The pressure drop measured under laminar flow for the structured microchannel reactor can be used to estimate the equivalent diameter of the channels based on the well-known relationship of Hagen-Poisieulle:

$$d_{eq} = \sqrt[4]{\frac{128 \cdot Q \cdot \mu \cdot L}{\Delta p \cdot \pi}} \tag{3}$$

Thus, the catalytic bed of 15 mm inner diameter consisting of about 46,000 filaments is equivalent to: N = 38,000 cylindrical microchannels of 70 μm diameter.

The estimated hydraulic diameter of 70 μm is in the same order of magnitude as for conventional multichannel microreactors. Other advantages of the filamentous packing are an easy manufacturing and simple handling.

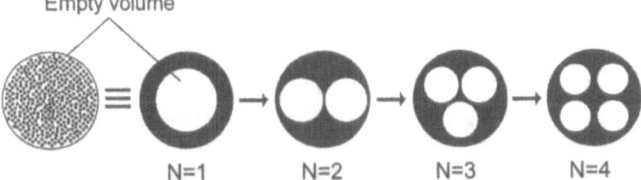

Figure 6: Simplified cross section of microstructured catalyst and representation of its empty volume by several cylindrical channels, N=number of channels

4.2 Catalyst testing

The propane dehydrogenation was first performed in a quartz tube reactor of 6 mm internal diameter. At 823 K and 0.14 MPa, propane conversion to thermodynamic equilibrium (22%) is reached. Observed by-products are methane, ethane and ethene. As shown on figure 7, the conversion is similar on powdered and on filamentous catalyst.

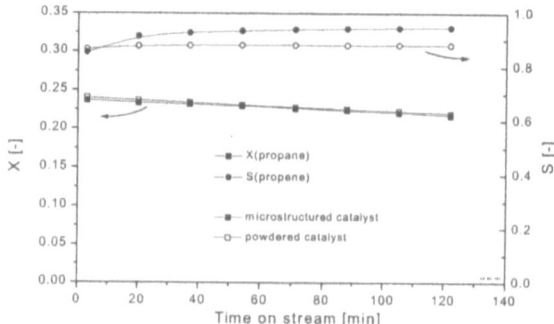

Figure 7: Equivalence of catalyst activity on filamentous and powdered support, and increase of propene selectivity with microstructured catalyst, 823 K, 0.14 MPa, GHSV = 1161 h^{-1} (τ = 3.1 s), m_{cat} = 0.375 g

Deactivation due to coke formation is slow in both cases. But the propene selectivity is higher over the microstructured catalyst: a maximum of 95% is reached instead of 88% over the powdered catalyst. This is explained by the narrower RTD obtained in the structured bed. It has been shown [13] that in this undesirable case, cracking reactions increase with propane conversion, diminishing selectivity towards propene.

Thus, in addition to lower pressure drop, the microstructured filamentous packing avoids undesired consecutive reactions due to the narrow RTD increasing the selectivity towards the target product.

The same reaction was carried out in the membrane reactor with microstructured catalytic bed installed in zone II (cf figure 2). The residence time was increased to 19 s to maximize propane conversion. Zone I was empty and supplied by air at 12.2 Nml/min (co-current). At the same temperature and pressure, the conversion (30%) has exceeded equilibrium value (22%) for the first 30 minutes, cf figure 8. According to the coke formation scheme of Larsson et al. [14], equilibrium between propene and coke precursors is shifted to the latter due to hydrogen removing out of the catalytic bed. This explains the faster deactivation of the catalyst in the membrane reactor as presented in figure 8.

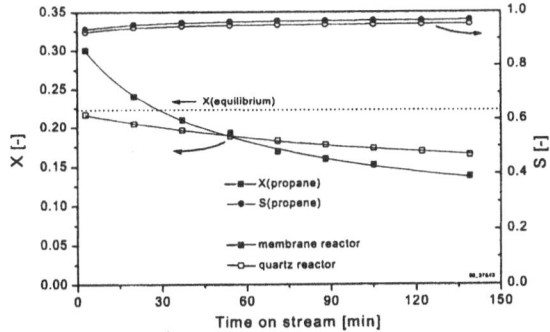

Figure 8: Enhancement of propane conversion over equilibrium and increase of propene selectivity in membrane reactor, 823 K, 0.14 MPa, GHSV = 189 h^{-1} (τ = 19 s), microstructured catalyst

Propene selectivity is enhanced in membrane reactor: the amount of by-products is twice less and propene selectivity reaches 97%. This performance is due to the absence of hydrogen in the gas phase reducing hydroisomerization and hydrogenolysis reactions [15]. The advantage of the microstructured packing in this case is that it allows radial hydrogen diffusion to the membrane even if longitudinal flow is highly preferred. H_2 can be efficiently removed using a macroscopic membrane with microfluidic channels.

5. Conclusions

A novel microstructured catalyst was developed. It is made of fine and long filaments with Pt-Sn as an active phase supported on γ-Al_2O_3. In this microstructured packing, gas flow in one direction contrary to isotropic beds where gas flow in many directions. Hydraulic diameter is about 70 μm and flow is laminar. This presents some of the most interesting advantages of microstructures, namely a narrow residence time distribution and low pressure drop. Compared to a packing made of spheres of 100-160 μm diameter, the pressure drop was diminished 5 times and the RTD was remarkably narrower.

This microstructured packing brings new opportunities to the heterogeneous catalysis domain because it can be used for many reactions improving highly the hydrodynamics and keeping an excellent contact with the reactants. Applied to non-oxydative propane dehydrogenation, it allowed a better control of reaction time due to the narrower RTD and increased propene selectivity from 88% to 95%.

In a reactor with a hydrogen permeable membrane, the propene selectivity was increased up to 97% due to the removal of H_2 suppressing hydroisomerization and hydrogenolysis reactions. But a remove of hydrogen in the gas phase increased coke formation and catalyst deactivation. Nevertheless, propane conversion has exceeded equilibrium conversion for the first 30 minutes.

200

Nomenclature

ASF	Alumina Silica Filament	p	Pressure [Pa]
d_{eq}	Equivalent diameter [m]	Q	Volumetric flowrate [m^3s^{-1}]
GHSV	Gas Hourly Space Velocity [h^{-1}]	RTD	Residence Time Distribution
ID, OD	Internal and outer diameter [m]	SSA	Specific Surface Area [m^2g^{-1}]
L	Length [m]	τ	Residence time [s]
μ	Viscosity [Pa s]	X, S	Conversion and selectivity [-]

Acknowledgment

The authors gratefully acknowledge the financial support of the Swiss National Science Foundation, Dr I. Youranov and A. Udriot (Institute of Chemical Engineering, EPFL) for the catalyst preparations.

References

1. A. Rouge, B. Spoetzl, K. Gebauer, R. Schenk, and A. Renken, Microchannel reactors for fast periodic operation: the catalytic dehydrogenation of isopropanol, *Chem. Eng. Sci.*, *in press* (2001).
2. D. Hoenicke and G. Wiessmeier, Heterogeneously Catalyzed Reactions in a Microreactor, *DECHEMA monographs*, 1996.
3. D. P. VanderWiel, J. L. Zilka-Marco, Y. Wang, A. Y. Tonkovich, and R. S. Wegeng, Carbon dioxide conversions in microreactors, *Proceedings of the 4th International Conference on Microreaction Technology (IMRET 4)*, W. Ehrfeld, U. Eul and R. S. Wegeng, Eds., AIChE, Atlanta, 2000.
4. S. P. Fitzgerald, R. S. Wegeng, A. Y. Tonkovich, Y. Wang, H. D. Freeman, J. L. Marco, G. L. Roberts, and D. P. VanderWiel, A compact steam reforming reactor for use in automotive fuel processor, *Proceedings of the 4th International Conference on Microreaction Technology (IMRET 4)*, W. Ehrfeld, U. Eul and R. S. Wegeng, Eds., AIChE, Atlanta, 2000.
5. E. A. Daym, D. P. VanderWiel, S. P. Fitzgerald, Y. Wang, R. T. Rozmiarek, M. J. LaMont, and A. Y. Tonkovich, Microchannel fuel processing for man portable power, *Proceedings of the 4th International Conference on Microreaction Technology (IMRET 4)*, W. Ehrfeld, U. Eul and R. S. Wegeng, Eds., AIChE, Atlanta, 2000.
6. D. Hönicke and G. Wiessmeier, Heterogeneously catalyzed reactions in a microreactor, *Microsystem technology for chemical and biological microreactors*, W. Ehrfeld, Ed., Dechema, VCH, Weinheim, 1995.
7. G. Wiessmeier and D. Hönicke, *Strategy for the development of micro channel reactors for heterogeneously catalyzed reactions*, 2nd International Conference on Microreaction Technology (IMRET 2), New Orleans, USA, W. Ehrfeld, I. H. Rinard and R. S. Wegeng, 1998, pp 24-32.
8. G. Wiessmeier, K. Schubert, and D. Hönicke, Monolithic microstructure reactors possessing regular mesopore systems for the successful performance of heterogeneously catalyzed reactions, *Proceedings of the 1st International Conference on Microreaction Technology (IMRET 1)*, W. Ehrfeld, Ed., Springer, Berlin, 1997.

9. O. Wolfrath, L. Kiwi-Minsker, and A. Renken, *Packed Bed which is Arranged in a Tubular Reactor Part*, No. P.7041, European Patent,,, 2000.

10. L. Kiwi-Minsker, I. Youranov, E. Slavinskaia, V. Zaikovskii, and A. Renken, Pt and Pd Supported on Glass Fiber as Effective Combustion Catalysts, *Catalysis Today*, *59*, 61-68 (2000).

11. I. B. Yarusov, E. V. Zatolokina, N. V. Shitova, A. S. Belyi, and N. M. Ostrovskii, Propane Dehydrogenation over Pt-Sn Catalysts, *Catalysis Today*, *13*, 655-658 (1992).

12. V. Hatzlantonlou, B. Andersson, and N.-H. Schöön, Mass Transfert and Selectivity in Liquid-Phase Hydrogenation of Nitro Compounds in a Monolithic Catalyst reactor with Segmented Gas-Liquid Flow, *Ind. Eng. Chem. Process Des Dev.*, *25*, 964-970 (1986).

13. M. Sheintuch and R. M. Dessau, Observation, Modeling and Optimization of Yield, Selectivity and Activity during Dehydrogenation of Isobutane and Propane in a Pd Membrane Reactor, *Chemical Engineering Science*, *51, No 4*, 535-547 (1996).

14. M. Larsson, N. Henriksson, and B. Andersson, Estimation of Reversible and Irreversible Coke by Transient Experiments, *Stud Surf Sci Catal*, *111*, 673-680 (1997).

15. T. Matsuda, I. Koike, N. Kubo, and E. Kikuchi, Dehydrogenation of Isobutane to Isobutene in a Palladium Membrane Reactor, *Applied Catalysis A: General*, *96*, 3-13 (1993).

Characterization of a Gas/Liquid Microreactor, the Micro Bubble Column: Determination of Specific Interfacial Area

V. Haverkamp, G. Emig*, V. Hessel, M.A. Liauw*, H. Löwe

Institut für Mikrotechnik Mainz GmbH, Carl-Zeiss-Str. 18-20, D-55129 Mainz, Germany
*   Lehrstuhl für Technische Chemie I, Universität Erlangen-Nürnberg, Egerlandstr. 3, D-91058 Erlangen, Germany

Motivation

The micro bubble column [1] was one of the first microstructured devices being specially designed for gas/liquid processing. Within a short time, the experimental performance of this microreactor was demonstrated for the fluorination of toluene using elemental fluorine, as an example of use with industrial relevance. However, the understanding of fundamental reactor characteristics had to be postponed to succeeding, more detailed investigations. In this context, the objective of the work described in this article is to determine specific interfacial areas as a main parameter of reactor performance and to develop on this basis a first reactor model for gas/liquid reactions in the micro bubble column. This reactor model enables then proper prediction of suitable candidate reactions for processing in this microreactor.

For this purpose, the specific interfacial areas of an aqueous and an organic fluid, acting as models for reacting fluids, were determined using a micro bubble column, which has been previously described [1]. These data are crucial for the development of the empirical reactor model. Using such a model, operating conditions for various mass transport limited reactions with known kinetics can be properly set. As a candidate, the homogeneously catalyzed oxidation of butyraldehyde to butyric acid was chosen, a gas/liquid process well described in literature [2]. By comparison of the conversion data gained by this reaction to calculated data, a validation of the reactor model is achievable.

Experimental Setup

The micro bubble column consists of a mixing and a reaction unit. For the experiments described in this abstract, the mixing units were equipped with 20 μm deep gas and liquid feeding channels, being 5 μm and 20 μm wide, respectively. The reaction unit comprises 32 channels of 300 μm x 100 μm cross-section, termed slit-like geometry. More details on design criteria and microreactor assembly are given in reference [1]. For all experiments, the reactor was equipped with an inspection glass. The parts of the micro bubble column and the mounted setup are shown in Figure 1.

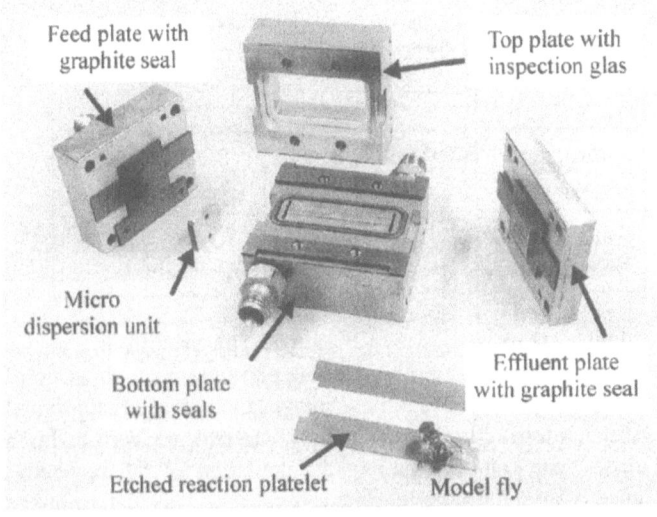

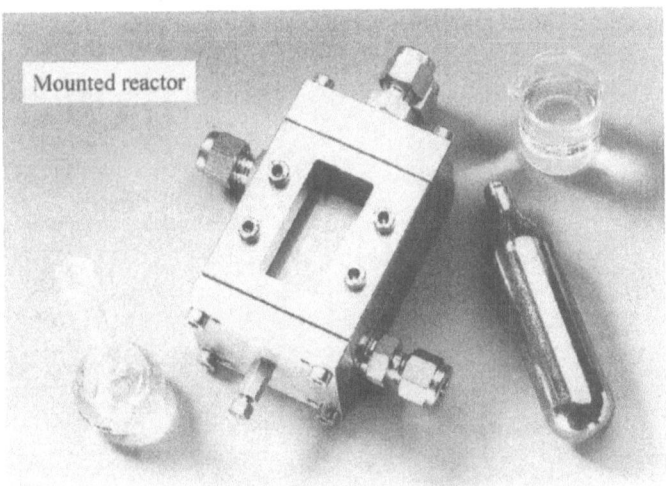

Figure 1: Top: parts of the micro bubble column with inspection glass (and model fly). Bottom: mounted reactor.

Setup for interfacial area measurements

The inert systems investigated were nitrogen/water and nitrogen/isopropanol. Experiments were carried out in a wide parameter range covering all flow patterns – bubbly, slug and annular flow - determined so far [3]. The volume flows of the gas were set between 1 and 180 mL/min and of the liquids between 0.11 and 0.86 mL/min. The range of superficial velocities w_L and w_G investigated is shown in Table 1.

Table 1: Experimental parameters for the optical interfacial area measurements.

Inert System	w_L (m/s)	w_G (m/s)
nitrogen/ water	0.006, 0.015	0.087 – 0.347
nitrogen/ isopropanol	0.002 – 0.015	0.087 – 1.74

The experimental setup is shown in Figure 2. The specific interfacial areas were measured using an optical method, by taking photographs with a microscope/CCD-camera/stroboscope system. From these images the characteristic dimensions of the disperse systems referring to the various flow patterns, concerning bubble sizes and film thickness, were gained, enabling a calculation of the respective interfacial area. For each volume flow set, an average area was determined out of four photographs, taken at different points of the micro bubble column. Therefore, the specific interfacial areas given below represent integral, rather than local values.

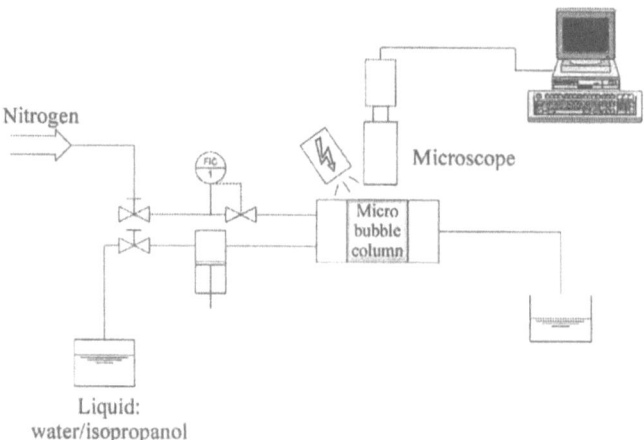

Figure 2: Experimental setup for the investigation of the specific interfacial areas.

Reacting model system

As reacting fluids, the well described system air/butyraldehyde in butyric acid with manganese acetate as catalyst was used [2]. As to include a broad range of flow patterns in the micro bubble column, two exemplary liquid volume flows, namely 20 mL/h and 50 mL/h, and a range of gas volume flows from 1.73 mL/min up to 172.8 mL/min were chosen. These fluid flows are indicated in the flow pattern map of the micro bubble column for the system nitrogen/isopropanol (Figure 3).

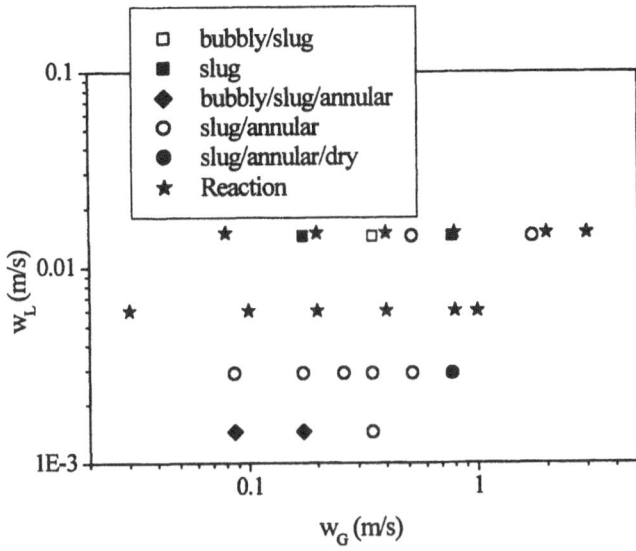

Figure 3: Flow pattern map for the inert system nitrogen/isopropanol with fluid flow parameters chosen for the reaction experiments.

The experiments were carried out in the same setup used for the measurement of the specific interfacial areas. A photograph of this setup is presented in Figure 4. The analytical determination followed the procedure described in [2]. The butyraldehyde was detected using gas chromatography, the butyric acid using iodometric titration.

Figure 4: Photograph of the experimental setup. The micro bubble column is surrounded by flow and pressure controlling equipment (left) and pumping and gas feeding systems (right).

Experimental approach

Specific interfacial areas and film thickness by an optical method

Interfacial areas usually are determined by optical or chemical methods. For some gas/liquid reactors, optical and chemical methods give nearly equivalent results, whereas in other cases optical methods turned out to give less reliable data [4]. Due to the broader applicability and lower technical expenditure, it was desired to use an optical method in the framework of investigations reported here. Consequently, it was verified at exemplary process parameters that the optical method chosen is in reasonable accordance to a chemical one (see below).

The optical method, used here, is simply based on the determination of the specific interfacial areas via the bubble geometry, in particular referring to the thickness of the liquid films on all four sides of the channel. In contrast to circular microchannels, the liquid films in rectangular channels are not constant, but differ distinctively, the opposite films being equal (see Figure 5). For the determination of interfacial areas, the length, width and height of the corresponding bubbles normally have to be known, the latter two parameters referring to the film thickness δ_{F1} and δ_{F2}, being actually measured (see Figure 5).

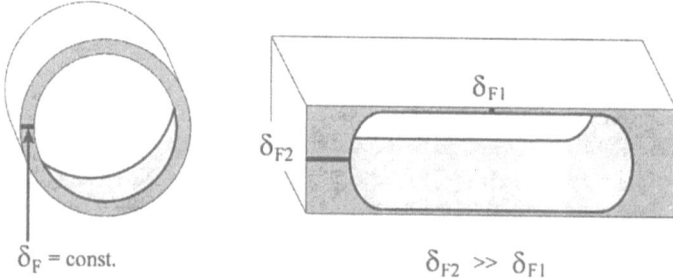

Figure 5: Film thickness in different microchannel geometries. Left: circular channel with constant film thickness. Right: slit like channel with two different film thickness δ_{F1} and δ_{F2}, with δ_{F1} being much smaller than δ_{F2}.

A first simplification of the above mentioned procedure relies on the neglect of the film thickness δ_{F1}, rendering the bubble height equal to the channel height. The reasons for that refers to the large difference between the film thickness δ_{F1} and δ_{F2}, the latter being only a few microns thick, as a consequence of the slit like channel geometry chosen (see dimensions in *Experimental setup*).

For reasons of fast data acquisition, it was further decided to measure the length of each individual bubble and to rely on mean values for the bubble width. This stems from the fact that there is a broad distribution of bubble geometries in the whole range of the superficial velocities, necessitating the second simplification. While the bubble length is easily accessible from photographs and the corresponding distribution hence can be gathered with reasonable efforts, determination of the bubble width, i.e. the respective film thickness, needs higher technical expenditure.

For nitrogen/water the mean film thickness amounts to 55 µm, with a scattering range of ± 10 µm, which corresponds to the measuring accuracy. The mean film thickness for the nitrogen/isopropanol system was determined to 64 μm, being slightly larger than the film thickness for nitrogen/water.

Experimental results

Specific interfacial areas

The such determined specific interfacial areas of the two model systems, nitrogen/water and nitrogen/isopropanol, are given in Figure 6 for different liquid flows, as a function of the ratio of the gas/liquid superficial velocities. The areas are increasing with increasing gas velocity, corresponding to a change of the flow pattern in the microchannels from bubbly/slug to annular flow. The measured maximum specific interfacial areas for the aqueous and organic model systems were 13,400 m^2/m^3 and 18,000 m^2/m^3, respectively. These values notably exceed the interfacial areas described for laboratory and technical equipment, being limited to about 2,000 m^2/m^3 [4].

The difference in specific interfacial area of 4,600 m^2/m^3 of the two systems may be explained by simple deductions based on the wetting behavior of isopropanol and water. The contact angle of isopropanol, when wetting the stainless steel reaction platelet, of about 6° is lower compared to that of water, being about 60°. It is assumed that the improved wetting of isopropanol at the inlet of the reaction unit should result in a more regular distribution of the gaseous and liquid flows. For more regular distributions, higher integral specific interfacial areas result.

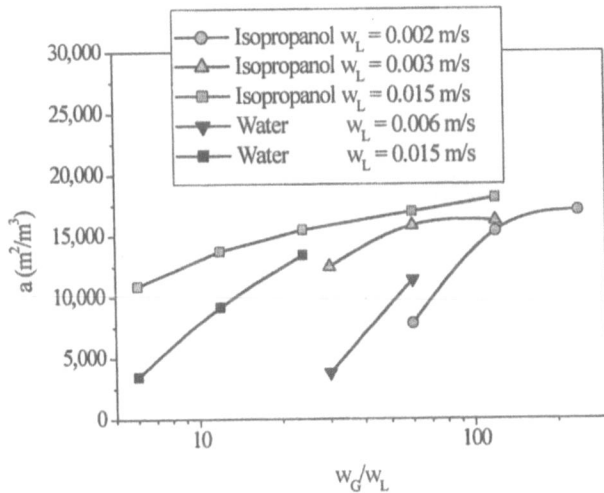

Figure 6: Specific interfacial areas in dependence on the ratio of gas to liquid superficial velocities w_G/w_L for two different model systems, nitrogen/isopropanol and nitrogen/water.

In order to allow predictions of interfacial areas for a wide range of fluid systems, it is intended in the following to extend this phenomenological description to a mathematical analysis. As main influencing parameters on the interfacial area, the gas superficial velocity, as evident from Figure 6, and fluid properties as density and surface tension, as suggested by the wetting behavior deductions, were considered. These data allow one to determine the interfacial area as a function of the Weber number as a dimensionless number. Hence this relationship encompasses different fluid systems with only one equation.

The Weber number is defined as

$$We = \frac{w_G^2 \cdot d_{eq} \cdot \rho}{\sigma}.$$

(1)

The interfacial data of both model systems are shown in Figure 7 with an interpolated curve that describes the dependence of the interfacial area on We

$$a = a_{max} + \frac{\left(C_1 - a_{max}\right)}{\left(1 + e^{\frac{We - C_2}{C_3}}\right)}$$

(2)

with the following fitted parameters: a_{max} = 16.900 m²/m³,

$C_1 = -45{,}721$ m²/m³; $\qquad C_2 = 0.3$; and $\qquad C_3 = 0.3$.

The good correspondence of the interpolated curve and the experimentally derived data proves that equation (2) provides a reasonable mathematical description for determination of the specific interfacial area.

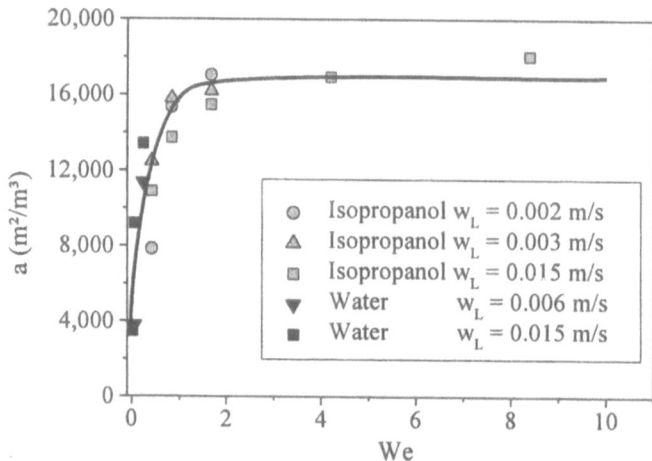

Figure 7: Specific interfacial area in dependence on the Weber number. The data correlate well with a function of the Weber number and the measured maximum specific interfacial area (equation 2).

Furthermore, it was aimed to judge the accuracy of the results gathered by the measurement of the bubble geometry, since this optical method, as discussed above, sometimes fails to describe the 'real' chemically effective interfacial areas (see also the discussion on the simplification of bubble geometry determination mentioned above). Figure 8 shows that the optically derived interfacial areas, gathered by simple means, are in acceptable accordance to data determined by a chemical method [1], namely the sulfite oxidation.

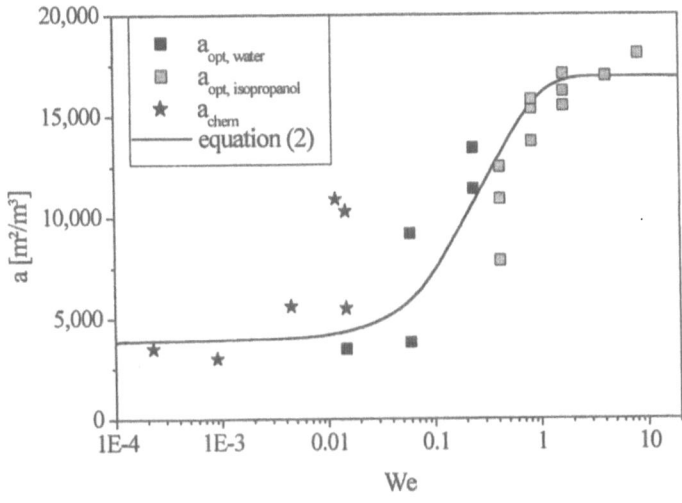

Figure 8: Comparison of specific interfacial areas of the fluid system nitrogen/water determined by an optical and a chemical method and fitted curve (equation (2)). The data of the two methods are in acceptable accordance.

Reacting system

Using the reaction system air/butyraldehyde in butyric acid, conversions of butyraldehyde up to 41% were realized in the micro bubble column (Table 2). First of all, the data suggest a strong dependence of conversion on superficial velocity ratio. Conversions are increasing up to factor of 4, e.g. from 10% for a ratio of 5 up to 41% for a ratio of 53 with increasing velocity ratio. This supports the basic assumption given above that the specific interfacial area ist the main parameter for reactor performance of the micro bubble column.

Table 2: Conversions of butyraldehyde in the micro bubble column for different superficial velocities.

w_L (m/s)	$\dot{V}_L$ (mL/h)	w_G (m/s)	$\dot{V}_G$ (mL/min)	w_G/w_L	X_{BA} (%)
0.006	20.74	0.03	1.73	5	14
		0.08	4.61	13	8
		0.1	5.76	17	2
		0.4	23.0	67	26
		0.8	46.1	133	27
		1	57.6	167	26
		3	173	494	28
0.015	51.84	0.08	4.61	5	10
		0.1	5.76	7	20
		0.2	11.5	13	19
		0.4	23.0	27	33
		0.8	46.1	53	41
		2	115	133	42
		3	173	200	41

In Figure 9, the conversions per mean residence time are plotted as a function of the Weber number, being a measure for the specific interfacial area. This clearly shows that conversion per residence time is strongly correlated to the specific interfacial area.

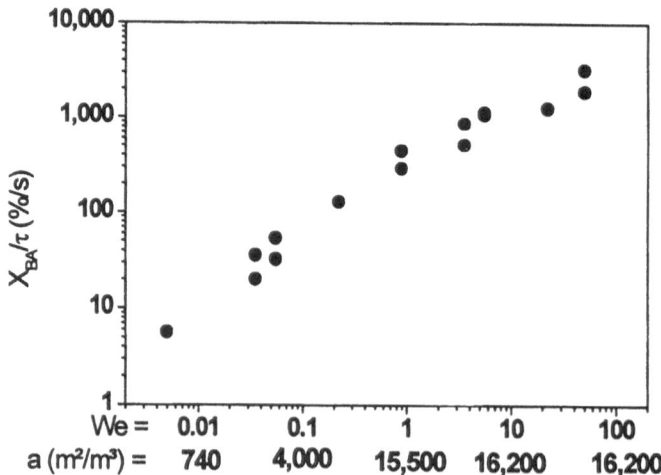

Figure 9: Conversions per mean residence time in dependence on the Weber number. Conversion per time increases strongly with rising We corresponding to increasing specific interfacial area.

Reactor model

Based on these main experimental results, a simple reactor model has been developed. Assumptions made for this reactor model are:
- system works isothermal,
- there exists only mass transport of the gaseous reactant A to the liquid phase; evaporation of the liquid component B is negligible, and
- the reaction carried out in the micro bubble column is fast, so that the concentration of gaseous reactant in the liquid phase is negligible.

To fulfill the latter assumption, the Hatta number has to be within the following limits [5]:

$$\text{Ha} > 3 \text{ and} \tag{3}$$

$$\text{Ha} << 1 + \frac{D_B \cdot c_{B,0}}{D_A \cdot c_{A,L}^*} . \tag{4}$$

Here, the Hatta number is defined as

$$\text{Ha} = \sqrt{\frac{2 \cdot D_A \cdot K_m \cdot \left(c_{A,L}^*\right)^{m-1}}{(m+1) \cdot k_L^2}} . \tag{5}$$

Within the limits expressed by the equations (3) and (4), the following simplification in describing mass transport over a gas/liquid interface with chemical reaction can be made:

$$R_A = E \cdot k_L \cdot \left(c_{A,L}^* - c_{A,L}\right). \tag{6}$$

Based on this equation and the mass balance over a volume element in one channel, the following differential equation for the concentration of liquid phase reactant B in dependence on channel length can be derived:

$$\dot{V}_L dc_{B,L} = -z \cdot a \cdot A_L \cdot \sqrt{\frac{2 \cdot D_A \cdot K_m \cdot \left(c_{A,L}^*\right)^{m-1}}{(m+1)}} \cdot c_{A,L}^* \, dx . \tag{7}$$

This equation is analytically solvable, and the solution for the reaction of butyraldehyde (BA) to butyric acid is:

$$c_{BA} = \frac{C_2 \cdot C_3 \cdot \exp(C_2 \cdot C_1 \cdot x)}{C_4 - C_4 \cdot C_3 \cdot \exp(C_2 \cdot C_1 \cdot x)} \tag{8}$$

with
$$C_1 = \frac{2 \cdot a \cdot A_L \cdot \sqrt{D_{O_2} \cdot K_3}}{\dot{V}_L}, \quad C_2 = c_{O_2,L0}^* - C_4 \cdot c_{BA,0},$$

$$C_3 = \frac{C_4 \cdot c_{BA,0}}{C_4 \cdot c_{BA,0} + C_2} \text{ and } C_4 = \frac{1}{2} \cdot \frac{\dot{V}_L}{\dot{V}_G} \cdot \frac{R \cdot T}{He} .$$

Comparison of the reactor model with experiments for oxidation of butyraldehyde

A comparison of conversions measured in experiments concerning butyraldehyde oxidation and conversions calculated with above reactor model shows that the model describes the reaction behavior in the micro bubble column well(see Figure 10). The maximum deviation in conversion is at a residence time of 0.2 s.

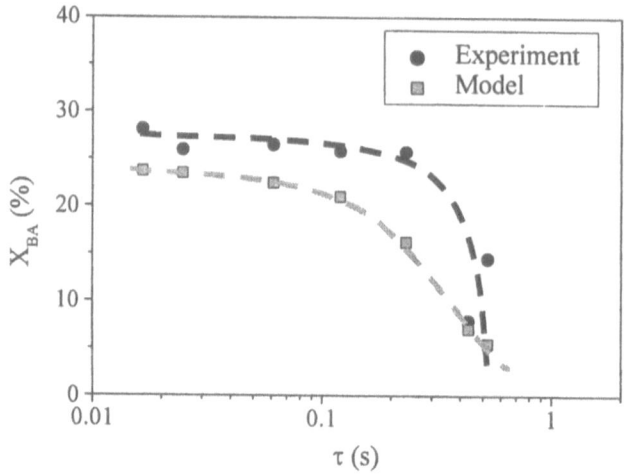

Figure 10: Measured and calculated conversions of butyraldehyde in dependence of residence time in the micro bubble column with a liquid volume flow of 50 mL/h. The model describes the experimental behavior with a deviation up to 10% in conversion.

Therefore, the model can be used to find optimum process conditions for this reaction. In Figure 11, conversions calculated with the reactor model are shown in dependence of specific interfacial area for a given volume flow of 50 mL/h. For selected data, the corresponding residence times are included. It becomes obvious that for the reaction with this specific liquid volume flow a single optimum point in dependence of specific interfacial area exists (reactions with more optimum points are known [6]).

This optimum is not related to the maximum specific interfacial area, but is determined by an interplay between high interfacial area for mass transport and sufficiently long residence time for the kinetics. For the liquid volume flow of 50 mL/h, optimum conversion is achieved at a specific interfacial area of 12,000 m²/m³ and at a residence time of 0.093 s.

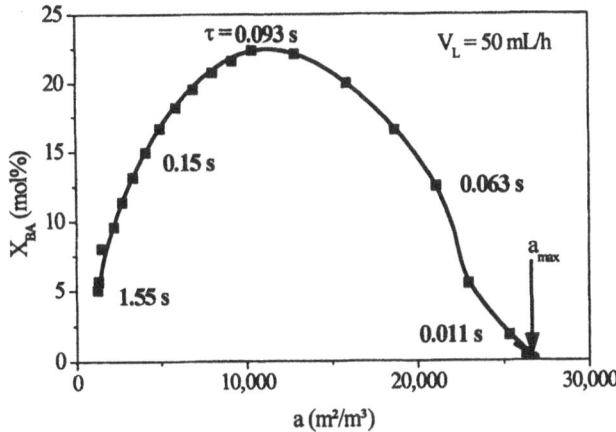

Figure 11: Calculated conversions in dependence of the specific interfacial area in the micro bubble column. The curve shows an optimum at an interfacial area of 12,000 m²/m³ and a residence time of 0.093 s for the oxidation of butyraldehyde.

Outlook

Today, it is common practice to apply reactions in microreactors without judging if the kinetics are sufficiently fast as compared to typical time-scales of fluid flow, due to the lack of reactor models. Considering the large range of processing times needed, ranging from milliseconds to hours, and comparing it to the smaller range provided by microreactors, it is obvious that such an undirected approach is not efficient, although currently unavoidable.

In the framework of the investigations of this article, a reactor model for gas/liquid reactions in a microreactor was developed for the first time. It is based on an empirical description with the specific interfacial area as a main parameter, rather than taking into account a more complex analysis of the relevant physical processes. This tool, although simple in structure, enables a first judgement of the suitability of the micro bubble column for different reactions.

Inevitably, kinetics are required for that purpose, hence limiting the use of the reactor model to only part of the reactions. Nevertheless, this paves the ground for a more directed approach in selecting suitable candidate reactions for microreactors. This allows a considerable reduction of the experiments needed and, consequently, notably enhances the probability to succeed in reaching the targets of a chemical engineering task.

214

Notation

Roman symbols:

a	:	specific interfacial area, m^2/m^3
A	:	channel cross-sectional area, m^2
c	:	concentration, mol/m^3
C	:	constant
d	:	diameter, m
D	:	diffusion coefficient, m^2/s
E	:	enhancement factor
Ha	:	Hatta number
He	:	Henry coefficient, m^3/mol bar
k	:	mass transport coefficient, 1/s
K	:	reaction rate constant
m	:	reaction order
R	:	gas constant, J/mol K
τ	:	residence time, s
$\dot{V}$	:	volume flow, m^3/s
w	:	superficial velocity, m/s
We	:	Weber number
x	:	channel length coordinate
X	:	conversion, mol%
z	:	stoichiometric factor

T	:	temperature, K

Greek letters:

δ	:	thickness, μm
ρ	:	liquid density, kg/m^3
σ	:	surface tension, N/m

Subscripts:

0	:	initial
*	:	equilibrium
A	:	fluid A
B	:	fluid B
BA	:	butyraldehyde
chem	:	measured with chemical method
eq	:	equivalent
F	:	liquid film
G	:	gas
L	:	liquid
m	:	reaction order
opt	:	measured with optical method
O2	:	oxygen

Literature

[1] V. Hessel, W. Ehrfeld, K. Golbig, V. Haverkamp, H. Löwe, M. Storz, Ch. Wille, A. Guber, K. Jähnisch, M. Baerns, Conference Proceedings of the 3rd International Conference on Microreaction Technology, April 18th-21st 1999, Frankfurt a.M., Germany, Springer-Verlag (2000), pp. 526-540

[2] M.E. Ladhabhoy, M.M. Sharma, Journal of Applied Chemistry (1969), pp. 274-280

[3] V. Hessel, W. Ehrfeld, V. Haverkamp, Th. Herweck, H. Löwe, J. Schiewe, Ch. Wille, Topical Conference Proceedings of IMRET 4, 4th International Conference on Microreaction Technology, March 5-9 2000, Atlanta, GA., The American Institute of Chemical Engineers (2000), pp. 174 – 186

[4] W.-D. Deckwer, Reaktionstechnik in Blasensäulen, Otto Salle Verlag, Frankfurt/M., Germany (1984)

[5] Fitzer, E., Fritz, W., Emig, G.: Technische Chemie, 4. Auflage. Springer-Verlag, Berlin Heidelberg New York (1995)

[6] Haverkamp, V., Dissertation Universität Erlangen-Nürnberg, 2001, to be published in VDI-Fortschrittsberichte

Visualization of Flow Patterns and Chemical Synthesis in Transparent Micromixers

T. Herweck, S. Hardt, V. Hessel, H. Löwe, C. Hofmann, F. Weise,
T. Dietrich*, A. Freitag*

Institut für Mikrotechnik Mainz GmbH, Carl-Zeiss-Str. 18-20, D-55129 Mainz, Germany
*mgt mikroglas technik AG, Galileo-Gallilei-Straße 28, 55129 Mainz

Abstract

The design of a glass made interdigital micromixer consisting of several functional layers is described. Its transparency allows the observation of the hydrodynamics directly in the mixing zone via optical imaging. By monitoring thin lamellae of dyed and pure water in the mixing zone, hydrodynamic and geometric lamellae focusing can be described. In the case of contacting immiscible fluids in the mixer, droplet formation is examined experimentally and compared to computational predictions. It is found that simulation data fit well with experimental results gained from the model system water / silicon oil. As a more realistic binary system of immiscible fluids, in the view of practical relevance in organic chemistry, concentrated sulfuric acid / toluene is investigated in a rectangular shaped glass micromixer. Droplet formation and quality of the emulsion are strongly influenced by the total flow rate. Subsequently, the hydrolysis of benzal chloride and one of its mono fluorinated derivates with concentrated sulfuric acid is examined in an interdigital micromixer. In comparison to results obtained from standard reaction tools like a stirred vessel and a mixing tee the dominance of the micro fluidic device can be demonstrated.

Motivation

Interdigital micromixers made of stainless steel, equipped, e.g., with LIGA-type metallic inlays, recently were utilized for mixing of miscible liquids, emulsion generation, gas / liquid dispersion, and forced precipitation of micro-scale solid particles [1,2]. Their performance as process optimization or production tools was outlined in several investigations by industry, e.g. being dedicated to extraction in miniplant mixer-settler systems [3], to premixing as a part of a tubular polymerization reactor [4], and to multiphase contacting in the explosive regime [5].

However, a fundamental understanding of the respective processes in the micromixers is still lacking. For instance, the knowledge about the formation of droplets within a liquid, as a first step in an emulsification process, is essential for a tailor-made design of the corresponding microsystem. In this context, experimental data are urgently required to validate first results obtained by simulation [6]. The direct observation of the hydrodynamics in micromixers, e.g. using colored solutions, is a simple method to get a first insight. For this purpose, mgt mikroglas technik AG and IMM used glass as a transparent material for the manufacturing of interdigital micromixers [7].

Glass made Micromixers

The transparent micromixers were fabricated of Foturan™, a photostructurable glass by means of UV-lithography and a subsequent chemical etching process. In contrast to the metal made IMM interdigital micromixers the glass made versions use a planar geometry, that means the channels for the inlet streams and the mixing zone for the product stream are arranged in one layer. Figure 1 shows two disassembled glass micromixers (Figure 1 left) and one assembled in a variable housing fitting for different types of glass made mixers (Figure 1 right).

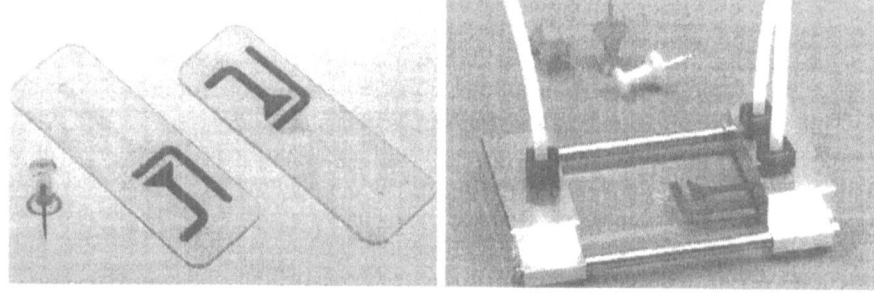

Figure 1: Disassembled glass interdigital micromixers (left) and assembled in a variable stainless steel housing (right).

Three designs were realized differing in their mixing zone comprising a triangular or rectangular shape as well as a more complex design (termed slit-shaped), which is similar to that of a stainless steel mixer, investigated earlier [2]. The devices are build up of four glass layers in a sandwich-like manner by an advanced glass-bonding process (Figure 2).

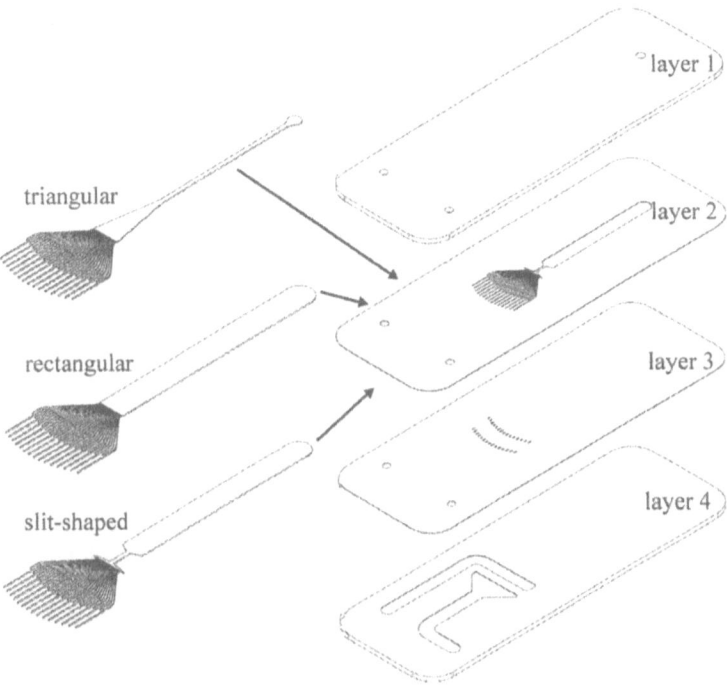

triangular

rectangular

slit-shaped

layer 1

layer 2

layer 3

layer 4

Figure 2: Schematic of an exploded view of a glass made micromixer and design variations of the mixing zone

The first layer from the upper side is a cover plate with 3 holes for inlet and outlet streams. The inlet streams are directed through the layers 2 and 3 to the channels in layer 4 and there fed to a diffuser. Layer 3, as an interface between the layers 2 and 4, contains 2 lines of 15 holes each covering the diffuser zones of layer 4. Through these holes the fluids reach the interdigitally arranged 2 x 15 microchannels with a width of 60 µm and 150 µm depth separated by 50 µm walls on layer 2. Adapted to the fluid distributing channels is the individually designed mixing channel with a cross section of 3250 µm x 150 µm where the fluids are contacted as a package of alternating thin fluid lamellae and directed to the outlet hole in layer 1. Fluid tightness up to about 6 bar is guaranteed by the bonding process.

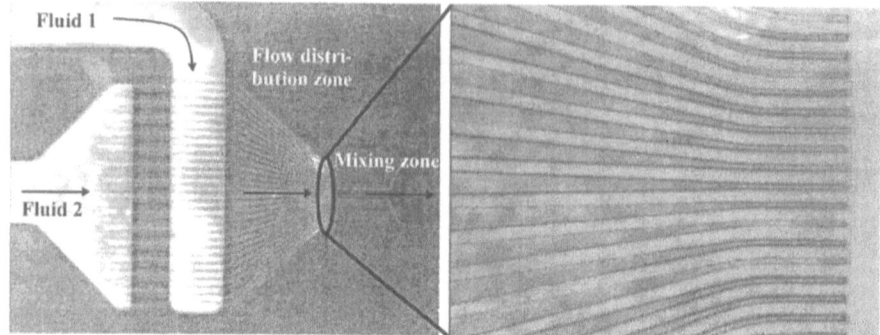

Figure 3: Photographic image of the microstructured fluid distributing area of a slit-shaped glass micromixer with a detailed view of the fluid contacting zone.

Figure 3 shows a photographical image of a slit shaped glass micromixer including two inlet channels for the different fluids, the interdigitally arranged microchannels for flow distribution and the mixing zone. Due to proper fluid equipartition the fluid spreading microchannels are very long compared to the slit-shaped part of the mixing zone. A detailed view of the orifice of the microchannels in the mixing zone in Figure 4 outlines the precision of the microstructuring process. The etching step creates a shell-like surface on the glass walls with a roughness of about 2 μm. This may influence the mixing or emulsification performance of these microfluidic devices and will be in the focus of future investigations.

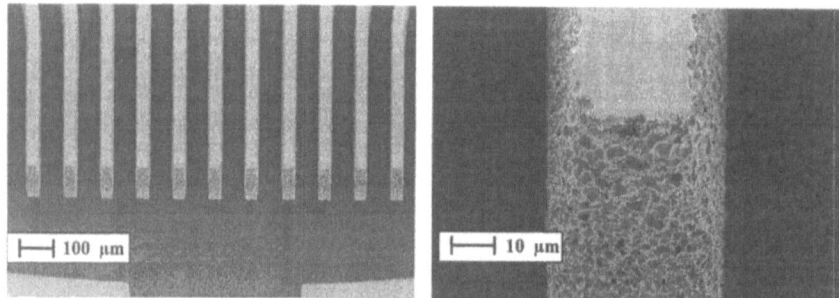

Figure 4: Etched glass walls from the multilamination zone of a micromixer with shell-like surface

Mixing of Miscible Liquids

The very promising results when using stainless steel interdigital mixers for chemical processing, mentioned above [1-5], may not exclusively be attributed to simple multilamination of fluid layers. For instance, parasitic mixing effects already were recognized in prior investigations [2], most likely corresponding to turbulent mechanisms.

Figure 5 shows these fundamentally different hydrodynamic features – pure multilamination and more complex flow patterns – which can be obtained by proper choice of the micromixing device. In a rectangular shaped glass made micromixer a uniform alignment of the fluid lamellae, reaching over a large area of the mixing channel (Figure 5 top). On the contrary, complex flow patterns were found when introducing colored and pure water streams in a slit-like glass interdigital mixer (Figure 5 bottom).

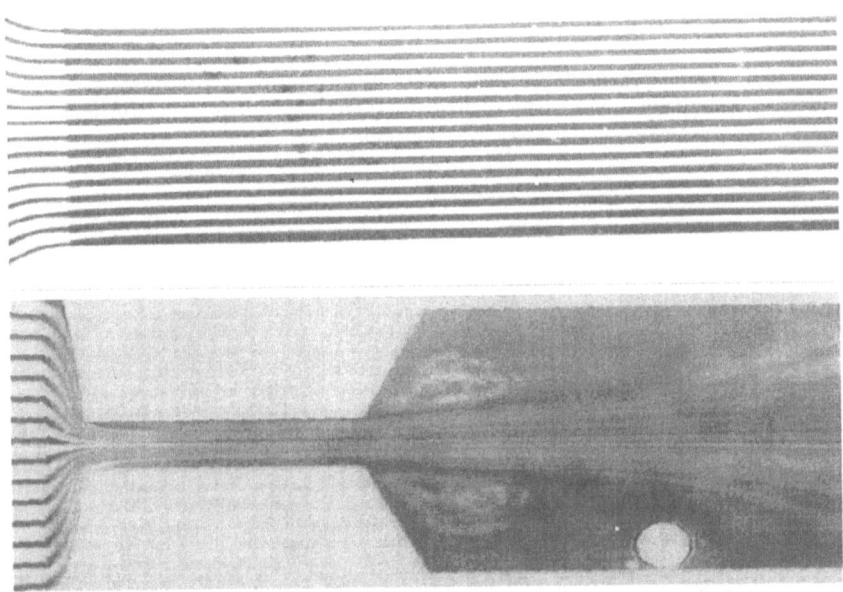

Figure 5: Flow platterns in interdigital glass micromixers: simple multilamination in the reactangular-shaped device (top image); more complex flow pattern in the slit-shaped (bottom image). Both devices were operated at a total flow of 1000 ml/h.

Please note that, particularly in the case of using the rectangular micromixer, no significant mixing effect is visible due to the too short residence time in the observed volume element. Calculations predict and experiments prove completion of mixing a few centimeters downstream, corresponding to about 0.1 seconds for the triangular mixer and 3 seconds for the rectangular mixer, respectively. This corresponds to the finding that the liquids in the slit-shaped system having a focusing zone, although still partially segregated, are much better mixed.

This stems from the fact of secondary mixing phenomena following the multilamination process. While in the front section of the mixing zone a focusing of the lamellae is achieved, thereby fastening mixing, a jet and surrounding eddies are found in the subsequent wide channel. The front section is connected to the wide channel by a remarkably thinner channel. Here, simulations predict a twisting of the lamellae finally leading to a spirally wound system of more and more thinned layers, when increasing the Reynolds number. Using a chemical reaction, the conversion of ferric(III) chloride with sodium rhodanide to a colored product, first evidence was gained for a possible existence of this type of flow behavior (for detailed information see Lit. [8]).

Two different focusing mechanisms of lamellae are exemplarily shown for the triangular shaped micromixer using again dyed and pure water as miscible liquids (Figure 6): geometric and hydrodynamic focusing. In the case of equal volume flows – e.g., a ratio of dyed water to pure water of 200 ml/h to 200 ml/h (see Figure 6 center image)- each lamella has a width of about 110 µm. The triangular structure of the mixing zone leads to a geometric focusing of these lamellae to a width of 17 µm each. Due to the laminar flow in the microstructure the only transport mechanism for mixing is diffusion. The shortened diffusion lengths by lamellae focusing therefore decreases the mixing time.

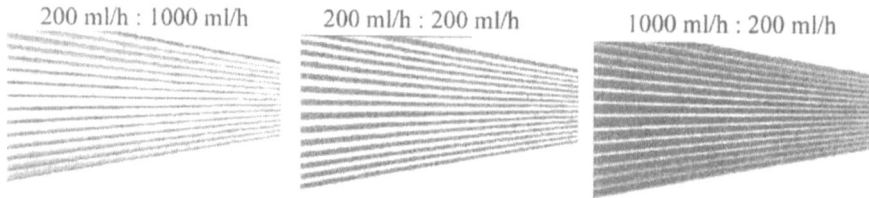

Figure 6: Geometric and hydrodynamic focusing of lamellae in a triangular shaped micromixer using dyed and pure water as fluids at various flow rates.

In addition to geometric focusing, lamellae will be hydrodynamically focused by applying different pressures of the two liquid flows, depending on the volume flow ratio of the fluids when entering the mixing zone [9]. The lamellae built by the fluid with the higher volume flow will compress the lamellae of the component with the lower flow rate. Figure 6 shows this effect by changing the ml/h-flow ratio of dyed to pure water from 200/1000 via 200/200 to 1000/200 (for detailed information Lit. [8]).

Mixing of Immiscible Liquids

In the following chapters, the emulsification behavior of three binary liquid systems, all being immiscible, (see Figure 7) in the interdigital mixer is investigated, ranging from a pure model (water / silicon oil) to a real reacting system (benzal chloride / conc. sulfuric acid).

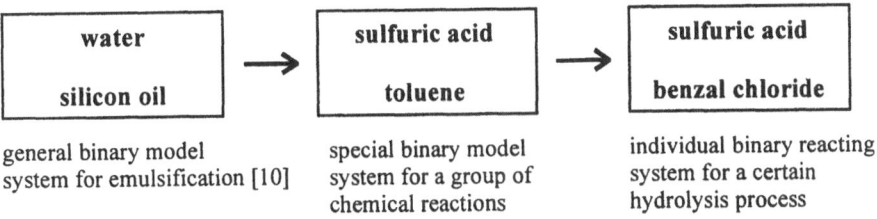

general binary model
system for emulsification [10]

special binary model
system for a group of
chemical reactions

individual binary reacting
system for a certain
hydrolysis process

Figure 7: Step-like investigation process of emulsification behavior of immiscible liquids in glass
made micromixers.

Water and Silicon oil

In the case of contacting immiscible liquids, the formation of droplets is
investigated. Using photographic imaging, well-resolved images, showing fine
details such as the breaking of cylinders into droplets, are obtained even at high
flow velocities of several m/s. While this feature already was anticipated before
getting experimental evidence [10], entirely new information is available now. The
location of droplet formation is influenced by the volumetric flow ratio of water
and silicon oil (η = 10 mPa * s) at room temperature. Figure 8 shows two images
of emulsion formation in a rectangular shaped micromixer both with a total liquid
flow of 800 ml/h. By running the system with a flow ratio of water to oil of 1:1 it is
proven that water and oil cylinders form after contact and decompose within a
short time, a few milliseconds, to a regular swarm of droplets of almost equal size.
When a flow ratio of 1:4 is used, droplet formation occurs immediately after
contact of the phases near the inlets of the mixing zone.

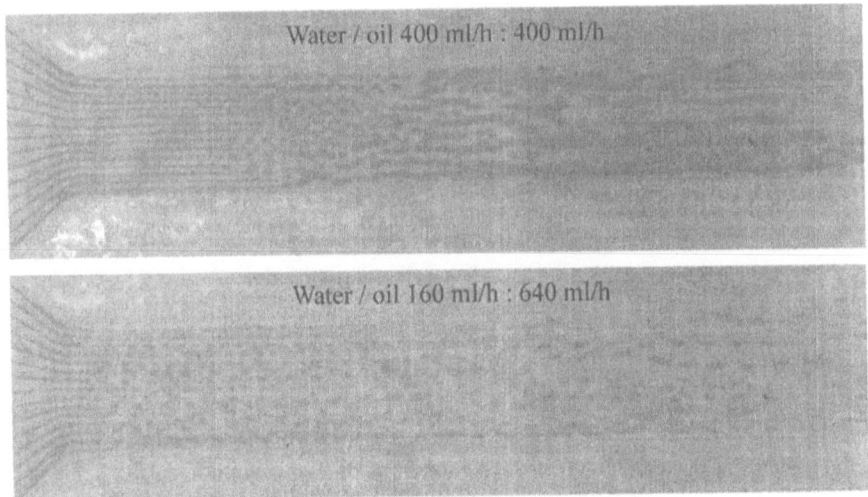

Figure 8: Photographic images of droplet formation and subsequent coalescence in a rectangular
shaped micromixer with a total volume flow of 800 ml/h and two different flow ratios
of dyed water to silicon oil.

Coalescene phenomena are visible as well, but will not be further discussed here. Certainly, they are favored by the large impact of surface forces in the tiny mixing chamber. An improved future design has to take account of enlarging the chamber following the contacting zone.

An analysis based on computational fluid dynamics of the droplet formation in the binary systems water / air and water / silicon oil underlines the experimental results [6]. A liquid cylinder at rest, consisting of water surrounded by air, decays from an initial configuration with small, sinusoidal modulations of the surface radius R. This so called Rayleigh-Plateau instability sets in for a perturbation wavelength greater than $2\pi R$. The fastest growing mode is predicted for a perturbation wavelength of 9.01R when inviscid irrotational flow is assumed. Even in the case of perturbations below $2\pi R$ the decay into droplets occurs via coupling to excitations of longer wavelengths in the range of the Rayleigh mode. Figure 9 shows the time evolution of a water/silicon oil system for an initial perturbation with a wavelength of 4.5R. The shape of the water cylinder is recovered after 0.5 ms which means that the subcritical perturbation is damped. Subsequently, a mode with a wavelength of about 9R is excited, initiating a decay of the water cylinder into droplets.

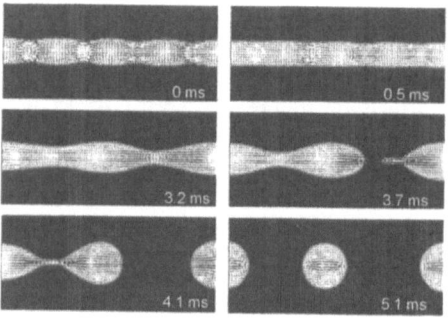

Figure 9: Decay of a water cylinder surrounded by silicon oil: time evolution of an initial perturbation with a wavelength of 4.5R.

Due to the non-zero velocity of the fluids, the simulation of lamellae decay in a micromixer is more complex as the above described model of a liquid cylinder at rest. A 3D computational model was set up for this problem and confirms the experimentally observed different droplet formation mechanisms for flow ratios of water to silicon oil of 1:1 and 1:4, as outlined in Figure 10 [6]. The black regions in the gray shaded water phase indicate a contact of the droplets to the glass walls of the mixing zone. Due to the small distance and the highly ordered row-to-row arrangement of the droplets, coalescence may occur and neighboring droplets frequently merge to a liquid column in the simulations. In reality, the arrangement of neighboring liquid cylinders is not absolutely uniform with regard to time and space, thus decreasing the extent of droplet coalescence.

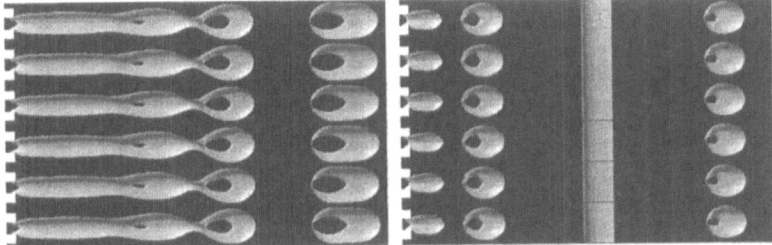

Figure 10: Prediction of droplet formation of a water/silicon oil system with total volume flow of 600 ml/h. Left side: flow ratio 1 : 1; right side: flow ratio 1 : 4.

The decay wavelength of the liquid lamellae was determined by measuring the spatial distance between subsequent droplets. Table 1 shows decay wavelengths and droplet diameters obtained from experiment and theory.

Flow ratio water : sil. oil	Decay wavelength [μm]		Droplet diameter [μm]	
	Exp.	Simul.	Exp.	Simul.
1 : 1	470	513	236	236
1 : 4	817	812	190	220

Table 1: Comparison of experimental and simulation results concerning decay wavelengths and droplet diameters. Notice: due to a mistake in Lit. [6] regarding the 1 : 4 flow ratio, the data listed in Table 1 differ from the literature source. Table 1 shows the corrected values.

It can be seen that at a flow ratio of 1 : 4 smaller droplets are formed as theoretically expected whereas the rest of the simulation results agree very well with experiments. In case of the 1 : 1 flow ratio a decay wavelength of 9.40R and 10.26R was obtained from experiment and simulations, respectively. These numbers underline the importance of the Rayleigh-Plateau instability for droplet formation in the micromixer.

Moreover, observation in transparent mixers now provides information on the size distribution of the droplets compared to that of the emulsion leaving the mixer. Hence dynamic, spatially resolved rather than only integral information was obtained, comprising the initial state of emulsion formation and subsequent processes such as droplet coalescence (see Figure 8).

Sulfuric Acid and Toluene

After discovering the emulsification processes of the model system water / silicon oil in transparent micromixers, the emulsification of chemically relevant fluids was investigated. As such a relevant model for a typical binary system of immiscible chemicals, sulfuric acid (conc.) and toluene was chosen. The reaction between sulfuric acid and functional groups of aromatic molecules is an often used step in organic synthesis routes, comprising e.g. the well-known hydrolysis of alkyl dihalides and the pinacol rearrangement. Hence, the model system described here should simulate the hydrodynamic behavior of such components without undergoing chemical reactions. The organic phase is dyed with iodine to gain better contrast of the image (Figure 11).

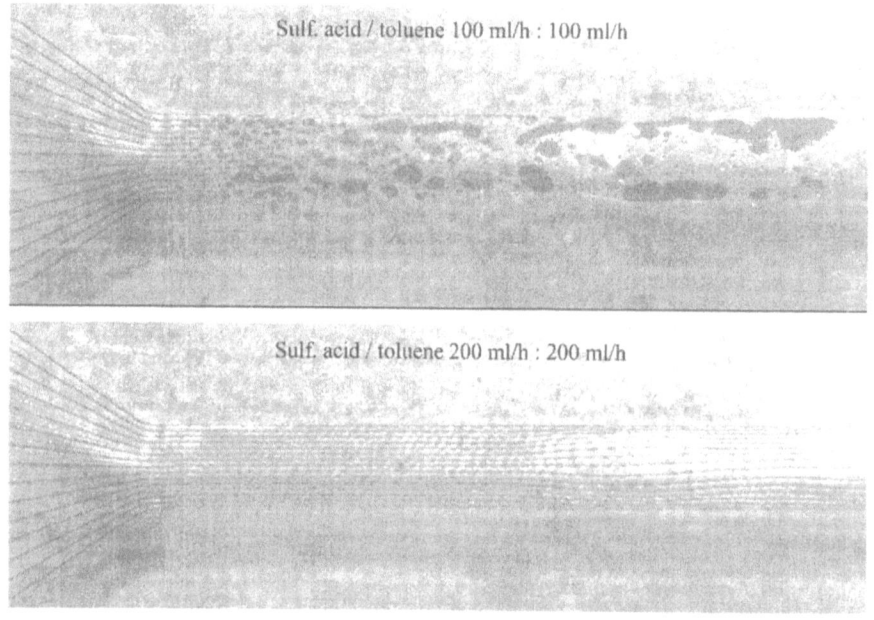

Figure 11: Optical images of contacting sulfuric acid with toluene (dyed with iodine) in a rectangular shaped micromixer at two different total flow rates.

Whereas in the case of a total flow rate of 400 ml/h the droplets formed look like pearl strings spread over the whole mixing channel, the hydrodynamics change totally at lower flow rates, e.g. 200 ml/h. In the latter case, droplet coalescence takes place shortly after phase contact. Although this comparatively low emulsion quality cannot compete with the finely dispersed system at the higher total flow rate, the droplets are still in the submillimeter range and exceed the dispersion quality produced in a standard reactor, like a stirred vessel (see below). This should have strong influences on reactions between two immiscible phases by means of conversion rates, selectivity, temperature control. To analysis the impact of this features, a real chemical reacting system was monitored.

Sulfuric Acid and Benzal chloride

As an example of use for liquid / liquid processing in micromixers, the synthesis of benzaldehyde from sulfuric acid and benzal chloride was carried out in a glass micromixer which was connected to a tubular reactor of about 6m length with regard to prolongation of residence times (Figure 12). The ratio of acid to substituted aromat was 5:1. This type of chemical reaction (liquid / liquid) is a standard lab synthesis, whereas at an industrial scale both the catalytic toluene oxidation (gas / liquid) and the benzal chloride hydrolysis using potassium carbonate solutions (liquid / liquid) are applied.

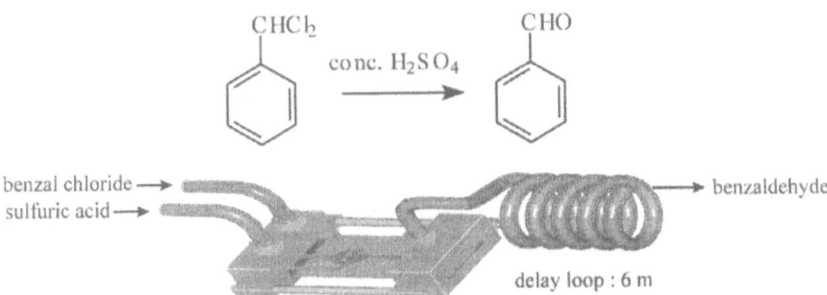

Figure 12: Sketch of the experimental set-up for hydrolysis of benzal chloride with conc. sulfuric acid (ratio 1:5).

The flow patterns of the reactants in a slit-shaped micromixer show the typical dependence on the total volume flow as expected from the hydrodynamic investigations of the model system sulfuric acid/toluene described above. Whereas at low flow rates large, separated acid zones are built through droplet coalescence, a fine dispersion is created at higher flow rates.

Using conventional methods, this hydrolysis requires intense cooling, namely operation at 0°C and stirring, while mixing of the highly viscous reaction solution tends out to be difficult. On the contrary, using the micromixer is a simple means to introduce the two reactant solutions at a defined concentration ratio into the reaction tube. In this manner, the benzal chloride hydrolysis can be performed even at temperatures of about 70°C, inevitably causing a bursting of any reaction mixture when carried out in a conventional vessel. The yield strongly depends on the temperature as outlined in Figure 13 and reaches a maximum of about 69%, which is slightly higher as compared to the classical approach (65%).

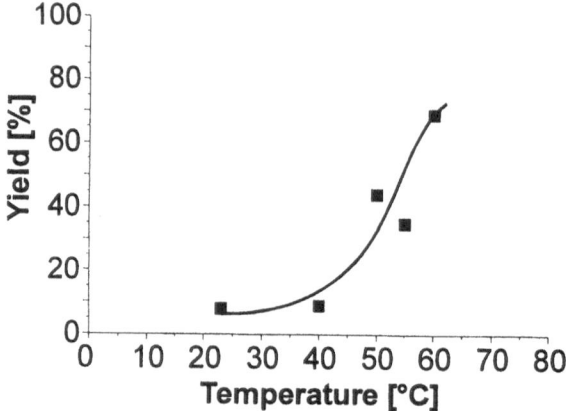

Figure 13: Hydrolysis of benzal chloride to benzaldehyde in a slit-shaped micromixer: yield against reaction temperature.

The dependence of the yield on residence time is characterized by a maximum at about 6 minutes, clearly demonstrating the need for correct setting of this parameter when using elevated temperatures. At longer residence times conversion increases to a maximum of about 100% but this as well leads to a decrease in selectivity by means of increasing side and follow-up reactions (Figure 14).

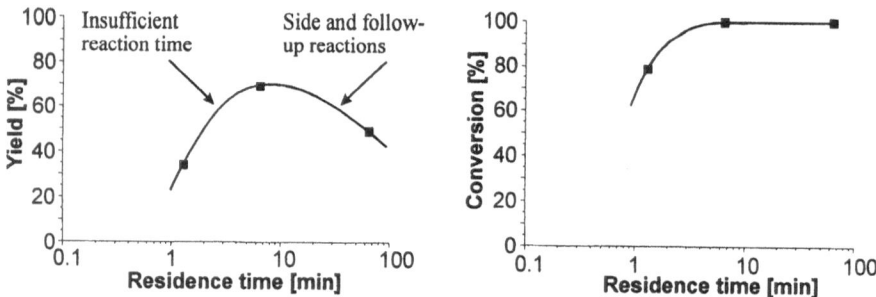

Figure 14: Hydrolysis of benzal chloride to benzaldehyde in a slit-shaped micromixer: yield against residence time (left diagram) and conversion against residence time (right diagramm).

Imaging of the flow patterns at low and high volume flows, i.e. different residence times, reveals a large difference in the respective mixing qualities, having an impact on the yield as well (see Figure 11).

As the micromixer is continuously working and the compared standard system is a batch reactor, it is useful to examine the reaction, in addition, in a mixing tee as a standard tool for continuously working systems. The mixing tee actually applied had an inner diameter of 1.5 mm. Both reactors, the interdigital mixer and the mixing tee, show increased selectivities at increasing conversion rates, but at different levels. As mentioned above, the slit interdigital micromixer produces at temperatures of about 60 °C a yield of 69% whereas the maximum yield of 33% shows, that the mixing tee cannot compete with the micro fluidic device (Figure 15).

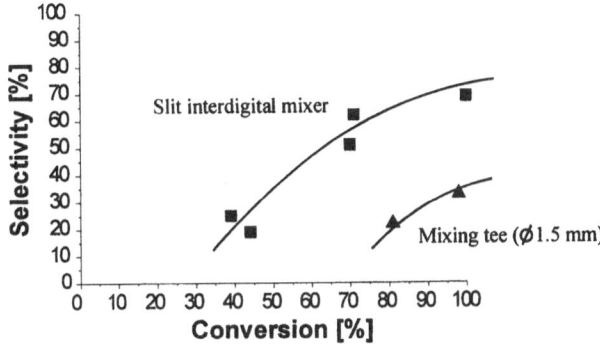

Figure 15: Comparison of the hydrolysis of benzal chloride to benzaldehyde in a slit-shaped micromixer and a mixing tee.

In a subsequent experiment, it was shown that similar features result for the synthesis of 4-fluoro benzaldehyde from the corresponding α,α-dichloromethyl compound. Using a slit-shaped interdigital steel micromixer the maximum yield at 100% conversion of about 50% can be reached at temperatures lower than those of the above described example (Figure 16).

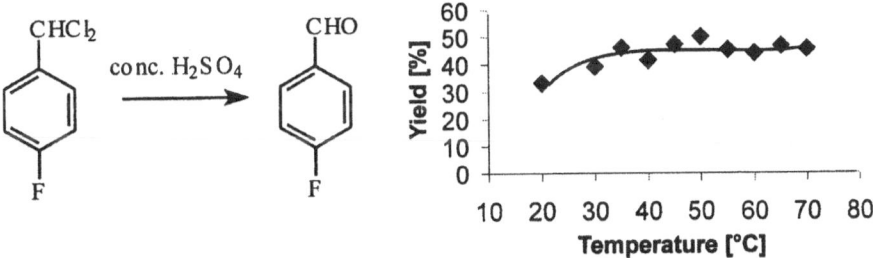

Figure 16: Hydrolysis of 4-fluoro-benzal chloride in a slit-shaped interdigital steel micromixer.

Conclusions and Outlook

A first analysis of hydrodynamics in three different transparent interdigital micromixers was made. On the one side, common hydrodynamic features were found for miscible and immiscible systems: namely, the formation of geometrically regular fluidic assemblies, being composed of either lamellae or cylinders, as well as their subsequent geometric and hydrodynamic focusing.

On the other side, some differences between the behavior of miscible and immiscible systems was apparent: For instance, in the larger zones of the mixing chamber of the slit-shaped micromixer, coexistence of hydrodynamic features typical for both, turbulent and laminar flow, namely stationary eddies, jets, wakes and multi laminated streams were observed. A striking feature of contacting immiscible liquids is the breakup of cylinders or direct droplet formation, dependent on the flow ratio of the two fluids. Similar phenomena occured also in real systems, but are expected to have additionally an interplay between mixing and reaction (see also the part *Discussion* in reference [11]).

Hopefully, these results may be the first step towards a vital interaction between experiment and simulation, as also presented elsewhere [6]. This ultimately will result in a more rationally based mixer design with well-directed adaptation to a specific processing function. The technological basis presented paves the ground for gathering much more information for a number of major research topics concerning microreactors in near future. This e.g. may include:

- Analysis of bubble formation and size distribution in gas / liquid systems [11]
- Micro-scale particle formation by forced precipitation [12]
- Fast realization of new customer-made devices

Acknowledgements

This work was financially supported by the "Volkswagen-Stiftung" ref. I/75 682.

References

[1] Ehrfeld, W., Hessel, V., Löwe, H.; *Microreactors;* VCH-Wiley, pp. 64-73, pp. 164-166 (2000)

[2] Ehrfeld, W., Golbig, K., Hessel, V., Löwe, H., Richter, T.; *"Characterization of mixing in micromixers by a test reaction: single mixing units and mixer arrays";* Ind. Eng. Chem. Res. **38**, 3 (1999), 1075-1082

[3] Benz, K., Regenauer, K.-J., Jäckel, K.-P., Schiewe, J., Ehrfeld, W., Löwe, H., Hessel, V.; *"Utilization of micromixers for extraction processes";* Chem. Eng. Technol. **24**, 1 (2001), 11-17

[4] Bayer, T., Pysall, D., Wachsen, O.; *"Micro mixing effects in continuous radical polymerization"*; Proceedings of the 3rd International Conference on Microreaction Technology, IMRET3, Frankfurt, Germany, pp. 165-170, Springer-Verlag, Berlin (2000)

[5] Löwe, H., Ehrfeld, W., Hessel, V., Richter, T., Schiewe, J.; *"Micromixing technology"*; Proceedings of the 4th International Conference on Microreaction Technology, IMRET 4, Atlanta, USA, pp. 43-44 (2000)

[6] Hardt, S., Schönfeld, F., Weise, F. Hofmann, Ch., Hessel, V., Ehrfeld, W.; *"Simulation of droplet formation in micromixers"*, Proceedings of the Conference on Modeling and Simulations of Microsystems 2001, MSM2001; Hilton Head Island, SC, March 19-21; pp. 223 - 226

[7] Freitag, A., Dietrich, T.R., Scholz, R., Hessel, V.; *"Glass as a material for microreaction technology"*; Proceedings of the Micro.Tec2000: World Micro-technologies Congress, VDI Verlag, Berlin, pp. 355-356 (2000)

[8] Hessel, V., Hardt, S., Herweck, T., Löwe, H., Schiewe, J., Schönfeld, F.; *"Hydrodynamics of liquid / liquid flow in interdigital micromixers"*; to be published in AIChE Journal

[9] Knight, J.B., Vishwanath, A., Brody, J.P., Austin, R.H.; *"Hydrodynamic focussing on a silicon chip: mixing nanoliters in microseconds"*; Phys. Rev. Lett. 80, 17 (1998), p. 3863

[10] Haverkamp, V., Ehrfeld, W., Gebauer, K., Hessel, V., Löwe, H., Richter, T., Wille, C.; *"The potential of micromixers for contacting of disperse liquid phases"*; Fresenius J. Anal. Chem. **364,** (1999) 617-624.

[11] Mathes, H., Plath, P.J.; "Generation of monodisperse foams using a micro-structured static mixer"; Proceedings of the Smart Systems and Devices, March 2001, in press

[12] Schenk, R., Donnet, M., Hessel, V., Hofmann, Ch., Jongen, N., Löwe, H.; *"Suitability of various types of micromixers for the forced precipitation of calcium carbonate"*; Proceedings of the 5th International Conference on Microreaction Technology, IMRET 5, Strasbourg, France, in this volume

Forced Periodic Temperature Oscillations in Microchannel Reactors

A. Rouge and A. Renken
Swiss Federal Institute of Technology, LGRC-EPFL
CH-1015 Lausanne, Switzerland

Abstract

A microstructured multichannel reactor allowing fast temperature changes and hence forced temperature oscillations is presented. Its thermal behaviour is discussed and a theoretical model is developed to describe the experimental results. The thermal response time in the centre of the microstructure is in the order of 3 s. The reactor can be used to investigate temperature cycling periods as short as about 20 s.

The catalytic dehydration of isopropanol to propene is used as model reaction. Whereas, the experimental results obtained at low frequencies (slow temperature variations) can be described by model calculations on the basis of a global kinetic model, the observed reactor behaviour at high frequency can only be explained by including detailed adsorption, desorption and surface reaction steps.

Keywords periodic operation, thermal behaviour, catalytic dehydration, isopropanol, propene

1 Introduction

Forced periodic operation of chemical reactors may lead to a considerable increase of selectivity and productivity [1]. The variables generally investigated are the fluid-flow and/or the inlet concentrations. Theoretical studies suggest that periodic changes of the reactor temperature can help to get useful informations on the catalyst behaviour as well as reaction mechanism and should be considered for reactor optimisation [2]. Investigation on the dehydration of ethanol to diethylether over cation exchange resins showed that for heterogeneously catalysed reactions the adsorption and desorption constant may influence greatly the transient reactor behaviour [3]. However, the published results highlight the difficulty for experimental investigations in this domain, due to the high inertia of conventional systems. Hence, no further studies were reported in the field.

Quiram et al. [4] showed that very fast temperature jumps can be obtained in single channel microsystems. Brandner et al. [5] studied the transient behaviour of an electrically heated microreactor and found a characteristic response time of about 30 s. In order to reduce the response time, thermo-fluids are used in this study for fast heating and cooling of the system. A proper design of the temperature control allows fast temperature changes or a periodic variation of the temperature in a multi-channel microreactor. The catalytic dehydration of isopropanol to propene was studied as a model reaction.

2   Experimental

The reactor used is based on a plate heat exchanger structure [6]. It contains 9 reaction plates coated with a catalytically active layer and 9 uncoated plates of stainless steel for the heat exchange (Fig. 2). They are stacked alternatively and assembled in a housing ensuring a sufficient tightness under present experimental conditions. Each plate is a square of 40x40 mm. It contains 34 rectangular channels of 300 μm width, 240 μm depth and 20 mm length. The total deposited catalyst mass is 110 mg, the measured reactor volume, V_R, is about 0.9 cm³. The catalyst, γ-alumina is deposited in the microchannels according to a special experimental procedure, resulting in a uniform catalytic layer in the channels [6]. The efficient use of the microchannel reactor for periodic concentration variations at cycle periods as short as 1s was evidenced experimentally [6].

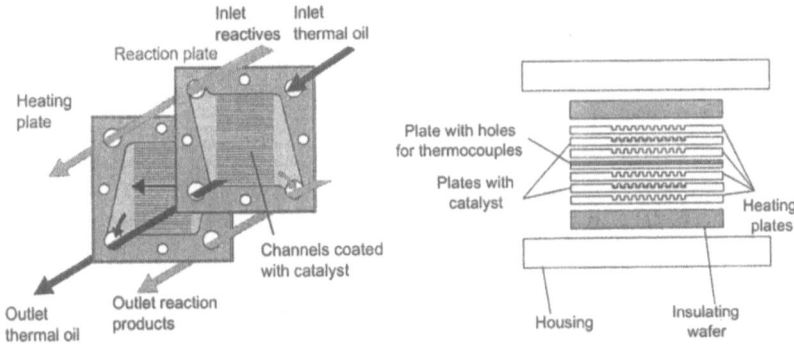

Figure 1 Schematic view of the set-up for temperature control

To reduce the thermal inertia of the system, insulating ceramic wafers (MACOR, Corning, Corning, NY, USA, 14 mm thickness) are inserted in-between the plates and the housing. Furthermore an unstructured plate is placed in the middle of the stack (Fig. 2 and 3). This plate contains openings in order to introduce thermocouples for monitoring the reactor temperature. Two oil circuits hold at different temperatures by the mean of two Juvo 500 thermostats (Karl Kurt Juchheim, Bernkastel, Germany) are used to control the reactor temperature. The switch is realised by acting simultaneously two 4-ways valves (Fig. 1).

Finally, for experiments under reaction conditions, isopropanol (>99.5%, Fluka chemie AG, Buchs, Switzerland) was provided through a temperature controlled bubble column fed by N_2 (>99.95%, Carbagaz, Lausanne, Switzerland). The outlet concentrations of reactants and products were measured online using a quadrupole mass spectrometer. A stable activity of the catalyst was obtained after one day on stream.

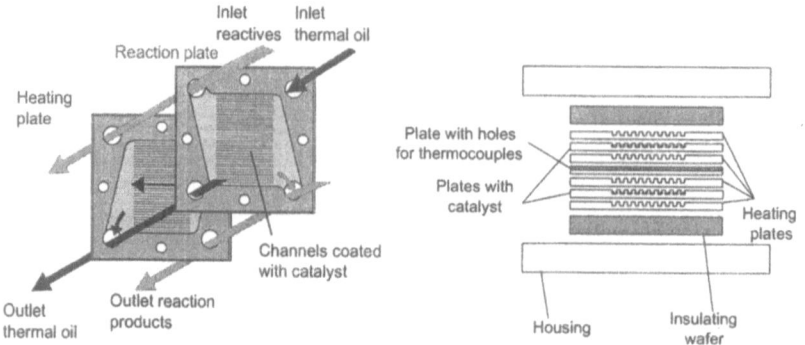

Figure 2 Structure of the stack of plates and and front view of the microreactor

3 Thermal response of the reactor

The temperature of the heating oil was switched from approximately 180 to 200°C and conversely from 200 to 180°C. No significant differences in the response between down- and up-steps were observed. The temperature of the oil at the inlet of the reactor and the reactor temperatures at different positions within the reactor (Fig. 3) were monitored.

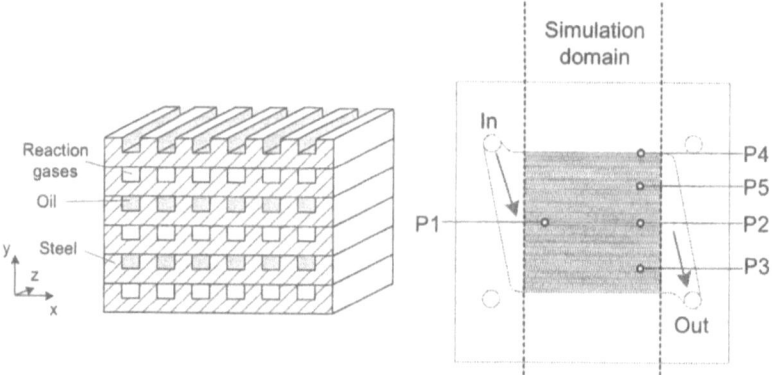

Figure 3 Geometry of channel array considered in the model for the thermal response and detail of one plate with position of the thermocouple and area considered for the mathematical model.

3.1 MODEL

The thermal behaviour of the reactor was simulated using a relatively simple model.
The key assumptions of this model are:

- There is no thermal flow perpendicularly to the plates. This implies a perfect isolation on each side of the stack.

- The temperature is locally homogeneous, i.e. the heat flow is neglected in the y direction and there is no temperature gradient in the metal between the vicinity of the oil channels and the bulk.

These hypotheses reduce the original situation to a 2-dimensionnal problem. Furthermore the simulation is limited to the regular domain of the plates, as shown in Fig. 3.

Within the channels, the flow is laminar (Reynolds-number, Re<100) The Nusselt-number is calculated using eq. 1, valid for laminar flow in a circular duct with constant wall temperature. This equation includes a correction for the average influence of the entrance effects [7]. The real value should be lower, since the channel is quadrangular and not circular. On the other hand, higher values can be expected due to the axial temperature profile. Probably, both effects compensate each other. Anyway, a small difference in Nu has a negligible effect on the model prediction.

$$ Nu = \left(3.66^3 + 0.7^3 + \left(1.615 \cdot \sqrt[3]{Re \cdot Pr \cdot d_h / l} - 0.7 \right)^3 \right)^{1/3} \tag{1} $$

The Prandtl-number ($Pr = \mu \cdot c_p / \lambda$) is 18 for the thermal oil used (oil viscosity: $\mu_{oil} = 0.001$ Pa.s, heat capacity: $c_{P,oil} = 1800$ J·kg^{-1}·K^{-1}, heat conductivity: $\lambda_{oil} = 0.1$ W·m^{-1}·K^{-1}), The hydraulic diameter of the channels is $d_h = 0.266$ mm and the channel length is $l = 20$ mm. Since the main limitation to heat transfer between the oil and the steel wall is on the oil side, the over all heat transfer coefficient $U,$, is given by :

$$ U \cong h_{oil} = \frac{Nu \cdot \lambda_{oil}}{d_h} \tag{2} $$

The heat released by the oil per unit of volume, q/V is equal to:

$$ \frac{q}{V} = (T_{st} - T_{oil}) \cdot U \cdot \frac{A}{V} \tag{3} $$

T_{st} is the local temperature of the reactor wall, T_{oil} is the oil temperature, U can be estimated from eq. 2 and the specific surface of the channels is given by the geometry: A/V=1833 m^2/m^3. Furthermore, the oil temperature can be calculated at each point of the reactor in the area covered by the channels:

$$ \frac{\partial T_{oil}}{\partial t} = u \cdot \frac{\partial T_{oil}}{\partial z} - \frac{q}{V} \cdot \frac{1}{c_{P,oil} \cdot \rho_{oil} \cdot \varepsilon_{oil}} \tag{4} $$

q/V can be estimated from eq. 3 where $\varepsilon_{oil} = 0.12$ is the fraction of the volume occupied by oil.

Finally the heat conductivity has to be considered to estimate the temperature in the stainless steel plate.

234

$$\frac{\partial T_{st}}{\partial t} = \left(\lambda_x^{eff} \cdot \frac{\partial^2 T_{st}}{\partial x^2} + \lambda_z^{eff} \cdot \frac{\partial^2 T_{st}}{\partial z^2} + \frac{q}{V} \right) \cdot \frac{1}{c_{P,st} \cdot \rho_{st} \cdot \varepsilon_{st}} \qquad (5)$$

$c_{p,st}$ =502 J·kg^{-1}·K^{-1} heat capacity of the steel, ρ_{st}=7550kg/m^3 its density and ε_{st}=0.76 volumetric fraction. λ_x^{eff} and λ_z^{eff} are the effective conductivities in the x and z directions. Heat conduction takes obviously only place within the solid material. Therefore, λ_z^{eff} is thus set to $\varepsilon_{st} \cdot \lambda_{st}$ with λ_{st}=15.0 W·m^{-1}·K^{-1} and ε_{st}=0.76.

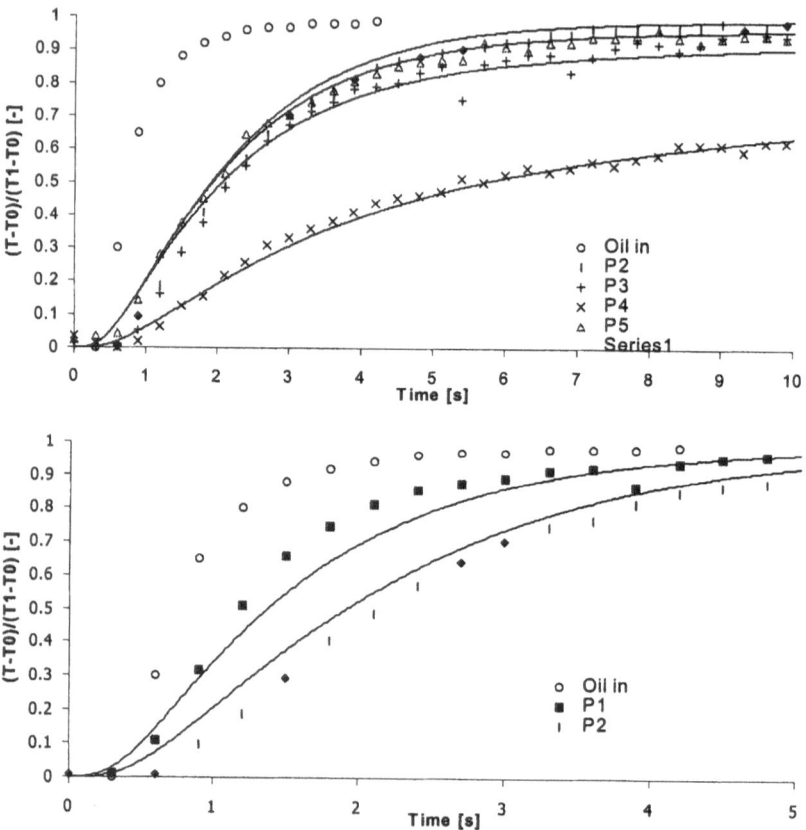

Figure 4 Normalised temperature response measured after a switch from one thermostatising circuit to the other (T-switches from 200 to 180°C). The location of the points is shown in Fig.1. Oil velocity in the channels: 0.35 m/s. Oil in: measured at the reactor inlet, lines: model predictions.

λ_x^{eff} was set to $0.52 \cdot \lambda_{st}$. It is supposed that the channels hinder the heat transfer and that the fins do not contribute to the transfer in the x-direction (see Fig. 3).

In the domain without channels, we have the normal behaviour of steel, λ_x^{eff} and λ_z^{eff} are equal to λ_{st} ($\varepsilon_{st} = 1$). Finally the inlet temperature of the oil changes according to the measured temperature curve at the reactor entrance; this curve is used for all the simulations.

3.2 RESULTS AND DISCUSSION

The actual flow velocity in the channels depends on different unknown parameters. As an estimation, a value of 7.7cm³/s, corresponding to a linear velocity of 0.35 m/s is taken. This corresponds to Re=75 and according to eq. 1 to Nu=4.6.

Experimental results (Fig. 4 and 5) show that 90% of the temperature change is reached within 3 s near the reactor entrance (point P1). This value is reached later at the end of the channels (Points P2-P5. The experimental results can be explained satisfactorily based on the developed model.

The delay in the temperature profile at the reactor outlet is due to the heat transfer in the microstructure. The oil releases its heat almost immediately at the reactor entrance, thus a pronounced axial temperature profile is developed. The axial temperature profile (z-direction) predicted by the model is less marked compared to the measured one (Fig. 5). This is due to the approximations introduced in the theoretical model. For example, a constant heat transfer coefficient between oil and the steel wall is assumed. In reality, due to the entrance effects, the heat transfer is maximal at the channel entrance and then diminishes. This causes simultaneously an underestimation of the temperatures near the inlet and an overestimation near the outlet.

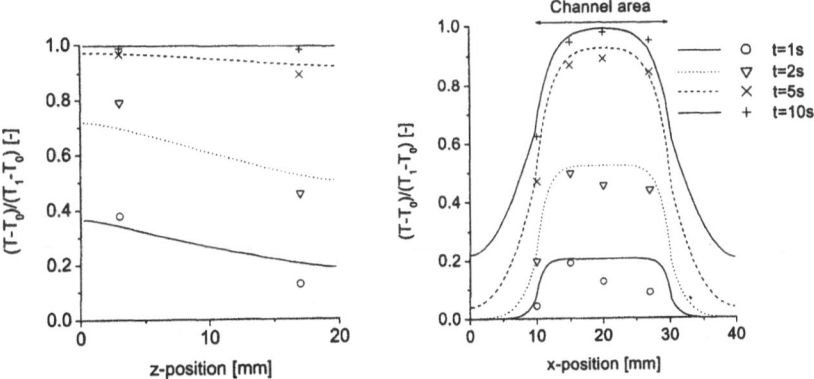

Figure 5 Temperatures in the reactor at different times after a temperature change. Left: Profile along the line P1-P2. Distance from the beginning of the channel. Right: Profile along P4-P3. Distance from the border of the plate.

There are larger temperature gradients perpendicularly to the channel structure (x-direction). The border region of the plate is heated or cooled only by heat conduction in the metal from the channel region. Therefore, the temperature changes only slowly at the edges of the plates. The border effect is seen clearly at point 4.

4   Reaction behaviour

The dehydration of isopropanol to propene and diisopropylether over γ-alumina was chosen as model reaction. The formation of propene is inhibited by the reactand and by water. This reaction has been studied previously [8, 9]. Quantitative investigations of the transient behaviour of this reaction under isothermal conditions have already been carried out in a similar microreactor [6, 10]. It was shown that the transient reactor response could be described by a mechanism implying two active sites (eq. 6). Isopropanol adsorbs on site S_1, expelling water in the process. This step can be reversed if water is present in the gas phase. The adsorbed iPrOH reacts to form propene, released in the gas phase and water, which remains on the catalyst surface. The reaction needs an empty S_2-site to take place. Isopropanol and water adsorb reversibly on S_2, thus inhibiting the reaction of propene formation.

Ether can also be formed, through the reaction between two isopropanol molecules adsorbed on S_1 and S_2. Since the selectivity to ether is low (<10%) and shows no particular transient behaviour, only the outlet concentrations of isopropanol and propene will be discussed.

$$A + WS_1 \xleftrightarrow{K_1} AS_1 + W$$
$$A + S_2 \xleftrightarrow{K_2} AS_2$$
$$W + S_2 \xleftrightarrow{K_{2W}} WS_2$$
$$AS_1 + S_2 \xrightarrow{k_3} P + WS_1 + S_2$$
$$AS_1 + AS_2 \xrightarrow{k_4} E + WS_1 + S_2$$

$A:$ *Isopropanol*
$W:$ *Water*
$P:$ *Propene*
$E:$ *Ether*
$S_1:$ *Site 1*
$S_2:$ *Site 2*

(6)

4.1   TEMPERATURE JUMPS

In this set of experiments, the temperature was suddenly increased or decreased from the initial to the final value. The cycle period was long enough to reach the new stationary state. Experiments were carried out with initial and final temperatures of 190 and 210°C, iPrOH concentration 0.92 mol/m³ (STP) and inlet flow ranging from 0.666 to 1.666 cm³/s (STP). The experimental temperature was continuously measured and checked in order to be sure that no temperature instabilities occur.

Qualitatively, the following effects are observed:

Immediately after the temperature change, there is a peak of iPrOH concentration in the outlet. This can be explained by the influence of the temperature on iPrOH adsorption equilibrium on S_2 sites. At high temperatures, the equilibrium of iPrOH adsorption on S_2 is shifted to the left, i.e. the surface coverage of iPrOH is lower. Therefore, by rising the temperature, a certain amount of iPrOH is immediately released in the gas phase, provoking the observed concentration peak. On the contrary, when the temperature decreases, the equilibrium is shifted in the other direction; a large amount of iPrOH is immediately adsorbed and causes the negative peak.

After this initial phenomenon, the rate increase of the surface reaction with temperature shows its effects, the iPrOH concentration tends to the new stationary value: Higher when the temperature increases, lower when it decreases.

The behaviour of propene is much more conventional. Its outlet concentration simply rises at high temperatures and decreases at lower ones.

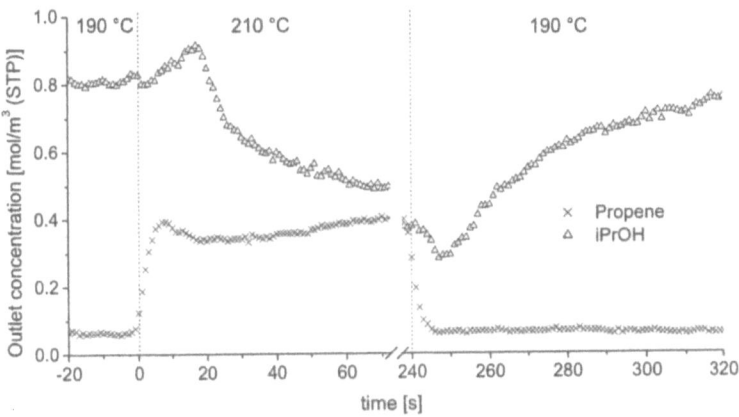

Figure 7 Outlet concentrations after a T-Jump in the microreactor with isolation: $C_{iPrOH,0}$=0.92mol/m^3 (STP), P=1.5 bar, Q=0.66 cm^3/s (STP).

The propene concentration tends rapidly to the new stationary value. This process is almost independent of the volumetric flow-rates. This implies that the characteristic response is mainly governed by the thermal behaviour of the reactor. During the ascending T-jumps, the propene production passes through a maximum and falls than to its stationary value. This can be rationalised by considering the adsorption equilibrium of iPrOH on sites S_1. At low conversions, the equilibrium is shifted to the right, and S_1 is is mainly occupied by iPrOH. With rising temperature the reaction rate increases, resulting in an important formation of water. The produced water shifts the equilibrium to the left, resulting in a decrease of iPrOH concentration on the surface and hence of propene formation.

238

4.2 PERIODIC TEMPERATURE VARIATIONS

The influence of periodic temperature variations is shown in Fig. 8 for a period of 20s. As the cycle period is short, the observed maximum production rate as observed in Fig. 7 can not be repeated. Characteristic is the shift of 5 s between the maximum propanol and propene concentrations, which correspond to 25% of the total period. But, due to the complex adsorption/desorption and the particular surface kinetics, a quantitative interpretation is difficult and not yet possible. As for periodic variations of the inlet concentrations, further optimisation of the frequency is necessary for maximum reactor performance.

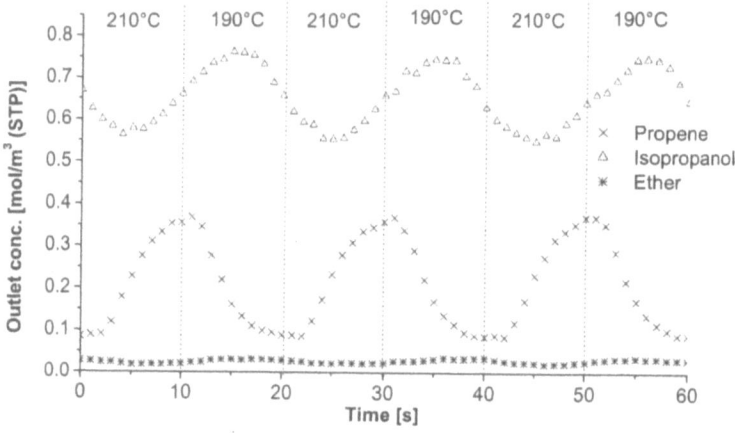

Figure 8 Periodic T-variation. Period 20s. $C_{iPrOH,0}$ 0.92 mol/m³ (STP), P 1.5 Bar, Q 0.66 cm³/s (STP).

5 Conclusion

A microreactor allowing fast temperature variations and hence fast forced temperature oscillations has been tested. The behaviour in the centre of the structure is satisfactorily, due to heat losses and the inertia of the housing, the transients at the border of the structure are prolonged. The measured transient temperature profiles could be simulated with a theoretical model. The model can be used to predict the influence of the construction material as well as the design of the microchannels on the thermal behaviour of the microreactor. Under reaction conditions, the evolution of reactand concentrations at transient temperatures were measured. The transient outlet concentration measured when the test reaction was carried out in this reactor highlighted the occurrence of particular effects short after the temperature jump. The observed reactor behaviour at high frequency can only be explained by including detailed adsorption, desorption and surface reaction kinetics. The periodic experiments showed that the reactor developed in this study can be used for temperature cycling with periods as short as 20 s.

Acknowledgement

The authors thank the Institut für Microtechnik Mainz (IMM) for manufacturing the microchannel reactor.

Financial support from the Swiss national science foundation is gratefully acknowledged.

6   Symbols

A	Isopropanol	T_0	Initial temperature
A	Exchange area	T_1	Final temperature
c_p	Heat capacity	u	Oil velocity
d_h	Hydraulic diameter	U	Global heat transfer coefficient
E	Ether	V	Volume
h	heat transfer coefficient	W	Water
l	channel length	x, y, z	Spatial directions
Nu	Nusselt-number $(h \cdot d_h / \lambda)$		
P	Propene		

<u>Greek letters</u>

P_1 to P_5	Points of T-measurements	ε	Fraction of total volume
Pr	Prandtl-number $(\mu \cdot c_p / \lambda)$	μ	Dynamic viscosity
q	heat flow	λ	Thermal conductivity
Q	Flowrate	ρ	Density
Re	Reynolds number $(u \cdot \rho \cdot d_h / \mu)$		

<u>Indices and superscripts</u>

S_1, S_2	Active sites on the catalyst	eff	effective
STP	Standard temperature and pressure (25°C, 1 bar)	Oil	related to the oil
		St	related to the steel
t	Time	x,z	in direction x or z
T	Actual temperature		

7   References

1. Silveston, P., R.R. Hudgins, and A. Renken, *Periodic Operation of Catalytic Reactors - Introduction and Overview*. Catalysis Today, 1995. **25**: p. 91-112.

2. Neer, F.J.R.v., A.J. Kodde, H. Den Uil, and A. Bliek, *Understanding of Resonance Phenomena on a Catalyst under Forced Concentration and Temperature Oscillations*. The Canadian Journal of Chemical Engineering, 1996. **74**: p. 664-673.

3. Denis, G.H. and R.L. Kabel, *The Effect of Temperature Changes on Tubular Heterogeneous Catalytic Reactors*. Chem. Eng. Sci., 1970. **25**: p. 1057-1071.

4. Quiram, D.J., K.F. Jensen, M.A. Schmidt et al. *Development of a Turnkey Multiple Microreactor Test Station*. in *IMRET 4*. 2000. Atlanta.

5. Brandner, J., M. Fichtner, and K. Schubert. *Electrically heated microstructure heat exchangers and reactors*. in *IMRET 3*. 1999. Frankfurt: Springer.

6. Rouge, A., B. Spoetzl, S. Schenk, K. Gebauer, and A. Renken, *Microchannel reactors for fast periodic operation: The catalytic dehydration of isopropanol*. Chem. Eng. Sci., 2001. **56**: p. 1419-1427

7. *VDI-Wärmeatlas*. 7., erweiterte Auflage 1994: VDI-Verlag.

8. Thullie, J. and A. Renken, *Model discrimination for reactions with a stop-effect*. Chem. Eng. Sci., 1993. **48**: p. 3921-3925.

9. Moravek, V., *Steady-state and transient kinetics of displacement adsorption and educt inhibition of alcohols on alumina*. Journal of Catalysis, 1992. **133**: p. 170-178.

10. Rouge, A. and A. Renken, *Performance Enhancement of a Microchannel Reactor under Periodic Operation*, in *Reaction Kinetics and the Development and Operation of Catalytic Processes*, G.F. Froment and K.C. Waugh, Editors. 2001, Elsevier Science B. V.: Amsterdam. p. in press.

Comparison of Ag/Al– and Ag/α-Al₂O₃ Catalytic Surfaces for the Partial Oxidation of Ethene in Microchannel Reactors

Ansgar Kursawe, Dieter Hönicke[1]

Technische Universität Chemnitz, Lehrstuhl für Technische Chemie
D-09107 Chemnitz / Germany

Abstract

The partial oxidation of ethene using Ag/Al and Ag/α-Al₂O₃ catalysts was performed in microchannel reactors applying high C_2H_4 and O_2 partial pressures within the explosion range. Recent investigations revealed different catalytic properties of bulk silver and silver coated aluminum. Thus, the effect of the Ag layer thickness on the selectivity and conversion of sputter-coated Ag/Al surfaces was investigated. Furthermore, the influence of the oxygen and ethene partial pressures on the selectivity and conversion was subject of concern, because it showed impossible to perform those experiments with fixed bed catalysts in tube type reactors. Finally, a comparison of both catalytic surfaces was carried out.

1 Introduction

Microchannel reactors are a new and efficient tool for the investigation of highly exothermic reactions. Their excellent heat transfer properties and their inherent safety allows the application of reaction conditions, which are not feasible in conventional, wall-cooled tube reactors with fixed bed catalysts. Due to the high heat conductivity of the metallic microchannel reactor framework, an occurance of hot spots is unlikely even at high specific heat production rates. Thus, an ignition of explosive mixtures by local hot spots is normally impossible and even if an ignition takes place, the result is due to the small reactor volume harmless as proved in H_2/O_2 oxidation experiments [1]. This proves that a microchannel reactor is a perfect tool for the investigation of partial oxidation reactions using very high oxygen partial pressures (or pure oxygen) and high hydrocarbon feed concentrations.

The partial oxidation of ethene to ethene oxide using silver or supported silver as catalyst is known since 1931 by the original patent by Lefort [2]. Nowadays, the worldwide annual production of ethene oxide using this catalyst is more than six million tons [3] with selectivities initially slightly above 90% using various promoters and oxidation inhibitors [4,5].

Parallel reaction steps, viz. the oxidation of ethene to ethene oxide on the one hand and to carbon dioxide on the other is commonly accepted as reaction scheme (fig. 1). The rate determining step is the reaction of absorbed oxygen with ethene [6]. It has been found that the consecutive combustion of ethene oxide to carbon dioxide is neglegible and much slower than the total combustion. Thus, the latter reaction path can be practically neglected under kinetic control [6,7].

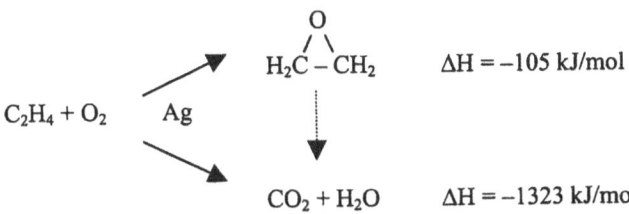

Figure 1: Reaction scheme of the partial oxidation of ethene to ethene oxide

In an industrial ethene oxide reactor, the choice of the reactand concentrations is strongly influenced by the explosion range of the ethene, oxygen and balance gas feed mixture [6,8]. Due to safety precautions, not every ethene / oxygen mixture can be applied and operation is performed either above or below the explosion range. Furthermore, high downstream ethene oxide concentrations are preferred [4]. Actually, the best industrial processes use high ethene concentrations of 20% to 40% and oxygen concentrations of about 8%, using methane as inert gas in order to avoid the explosion range [4]. It is known that a higher oxygen concentration generally enhances the selectivity towards ethene oxide. Thus, in any conventional reactor is a conflict whether to apply high ethylene or oxygen concentrations in order to avoid the explosion range.

Without gas-phase promoters, selectivities of about 50-70% are reported in open literature [9–13]. About 90% of the total heat production is caused by the deep oxidation of ethene to carbon dioxide and water. Even in an advanced industrial reactor having selectivities of about 80–90%, the total combustion contributes 50% to 75% to the total heat production.

The authors have chosen the ethene epoxidation as a model reaction for evaluating the performance of microchannel reactors, because those exothermic heat effects make experiments in conventional reactors dangerous, especially at high degrees of conversion and high ethene partial pressures within the explosion range. Furthermore, there should be no influence of heat effects on the selectivity and conversion behaviour, because unnoticed hot-spot formation in the catalyst bed can be excluded due to the high heat conductivity of the metallic microstructured parts.

2 Experimental

2.1 Catalyst preparation

As already mentioned, microchannel reactors exhibit excellent heat tranfer, because the catalytic active species are located within a very thin layer of only a few microns directly at the wall of the microchannel. But contrary, a time-consuming problem within the construction of a microchannel reactor is the immobilisation of catalytic active species at the walls. Immobilisation methods already described in numerous publications and patents are focused on the preparation of fixed bed catalysts. Thus, a method to immobilize the catalyst or catalytic active species at the walls of the microchannels had to be developed. Silver is known to be the only catalyst of industrial importance for the epoxidation of ethene to ethene oxide. The typical industrial catalyst is made of high calcinated, low surface area α-Al_2O_3, activated with silver and various promoters such as Cs, Ba, Re, K. Alpha-alumina with surface areas below typically 4 m²/g are used, because high surface area α-Al_2O_3 has a strong negative impact on the overall selectivity due to isomerisation of ethene oxide to acetaldehyde followed by combustion [13].

Traces of 1,2 dichlorethane (ppm range), as well as nitrogen dioxide (ppm) and carbon dioxide (up to 10 vol%) are typically used as inhibitors for the total combustion of ethene. Thus, the selectivity of the promoted catalyst can be improved by 20 to 25% depending on promotor concentration. In the recent studies the authors decided to use unpromoted catalysts, because the overall catalytic activity and therefore the heat production is reduced by promoters. If a catalyst performs well without any promoter, it is guaranteed that there will be no thermal problem when promotors are applied. Furthermore, the comparison of catalytic systems is easier, because different effects of promoters on the selectivity and conversion of different catalytic surfaces can be excluded.

In earlier work, the authors used microstructured silver wafers (bulk silver) and silver coated aluminum wafers in order to obtain catalytic active microchannels. Both catalytic systems allow the application of high ethene partial pressures such as 20% ethene in oxygen and high conversion degrees above 20% at selectivities up to 60% [14]. The silver-on-aluminum system was more active and selective than the bulk-silver microchannel reactor. Thus, an effect of the Ag layer thickness on the selectivity vs. conversion behaviour can be assumed and this dependence was investigated.

In order to study the effect of oxygen and ethene partial pressure on selectivity and conversion and to compare those Ag/Al catalysts with more commonly used Ag/α-Al$_2$O$_3$ catalysts, aluminum microstructures were coated with a thin α-Al$_2$O$_3$ layer by use of a sol-gel derived process and subsequently calcinated at about 773K. For covering the wafers with a such a ceramic layer, a boehmit sol was prepared by dropping aluminum-sec-butylate into an excess of water and following peptization of the resulting particles. Then, a nano-scaled α-Al$_2$O$_3$ powder was added to the sol. The homogeneous mixture was given drop wise on the wafer surface. Surplus sol was spun off at 2500 min^{-1} and the wafers were thermally treated on a hot plate at 550°C for 20 min. The resulting aluminum oxide layer with a thickness of slightly more than 1 μm consists predominantly of α-Al$_2$O$_3$. This porous layer was activated by a triple impregnation procedure using a silver oxide / lactate solution followed by calcination very similar to a method described elsewhere [8].

2.2 Microchannel reactor and wafer construction

Two different microchannel reactor concepts, a selfmade modular type and a commercial available one, were used for testing microstructured wafers.

The selfmade modular microchannel reactor was described earlier in detail [15]. This reactor grants easy access to the microstructured parts without destruction of the device. The microstructured wafers are fixed in a cylindrical inner housing having a rectangular millcut of 10 x 10 x 50 mm^3. Using this type of reactor, it is possible to manage heat production rates of about 30 Watts without external cooling or runaway / explosion when applying explosive reactand mixtures. Selfmade microstructured aluminum wafers as depicted in figure 2 were used in this modular reactor. The microstructures were manufactured by wire-EDM of the aluminum alloy AlMg3. Each wafer contains 14 parallel micochannels, each having a channel geometry of 300 μm x 700 μm x 50 mm.

Two different catalytic systems have been examined in this modular reactor. The effect of the silver layer thickness on the selectivity / conversion behaviour was examined by using three wire–electro discharge machined wafers coated with a thin silver layer by means of sputtering (MCR1). Furthermore, three wafers coated with α-Al$_2$O$_3$ and activated by threefold impregnation with silver-lactate (MCR2) as described have been tested in order to compare the selectivity / conversion behaviour with the former Ag/Al-system.

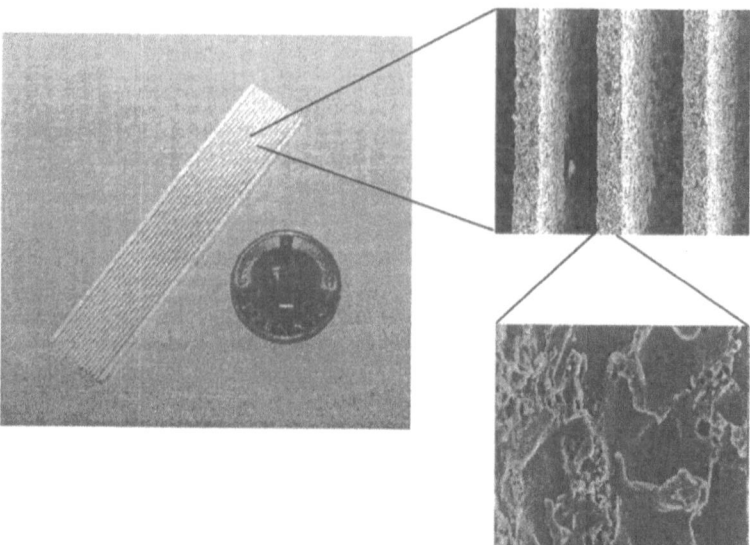

Figure 2: Microstructured wafer (50 x 10 x 1 mm³) having 14 channels of 300 x 700 μm made by wire-electro discharge machining and REM picture of the resulting surface (used in microchannel reactor MCR1 and MCR2)

It showed that the overall activity of the dense Ag/Al surface is much lower than the slightly porous Ag/α-Al$_2$O$_3$ surface. Thus, a microchannel reactor having a higher inner surface area and larger volume was required in order to achieve similar degrees of conversion. The third microchannel reactor (MCR3) used in the present study was made by the "Forschungszentrum Karlsruhe GmbH" by use of mechanical micromachining followed by a stacking and sealing process.

This commercial microchannel reactor consists of a stack of 26 microstructured metallic wafers of 10 x 50 x 0,3 mm³, each having 33 rectangular microchannels of 0,2 x 0,2 x 50 mm³. Two diffusors for reactand inlet and product outlet are attached to the stack on each side. A wafer in the middle of the stack with holes for thermocouples is provided. This reactor was catalytically activated by coating the microstructured aluminum wafers with a thin silver layer by means of PVD and sputtering techniques similar to the microchannel reactor MCR1. The layer thickness was adjusted to 1200 nm, controlled by a crystal oscillator in order to obtain an equal layer thickness on every single wafer. An overview about the microchannel reactors used in these investigations is given in table 1.

Table 1: Geometry and catalytic active coatings of microchannel reactors

	MCR 1	MCR2	MCR3
channel cross-section	300 x 700 μm^2	300 x 700 μm^2	200 x 200 μm^2
channel length	50 mm	50 mm	50 mm
number of channels	42	42	858
wafer material	AlMg3	AlMg3	AlMg3
catalytic active coating	50 to 1400 nm Ag by sputtering	1 μm layer of porous α-Al_2O_3 activated with silver-lactate	1200 nm Ag by sputtering / PVD

Similar investigations carried out for the partial oxidation of 1-butene indicate that the selectivity / conversion behaviour does not depend on the channel geometry for investigated channel diameters between 80 and 400 μm [16]. Thus, a change in channel geometry from 300 x 700 μm^2 to 200 x 200 μm^2 should not influence the selectivity at a certain degree of conversion.

3 Results

3.1 Reactor MCR1 (Ag/Al): Effect of the Ag-layer thickness

It is known that the Ag-particle size has a strong influence on the selectivity and activity of the silver particles used for ethene epoxidation [17,18]. Thus, the dependence of the Ag layer thickness on the catalytic performance was subject of the investigations.

Three microstructured aluminum wafers were initially coated with 50 nm Ag on the microstructured side by sputtering, mounted in the modular microchannel reactor and heated up to reaction temperature. The activation was performed under reaction conditions as long as the selectivity / conversion changed with time on stream. After typically 24 to 48h and a constant level of activity, the flowrate was reduced stepwise in order to get the selectivity / conversion behaviour of this reactor. Afterwards, the catalyst was removed from the modular microchannel reactor and coated with an additional amount of silver. Thus, a dense silver layer was added to an already activated catalyst and the procedure was performed again to monitor changes. REM examinations revealed that starting from an initially dense silver surface, small silver particles are formed by surface diffusion on the aluminum support leaving Al_2O_3 surface open.

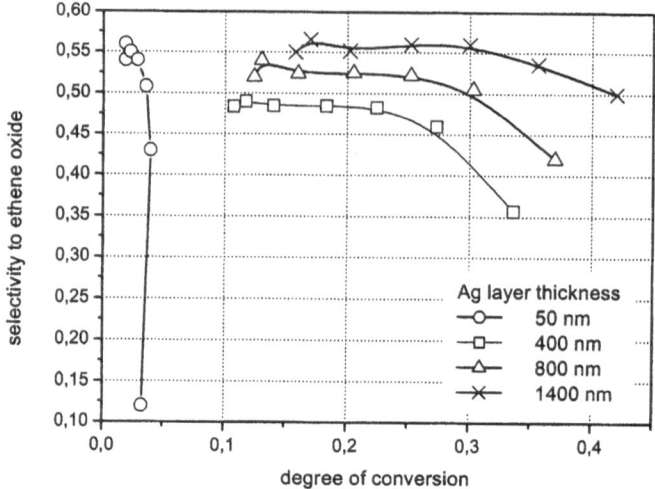

Figure 3: Selectivity and conversion degrees of the Ag/Al microchannel reactor MCR1 as a function of Ag layer thickness. Reaction conditions: T=523K, p=3 bar, 20% C_2H_4 in O_2, variation of residence time performed between 0.23 and 2 s

The selectivity / conversion behaviour of the Ag/Al reactor MCR1 is depicted in figure 3. It is noteworthy that at a very low layer thickness of about 50 nm and conversion degrees as low as 5%, the selectivity dropped down sharply with increasing degree of conversion. With an increasing layer thickness, much higher degrees of conversion up to 40% were attained. The maximum selectivity for each catalyst increased slightly with increasing layer thickness, the overall highest selectivity was provided by the thickest silver layer. Selectivities of initially 56% observed at 50 nm Ag and low degrees of conversion may be ascribed to a not sufficient initial activation period, because every catalyst showed the specific highest selectivies at the beginning of each aging / activation procedure, but before reaching its steady state. Furthermore, every catalyst showed nearly constant selectivities at low degrees of conversion. At a certain point, the selectivity dropped with increasing degree of conversion and for a layer thickness of 400 nm, this point was located at about 23%, for 800 nm at 27% and for 1400nm Ag layer thickness conversion degrees up to 32% were possible without losing selectivity.

3.2 Reactor MCR2 (Ag/α-Al₂O₃): Effect of C₂H₄ and O₂ partial pressure

The influence of the ethene partial pressure on the catalytic properties using an Ag/α-Al₂O₃ coated microchannel reactor was investigated by variation of the ethene concentration at a constant contact time using oxygen as balance and results are shown in figure 4. Selectivity increased slightly with higher ethene concentrations from 51.7% at 3.5% C_2H_4 to 54.6% at 60% C_2H_4 whereas the conversion degree decreased with increasing C_2H_4 concentration, from 21.3% at 3.5% C_2H_4 down to 1.52% at 60% C_2H_4. The highest ethene oxide concentrations were obtained at 30% ethene.

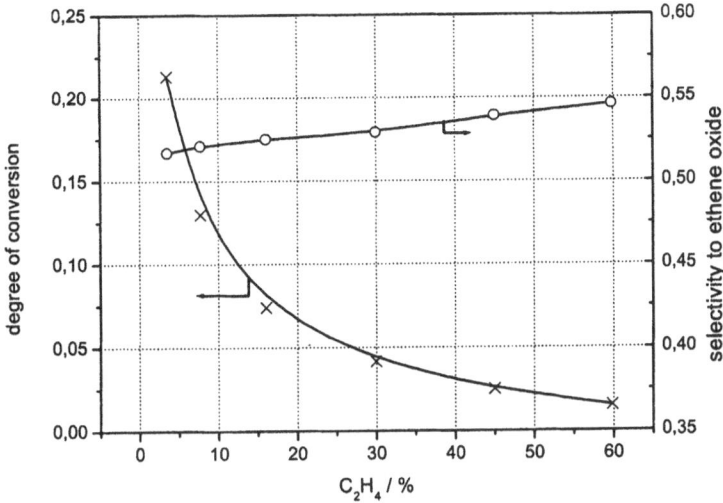

Figure 4: Selectivity and conversion degrees as a function of the C_2H_4 concentration using the $Ag/\alpha\text{-}Al_2O_3$ microchannel reactor MCR2. Reaction conditions: T=483K, p=3 bar, residence time 53 ms (STP), balance O_2 (no dilution by inert gas)

The dependence of selectivity and conversion degrees on the oxygen partial pressure were investigated in a similar way. In this experiment (fig. 5), the ethene partial pressure and the residence time were kept constant, but oxygen was stepwise replaced by methane. With increasing oxygen concentration, the selectivity improved by nearly 10% from 41% at 11.6% oxygen to 50.6% at 80% oxygen. The conversion degree was dramatically improved from 8.8% at 11.6% oxygen to 21.2% at 80% oxygen.

The low selectivity of only 40% at low oxygen partial pressures compared to approximately 60% selectivity reported in several publications indicates that this catalytic system has still about 20% improvement potential without application of promoters. This low selectivity may be ascribed to a not optimized and probably too porous $\alpha\text{-}Al_2O_3$ coating of the microstructured wafers. Nevertheless, 10% improvement in selectivity and an increase in conversion degree of factor 2.5 prove the advantage of using a microchannel reactor, because it showed impossible to apply comparable high ethene and oxygen concentrations to a conventional $Ag/\alpha\text{-}Al_2O_3$ catalyst without immediate explosion of the reactands.

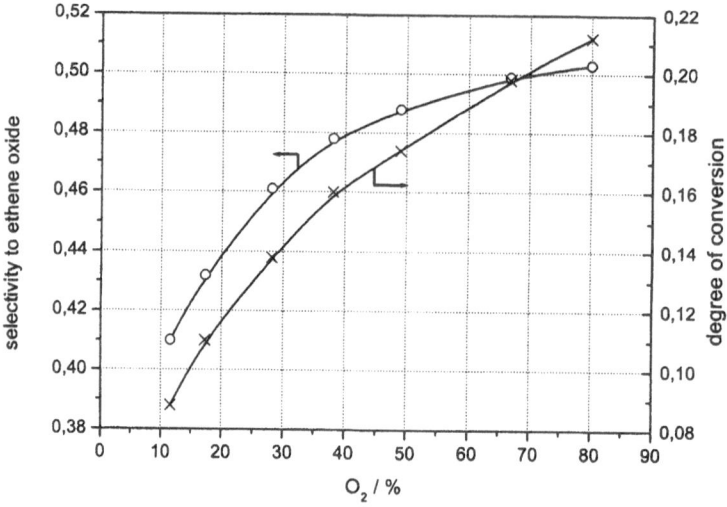

Figure 5: Selectivity and conversion degrees as a function of the O_2 concentration using the Ag/α-Al$_2$O$_3$ microchannel reactor MCR2. Reaction conditions: $c(C_2H_4)$ = 20%, $c(CH_4)$ = balance, T=483K, p=3 bar, residence time 300 ms (STP)

3.3 Comparison of Ag/α-Al$_2$O$_3$ (MCR2) and Ag/Al (MCR3) catalysts

The comparison of the systems Ag/Al and Ag/α-Al$_2$O$_3$ is performed using the selectivity / conversion behaviour as reference. Because of the low specific activity of the non-porous Ag/Al catalytic system, the commercially made microchannel reactor MCR3 having a higher channel volume and a higher geometric surface area was used instead of MCR1 in order to achieve comparable degrees of conversion. A comparison of both systems was performed by variation of the residence time and at temperatures of 503K in order to get higher conversion degrees (fig. 6).

The Ag/Al microchannel reactor MCR3 showed higher overall selectivities than the Ag/α-Al$_2$O$_3$ reactor MCR2. The maximum selectivity was 61% for the Ag/Al and 51% for the Ag/α-Al$_2$O$_3$ system. Both curves are initially horizontal and at their highest selectivities of 51% and 61% respectively, showing almost no dependence of the selectivity on degrees of conversion up to 20%. The more active Ag/α-Al$_2$O$_3$ microchannel reactor allows conversion degrees as high as 62% at selectivities as low as 15% without runaway or ignition of the explosive reactand mixture.

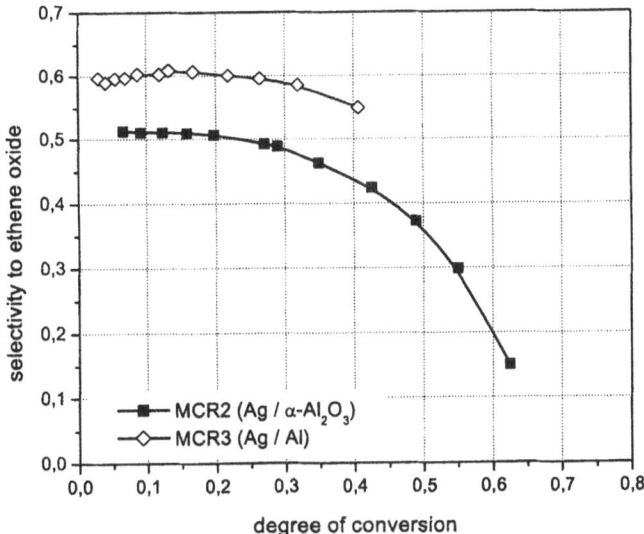

Figure 6: Selectivity and conversion degree of the Ag/-Al$_2$O$_3$ microchannel reactor MCR2 and the Ag/Al microchannel reactor MCR3. Reaction conditions: c(C$_2$H$_4$) = 20%, c(O$_2$) = 80%, T=503K, p=3 bar

4 Conclusions

The partial oxidation of ethene applying high ethene and oxygen concentrations in the explosion range was successfully performed in microchannel reactors using unpromoted Ag/Al and Ag/α-Al$_2$O$_3$ catalytic surfaces.

On the Ag/Al catalytic surface, selectivities of 60% were observed. A strong effect of the silver layer thickness on the selectivity/conversion behaviour has to be noted. Thick silver layers up to 1400 nm allow higher selectivities and activities than thinner ones, especially at high degrees of conversion. Conversion degrees up to 30% at 503K, 3 bar and 20% C$_2$H$_4$ in O$_2$ were achieved without any loss in selectivity.

The much more active Ag/α-Al$_2$O$_3$ surface allowed even higher degrees of conversion up to 63%, having lower selectivities of about 51%. At constant residence time, higher ethene concentrations allow modest gains in selectivity, improving the selectivity from 51.7% at 3.5% ethene to 54.6% at 60% ethene. The O$_2$ concentration proved to be crucial in order to achieve high catalytic activity and selectivity. With an increased oxygen concentration from 11% to 80%, the selectivity was improved by nearly 10% and the conversion by factor 2.4. It can be assumed that the observed improvements may persist when using an industrial catalyst in a microchannel reactor.

5 Acknowledgements

This study was supported by "Arbeitsgemeinschaft industrieller Forschungsvereinigungen Otto von Guericke e.V." (AiF[2]) under grant number 11146B/1 with financial support from the German Federal Ministry of Economics and Technology (BMWi). The commercial microchannel reactor was made in cooperation with Dr. Schubert[3] and Dr. Fichtner by the "Forschungszentrum Karlsruhe GmbH". The α-Al_2O_3 coating of the microstructured wafers was performed by Dr. Richter[4] from the „Hermsdorfer Institut für Technische Keramik".

6 References

[1] M.T.Janicke, H.Kestenbaum, U.Hagendorf, F.Schüth, M.Fichtner, K.Schubert; Journal of Catalysis 191 (2000), 282–293

[2] T.E. Lefort, French patent 729952, 1931, to Societe Francaise de Catalyse Generale

[3] R.Baratti, A.Gavriilidis, M.Morbidelli, A.Varma; Chemical Engineering Science 49 (1994), No. 12, p 1925–1936

[4] Information from CRI Catalysts, EO catalyst manufacturer

[5] G.L.Montrasi, G.C.Battiston; Oxidation Communication 3 (1983), p 259–267

[6] P.C.Borman, K.R.Westerterp; Industrial Engineering and Chemistry Research 34 (1995), 49–58

[7] E.P.S.Schouten, P.C.Borman, K.R.Westerterp; Chemical Engineering and Processing 35 (1996), 107–120

[8] D.M.Minahan, G.B.Hoflund; Journal of Catalysis 158 (1996), 109–115

[9] P.Yinsheng, Z.Shi, T.Liang, D.Jingfa; Catalyis Letters 12 (1992), 307–218

[10] A.Gavriliidis, A.Varma; ACS Symposium Series 523, 204th National Meeting of the American Chemical Society (Washington DC), 410–415

[11] E.P.S.Schouten, P.C.Borman, K.R.Westerterp; Chemical Engineering Science 49, No. 24a (1994), 4725–4747

[12] G.B.Hoflund, D.M.Minahan; Journal of Catalysis 162 (1996), 48–53

[13] C.F.Mao, M.A.Vannice; Applied Catalysis A 122 (1995), 61–76

---

[2] http://www.aif.de

[3] http://www.fzk.de

[4] http://www.hitk.de

[14]   A.Kursawe, D.Hönicke; IMRET 4 Conference Proceedings, AIChE Spring National Meeting 2000 (Atlanta), ISBN 0-8169-9882-5, 153–166

[15]   A.Kursawe, R.Pilz, H.Dürr, D.Hönicke; IMRET 4 Conference Proceedings, AIChE Spring National Meeting 2000 (Atlanta), ISBN 0-8169-9882-5, 227–235

[16]   S.Kah, D.Hönicke; „Selective Oxidation of 1-Butene to Maleic Anhydride [..]", IMRET 5 Conference Proceedings (this book), to be published.

[17]   V.I.Bukhtiyarov, A.F.Carley, L.A.Dollard, M.W.Roberts; Surface Science **381** (1997), L605–L608

[18]   S.N.Goncharova, E.A.Paukshtis, B.S.Bal'zhinimaev; Applied Catalysis A **126** (1995) 67–84

Monodispersed Droplet Formation
Caused by Interfacial Tension
from Microfabricated Channel Array

Shinji Sugiura[1, 2)], Mitsutoshi Nakajima[1)], Minoru Seki[2)]

1)   National Food Research Institute, Tsukuba, Ibaraki 305-8642, Japan

2)   Department of Chemistry and Biotechnology, The University of Tokyo,

Bunkyo-ku, Tokyo 113-8654, Japan

INTRODUCTION

Emulsions have been utilized in various industries.   Physical and qualitative stability depend on their size and size distribution.   Stability and resistance to creaming of emulsions are influenced by their size.[1, 2]   Furthermore, size and size distribution determine the characteristics and abilities of emulsions.   Rheology, appearance, chemical reactivity, and physical properties are influenced by both the average size and size distribution.[1, 3]   Monodispersed emulsions are applicable to valuable materials, such as drug delivery vehicles and as precursors of monodispersed beads for column chromatography.   Monodispersed emulsions are also useful for fundamental studies because the interpretation of experimental results is much simpler than that for polydispersed emulsions.[1]   However, size control of these MS is not so easy.

Membrane emulsification, in which the pressurized dispersed phase passes through a microporous membrane and forms emulsion droplets, is a promising technique for producing monodispersed emulsions with a coefficient of variation of approximately 10%.[4]   This technique is applicable to both oil-in-water (O/W) using hydrophilic membranes and W/O emulsions using hydrophobic membranes.[5, 6]   The emulsions are produced by pressurizing a dispersed phase into a continuous phase, and the emulsion droplet size is controlled by the membrane pore size.   This technique can be used to produce emulsions without strong mechanical stress.[7]   The porous glass membrane has been successfully applied in the production of monodispersed emulsions on a sub-micrometer or micrometer scale.   However, it is difficult to produce monodispersed emulsions on a ten-micrometer scale since the pore size distribution of the membrane is rather wide in this range.

Recently, the application of micromachining techniques has grown rapidly in various fields.   New applications are appearing in biotechnology and chemical reactions, such as polymerase chain reaction (PCR) on a chip,[8] DNA analysis,[9, 10] cell handling[11, 12] and micro-chemical reactors.[13, 14]   These microelectromechanical systems (MEMS) have the advantages of precise fabrication, replication by a photolithography, small volume, portability, small amounts of expensive reagents, short reaction and analysis times, and small volume of waste.   Hydrodynamics on a micrometer scale feature significant effects of viscosity and interfacial tension, laminar flow, fast diffusion, and so on.[13, 15]   Applying these features, microfabricated devices for pumping discrete droplets,[15] for measuring blood viscosity, [16] for micromixing,[17] and for diffusion-based separation and detection[18] have been developed.

Recently we proposed a novel microfabricated device for preparing super-monodispersed emulsion with coefficient of variation less than 5%.[19-22] This method is promising one not only for preparation of emulsions, but also various microspheres (MS), which need high monodispersity, for example, micro particles,[23] multiple emulsions, microcapsules and liposomes for drug delivery vehicle. Using this technique, monodispersed droplets are formed from microfabricated channel (MC) array. However, mechanism of droplet formation behavior from MC has not been clear because of very short time for droplet formation, less than 0.03 second (1 flame of video recorder). In this study, we used microscope high-speed camera system to analyze the droplet formation mechanism.

MATERIALS & METHODS
Materials

Triolein (purity >90 %) was obtained from Nippon Lever B. V. (Tokyo, Japan) was used as dispersed oil phase. MilliQ water was used as continuous water phase. Sodium dodecyl sulfate (SDS) was purchased from Wako Pure Chemical Ind. (Osaka, Japan) and used as the surfactant for emulsification.

Measurement and Analytical Method

The droplet diameter and interface area are determined from pictures obtained with the microscope video system described below. Winroof (Mitani Corporation, Fukui, Japan) software was used to analyze the captured pictures. The interfacial tension in each system was measured using a Full Automatic Interfacial Tensiometer (PD-W; Kyowa Interface Science Co., Ltd., Saitama, Japan).

MC Emulsification

Figure 1 shows the experimental setup and schematic flow through MC in the module. Experimental equipment consists of the MC module, the silicon MC plate and the liquid chambers supplying the both phases, and a microscope observation system. The silicon MC plate was tightly covered with a flat glass plate using an O-ring, in the module. During the operation of the MC emulsification, the applied pressure and the flow rate were regulated by changing the height of the chamber. Emulsification behavior was observed through the glass plate by microscope high-speed camera system, which can take 600 flames per second.

RESULTS & DISCUSSION

Droplet formation mechanism from MC

Sodium dodecyl sulfate (SDS) 0.3% aqueous solution was used as continuous phase. Sunflower oil was used as dispersed phase. Figure 2 shows the emulsification behavior observed through the glass plate at 3.5 kPa applied pressure. A monodispersed emulsion with an average diameter of 17.8 μm and coefficient of variation of 2.8% was prepared. The droplet formation behavior from MC was observed by microscope high-speed camera system. Dispersed phase passed through the MC inflated on the terrace having a disk-like shape in 0.1 s. At the time the dispersed phase reached at the end of the terrace, the dispersed phase flowed into the well part and was cut off in a moment, 0.01 s. From these, new droplet formation mechanism was proposed that distorted dispersed phase is sheared spontaneously into the spherical droplet by interfacial tension. Figure 3 shows schematic of droplet formation mechanism from MC. Dispersed phase passed through the MC inflated on the terrace having the disk-like shape. When dispersed phase inflated over the terrace end, the dispersed phase inflated having spherical shape in the well part. Here, the distorted dispersed phase on the terrace was transformed spontaneously into the spherical droplet by interfacial tension. This is the shearing mechanism of MC emulsification. Effect of interfacial tension forces is important at small scales relative to the other forces that influence liquid behavior. Shearing by this dominating interfacial tension force contributes to the stable droplet formation. MC emulsification is ingenious method taking advantage of the microfluidics phenomena that an effect of interfacial tension becomes larger at micrometer scale. The mechanism is shown to be an adequate model from the viewpoint of interfacial free energy.

Relationship between the interfacial tension and the breakthrough pressure

The relationship between the interfacial tension and the breakthrough pressure was examined by changing the surfactant concentration in the dispersed phase. For membrane emulsification the breakthrough pressure can be estimated from the capillary pressure.[24]

$$P_{bt} = \frac{4\gamma \cos\theta}{d} \qquad [1]$$

where P_{bt} is the breakthrough pressure, γ is the interfacial tension, θ is the interface contact angle with the wall of the channel and d is the channel diameter.

This equation originates from the Laplace equation expressing the relationship between the pressure inside the curved interface and the pressure outside the curved interface.[25]

$$P_{in} = P_{out} + \frac{2\gamma}{r}$$ [2]

where P_{in} is the pressure inside the interface, P_{out} is the pressure outside of the interface and r is the radius of droplet curvature.

Hydrogenated fish oil with Span40 was used as the dispersed phase, water was used as the continuous phase. Figure 4 shows the experimental result and the theoretical line calculated from Eq.[1]. For calculation of the theoretical line, an equivalent diameter of MC was taken as a cylinder diameter. From the experimental data, the breakthrough pressure was almost proportional to the interfacial tension and the experimental results had a good agreement with the theoretical line. It is suggested that the breakthrough pressure can be predicted from the interfacial tension and equivalent diameter of the MC.

CONCLUSION

During MC emulsification, distorted dispersed phase is cut off spontaneously into the spherical droplet by interfacial tension. MC emulsification is ingenious method taking advantage of the microfluidics phenomena that an effect of interfacial tension is significant at micrometer scale. The breakthrough pressure can be predicted from the interfacial tension and equivalent diameter of the MC.

258

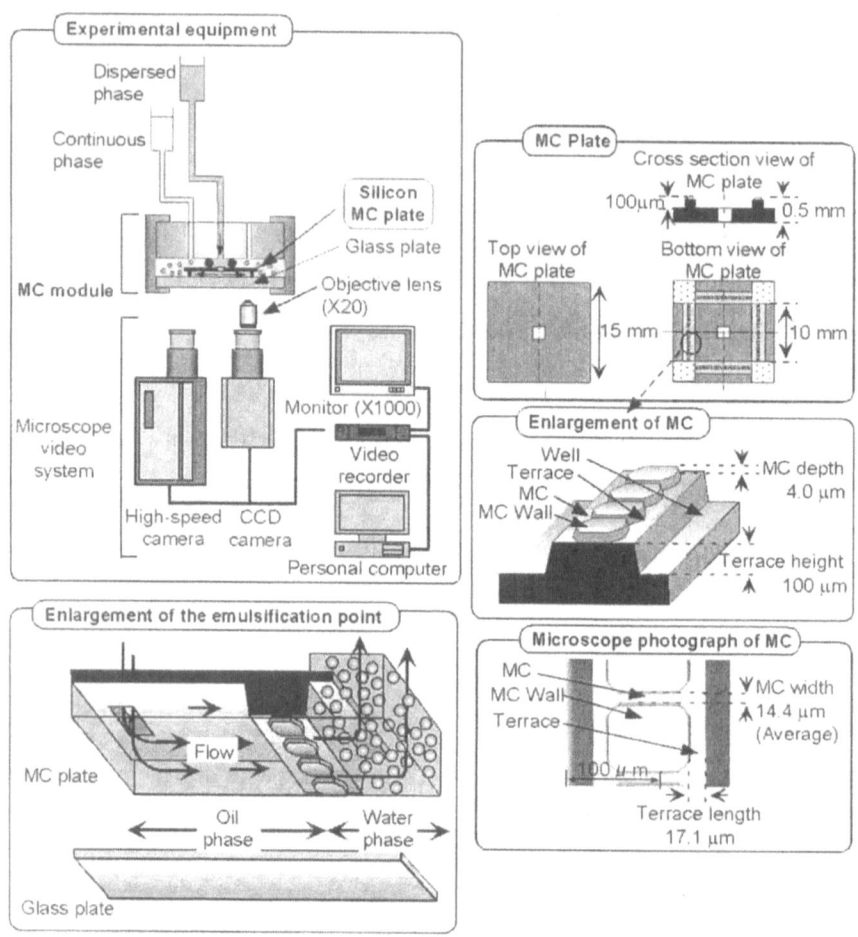

Fig. 1.   Experimental setup and schematic flow of dispersed oil phase

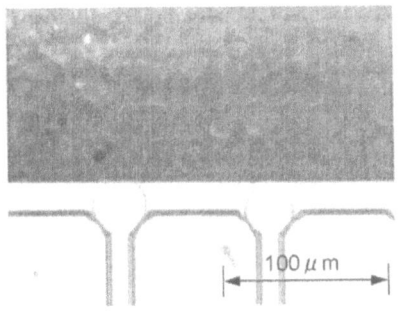

Fig. 2. Microscope photographs of MC emulsification process. The average diameter and coefficient of variation of prepared emulsion were 17.8μm and 2.8%.

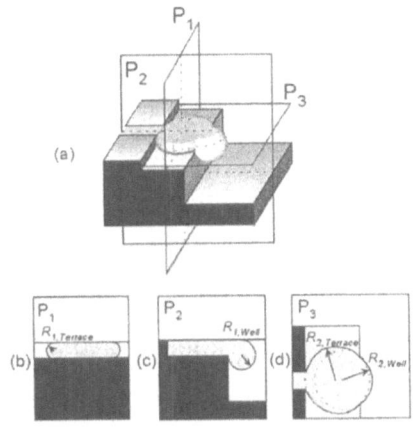

Fig. 3. Schematic of droplet formation mechanism from MC.

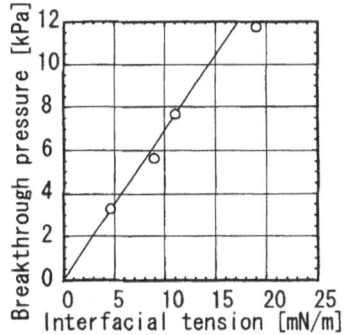

Fig. 4 Relationship between the interfacial tension and the breakthrough pressure

ACKNOWLEDGMENTS

This work was supported by Program for Promotion of Basic Research Activities for Innovative Biosciences (MS-Project).

260

REFERENCES

(1) McClements D. J., *Food Emulsions : Principles, Practice, and Techniques*; CRC Press: Boca Raton, FL, 1999.

(2) Mason, T. G.; Krall, A. H.; Gang, H.; Bibette, J.; Weitz, D. A., In *Encyclopedia of emulsion technology*, Becher, P. Eds.; Marcel Dekker : New York, 1996; Vol. 4, Chapter 6.

(3) Dickinson, E., *An Introduction to Food Colloids*; Oxford University Press: Oxford, 1992.

(4) Nakashima, T.; Shimizu, M.; Kukizaki, M., Key Engineering Materials, **1991**, 61&62, 513-516.

(5) Suzuki, K.; Shuto, I.; Hagura, Y., Food Sci. Technol., Int. **1996**, 2, 43-47.

(6) Suzuki, K.; Fujiki, I.; Hagura, Y., Food Sci. Technol. Int. Tokyo **1998**, 4, 164-167.

(7) Schröder, V.; Schubert, H., Colloids Surf. A, **1999**, 152, 103-109.

(8) Kopp, M. U.; de Mello, A. J.; Manz, A., Science, **1997**, 280, 1046-1048.

(9) Chou, H. P.; Spence, C.; Scherer, A.; Quake, S., Proc. Natl. Acad. Sci. USA, **1999**, 96, 11-13.

(10) Burns, M. A.; Johnson, B. N; Brahmasandra, S. N.; Hanhdique, K.; Webster, J. R; Krishnan, M.; Sammarco, T. S.; Man, P. M.; Jones, D.; Heldsinger, D.; Mastrangelo, C. H.; Burke, D. T., Science, **2000**, 282, 484.

(11) Takayama, S.; McDonald, J. C.; Ostuni, E.; Liang, M. N.; Kenis, P. J. A.; Ismagilov, R. F.; Whitesides, G. M., Proc. Natl. Acad. Sci. USA, **1999**, 96, 5545-5548.

(12) Li, P. C. H.; Harrison, D. J., Anal. Chem. **1997**, 69, 1564-1568.

(13) Jensen, K. F., AIChE J., **1999**, 45, 2051-2054.

(14) Srinivasan, R.; Hsing, I. M.; Berger, P. E.; Jensen, K. F., AIChE J., **1997**, 43, 3059-3069.

(15) Sammarco, T. S.; Burns, M. A., AIChE J., **1999**, 45, 350-366.

(16) Kikuchi, Y.; Sato, K.; Ohki, H.; Kaneko, T., Microvascular Res., **1992**, 44, 226-240.

(17) Ehrfeld, W.; Hessel, V.; Löwe, H., In Microreactors; WILEY-VCH: Weinheim, 2000; Chapter 3.

(18) Weigl, B. H.; Yager, P., Science, **1999**, 283, 346-347.

(19) Kawakatsu, T.; Kikuchi, Y.; Nakajima, M., J. Am. Oil Chem. Soc., **1997**, 74, 317-321.

(20) Kawakatsu, T.; Komori, H.; Nakajima, M.; Kikuchi, Y.; Yonemoto, T., J. Chem. Eng. Japan, **1999**, 32, 241-244.

(21) Kobayashi, I.; Nakajima, M.; Tong, J.; Kawakatsu, T.; Nabetani, H.; Kikuchi, Y.; Shono, A.; Satoh, K., Food Sci. Technol. Res., **1999**, 5, 350-355.

(22) Sugiura, S.; Nakajima, M.; Ushijima, H.; Yamamoto, K.; Seki, M., *J. Chem. Eng. Japan*, **2001**, in press.

(23) Sugiura, S.; Nakajima, M.; Tong, J.; Nabetani, H.; Seki, M., *J. Colloid Interface Sci.*, **2000**, 227, 95-103.

(24) Peng, S. J. and Williams, R. A., *Trans IchemE*. **1998**, 76(A), 894.

(25) Atkins, P. W., *Physical Chemistry*; Oxford University Press: Oxford, 1990.

INVESTIGATION OF MICROFLUIDICS AND HEAT TRANSFERABILITY INSIDE A MICROREACTOR ARRAY MADE OF GLASS

E. Marioth, S. Loebbecke, M. Scholz, F. Schnürer, T. Türcke, J. Antes, H.H. Krause

Fraunhofer-Institut Chemische Technologie ICT
P.O. Box 12 40, 76318 Pfinztal, Germany

N. Lutz

Fachhochschule Brandenburg, Magdeburger Str. 50, 14 770 Brandenburg/ Havel, Germany

Abstract

A microstructured glass reactor was investigated by different experimental and theoretical methods to characterize its heat flow and heat transport properties. Calorimetric studies were carried out to measure the heat flow from the microstructured reaction channels to the coolant channels being an integral part of the microreactor. Heat transfer coefficients and transferred heat flows were determined under different flow resp. thermal conditions. It was shown that the integrated cooling structure has a significant influence on the thermal control of the microreaction process. Heat transfer coefficients up to 10,000 W/m²K can be achieved.

Thermography in terms of an infrared imaging system was applied to achieve an online mapping of temperature distributions and temperature profiles over the whole microreaction process. Furthermore, thermography was a suitable tool for monitoring and optimizing process parameters and process conditions like flow rates, retention times, blockages, etc. which influence significantly temperature distributions and heat flow properties of the reactor.

CFD simulations of the glass reactor were carried out to confirm the influence of varied process conditions on the microfluidics and mixing performance of the microreactor, and therefore on the heat flow and heat transfer properties of the entire microreaction process.

Keywords

heat flow, heat transfer coefficient, glass reactor, exothermic reaction, calorimetry, thermal imaging, thermography, computational fluid dynamic (CFD), mixing performance

Introduction

In recent years microfluidic structures have received an increasing interest as reactors for highly exothermic chemical processes due to their high surface-to-volume ratios and good heat transfer characteristics.

Fraunhofer ICT has a particular interest in applying microreactors for strong exothermic nitrations both to reduce the hazardous potential of these reactions and to improve yield and selectivity for specific products [1 - 2]. For this purpose a microreaction device made of glass was developed in cooperation with mgt mikroglas technik AG, Mainz/Germany (Fig. 1).

This microreactor array consists of 20 reaction channels in parallel with individual educt mixing zones and an integrated cooling structure as well [1 - 3]. The channel dimensions (channel widths of 350 µm resp. 700 µm) were designed for relatively high throughputs. Glass was chosen as a reactor material for two reasons, its chemical inertness against highly corrosive nitrating agents and its optical transparency allowing a visual - and also spectroscopic - monitoring of chemical processes inside of the reactor.

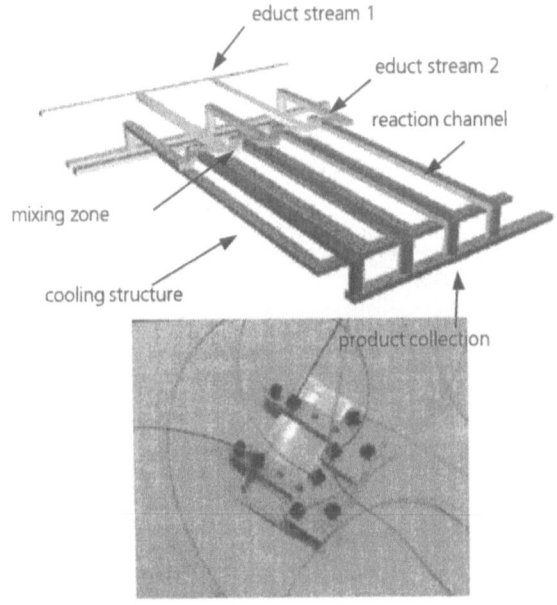

Fig. 1:   microreactor array made of glass

Experimental

Heat transfer properties of the microstructured glass reactor were investigated by three different methods. Calorimetric investigations were based on selective temperature measurements in microfluidic channels at the inlet and outlet of the reactor array [4].

264

Spatially resolved information on heat transferability and simultaneous mapping of the entire process were obtained by thermographic measurements of exothermicities in the microreactor array applying an infrared camera system. Finally, Computational Fluid Dynamic (CFD) simulations were carried out to investigate the influence of flow and mixing behavior on heat flow and heat transfer.

Calorimetric measurements

Calorimetric measurements allow to assess the heat transfer characteristic of the microstructured glass reactor. The experimental set-up is schematically shown in Fig. 2. Two reactant streams (index R) of different temperature and a defined coolant flow (index K) were continously fed into the microreactor while temperature was measured computer-controlled at selective positions inside of the reactor.

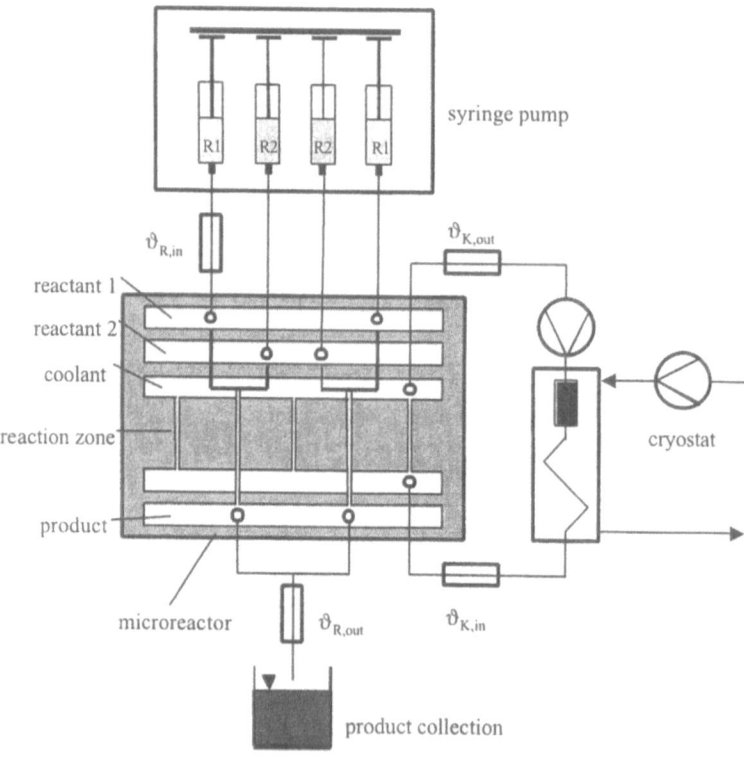

Fig. 2:   scheme of experimental set-up for calorimetric measurements

The heat transfer between the reaction channels and their vicinity was described by the heat transfer coefficient which was calculated from equation (1):

$$k = \frac{\left(\dfrac{\Delta Q}{\Delta t}\right)_{R-K}}{A \cdot \Delta T_m} \qquad (1)$$

with:

k: heat transfer coefficient [W/m²K]; $(\Delta Q/\Delta t)_{R-K}$: heat flow [W]; A: heat transfer surface [m²]; ΔT_m: mean logarithmic temperature difference [K]

The mean logarithmic temperature difference arises from the measured temperatures of the coolant and the reactants both at the inlet and outlet of the microreactor (equation 2).

$$\Delta T_m = \frac{\left(\vartheta_{R,out} - \vartheta_{K,in}\right) - \left(\vartheta_{R,in} - \vartheta_{K,out}\right)}{\ln\left(\dfrac{\vartheta_{R,out} - \vartheta_{K,in}}{\vartheta_{R,in} - \vartheta_{K,out}}\right)} \qquad (2)$$

As a first, most simple experiment the heat transfer potential of the glass reactor was investigated by using hot and cool water as the reactant resp. coolant stream. The heat flow and the heat transfer coefficient were determined under systematic variation of the flow conditions.

The transferred heat flow between the reactant and the coolant was calculated as the mean value of the heat flow released by the reactant stream and the heat flow absorbed by the coolant stream. Each heat flow was obtained as the product of the individual mass flow, specific heat capacity and temperature difference between in- and outlet of the reactant resp. coolant stream. Fig. 3 shows the transferred heat flow $(\Delta Q/\Delta t)_{R-K}$ as a function of the coolant flow rate. The coolant was conducted in counterflow.

The heat transfer coefficient calculated from equation (1) at different flow conditions is shown in Fig. 4. Both transferred heat flow and heat transfer coefficient show a linear dependence on the coolant flow rate.

For example, at a coolant flow rate of 34 mL/min and a reactant flow rate of 2 mL/min a heat transfer coefficient of k = 4,500 W/m²K was calculated. This value is twice the size of conventional heat exchangers having usual k-values in the range of 2,000 W/m²K. Even at lower flow rates of the reactant a heat transfer coefficient of k = 2,850 W/m²K was calculated going again beyond k-values of conventional heat exchangers.

In additional experiments the reactant and coolant flow rates were set equal, varying from 1 mL/min to 20 mL/min. The resulting heat transfer coefficients are plotted in Fig. 5 and compared with those coefficients which were calculated for a constant reactant flow rate of 1 mL/min.

Assuming a linear dependence of the heat transfer coefficient on the coolant flow rate k-values of more than 10,000 W/m²K seem to be achievable. Furthermore, the data also indicate that the heat transfer coefficient increases when the reactant flow rate is increased as well. Although the heat conductivity of glass or ceramic materials ($\lambda_w = 0.6 - 1.4$ W/m²K) is much lower than for metalic materials like copper ($\lambda_w = 394$ W/m²K), which are conventionally used materials for macroscopic heat

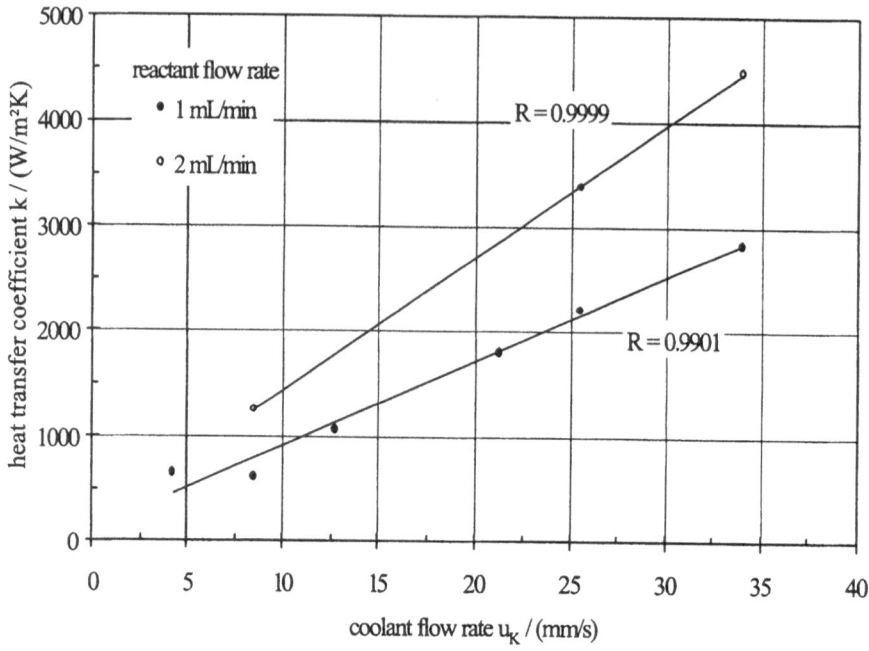

Fig. 3: transferred heat flow $\Delta Q/\Delta t_{R-K}$ at different flow rates of the coolant

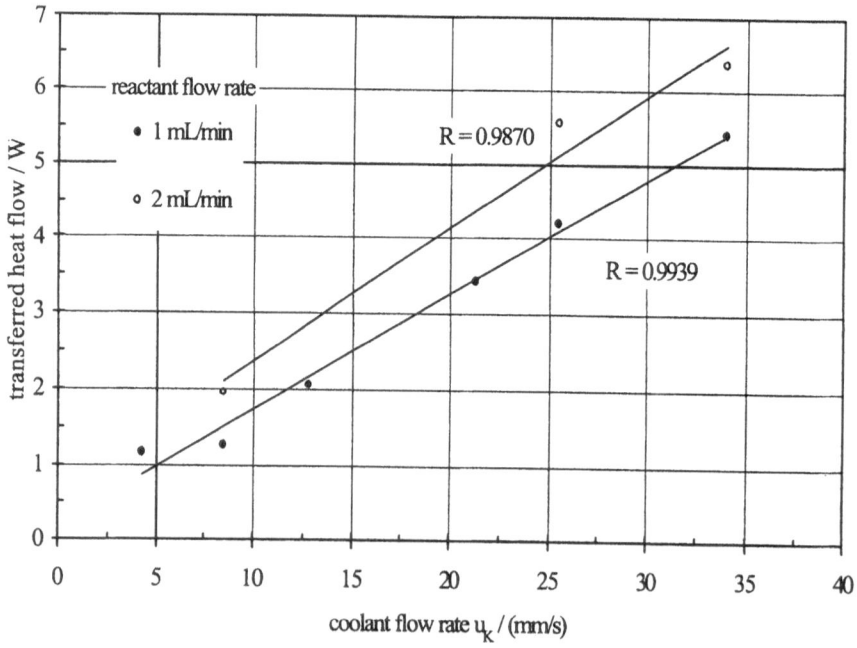

Fig. 4: heat transfer coefficient at different flow rates of the coolant

exchangers, they are sufficiently high for microscale applications. Studies of STIEF et al. [5] and BIER et al. [6 - 7] have recently shown, that the efficiency of heat transport in microstructured reaction/heat exchange units achieves its optimum at a heat conductivity of $\lambda_w = 0.66$ W/m²K due to the significant contribution of axial heat conductivity to heat transfer performances on a microscale. Hence, these studies confirm the good heat transport properties of the microstructured glass reactor applied in this work.

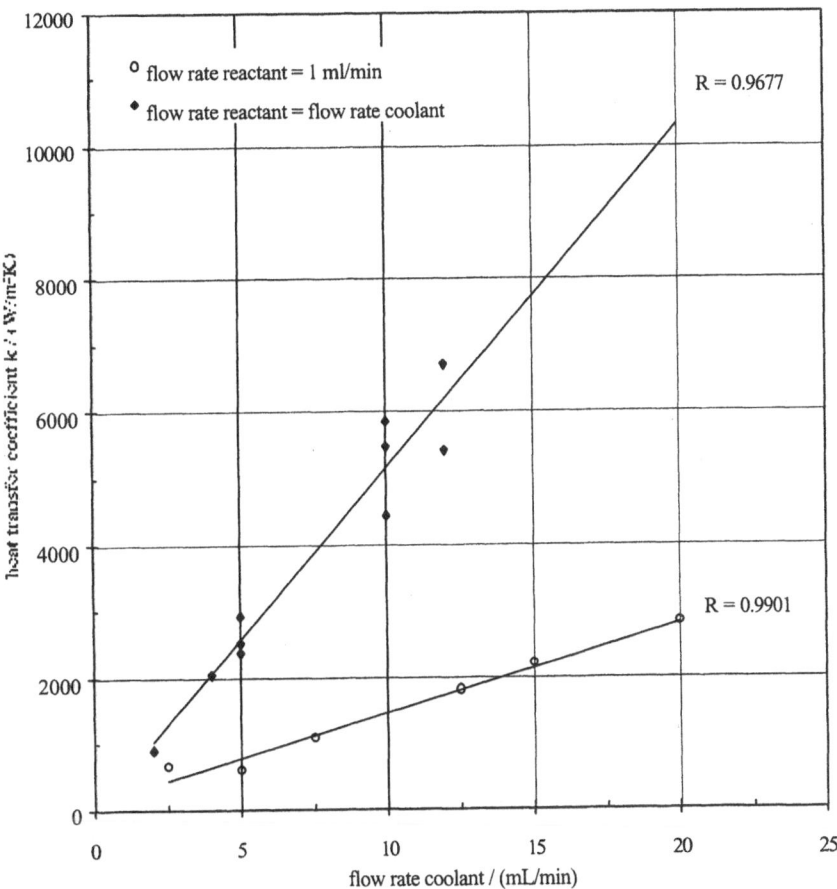

Fig. 5:   heat transfer coefficients for identical reactant and coolant flow rates

Beside water/water experiments heat transfer properties of the microreactor were also characterized by performing temperature controlled exothermic reactions. The dilution of concentrated sulphuric acid H_2SO_4 with water was chosen as an example for a strong exothermic process due to the released solvation enthalpy ΔH_s which enhances the heat content of the reactant streams. In the experiments the reactant flow rates of concentrated sulphuric acid and water were set equal while the coolant was conducted in counterflow.

The heat flow from the reaction channels to the cooling channels was again calculated on the basis of temperature measurements as described above. Heat flows for different coolant flow rates (11 mL/min and 21 mL/min) and different reactant flow rates (varied between 0.2 mL/min and 1.2 mL/min) were determined (Fig. 6).

Lower flow rates ensure a complete conversion in the reaction channels and thus a complete release of the solvation heat due to sufficient retention times of the reactants (also confirmed by CFD; see below).

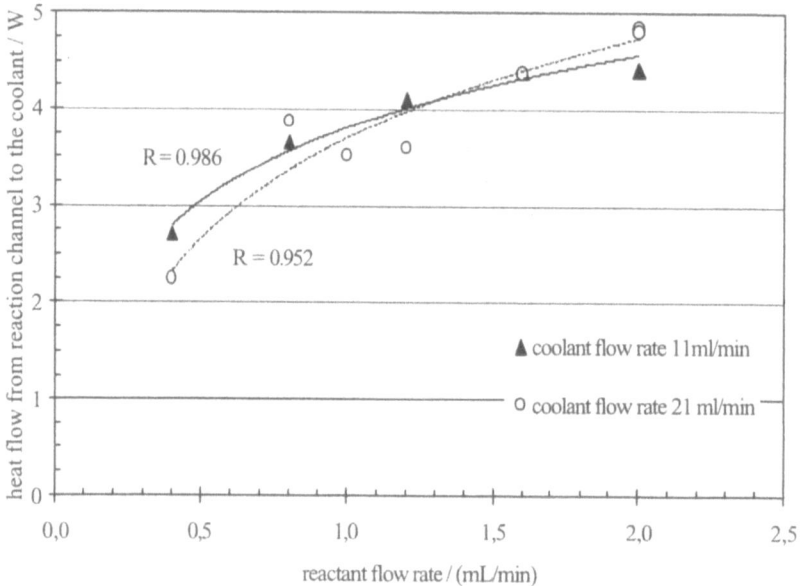

Fig. 6:   released heat flow to the coolant stream during dilution of concentrated sulphuric acid with water

The heat flow from the reactant stream to the coolant can be also theoretically assessed. The dilution of concentrated H_2SO_4 with water of same volume flow results in a molar ratio of 3.31 (water to sulfuric acid). For this dilution ratio a solvation enthalpy of $\Delta H_s = -49.6$ kJ per mol H_2SO_4 is known from literature [8]. On the basis of these data heat flows for different reactant flow rates can be calculated. Fig. 7 compares the calculated heat flows with those obtained by experimental approach. Deviations can be explained by heat flow into the vicinity of the reaction channels. However, differences between experiment and calculation are below 10%, thus confirming the good predictability of heat transport in the microstructured glass reactor.

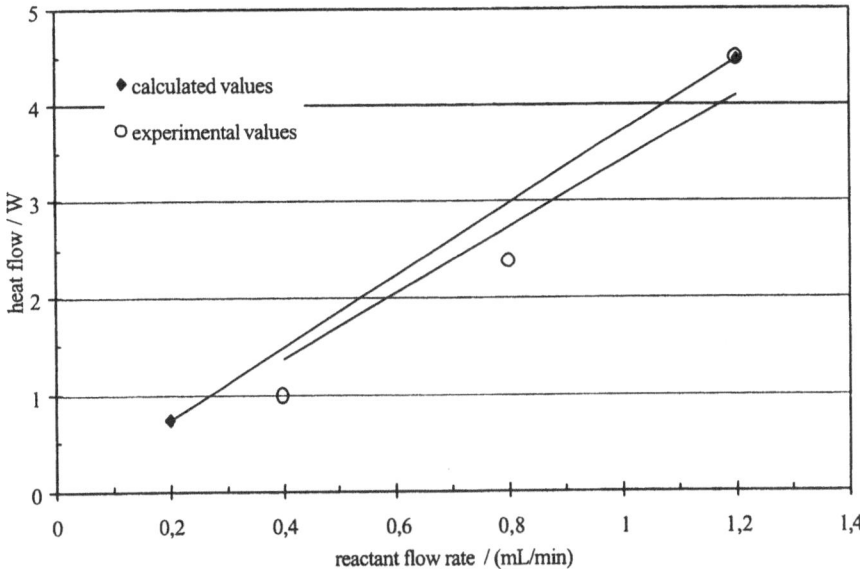

Fig. 7:   calculated and experimental values for released heat flows during dilution of sulphuric acid with water at different reactant flow rates (coolant flow rate: 11 mL/min)

Thermographic characterization of the microstructured glass reactor

In cooperation with the Fachhochschule Brandenburg, Germany a thermographic (thermal imaging) system was used to investigate heat transport in the microreactor.

Due to a sufficient transparency of the used FOTURAN® glass material in the IR spectral range between 3 and 5 µm such thermographic techniques can be applied. The striking feature of this technique is the visualization of the temperature distribution over the whole reactor surface. In this work the thermographic system was used for qualitative considerations only.

In a first experiment hot water of 80°C (flow rate: 10 mL/min) was pumped through the reactor without any active cooling to visualize the temperature distribution over the individual microfluidic channels. Fig. 8 shows the temperature distribution detected after 19 seconds (left) resp. 80 seconds (right).

The thermograms show the expected temperature increase inside and around the reaction channels. The individual reaction channels can be distinguished and for each channel a temperature profile can be determined along its longitudinal axis. Furthermore, the right picture in Fig. 8 shows also the appearance of a lower temperature zone on the reactor surface, clearly indicating an inhomogeneous flow distribution inside of the reactor due to a non uniform reactant supply. This example shows that thermographic measurements allow also some kind of control over the entire microreaction process, e.g. with respect to flow distribution, blockages inside of the microchannels, pump pulsation, or other experimental parameters.

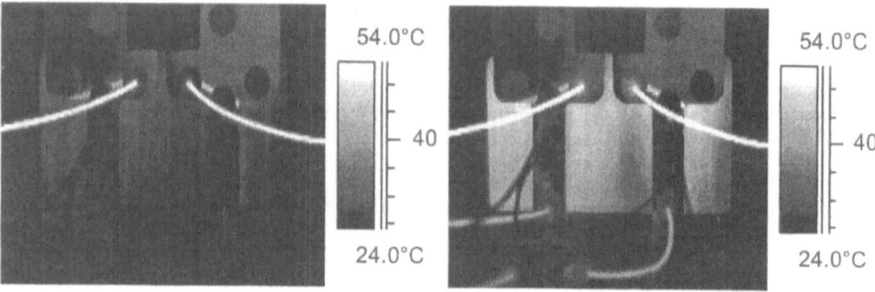

Fig. 8:   thermographic images of the microreactor after 19 s (left) resp. 80 s (right) pumping of hot water (80°C; 10 mL/min)

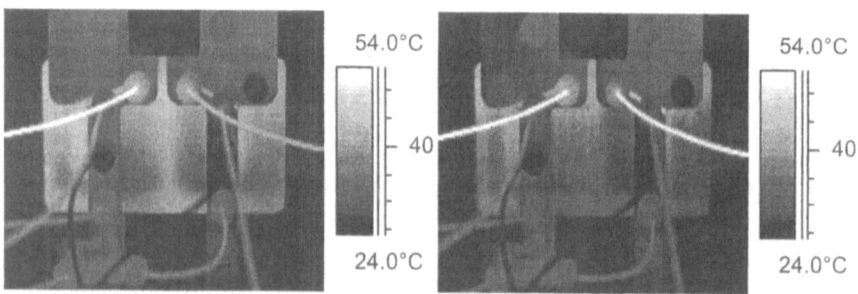

Fig. 9:   thermographic images of the microreactor after 250 s (left) resp. 330 s (right) feeding with hot water as "reactant" (80°C; 10 mL/min) and cold water as coolant (18°C; 20 mL/min)

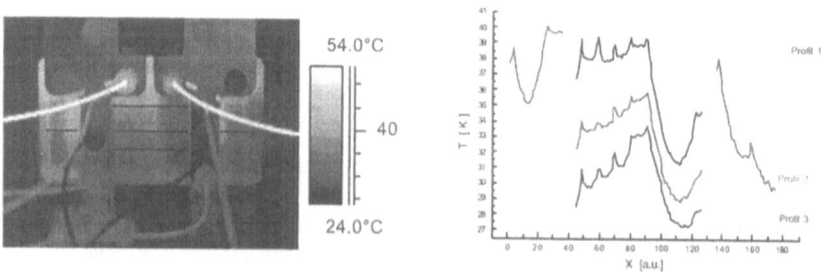

Fig. 10:   thermographic image of the microreactor and corresponding temperature profiles after 300 s feeding with hot water as "reactant" (80°C; 10 mL/min) and cold water as coolant (18°C; 20 mL/min)

After 215 s the temperature distribution reached steady state conditions and the coolant (water of 18°C; flow rate: 20 mL/min) was fed as a countercurrent flow. Fig. 9 shows the corresponding images of the temperature distribution after 250 s and 330 s. On these images the temperature gradient along the longitudinal axis of the reaction channels and thus the efficiency of the cooling structure can be clearly observed.

Beside the infrared mapping of the entire reactor surface thermographic measurements allow also discrete temperature measurements at any position of interest. Hence, qualitative and quantitative analysis of both spots and profiles is possible.

The thermographic system is extremely helpful for monitoring and evaluating the process conditions and parameters applied to the microreaction process. Suitable flow rates and retention times of the reactants can be easily determined. Fig. 11 shows the example of the exothermic dilution of concentrated sulphuric acid. In case of too short retention times the reaction will not take place in the microchannels anymore but in the main collecting channel near the outlet of the reactor. Hence, a thermal control of the reaction does not take place under these conditions. To achieve optimal process conditions a systematic variation of the applied flow conditions can be carried out and monitored simultaneously by thermographic techniques.

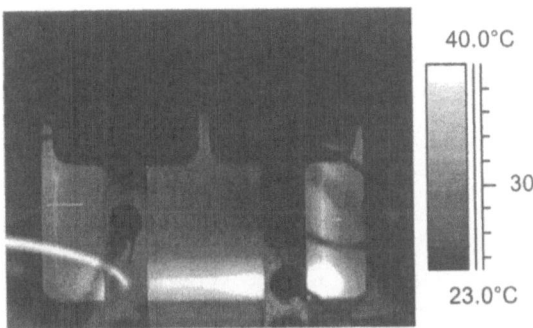

Fig. 11: thermographic image of the microreactor during the dilution of concentrated sulphuric acid and simultaneous active cooling applying too short retention times

Computational Fluid Dynamic (CFD)

CFD (ACE+ 6.0 from CFDRC, Huntsville Alabama, USA) simulations were carried out to investigate the mixing performance of the microstructured glass reactor and to determine suitable flow conditions with respect to an optimal mixing quality based on diffusion controlled mechanisms.

Fig. 12 shows simulation images with different boundary conditions leading to different mixing qualities at the end of a single reaction channel. The left simulation (boundary conditions: $\vartheta = 20°C$, volume flow rate at entrance: 3 mL/min, density of both reactants: 1 kg/m^3, cinematic viscosity 10^{-6} m^2/s) still contains laminar zones of

272

unmixed reactants indicated by dark grey zones (ideal mixing is indicated by a value of 0.5 resp. uniform light grey zone). In contrast to this incomplete mixing, the right simulation shows a good mixing quality with values between 0.4 and 0.6 due to different boundary conditions. The flow rate was decreased and the temperature increased (flow rate: 0.3 mL/min, ϑ = 50°C).

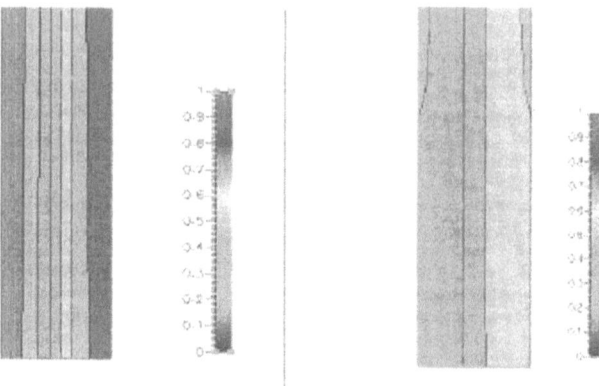

Fig. 12: CFD mixing and flow simulation at the outlet of a single reactor channel for flow rates of 3.0 mL/min at 20°C (left) and 0.3 mL/min at 50°C (right)

Fig. 13: images taken by light microscopy at the outlet of a single reaction channel indicating incomplete mixing performance (left image) at a flow rate of 3.0 mL/min and complete mixing performance (right image) at a flow rate of 0.3 mL/min

Neutralization experiments (HCl_{aq} + $NaOH_{aq}$) using bromothymol blue as an indicator confirm the CFD prediction of the mixing performance. Fig. 13 shows again the outlet of a single reaction channel observed under a light microscope applying the same mixing conditions as in the CFD simulations.

In accordance with the CFD simulations the mixing quality (indicated by dark grey of the bromothymol blue indicator) show significant differences due to the different process conditions applied.

These simultations point out that mixing quality as a function of process conditions has also a significant influence on the overall heat transport properties of a microreactor, not least because mixing quality determines conversion and thus heat flow of reactions.

Conclusions

The aim of this study was to test and apply different analytical methods to investigate the heat transfer and heat flow properties inside of a microstructured glass reactor. It could be shown by calorimetric measurements, that the integrated cooling structure has a significant influence on the thermal control of the microreaction process. Heat transfer coefficients up to 10,000 W/m²K can be achieved. Exemplary exothermic reactions (dilution of concentrated sulphuric acid) have shown that the measured heat flow is less than 10% lower than the theoretical value. A better insulation of the whole reactor unit may decrease this gap.

A new approach to characterize the heat transfer properties of the microreactor arised from the application of thermographic techniques as a mapping tool for the entire microreaction process. Temperature distribution and temperature profiles give an access to a spatially resolved thermal control over the whole reactor during operation. Furthermore, thermal imaging is a suitable tool to optimize process parameters and operation conditions, e.g. to visualize inhomogeneous flow conditions or insufficient retention times.

Finally, CFD simulations confirmed that process conditions have a significant influence on the flow behavior and mixing performance of the reactor, and thus also influence the heat flow and heat transfer properties of a microreaction process.

References

[1]   Loebbecke, S., Antes, J., Türcke, T., Marioth, E., Schmid, K., Krause, H.,
       Use of Microreactors for Nitration Processes,
       4th Int. Conference on Microreaction Technology (IMRET 4), 5-9 March 2000,
       Atlanta, GA, USA

[2]   Loebbecke, S., Antes, J., Türcke, T., Marioth, E., Schmid, K., Krause, H.,
       The Potential of Microreactors for the Synthesis of Energetic Materials,
       31st Int. Annu. Conf. ICT: Energetic Materials - Analysis, Diagnostics and
       Testing, 33, 27 - 30 June 2000, Karlsruhe, Germany

274

[3]   Freitag, A., Dietrich, T.R., Scholz, R.,
        Glass as a Material for Microreaction Technology,
        4th Int. Conference on Microreaction Technology (IMRET 4), 5-9 March 2000,
        Atlanta, GA, USA, 48-54

[4]   M. Scholz, diploma thesis 2001, Technical University Clausthal, Germany

[5]   Stief, T., Langer, O.-U., Schubert, K.,
        *Numerical Investigations on Optimal Heat Conductivity in Micro Heat
        Exchangers*,
        4th Int. Conference on Microreaction Technology (IMRET 4), 5-9 March 2000,
        Atlanta, GA, USA, 314-321

[6]   Bier, W., Keller, W., Linder, G., Seidel, D., Schubert, K.,
        *Manufacturing and testing of compact micro heat exchangers with high
        volumetric heat transfer coefficients*,
        ASME Symposium, Vol. DSC-19 (New York 1990), 189-197

[7]   Bier, W., Keller, W., Linder, G., Seidel,D., Schubert, K., Martin, H.,
        Gas to gas heat transfer in micro heat exchangers,
        Chemical Engineering and Processing, 32 (1993), 33-43

[8]   Sander, U., Rothe, U., Kola, R.,
        Schwefelsäure, Ullmanns Encyklopädie der technischen Chemie
        4., neubearbeitete und erweiterte Auflage, Band 21, S. 117 ff., Verlag Chemie,
        Weinheim (1982)

Part 5

Application of Microdevices for Production, Energy and Transportation Systems

Sub-watt Power Using an Integrated
Fuel Processor and Fuel Cell

Evan Jones, Jamie Holladay, Steve Perry, Rick Orth, Bob Rozmiarek,
John Hu, Max Phelps, Consuelo Guzman

Battelle
Richland, Washington, U.S.A.

Abstract

A sub-watt power system is being developed as an alternative to conventional
battery technology to better meet energy and power densities needed for operating
wireless electronic devices, such as microsensors and microelectromechanical
systems. This system integrates a microscale fuel processor, which produces a
hydrogen-rich stream from liquid fuels, such as methanol and butane, and a
microscale fuel cell, which uses the hydrogen as fuel to produce electric power.
Battelle, Pacific Northwest Division and Case Western Reserve University are
developing and demonstrating this technology for the Defense Advanced
Research Projects Agency. This paper describes work being performed by
Battelle on the fuel processor, in particular, catalyst and reactor design and testing.

The microscale fuel processor (integrated vaporizer/steam reformer/combustor)
assembled, fabricated, and tested during this study generated an equivalent power
level of 10 to 500 mW_e. This steam reformer test system has a reactor volume of
less than 5 mm^3. Catalyst testing achieved a near-maximum theoretical
conversion for methanol with <1% CO in product H_2 gas. High conversion and
H_2 selectivity was also achieved during catalyst testing with butane, but at higher
temperatures.

Introduction

The availability of onboard power, coupled with wireless data transmission, will
open numerous possibilities for autonomous devices for remote or difficult-to-
access locations. Current battery systems have two problems: excessive
weight/bulk, and reduced mission duration. Compact fuel cell systems operating
on liquid hydrocarbon fuels offer an efficient, light-weight alternative to batteries,
allowing for greater portability and longer mission lifetime. For instance, the
energy densities of diesel fuel and methanol are each at least an order of
magnitude greater than that of lithium-ion batteries. A hydrocarbon-based fuel

cell system operating at just 5% overall efficiency has a higher energy density than a lithium polymer battery

Under a program for the Defense Advanced Research Projects Agency, Battelle, Pacific Northwest Division (Battelle) and Case Western Reserve University (CWRU) are developing and demonstrating an integrated fuel cell and fuel processor for microscale (50- to 500-mW_e) power generation. This alternative power source has many potential advantages over conventional batteries for operating wireless electronic devices (e.g., microsensors and microelectromechanical systems), especially in terms of energy and power densities.

The technology consists of a microscale reformer for hydrocarbon fuels, based on a concept developed at Battelle, coupled to an elevated-temperature (150°C to 200°C) proton exchange membrane (PEM) fuel cell developed by CWRU. This paper describes the work being performed by Battelle on the fuel processor, in particular, catalyst and reactor design and testing.

The overall power system will eventually include several subsystems: fuel storage, fuel processor, synthesis gas treatment (optional), and fuel cell, along with associated peripherals such as pumps and control valves. The fuel processor contains a reforming reactor, combustion reactor, and heat exchangers. The fuel from storage is mixed with air and water in a fuel processor system that operates at 600°C to 700°C for butane and 250°C to 550°C for methanol. Heat generated in the combustion reactor is transferred to the endothermic reforming reactor to produce the H_2-containing synthesis gas. The synthesis gas may be processed directly in the fuel cell or treated to minimize the CO concentration, depending on the type of fuel cell. The fuel cell converts H_2 and O_2 (from air) to electrical power and water.

Experimental Results and Discussion

The testing described below was performed in different apparatuses. The catalyst testing was performed in dedicated catalyst test equipment while the micro-reactors were tested in a separate dedicated system. Since the catalyst performance and requirements are critical to final reactor design, the catalyst testing must be done first.

Reformer Catalyst Testing Results - Butane

A novel catalyst was used to examine reforming butane to hydrogen and carbon monoxide at 10 to 35 ms contact times in the temperature range of 500°C to 900°C. Figure 1 shows butane conversion (top) and hydrogen and methane selectivity (bottom) as a function of temperature and contact time. The data in

Figure 1 show that 100% butane conversion is achieved at approximately 600°C and at contact times of 25 and 35 mS. For a contact time of 10 mS, 100% butane conversion is not achieved until an operating temperature of over 750°C is reached. Longer contact times and higher operating temperatures also favor hydrogen selectivity. However, only at the highest temperatures does this selectivity approach that predicted by equilibrium.

The data show that approximately 10.5 moles of hydrogen can be produced per mole of butane reacted at the higher temperatures and longer contact times. In addition, at the longer contact times and higher temperatures, the hydrogen yield approaches the yield predicted at equilibrium. The catalyst volume of the reformer reactor operating at 100 ms contact time is still less than 5 mm^3.

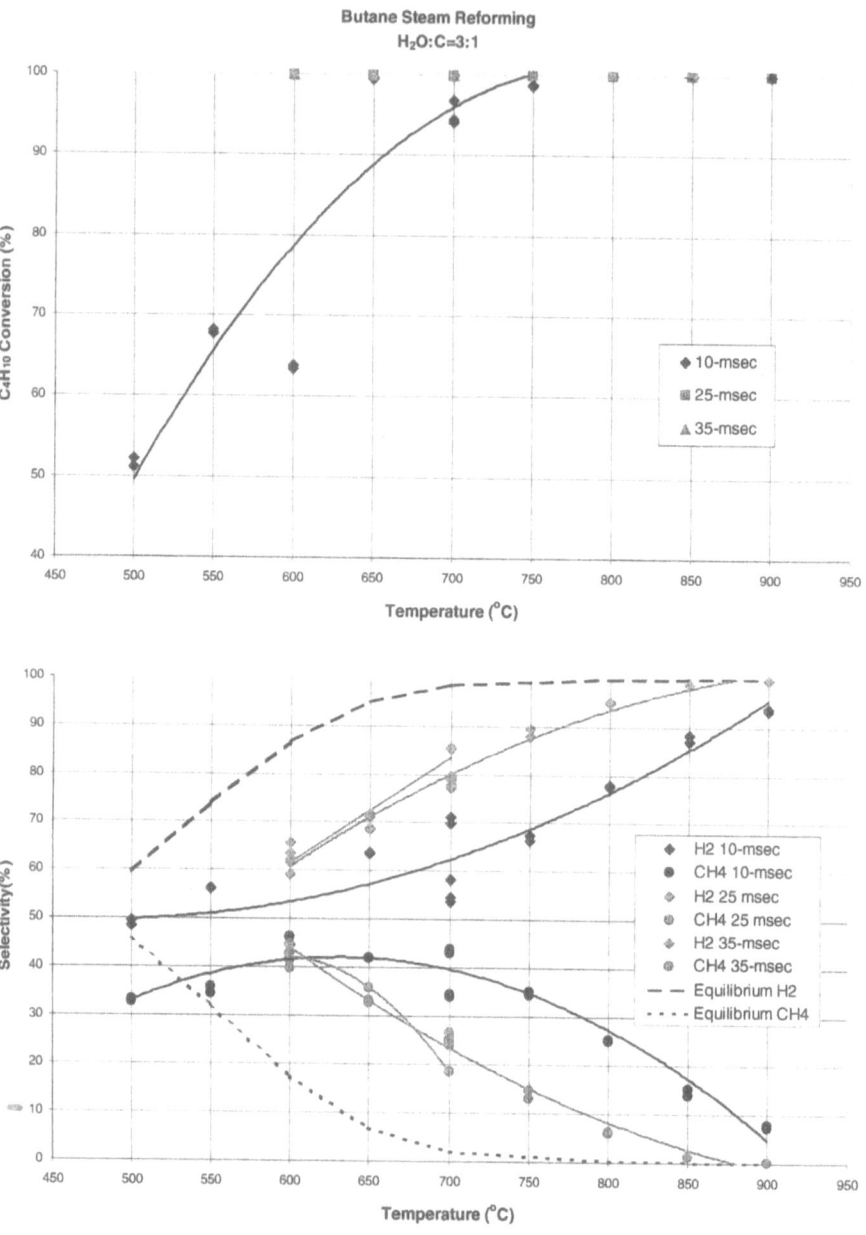

Figure 1. Butane Steam Reforming Catalyst Testing (3:1 Water: Carbon Molar Ratio). Butane conversion (top); selectivity (bottom).

Reformer Catalyst Testing Results - Methanol

Three types of catalysts were tested, labeled A, B, and C. Catalysts B and C were tested initially and performed well, but lower operating temperatures and lower carbon monoxide selectivity was desired. Consequently, a novel catalyst (labeled A) was tested at 300 msec contact times in the temperature range of 240°C to 400°C.

All testing reported here was conducted using a water:methanol molar ratio of approximately 1.8:1 in the incoming feed. Figure 2 shows methanol yield (left axis) and carbon monoxide yield (right axis) as a function of temperature. All three different catalysts were evaluated at these conditions. The data show the dramatic improvement in both methanol conversion and carbon monoxide content in the off-gas when going from catalyst C to catalyst A. For catalyst C, a methanol conversion of less than 80% was achieved at nearly 400°C, while nearly 100% methanol conversion was achieved at 360°C using catalyst B, and nearly 90% methanol conversion was achieved at temperatures as low as 280°C using catalyst A. For both catalyst B and catalyst C, the carbon monoxide concentrations in the off-gas were high, ranging from approximately 10 mol% to over 27 mol%. For catalyst A, on the other hand, the off-gas carbon monoxide concentrations were extremely low, less than 0.5 mol% for all of the conditions tested.

The hydrogen yield is shown in Figure 3 for these same catalysts. The data show that catalyst A is superior to the other two catalysts from this perspective as well. Hydrogen yields of nearly 3 moles hydrogen per mole methanol converted are achieved for catalyst A. This is very close to the theoretical yield based on the reaction given below:

$$CH_3OH + H_2O \longrightarrow CO_2 + 3H_2$$

The data are plotted in terms of equivalent electrical energy produced per methanol converted ("energy density") in Figure 4. In this figure, a fuel cell efficiency of 64% is assumed. The solid line shown in Figure 5 is the theoretical "energy density" assuming reaction (1) and a fuel cell efficiency of 64%. As can be seen, the performance of catalyst A comes very close to the theoretical "energy density."

This testing showed that at an operating temperature of approximately 280°C, catalyst A can convert over 90% methanol to yield nearly 3 moles of hydrogen/mole methanol converted. The concentration of carbon monoxide in the off-gas is less than 0.5 mol%. Testing is continuing develop catalyst with a lower operating temperature.

282

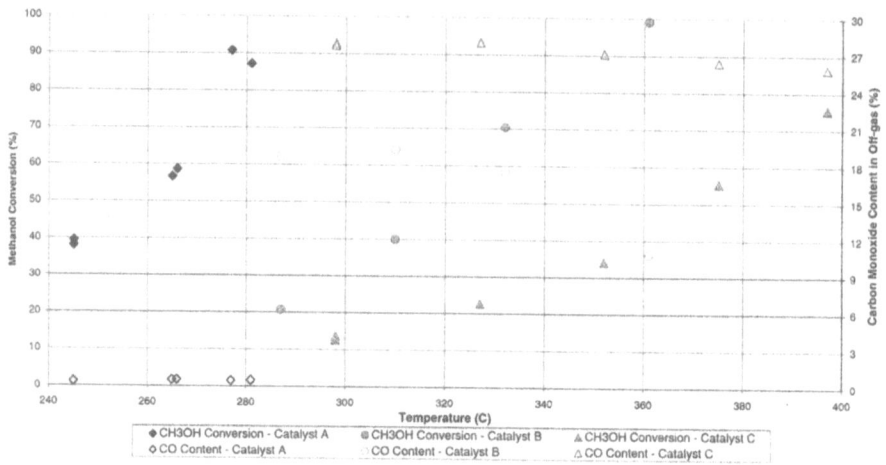

Figure 2. Methanol Conversion and Carbon Monoxide Content. Steam reforming test conditions: 1.8:1 water:methanol (mol:mol) feed, 300 msec contact time, three different catalysts.

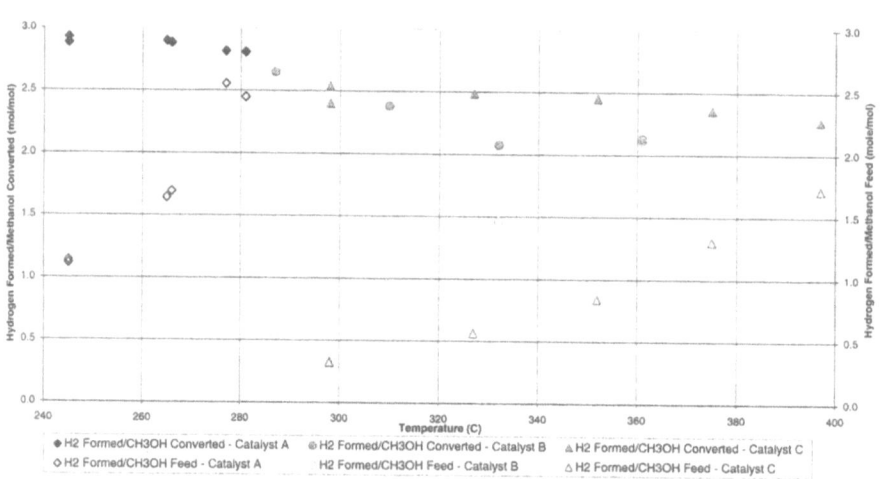

Figure 3. Hydrogen yield. Steam reforming test conditions: 1.8:1 water:methanol (mol:mol) feed, 300 msec contact time, three different catalysts.

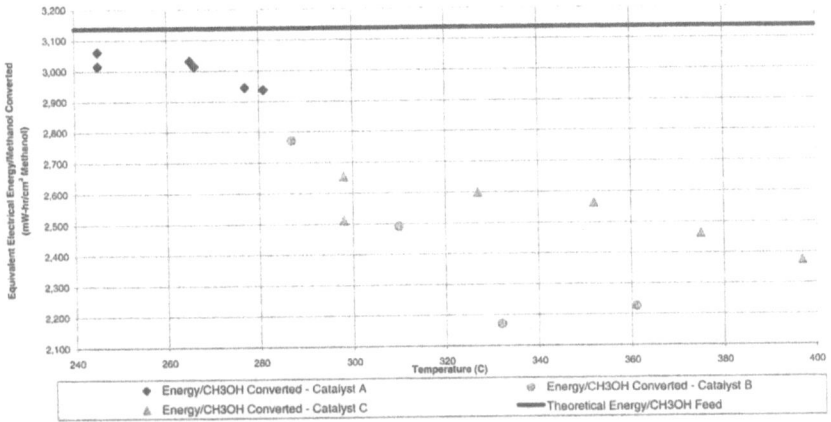

Figure 4. Equivalent Electrical Energy Production (64% fuel cell efficiency, methanol density = 0.79 g/ml assumed). Steam reforming test conditions: 1.8:1 water:methanol (mol:mol) feed, 300 msec contact time, three different catalysts.

Assembly and Fabrication of Steam Reformer/Combustion Test System

The 50- to 500-mW$_e$ integrated vaporizer/steam reformer/combustor test system is illustrated in Figures 5 and 6. Syringe pumps are used to supply the methanol/water mix (1.8:1 water:carbon molar ratio) to the steam reformer and pure methanol to the combustor. The methanol/water flow rates range from approximately 0.03 to 0.1 ml/hr (20°C basis). The mixed water and fuel then enter a vaporizer section in the integrated device and pass through the catalyst bed through a vapor/liquid separator section and exit to an online gas chromatograph for analyses. The steam reformer catalyst reactor volume is less than 5 mm^3.

Before the methanol was fed to the combustor, hydrogen (in the presence of air) was used as the fuel source to initiate the combustion. The hydrogen was fed to the reactor via a mass controller. The air was fed at flow rates ranging from approximately 8 to 14 sccm; the hydrogen was fed at rates from 1 to 5 sccm, and the methanol was fed at rates from approximately 0.05 to 0.2 ml/hr (20°C basis).

The experimental fuel processor below contains two reactors and a heat exchanger. The reformer has a volume of less than 5 mm^3 with a capacity of 200 mW. The combustor volume has a volume less than 1 mm^3 and a capacity of 3 W. The combustor capacity is larger than necessary so that a full range of conditions can be studied.

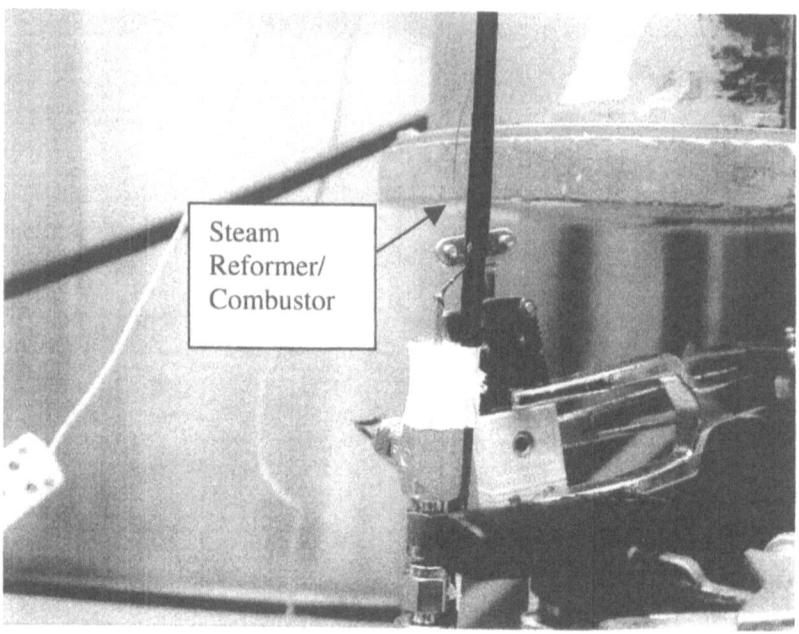

Steam
Reformer/
Combustor

Figure 6.   Above - Integrated Vaporizer/Steam
Reformer/Combustor Test System (50- to 500 mW$_e$). The integrated reactor is
mounted in a tube.

Below - Experimental Fuel Processor

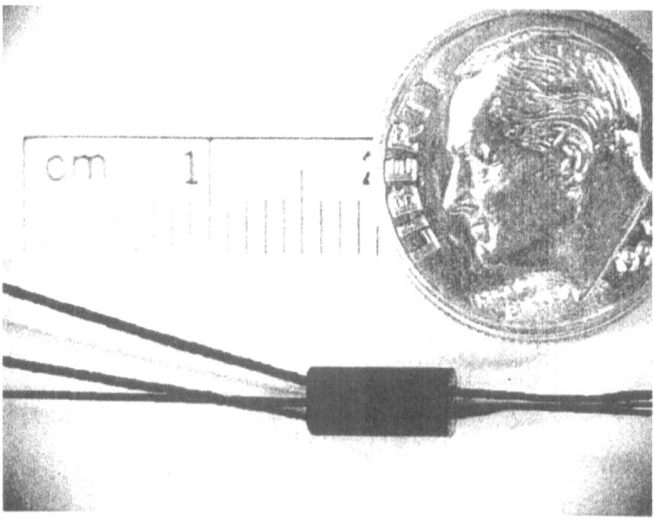

Summary and Conclusions

Catalyst test data from butane reforming show that 100% conversion is achieved at approximately 600°C and contact times of 25 and 35 mS. For a contact time of 10 mS, 100% butane conversion is not achieved until an operating temperature of over 750°C is reached. Longer contact times and higher operating temperatures also favor hydrogen selectivity. However, only at the highest temperatures does this selectivity approach that predicted by equilibrium. The catalyst volume of the reformer reactor operating at 100 ms contact time is 1 to 5 mm^3. Hydrogen yield data show that approximately 10.5 moles of hydrogen can be produced per mole of butane reacted at the higher temperatures and longer contact times.

Catalyst data from methanol reforming show that 100% conversion is possible with high H_2 selectivity and low CO concentration.

A 10- to 500-mW$_e$ steam reformer test system has been fabricated and assembled with a reactor volume of less than 5 mm^3. Shakedown and preliminary testing of the steam reformer reactor has been completed.

Catalytic Micro-Reactor Systems for Hydrogen Generation

Patricia M. Irving, W. Lloyd Allen, Todd Healey, Quentin Ming
InnovaTek, Inc. 350 Hills Street, Richland, WA 99352 USA
and
William J. Thomson
Department of Chemical Engineering, Washington State University
Pullman, WA 99164

Abstract

InnovaTek is developing a novel portable-sized fuel cell hydrogen generator by combining micro reactor technology with advanced catalysts and separations technology. Although fuel cells have been around for years, it wasn't until recently that advances, such as those made in micro-reactor technology, promise to make fuel cells economical and reliable enough for use in commercial applications.

The ultimate goal is the development of a micro channel catalytic reactor that produces pure hydrogen through catalytic reforming of methanol, diesel fuel and natural gas. Advanced membrane technology is incorporated to remove carbon dioxide, water and detrimental co-products from the reformate stream. Our technology provides a pure output stream of hydrogen that can be used in a compatibly sized PEM fuel cell for electrical generation.

A thermal and process system model that was developed as a system simulator was used to optimize the design of our micro channel reactor. With iterative testing and further refinement, the model will be used to provide a sound basis for improved reactor and process engineering.

On-going tests indicate that hydrogen production is maximized and CO production is minimized by proper selection of 1) temperature-dependent reaction equilibria, 2) ratio of fuel to steam, and 3) catalyst activity. The use of micro-reactor and micro-heat exchanger components help optimize these processes. Additional results include catalyst testing with sulfur present in the fuel and H-separation membrane development and testing, On going work includes mini-plasmatron development, and micro-fuel injection system development.

Introduction

Fuel Cells are electrochemical devices that convert hydrogen directly to electricity. Although fuel cells have been around for years, it wasn't until recently that advances, such as those made in micro-reactor technology, promise to make fuel cells economical and reliable enough for use in commercial applications.

The InnovaTek H2GEN™ hydrogen production system, currently under development, uses advanced technology to provide a pure hydrogen stream to fuel cells for stationary, portable, and mobile applications. The technology conveniently uses today's standard fuels such as natural gas, gasoline or diesel to generate hydrogen for clean on-site electrical power production.

Our portable power reformer (Fig. 1) can generate enough hydrogen for a 100 Watt fuel cell. Our development plans include system to generate hydrogen for equivalent sub-kW, kW, and multi-kW power production devices. We report here on our use of micro technology in the components of the H2GEN™ system and results from testing.

**Figure 1. Scale Model of the Components of InnovaTek's Proprietary
H2GEN™ Diesel Fuel Reformer**

System Components

InnovaTek's fuel processor is based on catalytic steam-reforming coupled with hydrogen separation membrane technology, and incorporates various proprietary and licensed components. The technology can reform gasoline, diesel, methanol and natural gas. We are developing and integrating the following critical enabling technologies into components (Fig. 2) that create a system offering significant advantages over traditional reactors.

- Sulfur-tolerant reforming catalyst that eliminates the requirement for extra components for sulfur removal
- Sulfur-tolerant H-separation membrane that yields 100% hydrogen product (no CO, H_2S or CO_2 to poison fuel cell or dilute hydrogen) producing higher fuel cell current densities
- Fuel Injector Micro-Nozzle that eliminates catalyst coking
- Micro-channel reactor and heat exchanger for compact high-efficiency system design
- Plasmatron for fast start-up and catalyst regeneration.

A thermal and process system model that was developed as a system simulator was used to optimize the design of our micro channel reactor. With iterative testing and further refinement, the model will be used to provide a sound basis for improved reactor and process engineering.

The process starts with water, air, and fuel, which are injected into two subcomponents – the burner unit and the vaporizor/fuel injector unit. The burner unit combusts the membrane reject gases to convert the water to steam and create enough heat for fuel vaporization and reforming processes.

Micro-channel heat exchangers transfer the energy to the catalytic micro-channels of the reformer where the vaporized fuel and steam are injected. The catalytic reaction occurs at about 800° C producing reformate that consists primarily of hydrogen (H_2), carbon monoxide (CO), and carbon dioxide (CO_2). Small amounts of hydrogen sulfide are produced from fuels with sulfur content.

The reformate is cooled through the use of microchannel heat exchangers and water is condensed and recycled. The dry reformate is heated to 450° C and then purified by the membrane component. Only hydrogen can pass through the membrane thereby producing a stream of pure hydrogen that is delivered to the fuel cell after additional cooling. The gas that does not pass through the membrane, known as the "reject stream" is sent back to the burner where the cycle continues.

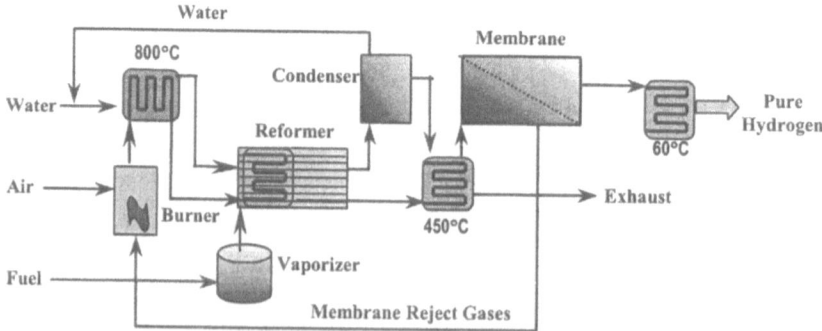

Figure 2. Process Flow Diagram For Hydrogen Generation System

Experimental Results

Catalyst Testing

The performance of InnovaTek's catalyst ITC-3 for reforming commercial-grade "regular" gasoline was evaluated using a tube reactor. Gasoline normally contains some sulfur compounds in the concentration range between 50 and 300 ppm, The results (Figure 3) indicate that the catalyst maintains its activity with no deactivation for 50 hours at which time the experiment was terminated.

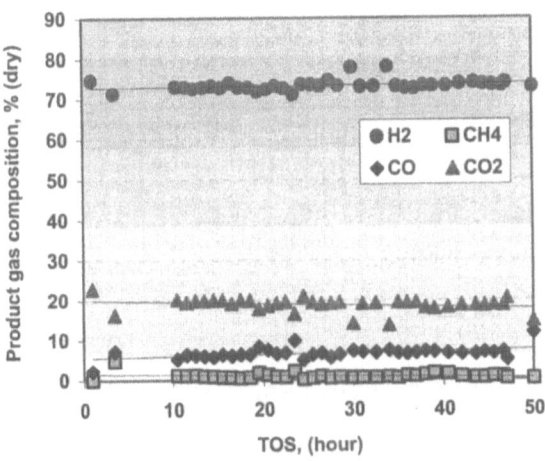

Figure 3. The product gas composition vs. the reforming time using InnovaTek catalyst ITC-3 for commercial gasoline; feed rate was about 0.1 g/minute; the ratio of steam:C varied from 5 to 8 and the temperature was 800°C.

Tests were also conducted using iso-octane feed containing 1000-ppm sulfur (Figure 4). The catalyst, ITC-3, has maintained its high activity and hydrogen selectivity for over 100 hours of testing; the reactor is still operating and data continues to be collected. The hydrogen concentration was maintained at about 70% during the testing period and no deactivation was observed. We believe the slight decrease (from 75% to 70%) after the first 30 hours was the result of some initial coking in our reactor that reduced the volume of active catalyst. The presence of H_2S in the reformate was detected (by lead acetate paper) shortly after the reaction started indicating that sulfur was reduced (and not absorbed by the catalyst bed, which would deactivate it).

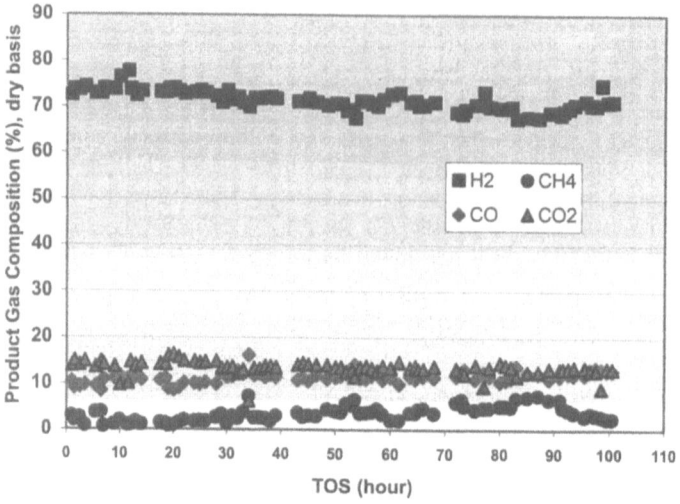

Figure 4. Product gas composition vs reforming time using InnovaTek catalyst ITC-3 for iso-octane with 1000 ppm sulfur; feed rate of iso-octane was about 0.1 gram/minute; steam/C ratio was about 4; and temperature was 800°C.

Microchannel Reactor Design and Fabrication

A micro-channel reactor was designed and fabricated from stainless steel and ceramic. The device consists of four layers performing separate functions: heat source (burner), fuel mixing, heat exchange, and catalytic reforming (Figure 5). The burner plate serves as the heat source for the reactor and the preheater for the fuel and water. The combustion of the fuel and air in the burner generates heat, a portion of which is transferred to the other plates by conductive heat

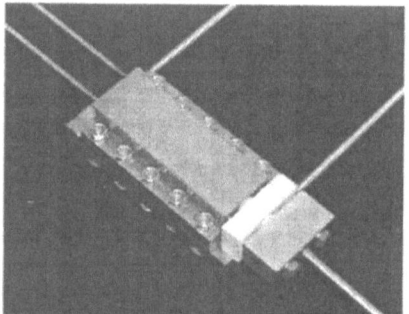

Figure 5. Integrated micro channel fuel reformer and burner 6"x 2.5"x 2".

transfer. Another portion of the heat is carried by the exhaust through micro channels generating convective heat transfer.

Both mixing and reactor plates (Fig. 6) have micro channels on top and bottom. This provides advantages in reducing mass and blocking unnecessary heat transfer to other regions. The mixing plate sits directly on top of the burner and the reactor plate is separated from the mixing channel by a thin stainless steel foil and graphite sheet. The top side of the reactor plate is enclosed by the cover plate. The plates and burner are fastened by bolts that prevent leakage but are removable for inspection of components or to install new catalyst.

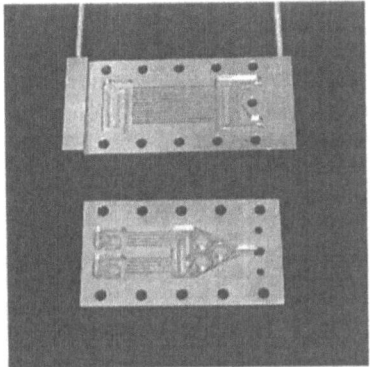

Tests were conducted with the catalyst packed into the micro-channel reactor that had heat supplied to it by an integrated micro-burner. The burner supplied heat, steam and vaporized fuel to the micro-channel reactor. A more complex fuel mixture consisting of isoctane, toluene, dodecane, and about 500 ppm sulfur was used to simulate diesel fuel. Our results for steam reforming indicate that the catalytic micro-reactor produced greater than 70% hydrogen at a constant level for 65 hours (Figure 7).

Figure 6. Catalytic reactor (top) and fuel mixer components (bottom).

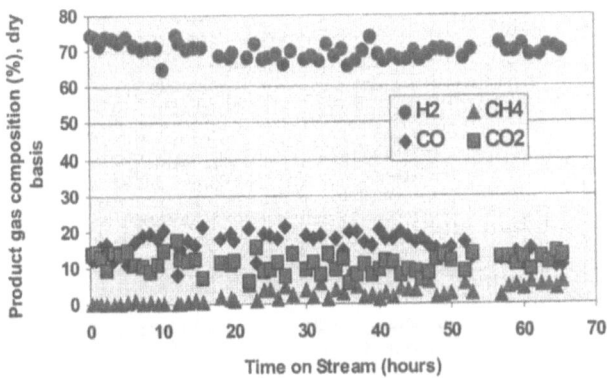

Figure 7. Steam reforming using InnovaTek Proprietary Catalyst ITC-2 in micro-channel reactor with integrated burner. Catalyst Amount: 3.75g; LHSV: 27hr $^{-1}$; Feed: 0.3g min $^{-1}$; Temp: 800°C; Steam/C: 4; Fuel Composition: 60% iso-octane, 20% toluene, 20% dodecane, and 476 ppm S.

Heat Exchange

Counter-flow micro-channel heat exchangers made of 316 SS (Fig. 8) were tested to determine efficiency and effectiveness. Heat exchanger size for a gas flow rate up to 9 LPM is approximately 12.3 x 1.4 x 0.9 cm. Pressure drop at 5 LPM was 0.6 psi. The core volume of the device is approximately 12 cm^3.

Results indicate that at 400° C heat exchange efficiency was greater than 80% and decreased to about 50% as flow rates were reduced to 2 LPM (Fig. 9). Room temperature (25° C) air was used for the counter-current side for these tests. Micro-channel heat exchangers will be used to maintain optimum temperature conditions for each stage of our fuel processing system (see Fig. 2). Tight temperature control is essential to maintaining maximum chemical conversion and thermal efficiencies in the system.

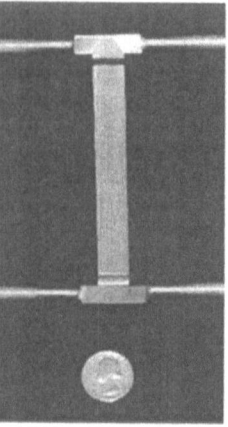

Figure 8. Micro-channel heat exchanger.

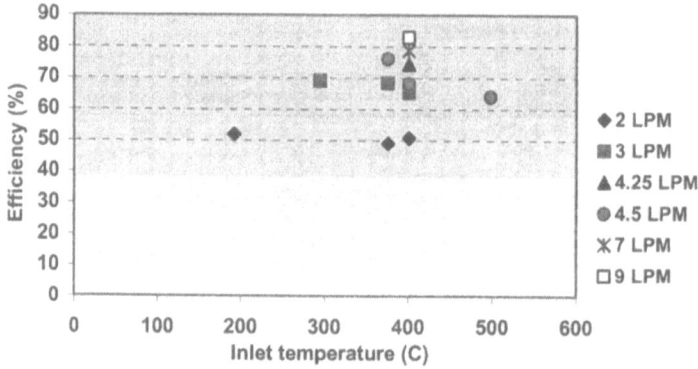

Figure 9. Heat Exchanger Efficiency as a Function of Inlet Temperature (hot side) and Flow Rate.

Hydrogen Separation

An apparatus to test the hydrogen membrane was constructed to measure performance under various temperature and pressure conditions. For simplicity a compressed cylinder of gas containing 65% H_2, 20% CO_2 and 15% CO (simulating our reformate composition) was used as the feed gas for separation tests. The membrane was fabricated on the inner surface of a support structure with 7 mm ID and 22 cm in length, with an effective surface of about 53 cm^2 and a membrane thickness of about 10 μm. Membrane development is continuing with the goal of further reducing membrane thickness and incorporating a composition that is sulfur tolerant.

Tests were conducted by changing the pressure (P_2) on the feed side of the membrane, while the pressure (P_1) at permeation side was at atmospheric pressure. As the pressure P_2 increases, the hydrogen permeation rate increases (Figure 10). Results are shown for varying pressures and temperatures.

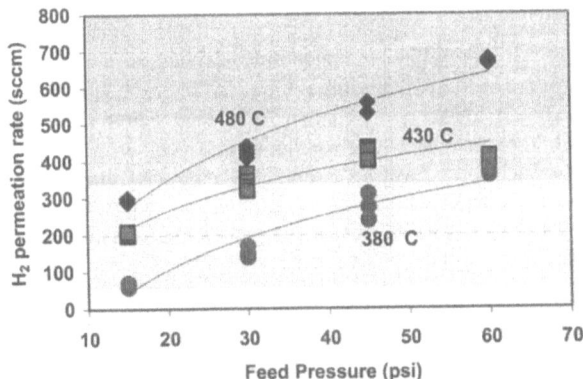

Figure 10. The hydrogen permeation rate vs. pressure of membrane feed gas at different testing temperatures; the pressure at permeate side (pure hydrogen) is atmospheric; the composition of feed gas is 65% H_2, 20% CO_2, and 15% CO.

Composition of the membrane permeate stream (which is the system output) is pure hydrogen with >80% recovery at a temperature of 450° C and pressure of 60 psi (Table 1). The reject gas stream is recycled to the system burner to vaporize fuel and water for the reformer and achieve the temperatures needed for catalytic reforming. This approach creates a very efficient system with little thermal and chemical losses.

Table 1. Composition and Flow Rate in Membrane Stream at 60 psi and 450 °C

Reformate Compound	% Composition Membrane Component		
	Feed	Reject	Permeate
Hydrogen	65	25	100
Carbon Dioxide	20	43	0
Carbon Monoxide	15	32	0
Hydrogen Recovery			82

Continuing Development

Further research is being conducted to develop additional micro components to optimize system design. These include micro-nozzles for fuel injection to reduce coking in diesel fuel systems and a micro-plasmatron for fast start-up and possible catalyst regeneration.

Summary

We have successfully engineered novel micro-technologies in developing a fuel processor that offers the following competitive advantages:

- Reforms multiple fuel types without the need for prior sulfur removal. This greatly expands the market potential to include 1) military logistical fuel and 2) those areas of the world with high sulfur content fuels.
- Produces nearly pure hydrogen output thereby enabling higher fuel cell voltages and power densities with no potential for electrode poisoning.
- Utilizes a steam reforming process that yields higher hydrogen product per volume of fuel consumed.
- Incorporates micro technology for reactor, heat exchanger, and fuel vaporizer components to improve system efficiency through optimized thermal management and fluid dynamics.

Acknowledgements

This work was funded by the U.S. Department of Energy Hydrogen Program under Contract DE-FC36-99GO, and the U. S. Army Communications-Electronics Command under Contract DAAD05-99-D-7014 DO 0007.

Palladium based Micro-Membrane for Water Gas Shift Reaction and Hydrogen Gas Separation

Sooraj V. Karnik[*], Miltiadis K. Hatalis[*], Mayuresh V. Kothare[**, 1]

[*]Department of Electrical and Computer Engineering, [**]Department of Chemical Engineering, Lehigh University, Bethlehem, PA 18015, USA.

Abstract

A novel palladium-based micromembrane has been fabricated which has the potential to be used for water gas shift reaction and hydrogen gas separation in catalytic microreaction systems for methanol fuel reforming. The membrane is supported by a copper film that has an array of patterned holes. Copper will also act as a catalyst in the shift reaction that will convert unwanted carbon monoxide gas produced earlier in the microreaction system during methanol fuel reforming into hydrogen which in turn will be separated by the membrane. The novelty of this structure is that we have integrated the water gas shift reactor as well as the hydrogen gas separator in the same structure. The microfabrication process allows for integration of heaters and temperature sensors into the device.

Introduction

The field of building miniature fuel processors for *in situ* hydrogen production is of particular interest in compact fuel cells, which are currently being considered as alternative energy sources for high-end portable devices such as cellular phones and laptop computers [1, 2]. Fuel cells, in general, transform the chemical energy of the reactants directly into a DC current. The basic chemical reaction is an oxidation-reduction reaction to form water molecules from their basic ingredients, namely oxygen and hydrogen in the presence of an electrolyte. Laptop computers, cellular phones and other portable applications require clean, efficient and most importantly, light and rechargeable sources of power. Besides the need for such miniature power sources, advances in large-scale fuel cells as well as microfabrication technology imply that such compact fuel cells can now be developed.

The fuel hydrogen is not conveniently available to recharge the compact fuel cells and its storage as a pressurized gas has also been a problem. Although it is intuitive to think of an on-line hydrogen generating fuel processor, its implementation at a micro scale remains conceptually difficult. This miniature fuel processor should produce pure-grade hydrogen from a hydrocarbon fuel via a catalytic chemical reaction and then provide it to the fuel cell. Due to process, design and environmental requirements, the miniature fuel processor must be

296

capable of separating hydrogen from other components while maintaining a low carbon monoxide concentration.

We propose to explore the use of a micro-reactor to produce hydrogen by the reaction of methanol (CH₃OH) with water. The schematic of our proposed fuel processor is shown in Figure 1. It will have the following four main components: (a) mixer/vaporizer of methanol and water; (b) catalytic steam reformer with copper catalyst where gaseous methanol and water react at 250 °C to produce carbon dioxide, carbon monoxide and hydrogen; (c) water gas shift reactor consisting of a perforated copper catalyst, where unwanted carbon monoxide produced earlier in the steam reformer reacts with water vapor in presence of copper catalyst at 280 °C and gets converted into hydrogen and a palladium micro-membrane for hydrogen separation; (d) integrated resistive heaters/sensors and control electronics. The fuel cell application dictates that the carbon monoxide level in the stream should be below 10 ppm for Polymer Electrolyte Membrane (PEM) fuel cells [3] since carbon monoxide is harmful to the fuel cell membrane catalyst. Hence, part (c), namely, the water gas shift reactor is necessary. Since reversible reactions are equilibrium limited, hydrogen removal promotes conversion of carbon monoxide in the shift reaction. This is accomplished by using a palladium membrane due to its excellent permselectivity to hydrogen [4].

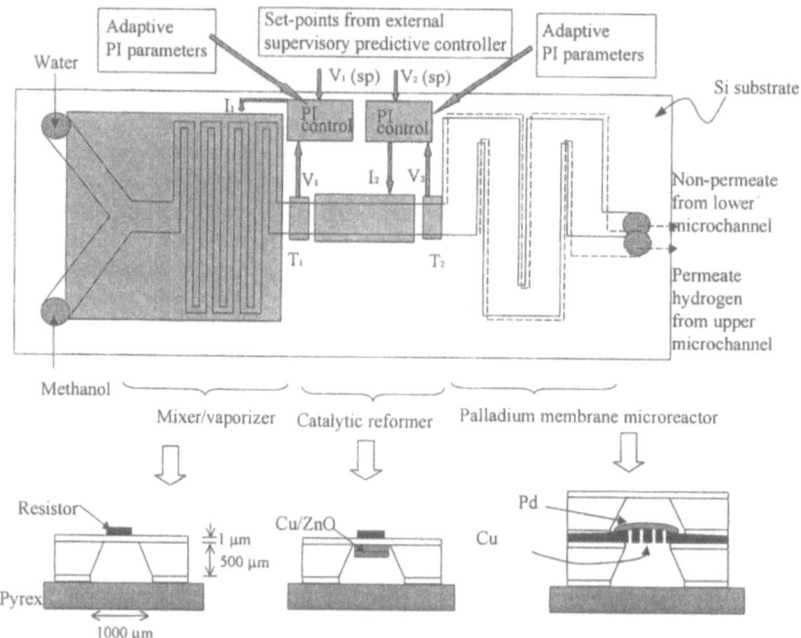

Figure 1: Schematic of the micro-reactor

This paper will focus on the design, fabrication and performance evaluation of a palladium based micro-membrane reactor for hydrogen separation as well as shift reaction [parts (c) and (d)] in a miniature fuel processor. Recently designs of micro-membranes based on palladium have emerged for selective separation of hydrogen [5, 6, 7]. The novelty of the structure proposed in this work is that it integrates hydrogen separation as well as water gas shift reaction in the same structure.

Fabrication

Integrated, palladium-based micromembranes have been successfully fabricated and tested for their mechanical strength. A schematic of the micromembrane is shown in Figure 2. The membrane is a composite of four layers: copper, aluminum, spin-on-glass (SOG) and palladium. Copper, aluminum and SOG layers have a pattern of holes etched into them, so as to have perforations. They serve as a structural support for the main element of the membrane, the palladium film. Copper can also act as a catalyst in the water gas shift reaction taking place in the same channel. The gases produced in part (b) of the micro fuel processor during the steam reforming reaction such as carbon monoxide, carbon dioxide and hydrogen with some remains of vapors of water and methanol will flow in the channel etched in the silicon substrate. Carbon monoxide at a temperature of 250 °C in presence of copper catalyst will get converted into more hydrogen gas that in turn gets selectively separated through the palladium membrane, thus shifting the reaction to product side. Separated hydrogen will pass through a channel etched in another silicon wafer bonded to the original silicon wafer on top of the palladium membrane.

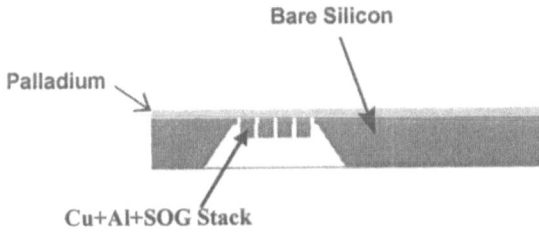

Figure 2: Pd-based shift reactor-separator

The micromembrane fabrication process steps are shown in Figure 3. We start with a 100 mm silicon wafer polished on both the sides. We deposit silicon nitride by plasma enhanced chemical vapor deposition (PECVD) method on both the sides of the wafer. We, then, sputter deposit 66 nm of copper followed by 50 nm of aluminum on the front side of the wafer. The aluminum film is added to provide mechanical strength to the free standing micromembrane. A pattern of holes is wet etched in this composite copper-aluminum film by using standard photolithography. We have investigated two different hole patterns. The density of hole pattern in one design is twice that in the other design, as shown in Figure 4. This is to see whether doubling the density of holes, that in turn doubles the area through which hydrogen gas can permeate, increases the rate of hydrogen separation and also how it affects the overall strength of the membrane. The front wafer surface is then planarised by spinning 400 nm of SOG. The SOG is cured at 250 °C for 30 minutes. A thin layer of aluminum followed by 50 nm of palladium are then sputter deposited on the front. The thin aluminum layer acts as an adhesion layer to palladium.

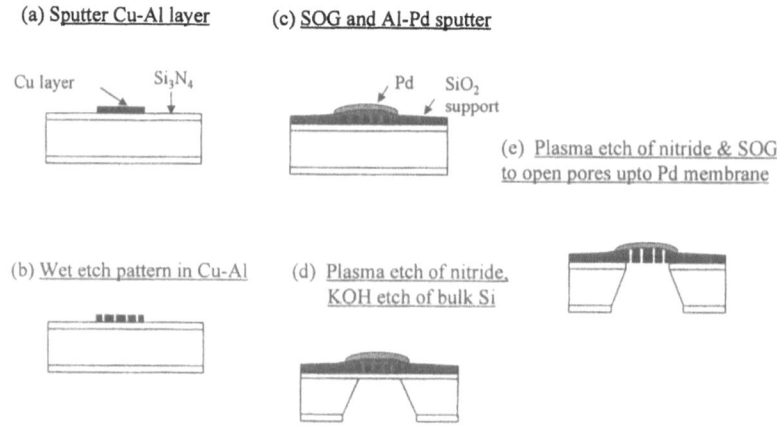

Figure 3: Fabrication steps

Photolithography is done on the back side of the wafer to obtain the channel pattern. Our process does not require expensive back side alignment equipment. Silicon nitride on the back side is then etched by plasma etch. This nitride acts as an etch stop for the KOH etching of silicon. Bulk silicon is etched in KOH from the back side until the silicon nitride layer on the front side is reached. The layers on the front side of the wafer are protected during KOH etching by a special arrangement. The palladium membrane is then released to the channel side by plasma etching the nitride and the SOG.

The microfabrication process described above enables use of arbitrarily thick or thin palladium films as a membrane. Thus, ultrathin films of Pd can be deposited, if their need arises in future. It also allows integration of resistive heaters and sensors. Figure 5 shows the design of heater and sensor lines with channel. Natural convection heat loss and power requirement were estimated for the resistor design. Resistive heaters are necessary to raise the temperature of the device to 250 °C in order for the water gas shift reaction to occur as well as hydrogen gas to permeate.

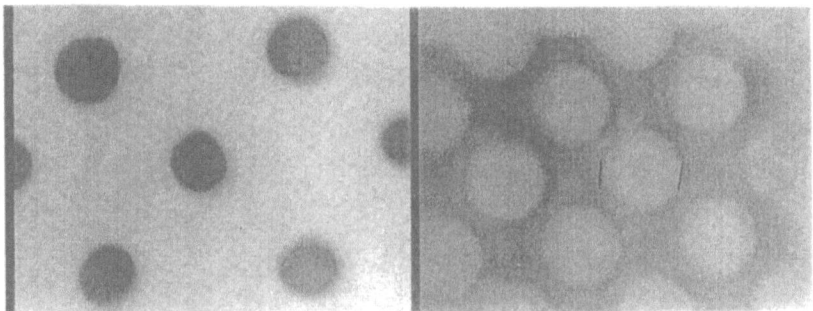

Figure 4: Two designs of holes etched in copper film

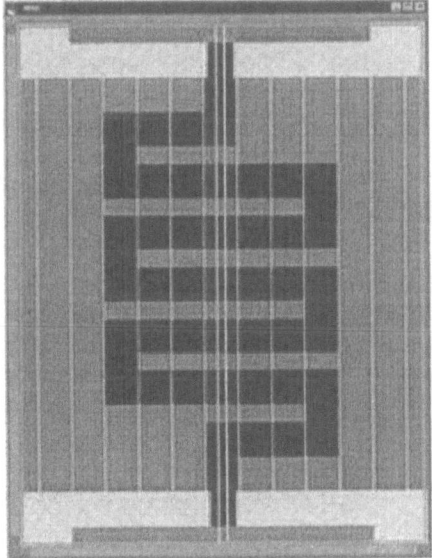

Figure 5: Design of heater and sensor lines with channel.

300

Fabricated Device & Characterization

A bottom view of the micromembrane device is shown in Figure 6. It shows the composite membrane with perforations in Cu-Al-SOG layers and palladium film on the top. The slanted walls of channel in silicon wafer are also visible. Figure 7 shows an SEM image of the free standing palladium membrane inside perforations in copper, aluminum and SOG layers.

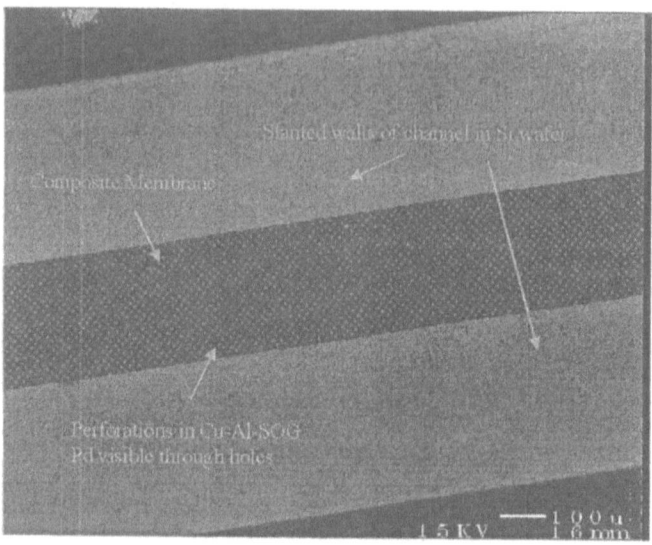

Figure 6: Bottom view of the Pd membrane device in a channel etched in Si

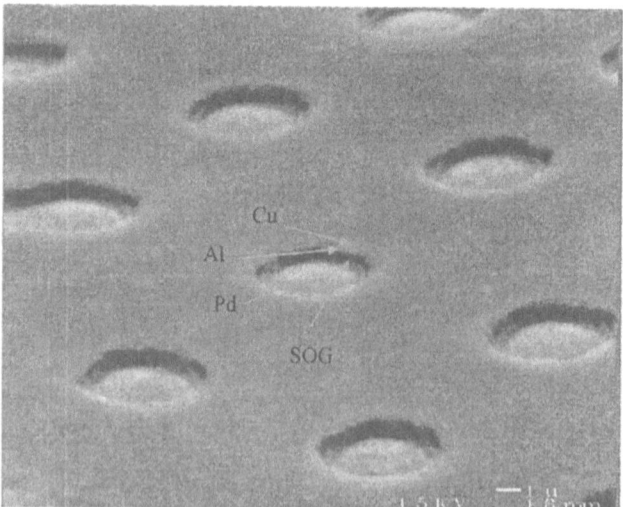

Figure 7: SEM image of free standing Pd inside perforations in Cu, Al and SOG layers

The mechanical strength of the microfabricated free standing membrane was tested. The membrane strength is very important, since passage of hydrogen through the membrane is driven by a pressure gradient across the membrane. The strength of the membrane at room temperature was measured, with the membrane pressurized from the silicon channel side and the front side of the membrane kept at atmospheric pressure. A micromembrane with 5.5 µm diameter perforations in Cu-Al-SOG stack, withstood a pressure gradient of 10 psi without breaking.

Conclusions

The overall goal of our study is to explore the use of a micro-reactor to produce hydrogen by the reaction of methanol with water. This paper concentrated on the design, fabrication and performance evaluation of a palladium based micro-membrane reactor for hydrogen gas separation as well as shift reaction in this miniature fuel processor. Integrated, palladium-based micromembranes have been microfabricated and tested for their mechanical strength. The device can be used for water gas shift reaction and hydrogen gas separation in catalytic microreaction systems for methanol fuel reforming. The novelty of this structure is that we have integrated the water gas shift reactor as well as the hydrogen gas separator in the same structure. This is because copper can act as a membrane support as well as a catalyst in the water gas shift reaction. A currently unoptimized design of micromembrane was capable of withstanding a pressure gradient of 10 psi without breaking.

Acknowledgements

We acknowledge financial support for this project from the U.S. National Science Foundation under the "XYZ-on-a-Chip" grant CTS-9980781, the Pittsburgh Digital Greenhouse and Sandia National Laboratories. Help from Chemical Engineering undergraduate students, Jake Towne and Josh Tilghman, and Electrical Engineering undergraduate students, Susan Alexander and Pete Irvin, is also thankfully acknowledged.

References

1. The Knowledge Foundation, Inc. The latest developments in portable power: Advances in R & D for the commercialization of small fuel cells and battery technologies for use in portable applications, Bethesda, MD, April 29-30 1999.
2. C. Hebling, A. Heinzel, D. Golombowski, T. Meyer, M. Muller and M. Zedda, "Fuel cells for low power applications," in *the 3rd International Conference on Microreaction Technology*, Frankfurt, Germany, April 18-21 1999. DECHEMA.
3. M. Datta (chair) and N. Smilanich (co Chair), "Technology Roadmaps: Energy Storage Systems," National Electronics Manufacturing Technology Roadmaps, December 1996.

4. A. G. Knapton, "Palladium Alloys for Hydrogen Diffusion Membranes," *Platinum Met. Rev.*, **21**, pp. 44-50, (1977).

5. A. J. Franz, K. F. Jensen and M. A. Schmidt, "Palladium based micromembranes for hydrogen separation and hydrogenation/dehydrogenation reactions," IEEE MEMS, pp. 382-387 (1999)

6. A. Zheng, F. Jones, J. Fang and T. Cui, "Dehydrogenation of cyclohexane to benzene in a membrane microreactor," in the proceedings of 4th Annual Conference on Microreaction Technology, March 5-9, 2000, pp. 284-292, Atlanta, GA, 2000.

7. S. V. Karnik, M. K. Hatalis and M. V. Kothare, "Issues Involved in Fabrication of a Palladium based Micro-Membrane for Hydrogen Gas Separation," Presented at American Institute of Chemical Engineers (AIChE) annual meeting, November 12-17, 2000, Los Angeles, CA.

Demonstration of Energy Efficient Steam Reforming in Microchannels for Automotive Fuel Processing

G. A. Whyatt, W. E. TeGrotenhuis, J. G. H. Geeting, J. M. Davis, R.S. Wegeng, L. R. Pederson

Pacific Northwest National Laboratory, USA

Abstract

A compact, energy-efficient microchannel steam reforming system has been demonstrated. The overall volume of the reactor is 4.9 liters while that of the supporting network of heat exchangers is 1.7 liters[1]. The reactor contains alternating reaction and combustion gas channels, arranged in crossflow, to provide heat to the reaction. Use of a microchannel configuration in the steam reforming reactor produces rapid heat and mass transport which enables fast kinetics for the highly endothermic reaction. The microchannel architecture also enables very compact and highly effective heat exchangers to be constructed. A network of microchannel heat exchangers allows recovery of heat in the reformate product and combustion exhaust streams for use in vaporizing water and fuel, preheating reactants to reactor temperature and preheating combustion air. As a result of the heat exchange network, the system exhaust temperatures are typically ~50°C for the combustion gas and ~130°C for the reformate product while the reactor is operated at 750°C. While reforming isooctane at a rate sufficient to supply a 13.7 kWe fuel cell, the system achieved 98.6% conversion with an estimated overall system efficiency after integration with WGS and PEM fuel cell of 44% (electrical output / LHV fuel). The efficiency estimate assumes integration with a WGS reactor (90% conversion CO to CO_2 with 100% selectivity) and a PEM fuel cell (64% power conversion effectiveness with 85% H_2 utilization for an overall 54% efficiency) and does not include parasitic losses for compression of combustion air.

Introduction

A fuel cell powered car provides the potential of greater fuel efficiency and less pollution compared to conventional internal combustion engines. Although on-board hydrogen storage technology is being pursued, an alternative approach

---

[1] The heat exchanger volume of 1.7 liters includes a low-temperature air preheater (0.7 liter) although this exchanger may not be required upon integration within a fuel cell system where cathode air is combusted. The volume excludes the low temperature water preheater (~0.1 liter) which was not utilized in generating the data presented.

is to process a liquid hydrocarbon fuel to produce a hydrogen-rich gas stream suitable for consumption by a fuel cell. Conventional technology for hydrocarbon steam reforming experiences heat transfer limitations resulting in long residence times and large equipment. As a result, most automotive fuel reforming efforts have targeted partial oxidation (POX) and autothermal (ATR) reforming approaches that provide the heat by injecting air along with the reactants. However, steam reforming offers several potential advantages over the POX or ATR approaches, including

1) The hydrogen content in the reformate stream is higher because it is not diluted by nitrogen.
2) High reformate pressures can be efficiently generated by pumping liquid fuel and water as liquids without the need to compress air to the reaction pressure.
3) Steam reforming can combust waste anode gas as fuel to provide the necessary heat input, allowing it to be more efficient. POX and ATR, which are thermally neutral or exothermic, cannot use the waste anode gas in this way.

In May 1999, PNNL successfully demonstrated rapid kinetics for steam reforming in heated microchannels. Based on this observation, a microchannel steam reforming system was designed and built with the key objectives being to demonstrate productivity sufficient to support a 10 kWe PEM fuel cell[2], to demonstrate conversions >90% and to demonstrate a reformer efficiency sufficient to support a >40% efficiency for an integrated PEM fuel cell system[3].

The experimental system demonstrated a capacity of 13.7 kWe capacity at 98.6% conversion with an estimated 44% overall system efficiency, exceeding the design objectives. In addition, productivity >20 kWe was demonstrated while maintaining conversion >90% and estimated system efficiency >40%. This paper describes the system and the results obtained.

System Configuration

The experimental system prior to application of insulation is shown in Figure 1. The Figure 1 caption explains the function of the various exchangers. A schematic of the system flows is shown in Figure 2 for the combustion side and in Figure 3 for the reaction side. The controls required to operate the system include metering pumps for each isooctane and water inlet, a controller to

---

[2] The kWe capacity is calculated using an assumed WGS reactor performance (90% conversion, 100% selectivity to CO_2) and a 54% efficient PEM fuel cell (64% fuel conversion effectiveness with 85% fuel utilization).

[3] Efficiency is defined as (electrical output from fuel cell)/(lower heating value of input fuel).

maintain the desired air flow rate on the combustion side and 4 temperature controllers tied to hydrogen flow controllers used to control combustion temperatures at the primary combustor and each of the 3 secondary combustion points just prior to each of the reactor cells. In addition, there is a manual valve that allows incoming combustion air to be diverted around the high temperature recuperator if more heat is needed in the water vaporizor. This is used to operate at higher capacities when a proportional increase in combustion air flow is not available.

Fabrication Approach

The individual microchannel components are fabricated from 316L stainless steel in a process that includes photo-chemical etching of thin metal sheets which are then stacked and diffusion bonded to form a laminated structure with microchannels. The laminated fabrication approach is discussed in more detail in Reference 1.

Experimental Results

A summary of operational data for 2 steady-state operating points is shown in Table 1. The reforming system is capable of achieving 98.6% conversion operating at 13.7 kWe capacity. In addition, the system can exceed 20 kWe capacity with conversions greater than 90%. The appropriate operating capacity of the system will depend on the level of conversion required to avoid problems due to unconverted fuel in components downstream from the reformer. The energy efficiency values in Table 1 take no credit for the fuel value of unconverted multi-carbon components that may show up in the anode gas (effectively assuming they are removed prior to the fuel cell). If unconverted fuel components can pass through downstream components and return with the waste anode gas the high output energy efficiency would be improved and the system could operate efficiently at the high end of its capacity. Alternatively, it could be smaller at a lower capacity level.

A plot of data for various operating points is shown in Figure 4. The reaction rate has an approximate Arrhenius temperature dependence which allows a simple model of the reactor to be formulated to predict productivity and conversion as a function of the reformate exit temperature. For a fixed level of conversion, the plot indicates a strong dependence of productivity on temperature. Alternatively, at a given hydrocarbon feed rate the conversion level is strongly dependent on temperature.

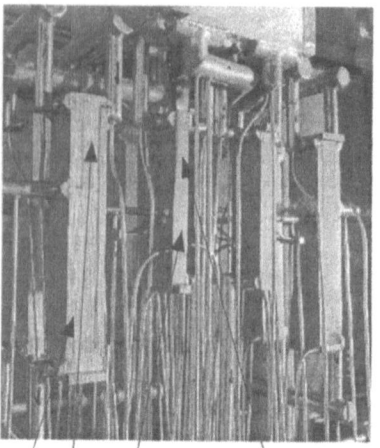

Reformate Recuperator

Fuel Vaporizer

High-T Air Recuperator

Water Vaporizer

Figure 1. 10 kWe Microchannel Steam Reforming System. The picture on the left shows the reforming system with 4-cell reactor at top, and the low-temperature air preheater in front. The low-temperature air preheater uses the combustion exhaust leaving the water vaporizer to preheat incoming combustion air. This exchanger may not be necessary after system integration if air enters from the cathode of the fuel cell. The picture at right provides a closeup of the multistream exchangers directly below the reactor. To the outside are exchangers that include (in a single block) a high-temperature air preheater and water vaporizer. This exchanger takes the combustion exhaust from the reactor and uses it to first preheat incoming combustion air from the low-temperature air preheater and then to vaporize water. Toward the inside are exchangers that include (in each block) a reformate recuperator, a fuel vaporizer, and a water preheater. These exchangers take the hot reformate product from the reactor and use it to first heat vaporized water and fuel to near reactor temperature and then to vaporize the liquid fuel. The exchangers also include a section for preheating water when operating at high pressure. The vertically orientated exchangers are designed so that they can all be combined into a single integrated, multi-stream exchanger. However, to accommodate instrumentation the exchangers were bonded as separate units. The large number of small diameter tubes provide for detailed monitoring of temperature and pressure. When opened, the manual valve (left picture) allows some fraction of incoming air to bypass the high temperature combustion gas recuperator, thus leaving more combustion gas heat for use in the water vaporizer.

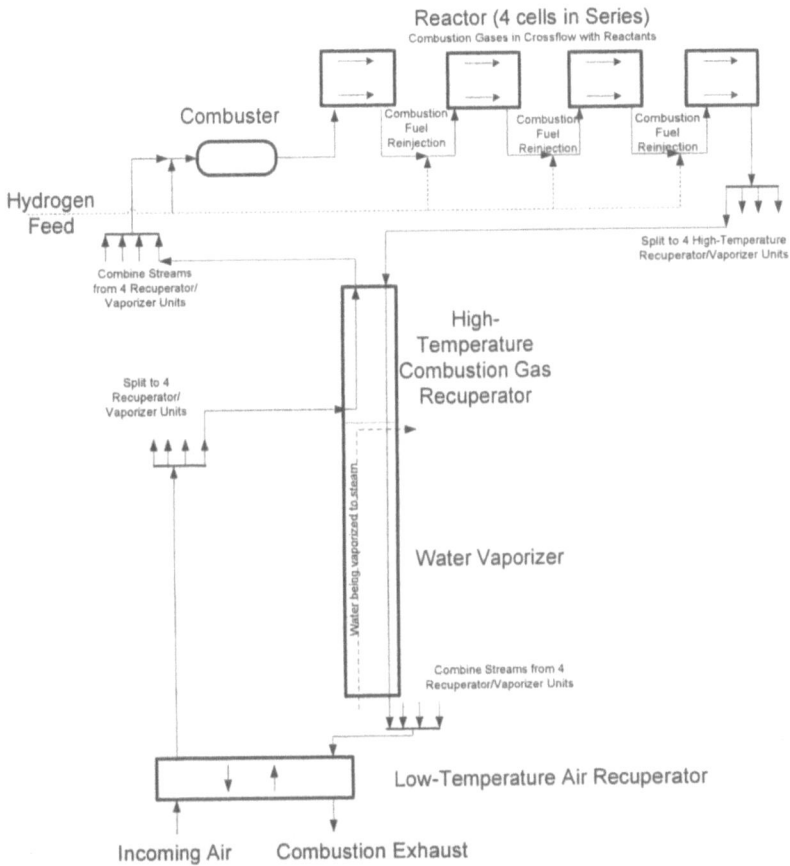

Figure 2. Flow Schematic for Combustion Gas Side of Steam Reformer System. Combustion gas flows through the four reactor cells in series and then through the four high temperature air recuperator/water vaporizors in parallel. The combustion gas then recombines before passing through the single low temperature air preheater. In this system, hydrogen is burned for fuel since the system is not yet integrated with a fuel cell and waste anode gas is not available. Most of the combustion occurs in the main combustor. However, additional fuel is burned after each pass through the reactor to restore the desired temperature before reentering the reactor.

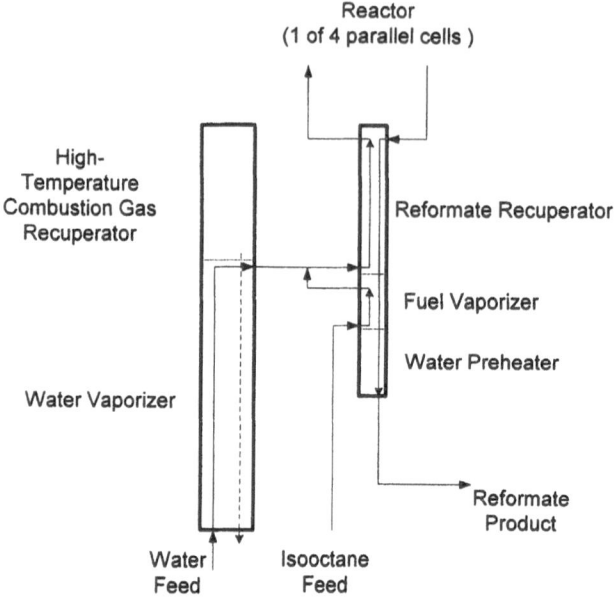

Figure 3. Schematic of Reaction Side Flows. There are 4 independent reaction cells in the reactor that operate in parallel. Each cell has a dedicated water and fuel vaporizer and reformate recuperator. Since each cell is independent, the processing rates of each cell can be independently varied. The water preheater is only used at high pressure. Near ambient pressure, the water preheater is bypassed and water is fed directly to the vaporizer as illustrated in the figure.

Table 1. Steam Reformer System Performance at Two Conditions

Productivity[1]	13.7 kWe	22.2 kWe
Fuel Conversion to C1	98.6%	93.6%
Estimated System Efficiency[2,3]	44%	41%
Power Density[4]	2100 We/L	3500 We/L
Combustion Temperature	750°C	775°C
Combustion Exhaust Temperature	43°C	50°C
Reformate Exit Temperature	129°C	115°C
Dry Gas Composition	70.6% H_2	69.7% H_2
	14.6% CO	16.1% CO
	13.7% CO_2	12.3% CO_2
	0.9% CH_4	1.3% CH_4

1) Calculated potential power output from a PEM fuel cell is based on assuming 90% CO conversion and 100% selectivity to CO_2 in a downstream water gas shift reactor and a fuel cell with 54% efficiency (85% H_2 utilization and 64%

fuel conversion efficiency). These assumptions are also used to calculate system efficiency.

2) Efficiency is calculated as electrical output from the fuel cell divided by the lower heating value of the fuel fed to the system (both for reforming and combustion). Unutilized H_2 and CH_4 in the fuel cell waste anode gas are assumed to be combusted to provide heat for the system.

3) The PNGV 2000 target for system efficiency at 25% of system capacity is 40%. The 2004 goal is 48%.

4) The PNGV 2000 target for power density in the fuel processor is 600 We/L. The 2004 goal is 750 We/L. The goals include the volume of the water gas shift and CO cleanup as well as the reformer. However, the current reformer consumes only a fraction of the volume goal making it likely that an integrated system including WGS and CO cleanup can be made to meet the goal.

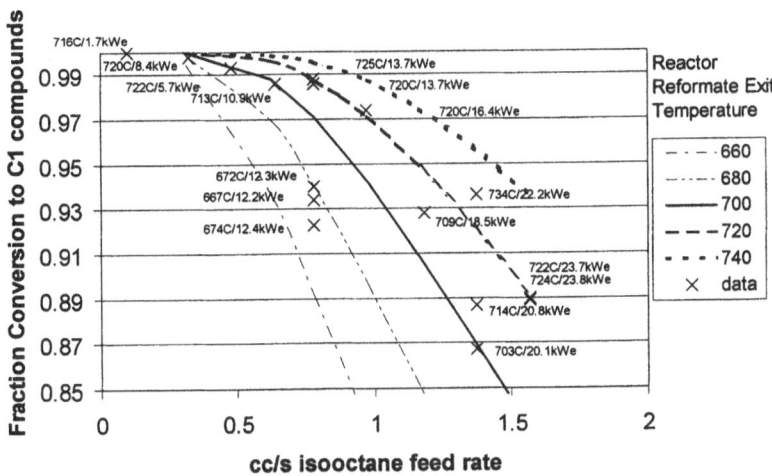

Figure 4. Relationship Between Fuel Reforming Rate and Conversion at Various Temperatures. The horizontal axis indicates the rate at which isooctane is being fed to the system while the vertical axis indicates the fraction converted to single carbon species. The curves show model predictions for conversion as a function of isooctane feed rate and reformate outlet temperature. Actual observed operating points are shown with labels indicating the reformate outlet temperature and calculated productivity. The productivity ratings in kWe are calculated as described in Table 1, footnote 1. The data plotted above are restricted to a steam to carbon ratio of 3 and to operating points where reaction pressure was not intentionally increased.

Thermal Efficiency vs. Processing Rate

In Table 1, a 44% efficiency is estimated for a system consisting of the steam reformer plus an assumed water gas shift reactor, CO cleanup and PEM fuel cell. The steam reformer within the integrated PEM system increases system efficiency because roughly half of the heat input required by the reformer is provided by the fuel value of the waste anode gas. However, this efficiency value tends to mask the efficiency of the network of microchannel heat exchangers in the reformer system. A theoretically obtainable efficiency for the reformer system can be calculated by summing the lower heating value (LHV) of the hydrogen and carbon monoxide leaving in the reformate for 100% conversion and equilibrium gas composition and dividing this quantity by the LHV of the isooctane fed to the reformer plus the net heat input required to sustain the reaction, vaporize fuel/water, and preheat reactants while recovering heat from the reformate sufficient to cool it to its dewpoint[1]. By replacing the net heat term in the denominator with the actual LHV of hydrogen burned as fuel to provide heat during an experiment the percent of theoretically obtainable efficiency can be evaluated. This quantity is shown in Figure 5 as a function of the processing rate of the system. The system is capable of >90% of theoretical efficiency over a range of 5 to 15 kWe capacity. At high processing rates the efficiency decreases due to lower conversion. At low processing rates longitudinal conduction in the heat exchangers begins to reduce efficiency.

Heat Exchanger and Vaporizer Heat Transfer Densities

The efficient operation of the steam reforming system depends on a number of diffusion bonded microchannel heat exchangers. The heat exchanger network was sized to operate up to a processing rate of ~20 kWe in order that the heat exchangers would not limit the system productivity. The heat exchanger volumes and observed heat transfer duties and densities for the conditions reported in Table 1 are summarized in Table 2.

Transient Response

The system has a rapid warm transient response due to the very short residence time. The current manually adjusted metering pumps limit the rate at which process changes can be made. However, doubling the reformate output can be accomplished within 20 seconds, limited by the pump adjustment. The 2004 PNGV goal for warm transient from 10 to 90% is 1 second.

---

[1] This definition penalizes the theoretical efficiency value for the heat required to vaporize water, which is needed as vapor downstream in the WGS reactor in any case. However, theoretical obtainable efficiency is only used as a point of comparison to evaluate the relative efficiency of the reforming system.

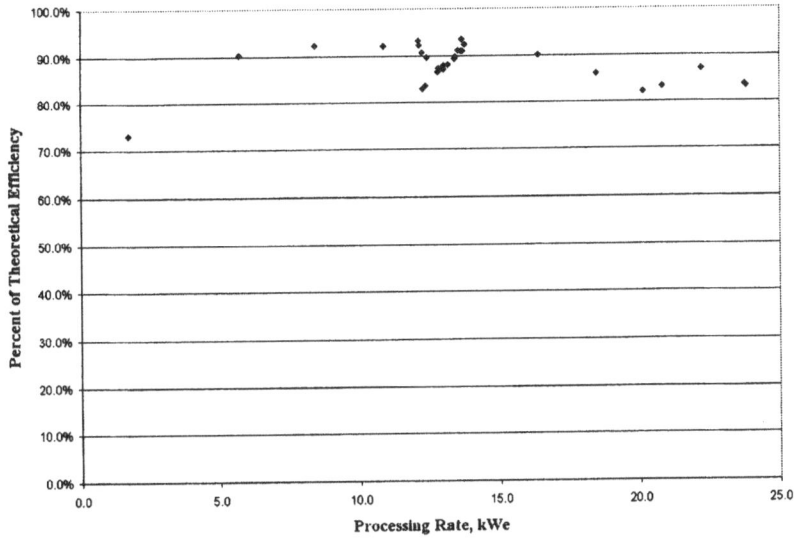

Figure 5. Fraction of Theoretical Efficiency Observed in Various Experiments. Test conditions shown are limited to 3:1 S:C and reformate exit temperature between 650 and 750°C. A value of 100% would indicate complete conversion of fuel to single carbon compounds, zero heat loss, recovery of heat from reformate stream sufficient to bring it to the dewpoint, and complete utilization of the LHV of the fuel combusted for heat.

Table 2. Heat Exchanger Duties for Conditions Reported in Table 1.

Exchanger	Bonded Stack [a]	Heat Exchanger Duty [b]		Bonded Stack Heat Transfer Density	
	cm^3	13.7 kWe Reformate Watts	22.2 kWe Reformate Watts	13.7 kWe Reformate W/cm^3	22.2 kWe Reformate W/cm^3
Reformate Recuperator	41.3	705	1087	17	26
High-T Air Preheater	74	1517	1119[c]	21	15[c]
Low-T Air Preheater	726	1228	1346	1.7	1.9
Fuel Vaporizer	10.3	55	98.5	5	10
Water Vaporizer	125	1346	2344	11	19

See next page for footnotes

TABLE 2 FOOTNOTES

a- Volume within integrated exchangers apportioned between different heat exchange functions. The reactor system contains 4 of each heat exchanger with the exception that there is only 1 low temperature air preheater. Volume is taken as the volume of the diffusion bonded stack which includes the volume of the internal flow distribution headers.

b- Duties are for a single exchanger. For all units except the low temperature air preheater the total system duty can be determined by multiplying by 4.

c- The air flow available to the system was limited and was only increased by 12% in going from the 13.7 kWe condition to the 22.2 kWe condition. A control valve intentionally bypassed incoming air around the high temperature air recuperator to reduce its duty and increase heat available in the vaporizer section.

Conclusions

The current steam reformer system has successfully demonstrated that a microchannel steam reformer can achieve a significant capacity in a compact package while achieving high conversion of the hydrocarbon feed. In addition, a highly integrated system of microchannel heat exchangers and vaporizors has been demonstrated to allow the system to operate in an energy efficient manner. Although work remains to develop the system sufficiently for operation in an automotive setting, the current demonstration is a major step towards that goal.

Acknowledgement

The work described here was funded by the U.S. Department of Energy, Office of Transportation Technology as part of the OTT Fuel Cells Program.

References

1) Matson, D. W., P. M. Martin, D. C. Stewart, A. Y. Tonkovich, M. White, J. L. Zilka, G. L. Roberts. *Fabrication of Microchannel Chemical Reactors Using a Metal Lamination Process*. Microreaction Technology: Industrial Prospects. IMRET 3: Proceedings of the Third International Conference on Microreaction Technology. W. Ehrfeld, ed.

A Microstructure Reactor System for the Controlled Oxidation of Hydrogen for possible Application in Space

K. Haas-Santo[1], O. Görke[1], K. Schubert[1], J. Fiedler[2] and H. Funke[2]

[1] Forschungszentrum Karlsruhe GmbH, Postfach 3640, 76021 Karlsruhe, Germany
[2] Astrium GmbH, 88039 Friedrichshafen, Germany

1. Introduction

Microchannel reactors and heat exchangers may have distinct advantages in comparison to conventional reactor designs. The advantages of the microstructure reactors include their inherent safety as flame arresting capability and high heat transfer rates. The small size of the channels results in dimensions that are less than the quench distances for explosions, ultimately providing an opportunity to perform hazardous reactions safely. The hydrogen/oxygen reaction is particularly demanding as the reaction is highly exothermic and mixtures of H_2 and O_2 are extremely explosive in a wide range of gas concentrations.

In space applications electrolytic cells powered by solar cells are used to produce hydrogen and oxygen for various purposes. When hydrogen is not needed, it is released into space. An alternative possibility would be to convert hydrogen and oxygen to water again for further use inside the space lab. Therefore the aim of this work was to provide a reaction system to convert explosive mixtures of hydrogen and oxygen completely and safely to condensed water which can be recycled. For this purpose a catalytically active microstructure reactor system had to be prepared.

For this reactor system existing standard reaction modules developed at the Forschungszentrum Karlsruhe [1] like micromixer, microreactor and micro heat exchanger have been used. As catalyst support inside the microchannels of a micro heat exchanger an alumina coating, obtained by sol/gel method, was developed. The coating has been impregnated with a palladium catalyst.

2.   Microchannel reactor configuration

The microstructure reaction system itself consists of a micromixer, the catalytic microreactor with a reaction and a cooling passage and a micro heat exchanger to cool the reaction product, so that condensed water is received. Between the outlet of the microreactor and the inlet of the micro heat exchanger a tube containing catalytic material to guarantee complete conversion of the hydrogen/oxygen mixture is placed. A picture of the microreaction system is shown in figure 1. In the second passage of the reactor and the heat exchangers air is used as coolant fluid. Gas temperatures are measured with thermocouples in the inlet and outlet of the reaction and coolant passages.

The microstructure reactor and micro heat exchanger used for this work were 1 cm^3 cross-flow devices made of stainless steel similar to the device already described in a previous paper [2]. The stack of foils is diffusion bonded and welded into a stainless steel housing with standard tube connections. The assembly is vacuum tight from passage to passage and can handle high pressure. The stainless steel foils had channels with dimensions of 200 x 70 microns separated by fins of 100 microns.

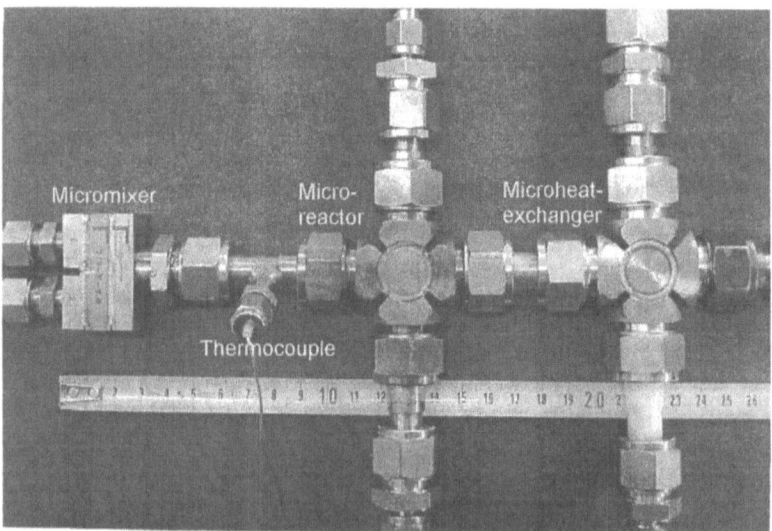

Figure 1: Picture of the microstructure reaction system consisting of a micromixer, a micro reactor and a micro heat exchanger.

3.   Catalyst preparation

To provide a support for the catalyst the channels of one passage of the microstructure reactor was coated with porous aluminum oxide by the sol/gel method. The sol was prepared using aluminum-sec.-butylate as metal alkoxide and ethanol as solvent. After preparation the sol was aged at room temperature for several days before use. The coating of the inner channel walls was done as follows: The microstructure reactor was placed in such a way that the channels were vertical. The sol was pipetted into the openings of the channels. The viscosity of the sol was so low that it readily flowed into the channels. After a definite time the superfluous sol was removed by air. Afterwards the thin gel layer was treated at a calcination temperature of 500 °C.

The resulting alumina oxide layer was examined by microscopy. The alumina layer showed a constant thickness of about 2-3 μm (figure 2).

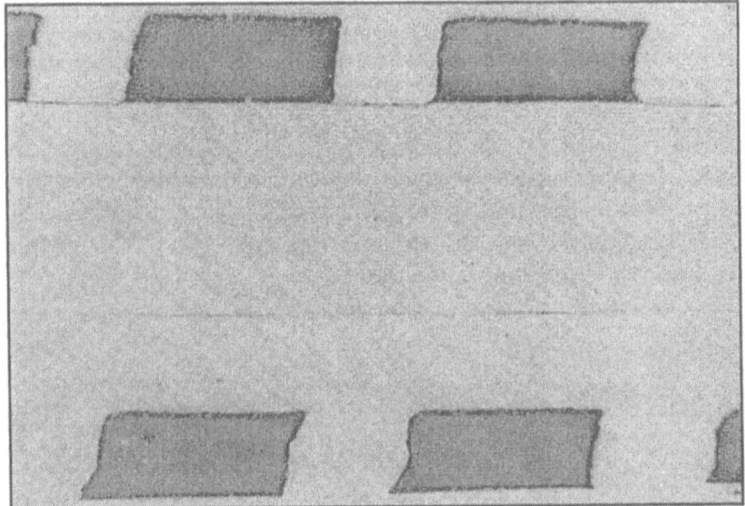

Figure 2: Microscopic picture of a cross section of a microstructure reactor coated with alumina oxide.

The specific surface area and pore volume distribution of the coatings were determined by physisorption using nitrogen as adsorptive. The sorption isotherm is shown in figure 3. The shape of the isotherm represent typical nitrogen sorption isotherms for Al_2O_3 made of aluminium-sec.-butylate, likely IUPAC type IV isotherms indicating a mesoporous solid. The diameter of the mesopores was calculated to be between 10 and 50 Å with a maximum at 40 Å by applying the BJH model for the desorption curve. For comparison a surface enhancement factor was defined as the quotient of surface area estimated by physisorption and geometrical surface area of the microchannels. In the experiments an enhancement factor of 430 m^2/m^2 (film thickness = 2 μm) was reached.

316

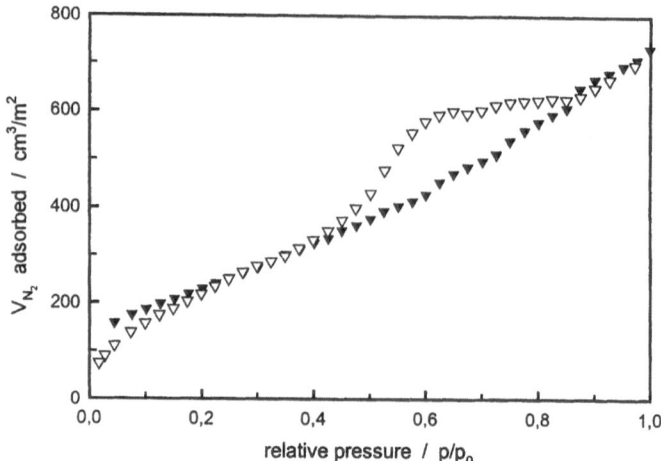

Figure 3: Nitrogen sorption isotherm of an alumina oxide coating in microchannels (filled symbols indicate values of the adsorption curve, light symbols those of the desorption curve).

The noble metal catalyst was prepared by wetness impregnation. Palladium as noble metal catalyst was incorporated from an aqueous solution of palladium nitrate. To obtain an optimum adsorption, impregnation was carried out at 50 mbar for 30 min. The catalyst was activated by drying at 80 °C for 1 h and following reduction with hydrogen in nitrogen as carrier gas.

4. Experimental Set-up

Two different experimental set-ups have been used. First the reaction experiments have been performed in an existing testrig at Karlsruhe Research Center, which consists of a gas dosing section, the microstructure reaction system and an analysis section. Here the generation of model gas mixtures containing various amounts of hydrogen and oxygen was possible. Oxygen was added in the most experiments as air, thus nitrogen as inert gas was present in these experiments. A stand alone microreactor and the microstructure reaction system were tested under various reaction conditions, for a flow schema see figure 4. The flow rate of hydrogen was varied in the range of 0.05 to 1.0 Nl/min, air was added in such an extent that oxygen was in stoichiometric amount. Reaction pressure and the cooling gas flow have been further parameters. As cooling fluid air was used. The conversion of the hydrogen/oxygen mixture was determined by measuring the concentration of the remaining oxygen in the product gas.

After these experiments the microstructure reaction system was tested at Astrium GmbH in the off stream of the electrolytic cell with hydrogen and oxygen throughputs to be expected at space lab conditions. The total gas flow was in the range 1.4 to 1.74 Nl/min. The total gas flow effusing from the electrolysis cell stack is dependent on the number of operated cells. The microstructure reaction system was tested at various reaction parameters. Conversion was calculated by weighing the amount of the produced water.

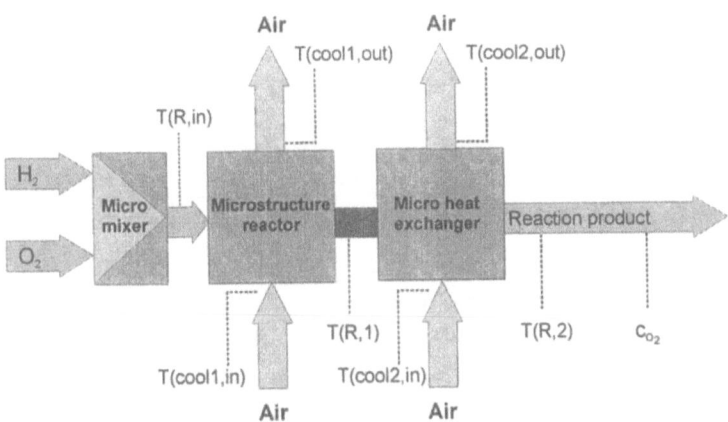

figure 4: Flow scheme of the microstructure reaction unit. Hydrogen and air/oxygen are provided by gas cylinders (experiments at FZK) or by the electrolysis cell (experiments at -Astrium).

5.   Results

5.2   Experiments with separate microstructure reactor

The starting behavior and conversion of the hydrogen/oxygen mixture was first tested with a single microstructure reactor placed after the micromixer. Figure 5 shows the decrease of the oxygen concentration directly at the beginning of the reaction and at the same time an increase of the cooling gas temperature. This indicates an ignition of the reaction directly at the beginning. After 150 seconds the concentration of oxygen was zero, indicating complete conversion of the stoichiometric hydrogen/oxygen mixture. Steady state was reached after 320 sec. There the measured temperatures are not changing any more. But the temperature of the effluent reaction gas reached 130 °C. This product temperature should be reduced to obtain water with a temperature of about 50 °C. Therefore in the next experiment a microstructure reaction system consisting of a micromixer, a microstructure reactor and a micro heat exchanger was used.

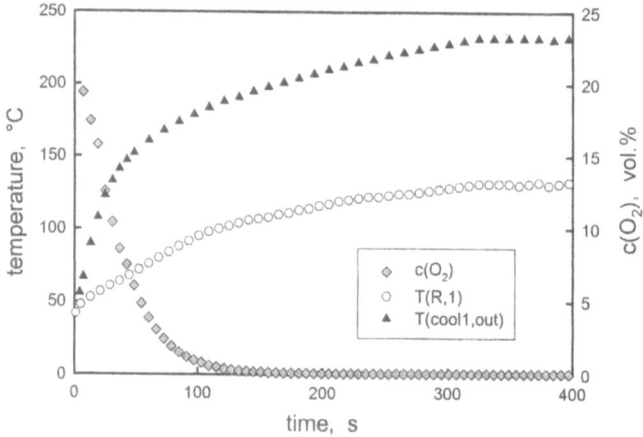

Figure 5: Temperatures and oxygen concentration as a function of time (hydrogen flow 1,0 Nl/min; cooling air flow 20 Nl/min).

5.2   Experiments with the microstructure reaction system

Figure 6 shows as example the oxygen concentration and the temperatures for a stoichiometric hydrogen/oxygen mixture (hydrogen flow 1,0 Nl/min, air flow 2,5 Nl/min), cooling air flow was 20 Nl/min. As an immediate raise of the temperature of the cooling air of the microreactor could be observed, the reaction started at about 25 °C, thus indicating the direct ignition of the reaction. Complete conversion was attained after 120 seconds, as the oxygen concentration in the product gas reached zero. This could be observed for various total gas flows with a hydrogen/oxygen ratio of 2.

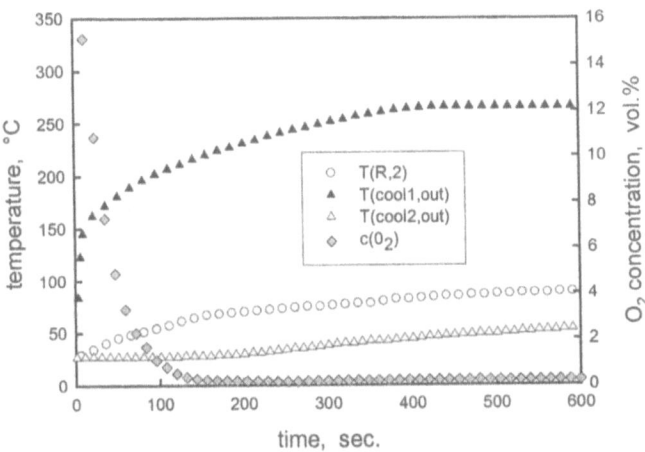

Figure 6: Temperature and concentration as a function of time (hydrogen flow 1,0 Nl/min; cooling air flow each 20 Nl/min; T(cool 1,in) = 30 °C).

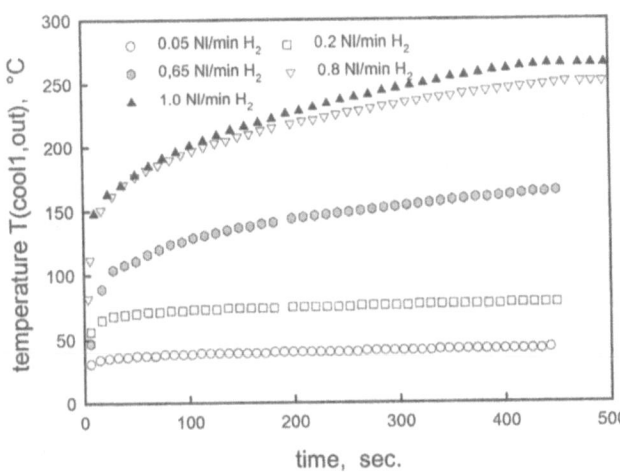

Figure 7: Temperature of the cooling outlet gas of the microreactor at different hydrogen flows (cooling air each flow: 20 Nl/min).

A steep increase of the temperature of the cooling gas 2 at the outlet can be observed during the first 20 seconds of the reaction, afterwards a constant temperature is approached asymptotically after 350 seconds for 1 Nl/min

320

hydrogen, as can be seen in figure 6. This indicates that the reaction system could be operated at a stationary overall temperature, without overheating the system. The temperature of the effluent gas was about 70 °C, so that liquid water was obtained. Also, even after 200 h no deactivation of the catalyst activity was observed.

The influence of different hydrogen flows on the heat production was investigated. In figure 7 the time dependence of the temperature of the cooling gas of the microreactor is shown, the achieved temperature is dependent on the hydrogen gas flow. The ignition of the reaction is not dependant on the hydrogen gas flow as in all experiments the oxygen concentration decreased at the beginning to zero (not shown).

In the next step the microstructure reaction system was placed in the off-stream of a electrolysis cell stack. This implies that pure hydrogen and oxygen was feed into the catalytic reactor. The number of cells of the electrolysis cell was varied during the experiment, resulting in hydrogen flows of. 0,95 Nl/min for 3 cells and 1,16 Nl/min for 5 cells. Figure 8 shows the temperatures of the cooling gases (T(cool1,out) and T(cool2,out)) and of the reaction product water (T(R,2)). At the beginning of the experiment the gas flows of hydrogen and oxygen are raising slowly due to the start of the electrolysis cell. The reaction ignites spontaneously, resulting in raising temperatures of the cooling gases. After changing the number of electrolysis cells, a steady state is reached after 20 minutes. The temperature of the reaction product water could be kept at about 40 °C, so that liquid water was obtained. Hydrogen could be converted to a full extent into water.

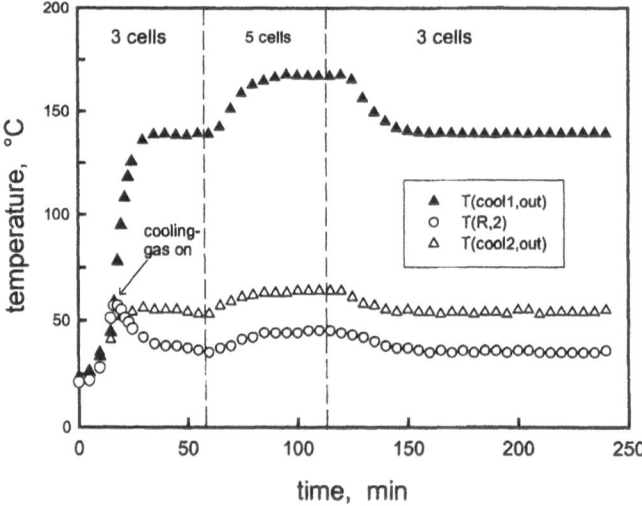

figure 8: Temperatures of the cooling gases and the reaction outlet on two different electrolysis cell parameters (cooling air flow 1 =10 Nl/min and cooling air flow 2 = 30 Nl/min).

6. Summary and conclusion

A microstructure reaction system was developed for the catalytic reaction of explosive hydrogen/oxygen mixtures. This reaction system consisted of a micromixer, a microstructure reactor and a micro heat exchanger. The catalytic coating of the microchannels of the microstructure reactor was manufactured by a sol/gel process, followed by a wet impregnation procedure with diluted palladium nitrate. The catalytic microstructure reaction system showed a high catalytic activity for hydrogen/oxygen mixtures with inert gas (tests in Karlsruhe using air as oxygen source) and without (tests at Astrium). The reaction in the microstructure system results in liquid water at 40 to 50 °C, that can be recycled.

This catalytic microstructure reaction system showed a possible option to be used in space applications for surplus products of the electrolysis cells.

[1] K. Schubert, W. Bier, J. Brandner, M. Fichtner, C. Franz, G. Linder. Realization and Testing of Microstructure Reactor, Micro Heat Exchangers and Micromixers for Industrial Applications in Chemical Engineering. 2nd International Conference on Microreaction Technology, March 09-12, 1998 New Orleans, Louisiana, pages 88-95

[2] M. Janicke, H. Kestenbaum, U. Hagendorf, F. Schüth, M. Fichtner and K. Schubert, *The Controlled Oxidation of Hydrogen from an Explosive Mixture of Gases using a Microstructured Reactor/Heat Exchanger and Pt/Al$_2$O$_3$ catalyst*, J. Catal. **191** (2000) 282 – 293.

Catalyst Coating in Microreactors for Methanol Steam Reforming: Kinetics

Pierre Reuse, Pascal Tribolet, Lioubov Kiwi-Minsker and Albert Renken
Swiss Federal Institute of Technology, EPFL – LGRC
CH-1015 Lausanne, Switzerland

Abstract

The intrinsic kinetics of the methanol steam reforming over a copper-based catalyst (Süd-Chemie) was investigated. A higher hydrogen inhibition for the catalyst coated on a microstructure compared to the fixed bed reactor was noticed. With no hydrogen in the inlet feed, the rate of methanol disappearance is on average 34% higher at 200°C with the coated catalyst.

Keywords: methanol steam reforming, hydrogen production, kinetics, washcoating

Introduction

One of the main problems in the use of microreactors for heterogeneous catalytic reactions is the development of catalytically active microporous wall materials. Several solutions have been proposed to overcome this problem. Filling the channels with the catalyst – scale down of industrial reactors – has been described earlier [1]. Due to high pressure-drop and the risk of flow maldistribution other techniques have been investigated. Wiessmeier and Hönicke proposed anodic oxidation of microstructured aluminium foils [2]. By this mean, the microchannel's surface is converted to a thin shell of α-alumina. The porous layer can then be impregnated with active components. Washcoating is an other useful tool to create a catalytically active layer on the wall of the microchannels. This method is not restricted to aluminium. Rouge et al. [3] have shown that it is possible to create a uniform catalytic layer within the microchannels. By measuring the residence time distribution they confirmed that maldistribution within the reactor could be efficiently avoided.

Pfeifer et al. [4] used commercially available nanocrystalline powders to coat the microchannels with a porous layer containing active components or an inert layer for wet impregnation.

In the present paper the kinetics of methanol steam reforming over a low temperature shift catalyst placed in a fixed bed and coated on microchannels will be discussed.

Methanol Steam reforming

The continuous expansion of roads traffic causes ongoing demand for reducing pollution from automotive engines [5]. A promising solution to address this problem is the use of hydrogen fuel cell powered cars [6]. The hydrogen could be produced on-board through the steam reforming of carbon sources such methanol, coal, gasoline... The advantages of methanol stem from the fact that it can be produced from renewable resources (e.g. biomasses).

$$CH_3OH + H_2O \leftrightarrows CO_2 + 3 H_2 \qquad (1)$$

Methanol steam reforming is well known being catalysed by noble metals or by copper-based catalysts. In the first case, decomposition of methanol to H_2 and CO, followed by the "water gas shift" reaction have been shown to be the reaction pathway [7].

$$CH_3OH \leftrightarrows CO + 2 H_2 \qquad (2)$$
$$CO + H_2O \leftrightarrows CO_2 + H_2$$

On copper-based catalysts, the reaction mechanism does not involve carbon monoxide. According to numerous authors, the following reactions occur on this type of catalyst [8-13].

$$2 CH_3OH \leftrightarrows HCOOCH_3 + 2 H_2 \qquad (3)$$
$$HCOOCH_3 + H_2O \leftrightarrows HCOOH + CH_3OH \qquad (4)$$
$$HCOOH \leftrightarrows CO_2 + H_2 \qquad (5)$$

Only small amounts of carbon monoxide are produced through methanol decomposition.

$$CH_3OH \leftrightarrows CO + 2 H_2 \qquad (6)$$

To counteract this reaction, it is beneficial to increase the steam to methanol ratio. This improves the hydrogen selectivity and reduces the formation of CO. The benefit of this effect is somewhat offset by the increased heating requirement for vaporizing the additional water in the feed [14].

Experimental

Fixed bed reactor

30 mg of copper-based catalyst (Süd-Chemie), with particle diameter between 100 – 250 μm, have been packed into a tubular quartz reactor with inner diameter of 5 mm. The temperature inside the catalytic bed was measured with a K-type thermocouple (Thermocontrol GmbH) and the reactor was heated with an electrical tape.

Microstructured plates

Plates of FeCrAlloy (Al 4.8 %, 20.5 % < Cr < 23.5 %, Fe balance) with dimension 20 mm x 20 mm x 200 μm with microchannels hollowed out – 20 mm x 200 μm x 100 μm – have been coated with a suspension of the catalyst in isopropanol. After drying at room temperature, the catalyst was calcinated at 500 °C for two hours to burn away any organic contaminants. Careful measurements of plate's weight have been performed in order to determine the catalyst layer thickness.

To characterize the activity of the coated catalyst, a special reactor has been designed. The test unit allows to test up to twenty plates at the same time under reproducible conditions.

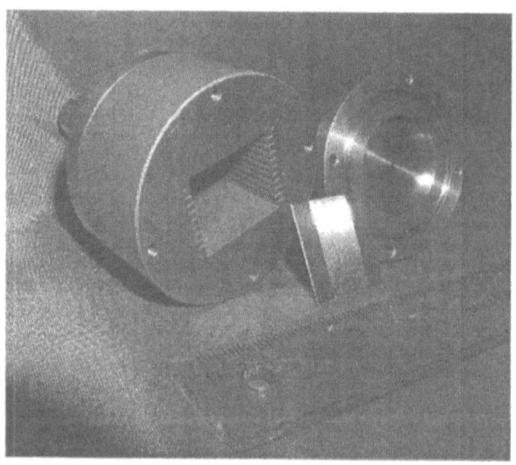

Figure 1: Test unit for the coated microplates.

Experimental set-up

The feed section of the installation provides methanol (> 99.5 % Romil, Fluka Chemie AG) and deionised water through two bubble columns. Argon (> 99.998 Carbagas, Lausanne, Switzerland) was chosen as carrier gas. The gas flow is controlled by mass flow controllers (Bronkhorst high-Tech B.V., Ruurlo, The Netherlands).

An operating pressure of 1.5 bar was used. Temperature of 200°C was chosen to ensure that the "water gas shift" reaction is negligible.

Hydrogen is detected by gas chromatography using a Carboxen 10-10 column in a Shimazu GC-14 A with a TCD detector. Unreacted methanol is determined using a HP-5 MS column with a FID detector. Carbon oxides were measured by infrared spectrometry (Ultramat 22, Siemens).

Catalyst activation

Prior to the reaction, the catalyst was oxidised in oxygen – 160 ml(STP)/min, 10% O_2 (> 99.95 Carbagas, Lausanne, Switzerland) – for 30 minutes at 200 °C. Reduction has been performed under the same conditions – 160 ml(STP)/min, 10% H_2 (> 99.995 Carbagas, Lausanne, Switzerland).

Reaction conditions

Reaction kinetics were determined under differential conditions which have been investigated. Volumetric flow rates were varied from 80 ml(STP)/min to 270 ml(STP)/min.

Inlet concentrations of methanol, water and hydrogen were fixed at : 2.0 %$_{mol}$, 5.0 %$_{mol}$, 9.0 %$_{mol}$ and 12.0 %$_{mol}$.

In addition measurements without hydrogen were performed to simulate the "real" feed of the future microreactor and to compare the activity of the coated catalyst with the original preparation in the fixed bed reactor.

At a temperature of 200°C, catalyst activity was sufficiently stable throughout the course of experiment.

326

Results and discussion

Differential conditions

Figure 2 shows the dependence of the conversion on the modified residence time. As a linear relationship is obtained in the working range, differential conditions could be assumed. This shows that the measured kinetics are representative.

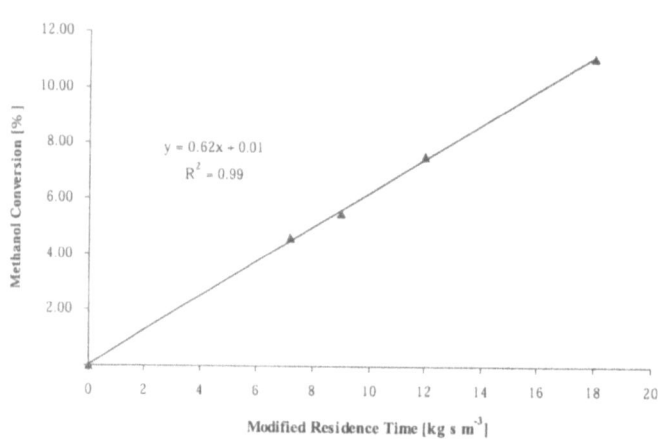

Figure 2: Methanol conversion as a function of the modified residence time for the coated catalyst, illustrating the differential condition, proper to perform a kinetic analysis.

Transfer limitations

Based on the Mears criteria [15], the internal transfer limitation was excluded.

The characteristic length, used to estimate the second number of Damköhler for the Mears criteria, is not well defined for the microplates. Therefore, the apparent activation energy was determined for the temperature range of 175°C to 250°C with an Arrhenius plot.

The estimated activation energy of 56 kJ mol^{-1} indicates that external mass transfer limitation can be neglected.

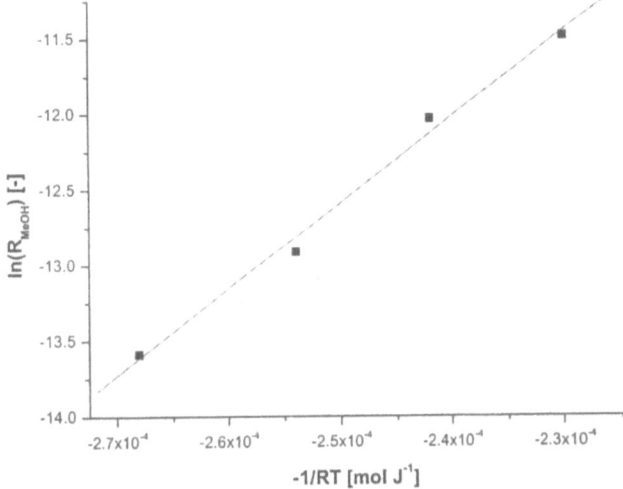

Figure 3: Determination of the activation energy for the methanol steam reforming reaction.
P_{MeOH} 0.12 bar, P_{H2O} 0.15 bar, P_{tot} 1.50 bar (argon as diluent), F 160 ml(STP) min^{-1}, Temperature of reaction 250°C, 225°C, 200°C, 175°C; 30 mg of catalyst

Kinetic model

The formal kinetics is described by the following empirical relationship.

$$-R_{MeOH} = k_0 \exp\{-E_a/RT\} \, p_{MeOH}^{m} \, p_{H2O}^{n} \, p_{H2}^{o} = k \, p_{MeOH}^{m} \, p_{H2O}^{n} \, p_{H2}^{o} \qquad (7)$$

To determine the partial order for component i, its concentration was varied while the concentrations of the two remaining components have been kept constant. Linear regressions of $\ln(-R_{MeOH})$ versus $\ln(p_i)$ were used to determine the partial order of the constituent i. Figure 4 illustrates typical results for the determination of the partial order of methanol.

328

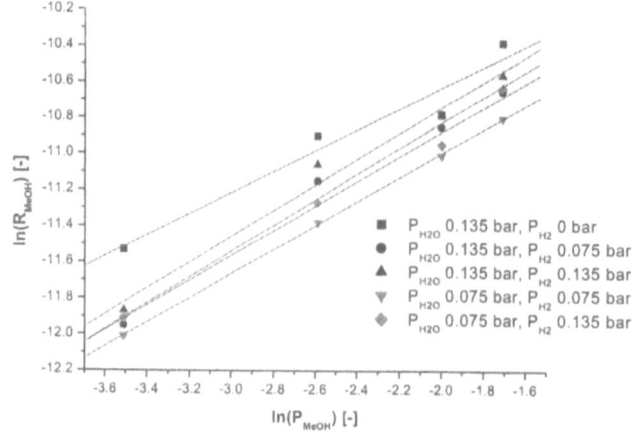

Figure 4: Determination of the partial order for the methanol for different constant concentrations of water and hydrogen, Ptot 1.5 bar (Argon as diluent), Temperature 200°C, 30mg catalyst.

The following kinetic model has been established for the original catalyst in the fixed bed:

$$-R_{MeOH} = (7.8 \ 10^{-5} \pm 0.9 \ 10^{-5}) \ p_{MeOH}^{\ 0.7\pm0.02} \ p_{H2O}^{\ 0.1\pm0.04} \ p_{H2}^{\ -0.1\pm0.1} \qquad (8)$$

A similar expression for the reaction rate with the catalyst coated on microchannels has been determined. The results are summarized in Table 1.

Table 1 : Summary of the different partial order for the components of the kinetic analysis.

	Catalyst in fixed bed Reuse et al.	Catalyst coated on microchannels Reuse et al.	Jiang et al. [10]	Amphlett et al. [16]
n_{MeOH}	0.70 ± 0.02	0.70 ± 0.1	0.26	0.62
n_{H2O}	0.1 ± 0.04	0.0 ± 0.1	0.03	0
n_{H2}	-0.1 ± 0.1	-0.2 ± 0.1	-0.2	-0.66
k (mol g^{-1} s^{-1} bar^{-1})	$7.8 \cdot 10^{-5} \pm 0.9 \ 10^{-5}$	$4.8 \cdot 10^{-5} \pm 0.6 \ 10^{-5}$	$4.1 \cdot 10^{-6}$	$1.7 \cdot 10^{-5}$

Jiang et al. and Amphlett et al. have studied similar copper based catalysts. Jiang et al. used a BASF catalyst (S3-85); Amphlett et al. reported results obtained for the catalyst C18HC from United Catalyst.

According to our expectations, a positive partial order for methanol is obtained. The value is slightly higher compared to the one measured by Amphlett et al. [16]. These results also show that the partial pressure of water has no influence on the rate of the reaction.

The study also demonstrates that the inhibition due to hydrogen on the coated plates is more important than in the fixed bed. The apparent rate constant is therefore lower in the expression of the reaction rate on the coated plates than in the fixed bed.

With no hydrogen in the inlet feed, the catalyst coated on the microplates appears to be more active than the original catalyst used in the fixed bed. Figure 5 illustrates this fact, comparing the consumption rate of methanol for different inlet concentrations of water and methanol.

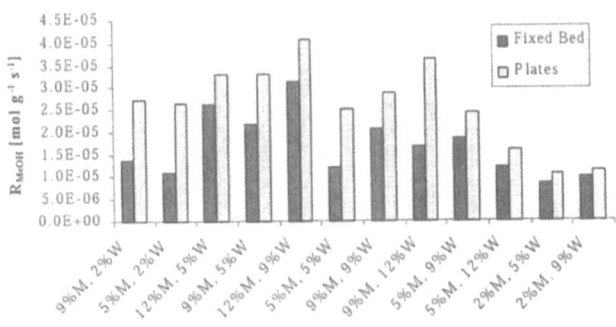

Figure 5 : Rate of methanol conversion for the fixed bed reactor and the coated plates. x_{MeOH} as described above, x_{H2O} as described above, P_{tot} 1.50 bar (argon as diluent), F 160 ml(STP) min^{-1}, Temperature of reaction 200°C, 30 mg of catalyst.

In all case, the rate of methanol disappearance on the coated catalyst is greater than on the raw catalyst. Any correlations between the inlet feed composition and the enhancement of activity is still undetermined. The mean activity is, at 200°C, 34 % higher on the coated catalyst. This beneficial effect may be attributed to the catalyst thin layer.

To evaluate the thoroughness of our results, a parity plot of the measured rates versus calculated ones are shown for two sets of experiments (Fig.6).

For the catalyst in the fixed bed reactor, the majority of measurements are well described by the determined kinetic law. All points are within 15% of the predicted value. With the coated catalyst, the results are less accurate; the larger dispersion of these measurements. hydrodynamic of the test unit – for example dead volumes – could explain the larger dispersion of the measurements.

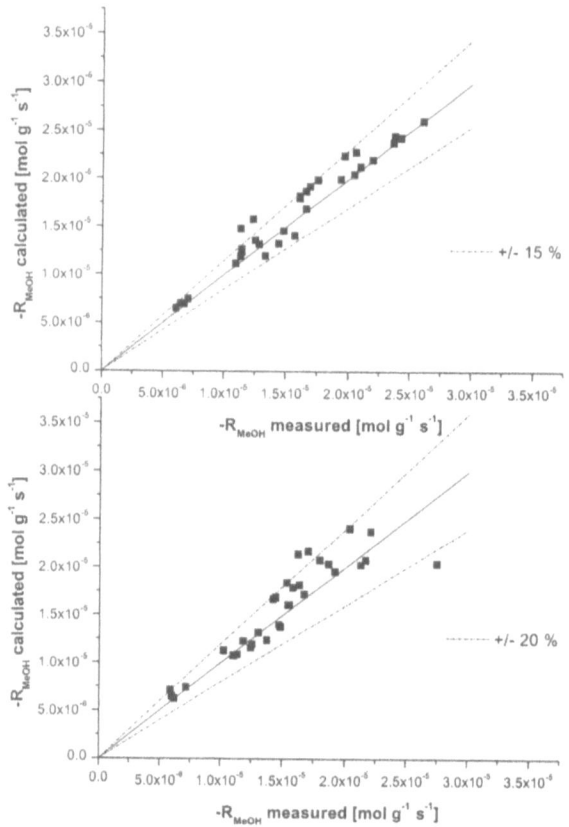

Figure 6 : Plot of the measured rate of methanol decomposition versus the theoretical value obtained through the kinetic law

Conclusion

This copper-based catalyst (Süd-Chemie) exhibits a higher rate of methanol decomposition once deposed on the microchannels. Also, it presents a stronger inhibition for the hydrogen.

Acknowledgments

The authors gratefully acknowledge the "Bundesministerium für Bildung und Forschung" (Germany) for financial support. We would like also to thank Süd-Chemie for providing the catalyst.

References

1. Tonkovitch, A.L., et al. *The catalytic partial oxidation of methane in a microchannel chemical reactor.* in *Process Miniaturization: 2nd International Conference on Microreaction Technology.* 1998. New Orleans: AiChE.

2. Wiessmeyer, G. and D. Hönicke. *Microreaction Technology: Development of a micro channel reactor and its application in heterogeneously catalyzed hydrogenations.* in *Process Miniaturization: 2nd International Conference on Microreaction Technology.* 1998. New Orleans: AiChE.

3. Rouge, A., et al., *Microchannel reactors for fast periodic operation: The catalytic dehydration of isopropanol.* Chem. Eng. Sci., 2001, 56(4), p. 1419 -1427.

4. Pfeifer, P., et al. *Microstructured catalysts for methanol-steam reforming.* in *Third International Conference on Microreaction Technology.* 1999. Frankfurt: Springer.

5. Borthwick, W.K.D., *The European Union approach to fuel cell development.* Journal of Power Sources, 2000. **86**: p. 52-56.

6. Sattler, G., *Fuel cells going on-board.* Journal of Power Sources, 2000. **86**: p. 61-67.

7. Takahashi, H., K. Takahashi, and N. Takezawa, *Steam reforming of methanol over group VIII metals supported on SiO2, AL2O3 and ZrO2.* React. Kinet. Catal. Lett., 1994. **52**(2): p. 303-307.

8. Peppley, B.A., et al., *Methanol.steam reforming on Cu/ZnO/Al2O3. Part 1: The reaction network.* Applied Catalysis A: General, 1999. **179**: p. 21-29.

9. Peppley, B.A., et al., *Methanol steam reforming on Cu/ZnO/Al2O3. Part 2: A comprehensive kinetic model.* Applied Catalysis A: General, 1999. **179**: p. 31-49.

10. Jiang, C.J., et al., *Kinetic study of steam reforming of methanol over copper-based catalysts.* Applied Catalysis A: General, 1993. **93**: p. 245-255.

11. Jiang, C.J., et al., *Kinetic mechanism for the reaction between methanol and water over a Cu-ZnO-Al2O3 catalyts.* Applied Catalysis A: General, 1993. **97**: p. 145-158.

12. Breen, J.P. and J.R.H. Ross, *Methanol reforming for fuel-cell applications: development of zirconia-containing Cu-Zn-Al catalysts.* Catalysis Today, 1999. **51**: p. 521-533.

13. Breen, J.P., F.C. Meunier, and J.H. Ross, *Mechanistic aspects of the steam reforming of methanol over a CuO/ZnO/ZrO2/Al2O3 catalyst.* Chem. Commun., 1999. **1999**: p. 2247-2248.

14. Amphlett, J.C., et al., *Hydrogen production by steam reforming of methanol for polymer electrolyte fuel cells.* Int. J. Hydrogen Energy, 1994. **19**(2): p. 131-137.

15. Mears, D.E., *Tests for transport limitations in experimental catalytic reactors.* Ind. Eng. Chem. Process Des. Develop., 1971. **10**: p. 541.

16. Amphlett, J.C., R.F. Mann, and B.A. Peppley. *Performance and operating characteristics of methanol steam-reforming catalysts for on-board fuel-cell hydrogen production.* in *Hydrogen energy progress XI.* 1996. Stuttgart, Germany: International Association for Hydrogen Energy.

17. Ledjeff-Hey, K., et al., *Compact hydrogen production systems for solid polymer fuel cells.* Journal of Power Sources, 1998. **71**: p. 199-207.

18. Werner, H., et al., *Reaction pathways in methanol oxidation : kinetic oscillations in the copper/oxygen system.* Catalysis Letters, 1997. **49**: p. 109-119.

A microreactor for in-situ hydrogen production by catalytic methanol reforming

Ashish V. Pattekar and Mayuresh V. Kothare[*]
Department of Chemical Engineering, Lehigh University, Bethlehem, PA18015, USA.
Sooraj V. Karnik and Miltiadis K. Hatalis
Department of Electrical Engineering, Lehigh University, Bethlehem, PA18015, USA.

Abstract

A detailed study of the theoretical and experimental issues involved in the design and operation of a silicon-based methanol microreformer is presented in this paper. Thermal simulations for heat loss from the reformer chip and calculations involving the endothermic heat effect of the reforming reaction were carried out to estimate the total power requirement for continuous operation of the reformer. Micromachining technology was utilized to fabricate a prototype silicon microchannel based reformer of channel cross-section $1000\mu m$ x $230\mu m$ with a copper layer of thickness ~ $33nm$ as catalyst. The reactor chip was interfaced with the tubing for reactant and product transport using a stainless steel housing machined to exact dimensions and flexible graphite pads to provide gas-tight seals and excellent thermal contact between the chip and the housing for external heating of the reformer. Runs of this prototype microreformer were carried out using a custom made experimental setup for generation of the reactant mixtures, temperature control of the microreactor and online measurement of the product gas composition. Results from test runs of the microreformer are presented and strategies for improving the hydrogen yield are discussed.

1. Introduction

The use of microreactors for in-situ and on-demand chemical production is gaining increasing importance as the field of microreaction engineering matures from the stage of being a regarded as a theoretical concept to a technology with significant industrial applications. Various research groups have successfully developed microreactors for chemical processing applications such as partial oxidation of ammonia [10], nitration [2] and chemical detection [4]. The objective of the research effort at the Integrated Microchemical Systems Laboratory at Lehigh University's Chemical Engineering Department is to demonstrate a working microreaction system for use as a sustained source of hydrogen fuel for proton exchange membrane (PEM) fuel cells through catalytic steam reforming of methanol. The complete reformer-fuel cell unit is proposed as an alternative to conventional portable sources of electricity such as batteries for laptop computers and mobile phones due to its ability to provide an uninterrupted supply of electricity as long as a supply of methanol and water can be provided. Though considerable work already exists in literature on the catalytic steam reforming of methanol for production of hydrogen using conventional reactors [1,7,8], the use of microreactors for in-situ methanol reforming is a relatively new idea [3,6,11,12]. Literature on the macro-scale steam reforming of methanol includes analysis of the reaction thermodynamics [1] for prediction of optimum reactor temperature and feed compositions, catalyst characterization studies [7], and experimental studies on macro-scale pilot reactors [8]. Results obtained in the study of methanol reforming in these conventional reactors form a good background for the development of microreactors for this purpose. Silicon is considered a good material for fabrication of microreactors due to the high strength of the Si-Si bonds which results in the chemical inertness and thermal stability of silicon. Well established silicon micromachining techniques commonly used in the microelectronics industry facilitate easy fabrication of microchannels and other desired features on silicon substrates thus making silicon the 'material of choice' for microreactor fabrication.

In the following sections we discuss the theoretical and experimental issues in the development of a prototype silicon chip based microreformer. Preliminary calculations giving an idea of the power required to operate the microreformer are presented followed by a description of the microreformer fabrication procedure. The experimental setup for carrying out test runs of the microreformer is described and results from the test runs are presented.

2. Preliminary Simulations

Preliminary simulations on a Silicon chip of size $2cm$ x $2cm$ were carried out using the MEMCAD design and simulation software [9] to study the effects of fluid flow and heat loss from a Silicon chip operating at the reforming temperature of 500K. Figure 1a shows the velocity profile (velocity units: $\mu m/sec$) of the fluid (water) flowing through microchannels of cross section $1000\mu m$ x $500\mu m$. The channel geometry is such that the fluid enters along the negative z axis direction through the two inlets, mixes at the intersection of the two inlets and leaves along the positive z axis direction at the outlet. The boundary condition used for the flow simulation was a constant flow rate of 0.1 cc/sec at each inlet with the outlet being at atmospheric pressure, with no slip at the channel walls. Figure 1b shows the results of a thermal simulation on the same chip with a temperature boundary condition of 500K at the bottom wall of the microchannels in addition to the flow boundary conditions described above. The fluid inlet temperature was taken as the room temperature (300K) and the heat loss from the rest of the chip surface was governed by natural convection to ambient air at 300K.

334

Figure 1(a):
Flow through a simple microchannel

Figure 1(b):
Temperature gradients within bulk Si

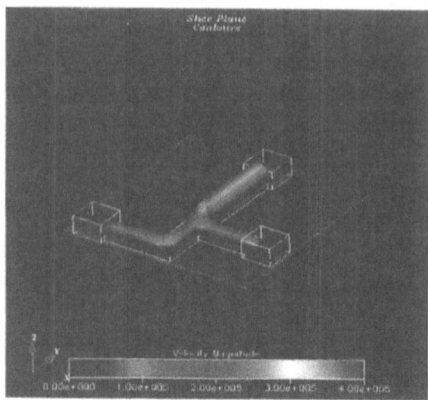

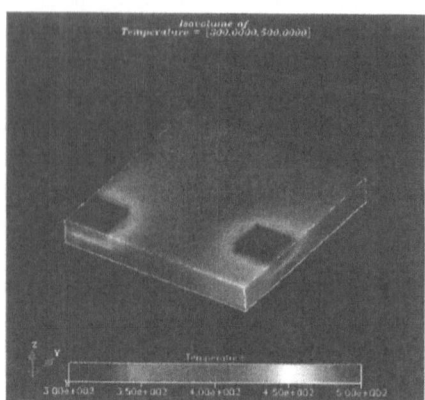

As can be seen in Figure 1b, most of the chip surface reaches a temperature very close to the maximum of 500K within a short distance from the inlets, which are at 300K due to the incoming fluid being at room temperature. The excellent thermal conductivity of bulk silicon results in close to uniform temperature distributions within the bulk with most of the chip temperature being very close to the maximum (500K for this calculation) when the heat loss from the chip to the surrounding is governed by relatively weaker modes of heat transfer such as natural convection. This conclusion is especially relevant for the operation of high temperature microreactors since this means that Silicon based microreactors for high temperature reactions will have most of their surface at the high reaction temperature resulting in significant heat losses to the surrounding unless sufficient thermal isolation from the ambient is provided. Since the optimum temperature for the methanol reforming reaction is about 250 °C [1], it follows that heat integration and thermal isolation issues will be extremely important for efficient operation of the microreformer. For example, the heat loss from a Silicon chip of size $2cm$ x $2cm$ operating at 250 °C is about 4.5 *Watt* as governed by natural convection to air at 25 °C with a natural convection heat transfer coefficient of 25 W/m^2K. A hydrogen flow rate of 0.0233 gmol/hr is needed for a 20-Watt fuel cell assuming 80% hydrogen utilization, which translates into a methanol flow rate of 0.0129 gmol/hr at 60% conversion inside the reformer. Then the power required for the endothermic reforming reaction (with the reformer operating at 250 °C) at this flow rate is about 0.127 Watt as shown in the calculations below using data on the heat of reaction and specific heats of the components of the reaction mixture[5].

The steam reforming of methanol for hydrogen production in the presence of Cu/ZnO catalyst involves the following reactions at 250 °C:

Primary reactions:

$$CH_3OH + H_2O \Leftrightarrow CO_2 + 3H_2 \qquad (\Delta H_{298K} = 48.96 \text{ KJ/mol}) \qquad (1)$$
$$CH_3OH \Leftrightarrow CO + 2H_2 \qquad (\Delta H_{298K} = 90.13 \text{ KJ/mol}) \qquad (2)$$

Secondary Reaction:

$$CO_2 + H_2 \Leftrightarrow CO + H_2O \qquad (\Delta H_{298K} = 41.17 \text{ KJ/mol}) \qquad (3)$$

Considering only the main reforming reaction,

$$\Delta H_{298K} = 48.96 \text{ KJ/mol}$$

$$\Rightarrow \Delta H_{250C} = \Delta H_{523K} = \Delta H_{298K} + \int_{298}^{523} \Delta Cp(T)dT = 59.17 \text{ KJ/mol}$$

$\Rightarrow$ Heat of reaction requirement for endothermic reforming reaction
$$= 59.17 \times 0.0129 \times 0.6 / 3600 \text{ KJ/sec} = \textbf{0.127 Watt}$$

Thus a large part of the power required to run the reformer is due to heat loss from the reformer chip to the surrounding and can be reduced by effective packaging and insulation of the microreformer.

3. Microreformer Fabrication

Fabrication of reactor microchannels of cross sectional dimensions $1000 \mu m$ x $230 \mu m$ on the prototype silicon reformer chips was carried out using standard silicon micromachining techniques such as Photolithography and KOH etching. We used a single mask process to get four identical microreformers from a single $100 mm$ silicon wafer polished on both the sides. Silicon nitride was deposited by plasma enhanced chemical vapor deposition (PECVD) on both sides of the wafer. Photolithography was done on the back side of the wafer to obtain the microchannel pattern. Silicon nitride on the back side was then etched by plasma etch to get pattened nitride which acts as an etch stop during KOH etching of silicon. Bulk silicon was etched in KOH from the back side to obtain microchannels of desired depths. About 33 nm of copper (catalyst layer) was then sputter deposited. Figure 2 illustrates the major steps involved in the fabrication of the microreformer chips. As will be discussed in the section on the experimental set-up and external interfacing of the microreformer, one end of the channel needs to have a hole in the silicon substrate for the dosing of reactants into the channels. This hole was made after fabrication of the channels by etching through the wafer with KOH. A protective coating of black wax prevented etching of the rest of the wafer. This technique exposes only one end of the channel (an area of approximately $1 mm^2$) to the etchant, which results in the desired hole after completion of the etching process. This step is illustrated in Figure 2(d). Figure 3 shows a schematic of the final reformer chip.

Figure 2: Microreformer fabrication steps

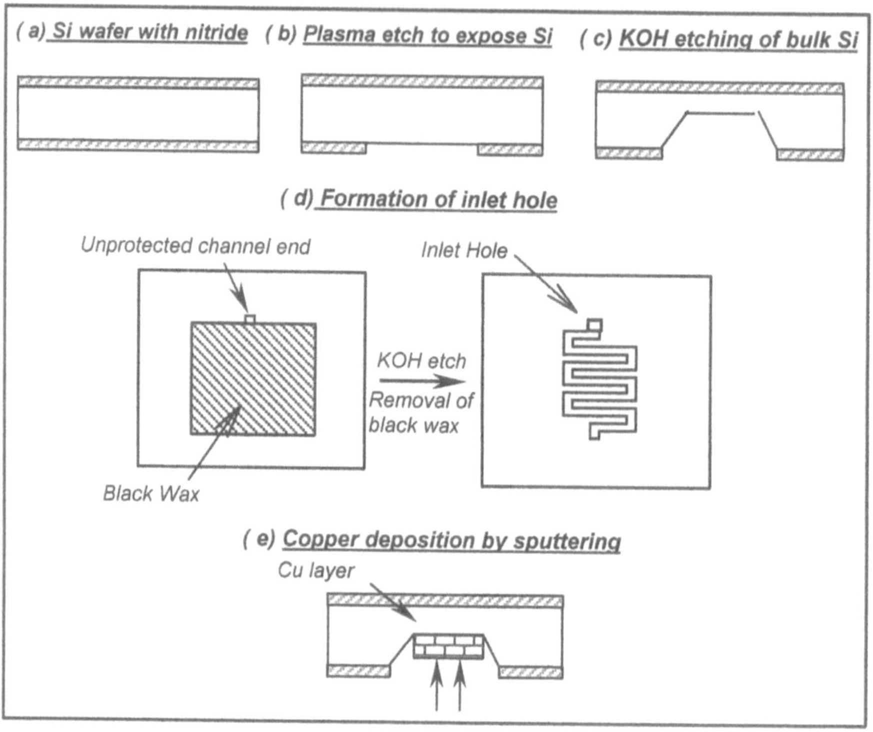

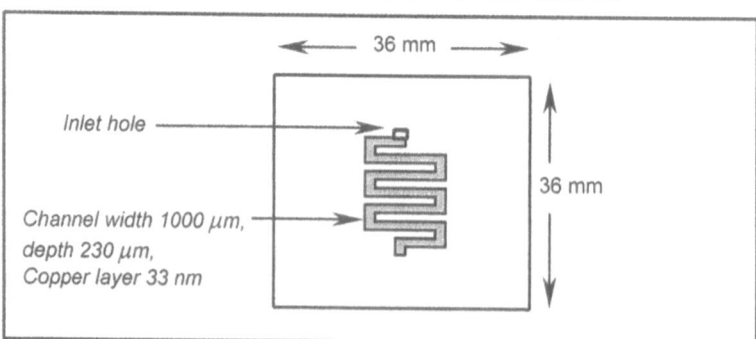

(a) *Si wafer with nitride*   (b) *Plasma etch to expose Si*   (c) *KOH etching of bulk Si*

(d) *Formation of inlet hole*

Unprotected channel end

Inlet Hole

KOH etch
→
Removal of
black wax

Black Wax

(e) *Copper deposition by sputtering*

Cu layer

Figure 3: The fabricated microreformer

← 36 mm →

Inlet hole

36 mm

Channel width 1000 μm,
depth 230 μm,
Copper layer 33 nm

4. <u>External Interfacing</u>

The interfacing of the microreformer chip to tubing for reactant and product gas transport was done using a custom made stainless steel housing shown in Figures 4 and 5. The steel housing blocks have ridges of appropriate length connecting the central bore to appropriate points on the reformer. The ridge on one of the steel blocks was machined to face the reactant inlet hole on the reformer chip when the chip was properly aligned with the housing. Similarly, the product gases exit from the other end of the channel through a similar ridge in the housing block machined to face the other end of the channel when aligned correctly.

Figure 4: Schematic of microreformer housing

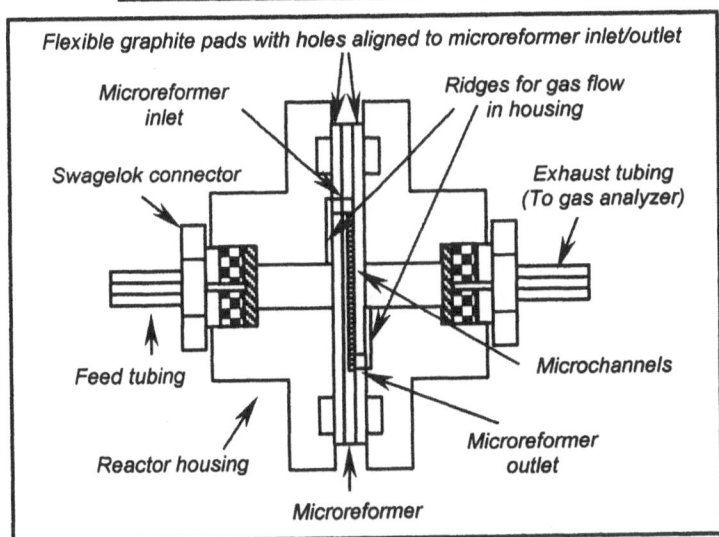

Leak-proof sealing between the reformer chip and the steel housing was ensured by using flexible graphite pads acting as gaskets with holes drilled at the right positions for reactant and product transport. Thus the reformer chip was covered from either side by a graphite covering pad to provide a leak-proof conduit for reactant and product gas transport between the housing and the reformer. The tubing for gas transport was connected to both sides of the housing via standard Swagelok connectors so that the reactants entered the housing-microreformer assembly from one side, passed to the microreformer through the hole in the silicon substrate and exited the reformer through a similar hole in the covering graphite pad at the other end of the microchannel after flowing over the deposited catalyst. The advantage of using flexible graphite pads for sealing purposes was that graphite being a good conductor of heat, the reactor temperature could be maintained at 250 °C by heating the steel housing using heating tapes and suitable temperature controllers. This allowed testing of the reformer chip without having to fabricate on-chip resistive heaters and temperature sensors, which will be integrated with the microreformer in future.

Figure 5: Microreformer housing

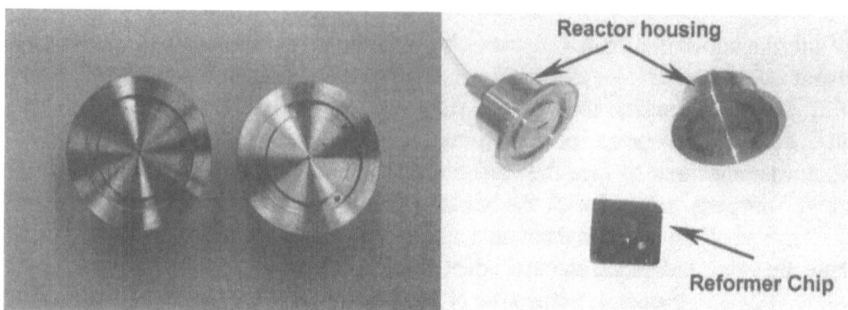

5. Experimental set-up

The operation of the microreformer involved:

1. Setting up of the methanol vapor and steam source
2. Interfacing of the microreformer chip to external tubing for transport of reactants and products as discussed in Section 4, and
3. Connection of the reactor exhaust tubing to the gas analysis equipment (a quadrupole mass spectrometer) for online analysis of the product gas composition.

A gas-tight sample cylinder containing a liquid methanol-water mixture immersed in a constant temperature hot water bath maintained at temperatures in the range 80 °C – 95 °C was used for generation of methanol vapor-steam mixtures of desired compositions by manipulating the composition of the liquid mixture in the sample cylinder on the basis of methanol-water vapor-liquid equilibrium (VLE) data. Analysis of the VLE data for the mixture and pressure drop calculations from simulations using MEMCAD confirmed that enough pressures could be generated in this set-up to drive the flow of the reactants through the reformer. The product gas tubing coming from the microreformer housing connected to the mass spectrometer, which gave an online analysis of the composition of the product gases. Figure 6 shows a schematic of this set-up. Electric heating tapes and simple on-off controllers with thermocouple probes were used to maintain the temperature of the connecting tubing above 100 °C to avoid condensation of the vapors during flow. A pressure transducer-strain gage meter combination was used to continuously monitor the pressure in the sample cylinder. The reactor housing was also maintained at the temperature required for the reforming reaction (around 250 °C) using electric heating tapes and an auto-tuning PID controller with a thermocouple sensor which maintained the reactor temperature to within 1 °C of the setpoint.

Figure 6: Schematic of experimental set-up

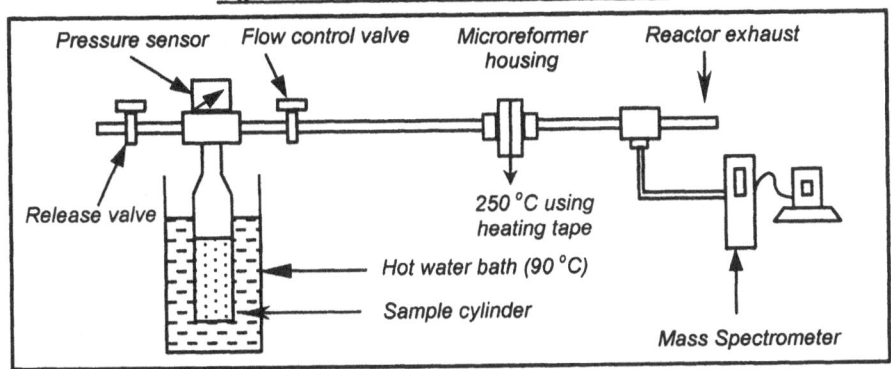

The gas analyzer used for analysis of the product gas composition was a quadrupole ion-trap mass spectrometer based residual gas analyzer (RGA) with a pressure-reducing inlet allowing the sampling of gases at atmospheric pressure. The pressure reducing inlet capillary of the analyzer was also maintained at above 100 °C using heating tapes to avoid any condensation before the pressure was sufficiently reduced to obviate condensation of water and methanol at room temperature inside the capillary. The mass spectra were collected at short intervals and the peak amplitudes were converted to mole fraction data for conversion calculations.

6. Results and Discussion

I. Test Runs

Test runs of the microreformer were carried out to verify hydrogen production capability inside the microreformer and to get an idea of the conversion possible using this set-up. Though the reformer design wasn't optimized from the point of view of conversion or hydrogen yield, the preliminary runs gaveimportant insights into the operation of the microreformer in this set-up. For the test run reported here a 50:50 mixture of methanol and water was fed into the sample cylinder maintained at 85 °C using the hot water bath shown in Figure 6. Using methanol-water VLE data at 85 °C obtained using the UNIFAC method in Aspen Plus, the vapor mixture at equilibrium with this liquid will have methanol vapor mole fraction ($Y_{methanol}$) of about 0.76. The methanol vapor-steam mixtures generated in this manner were passed through the reformer maintained at 250 °C and the product gas mass spectra were obtained at regular intervals during the run. The mass spectra were converted to mole fraction data using the library for molecule fragmentation inside the mass spectrometer ionizer. A plot of the exit gas composition (*mole fraction*) v/s time (*sec.*) for a microreformer test run of 500 sec is shown in Figure 7. As can be seen from the plot, sufficient amount of hydrogen was detected at the reactor exit to confirm that reforming was indeed taking place in the microreactor.

Figure 7: Microreformer product gas composition (mole fraction) v/s time (sec)

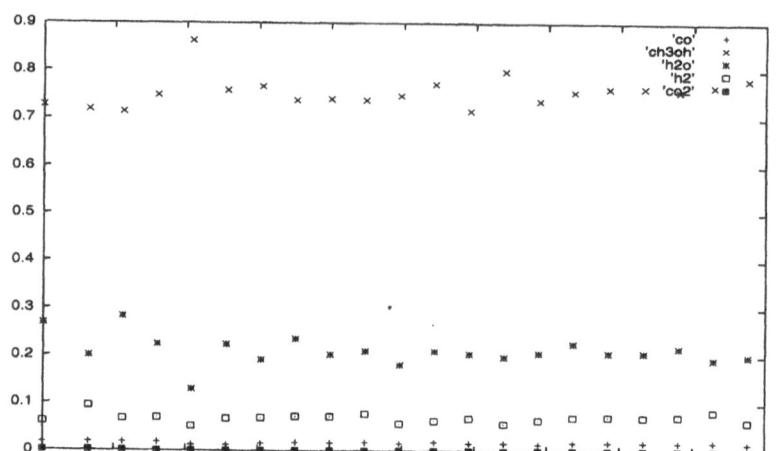

Average values of the mole fraction of each component in the product gas for this run can be used to get a crude estimate of the hydrogen yield. These values were 0.755 for methanol, 0.21 for water, 0.070 for hydrogen, 0.00129 for carbon dioxide and 0.017 for carbon monoxide. Though the hydrogen yield obtained in this prototype unoptimized microreactor was not significant (approximately about 0.092 moles H_2 produced per mole CH_3OH fed into the reformer with an average flow rate of about 2.36×10^{-3} gm/sec of methanol-steam mixture at the microreformer inlet with inlet methanol mole fraction of 0.76) these results demonstrate the capability to carry out methanol reforming in a microreactor and to test the microreactor operation at different operating conditions. Future work will involve attempts to improve the hydrogen yield by increasing the catalyst contact area per unit channel length and optimizing microchannel geometries for more efficient utilization of available total chip area. Experimental runs with reformer chips having the composite copper-zinc oxide catalyst instead of using only a copper layer will also be carried out to study the effect of catalyst properties on reaction extent.

II. Catalyst Deactivation

Referring to the chemical reactions (equations 1, 2 and 3 in Section 2) it should be noted that the amount of CO and CO_2 produced in the test runs as obtained from the mass spectra was less than what would be expected from the reaction stoichiometry for the observed production of H_2. One possible reason for this could be the deposition of minute quantities of carbon during the reaction in the microchannels which would reduce the amount of CO and CO_2 present in the exit gases [1]. Carbon deposition on the catalyst surface can have several undesirable

341

effects affecting the product purity. It was also observed that the copper layer in the microreformer degraded and turned blue (as shown in Figure 8) after several test runs probably due to the formation of some compound of copper in the presence of the reaction mixture at the high reforming temperature of 250 °C. These issues will be important in refining the design of the microreformer since catalyst deactivation will lead to lower yields and some form of catalyst regeneration will have to be incorporated for restoring good efficiency of the reformer operation.

Figure 8: Degradation of copper layer in the microreformer

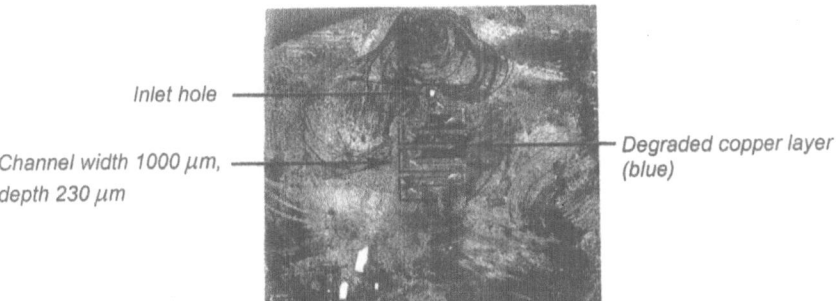

Inlet hole —

Channel width 1000 µm, —
depth 230 µm

— Degraded copper layer (blue)

III. Conclusion
The main contribution of this work is the demonstration of a microchannel based methanol reformer for small-scale hydrogen production and the development of a set-up for carrying out experimental runs of the microreformer. A prototype silicon based microreformer was successfully fabricated and tested using a setup designed for easy operation and replacement of microreformer chips and testing of different microreformer designs. Preliminary results from the test runs confirmed the presence of hydrogen at the reformer exit and revealed the need for extensive optimization of the reformer design for obtaining good hydrogen yields. Catalyst degradation in the microchannels was also observed. The optimization of the reformer design and methods to deal with catalyst deactivation issues will be an important part of future work in this project.

Acknowledgment: Financial support for this project from the U.S. National Science Foundation under the 'XYZ-on-a-chip' grant CTS-9980781, the Pittsburgh Digital Greenhouse and Sandia National Laboratories is gratefully acknowledged.

We also acknowledge the assistance of undergraduate Chemical Engineering student Joshua Tilghman in machining of the microreactor housing and setting up of the experimental apparatus for testing the microreformer.

References

[1] Amphlett, J., M. Evans, R. Jones, R Mann and R. Weir. Hydrogen production by the catalytic steam reforming of methanol Part 1: The thermodynamics. *Can. J. Chem. Engng.* 59:720-727, 1981.

[2] Antes, J., T. Tuercke, E. Marioth, K. Schmid, H. Krause and S. Loebbecke. Use of microreactors for nitration processes. *Proceedings of the Fourth International Conference on Microreaction Technology.* 194-200. Atlanta, GA, 2000.

[3] Fitzgerald, S., R. Wegeng, A. Tonkovich, Y. Wang, H. Freeman, J. Marco, G. Roberts and D. VanderWeil. A compact steam reforming reactor for use in an automotive fuel processor. *Proceedings of the Fourth International Conference on Microreaction Technology.* 358-363. Atlanta, GA, 2000.

[4] Floyd, T., K. Jensen and M. Schmidt. Towards integration of chemical detection for liquid phase microchannel reactors. *Proceedings of the Fourth International Conference on Microreaction Technology.* 461-466. Atlanta, GA, 2000.

[5] Perry, R., and C. Chilton. Chemical Engineers' Handbook. *Mc.Graw-Hill Book Company.* 1973.

[6] Pfeifer, P., M. Fichtner, K. Schubert, M. Liauw and G. Emig. Microstructured catalysts for methanol steam reforming. *Proceedings of the Third International Conference on Microreaction Technology.* 372-382. Frankfurt, Germany, 1999.

[7] Raphel, O., and N. Bakhshi. Production of Hydrogen from Methanol. 1. Catalyst characterization studies. *Ind. Eng. Chem. Res.* 33:2047-2055, 1994.

[8] Raphel, O., and N. Bakhshi. Production of Hydrogen from Methanol. 2. Experimental studies. *Ind. Eng. Chem. Res.* 33:2056-2065, 1994.

[9] Senturia, S., R. Harris, B. Johnson, S. Kim, K. Nabors, M. Shulman and J. White. A computer-aided design system for microelectromechanical systems (MEMCAD). *Journal of Microelectromechanical Systems.* 1:3-13, 1992.

[10] Srinivasan, R., I. Hsing, P. Berger, K. Jensen, S. Firebaugh, M. Schmidt, M. Harold, J. Lerou and J. Ryley. Micromachined reactors for catalytic partial oxidation reactions. *AIChE Journal.* 43:3059-3069, 1997.

[11] Tonkovich, A., S. Fitzgerald, J. Zilka, M. LaMont, Y. Wang, D. Vander-Wiel and R. Wegeng. Microchannel chemical reactors for fuel processing applications. *Proceedings of the Third International Conference on Microreaction Technology.* 364-371. Frankfurt, Germany, 1999.

[12] Zilka-Marco, J., A. Tonkovich, M. LaMont, S. Fitzgerald, D. VanderWiel, Y. Wang and R. Wegeng. Compact microchannel fuel vaporizer for automotive applications. *Proceedings of the Fourth International Conference onMicroreaction Technology.* 301-307. Atlanta, GA, 2000.

Applications for Micro Chemical and Thermal Systems

David L. Brenchley

Pacific Northwest National Laboratory, Richland, Washington 99352
Prepared for presentation at IMRET 5, May 27-30, 2001

Abstract

Pacific Northwest National Laboratory (PNNL) is developing a variety of applications using Micro Chemical and Thermal Systems (MICRO-CATS). The products represent a decade of research and development by PNNL in miniature systems. Miniature devices include fuel vaporizers, fuel processors, fuel cells, heat pumps, heaters, power generators, heat exchangers, and chemical processing units. The sponsors for their development include the Office of Transportation (OTT) in the U.S. Department of Energy (DOE), National Aeronautics and Space Administration (NASA), Department of Defense (Army), the Defense Advanced Research Projects Administration (DARPA) and others. OTT wants to have fuel cells in automobiles. DARPA needs miniature heating, cooling and power generation units for soldiers in the field. NASA wants miniature chemical plants to produce oxygen, fuel, and other things from the resources available in outer space. For these microtechnology users, the status of current development and the promise for future applications of miniature systems are shown.

Pacific Northwest National Laboratory

Pacific Northwest National Laboratory (PNNL) is a pioneer in developing Micro-Chemical and Thermal Systems (MICRO-CATS™). The range of research, development, and products are illustrated on our MICRO-CATS web site (http://www.pnl.gov/microcats) and shown in Figure 1.

MICRO-CATS Micro Chemical and Thermal Systems

R&D 100 Winners
(select picture above to see paper)

Why Microtechnology?

More About Us

- Technical Staff
- Research Projects
- Capabilities
- Publications
- Doing Business With Us
- Conferences
- Search

Contact Us

- More Information
- Your Interests

What's New

Related Websites

Microtechnology Menu	Industrial Chemical Processing
Buildings	Military
Carbon Management	Space Exploration
Environmental Restoration	Transportation

Microtechnology doesn't work for us until it works for you.

Welcome to the Micro Chemical and Thermal Systems (MICRO-CATS™) website. The Pacific Northwest National Laboratory, operated by Battelle, is a recognized leader in the development of MICRO-CATS for a variety of applications. Wherever chemical reactions and heat are used, our microsystems can give you important advantages. We invite you to browse our site and contact us about your needs. Let's talk.

Figure 1. MICRO-CATS Web Page

Microtech-Based Applications

The potential exists for every chemical engineering unit operation to be miniaturized. When thermal and chemical systems are miniaturized, their overall efficiency and productivity improve. PNNL uses miniaturization to develop products with the following inherent microtech advantages:

- Lightweight and compact
- Rapid heat and mass transport
- Extremely precise control of process conditions
- High performance (i.e., high throughput per unit hardware volume)
- Cost economies through mass production
- Available for distributed or mobile applications
- Short lead times for development and deployment.

Just as rapid advances in microelectronics have revolutionized computers, appliances, communication systems, and many other devices, PNNL's efforts in creating MICRO-CATS seek to revolutionize heat exchangers, heat pumps, combustors, gas absorbers, solvent extractors, fuel processors, and many other devices. These lightweight, compact, high-performance systems can work in many applications in transportation, buildings, military, environmental restoration, space exploration, carbon management, and industrial chemical processing.

PNNL's patented "sheet architecture" [U.S. Patent No. 5,611,214 (issued March 18, 1997) and U.S. Patent No. 5,811,062 (issued September 22, 1998)] provides a format for mass production of chemical and thermal systems. In addition to the "sheet architecture," PNNL has pioneered microfabrication methods for microdevices. We make microdevices with a variety of fabrication methods using copper, aluminum, stainless steel, high-temperature alloys, plastics, and ceramic materials. With all of our clients we stress that "Microtechnology doesn't work for us until it works for you."

Microthermal Systems Applications

Microthermal systems focus on devices such as heat exchangers and heat pumps. Using its "sheet architecture" and microfabrication capabilities, PNNL crafts novel microthermal systems. Two examples of this strategy won R&D 100 Awards: the Compact Microchannel Fuel Vaporizer and the Microheater. Both of these products target applications that give the microtech-based devices considerable market advantages.

The compact microchannel fuel vaporizer (CMFV) (Figure 2) is a breakthrough in miniaturizing process technology and a key component of a microchannel fuel processor that will enable fuel cell-powered vehicles. Fuel cells are a high-efficiency, low-emission alternative to the internal combustion engine and run on hydrogen. Currently, no infrastructure exists to deliver either gaseous or liquid hydrogen to fuel cell-powered automobiles, but a gasoline-distribution infrastructure does exist. The CMFV can vaporize the requisite amount of gasoline needed to power a 50-kWe fuel cell and is small enough (0.3-L) for portable or automotive applications.

The Microheater (Figure 3) is a microscale combustion system (the palm-size combustion unit weighs less than 0.2 kg [5 oz]). It can provide heat for portable personal heating/cooling devices, indoor heating devices such as baseboard heaters, in-line water heaters, and fuel cell systems. It is 10 times smaller and lighter than conventional combustors. The Microheater can produce 30 W of thermal energy per square

Figure 2. CMFV

Figure 3. Microheater

centimeter of external combustor area. One module can power a personal, portable heater for eight hours on little fuel or provide instantaneous in-line water heating; an array of modules will heat a house efficiently and reduce ducting and zoning thermal energy losses by 45%. This technology is the first application of enhanced microscale heat and energy mass-transfer to a combustion process.

Micro Chemical Systems Applications

Micro chemical systems focus on such devices as reactors and chemical separations. Miniaturization of chemical processes increases the process intensity. Examples of PNNL's use of microreaction technology include fuel processor development for the U.S. Department of Energy (DOE), man-portable power development for the Army, milliwatt power development for DARPA, and propellant processing for NASA.

Microtechnology is being used to develop a compact hydrogen generator for fuel cells that can generate 50 kW-electric for automotive use. This effort, funded by DOE's Office of Transportation Technology, is now demonstrating high performance in compact units. Over the past year, the project has concentrated most of its effort on demonstrating an overall microchannel steam reforming system, including four microchannel steam reformers and more than 24 microchannel heat exchangers, which, as a system, are intended to provide both high energy-efficiencies and high power densities. Work is also under way on other microchannel components that may ultimately find value within an automotive fuel processing system or within distributed power systems.

PNNL is developing a man-portable power system that will continuously provide 10 to 100 watts of base-load electric power for weeks or months using a microtechnology-based fuel processor. This lightweight, compact system is suitable for long-duration missions. In the fuel processor, hydrogen for the fuel cell is produced from liquid hydrocarbon fuel. The fuel processor consists of a primary fuel-reforming reactor (endothermic), a water-gas shift reactor (exothermic), and a method of removing carbon monoxide. The power generation module consists of liquid hydrocarbon fuel storage, a microchannel fuel processor, and a microscale fuel cell. A lithium polymer battery could be included in the system for startup and peak power.

Under a program for the Defense Advanced Research Projects Agency (DARPA), Battelle, Pacific Northwest Division (Battelle) and Case Western Reserve University (CWRU) are developing and demonstrating an integrated fuel cell and fuel processor for microscale (10- to 500-mWe) power generation. The system includes a microscale fuel processor, which produces hydrogen from liquid fuels such as methanol, butane, JP-8, or diesel; and a microscale fuel cell that will use the hydrogen as fuel to produce electric power. This alternative power source has many potential advantages over batteries for operating wireless electronic devices (e.g., microsensors and microelectromechanical systems), especially in terms of energy and power densities. This technology broadens the possibilities for using self-sustaining devices in remote or difficult-to-access locations. In this application, the fuel from storage is mixed with air and water in a fuel processor system that operates at 600° to 700°C for butane and 250° to 550°C for methanol. The overall power system is composed of several subsystems: fuel storage, fuel processor, synthesis gas treatment (optional), and fuel cell, along with associated peripherals such as pumps and control valves. The fuel processor contains a reforming reactor, combustion reactor, and heat exchangers. Heat generated in the combustion reactor is transferred to the endothermic reforming reactor to produce the H_2-containing synthesis gas. The synthesis gas may be processed directly in the fuel cell or treated to minimize the CO concentration, depending on the type of fuel cell. The fuel cell converts H_2 and O_2 (from air) to electrical power and water.

The NASA In Situ Resource Utilization (ISRU) program plans to include chemical processes for converting the carbon dioxide and possibly water from the environment on Mars into propellants, oxygen, and other useful chemicals. The use of such indigenous resources significantly reduces the size and weight of the payloads that need to be lifted off from Earth. The in situ propellant production involves collecting and pressurizing the atmospheric carbon dioxide, conversion reactions, chemical separations, heat exchangers, and cryogenic storage. PNNL's systems study demonstrated that microtechnology could provide significant size, weight, and energy-efficiency gains. In this system, energy management is very important. First of all, energy is a scarce resource. Secondly, heat rejection is a problem because the low-pressure atmosphere makes convective heat transfer ineffective. Mictrotechnology-based systems are attractive because processing fluids in microchannels can change the governing physics dramatically. The gravitational forces are secondary to surface, interfacial, and hydrodynamic forces. Therefore, microscale technology very naturally overcomes the challenges and limitations of operating in reduced gravity and micro gravity environments.

Revolutionary Science and Technology

Nanotechnology will have a significant impact on microscale technology of the future (see Figure 4). Nanotechnology is changing fundamentally the way materials and devices will be produced. With the ability to build things atom-by-atom and molecule-by-molecule, there will be new classes of structural materials.

Figure 4. Molecular Beam Epitaxy

PNNL is spearheading R&D in nanoscience and nanotechnology, especially in the areas of nanomaterials and nanobiology. The PNNL NANO website (http://www.pnl.gov/nano) tells how revolutionary new materials with enhanced surface properties will greatly improve microscale devices and systems. PNNL is in the business of discovering fundamental phenomena and applying this knowledge to the development of commercial products. Nanotechnology will enable products to be lighter, stronger, smarter, cheaper, cleaner, and more precise. All of these advances will provide many benefits to future microscale systems. PNNL seeks to make revolutionary strides in putting nanotechnology to work for the benefit of humankind.

Acknowledgment

Pacific Northwest National Laboratory is operated by Battelle for the U.S. Department of Energy under Contract DE AC06-76RLO1830.

MIXING SIMULATION OF A ZIGZAG MICROCHANNEL:

A step towards the methoxylation of methyl-2-furoate

Virginie Mengeaud[1], Rosaria Ferrigno[2], Jacques Josserand[1] and Hubert H. Girault[1]

[1] Laboratoire d'Electrochimie, Ecole Polytechnique Fédérale de Lausanne
1015 Lausanne, Switzerland
[2] Department of Chemistry and Chemical Biology, Harvard University, 12 Oxford Street, Cambridge, MA 02138, USA

Abstract

Numerical simulations of laminar flows are a useful tool for designing effective mixing devices for microsystems. A finite element model has been used in order to examine the channel geometry for a zigzag micromixer integrating a ~Y" junction. The results illustrate the effects of flow rate and channel geometry on the hydrodynamics and mixing efficiency. Finite element simulations illustrate the mixing effect of laminar flow recirculations.

Keywords

finite element, simulation, mixing, zigzag channel

1 - Introduction

Downsizing reactor systems represents a novel approach for carrying out chemical reactions, since significant improvement of mass and heat transfer properties are achieved. So far, most of recent developments have been in the area of micro total analytical systems (μTAS) [1, 2]. The same idea adapted to synthesis has opened new horizons with the ~Plant-on-a-chip" concept involving the integration and miniaturisation of various independent units used in a single flow configuration such as for instance mixers, heat exchangers [3, 4], pumps and valves. From an industrial standpoint, this approach can offer a fast and low cost scale-up step via replication of the microreactor developed on bench scale [5]. Chemical microreactors incorporating a micro-heat exchanger and a micromixer unit have already demonstrated their efficiency compared to classical reactors. As far as the mixing step is concerned, the flow remains laminar in these systems, therefore turbulence does not play any role as in the case of macrosystems. Thus, layers of different concentrations mix only by molecular diffusion. The absence of turbulence introduce some limitations in the mixing efficiency. However these limitations are compensated by the small dimensions of the device. Due to this size effect, all the fluid is then submitted to the concentration gradient (no region of concentration stagnation).

Concerning others operation units, electrochemical and distillation processes are among the few examples that have been reported in literature [6]. In the electrochemistry approach, the decrease of the characteristic dimension of the electrodes (inter-electrode gap) has already shown some advantages. The reduction of the inter-electrode ohmic losses allows a better control of the current and potential distributions improving the selectivity and the energy consumption. For instance, the interdigitated band electrode configuration incorporated in a microcell have shown a high selectivity of the electrochemical methoxylation of furan [7]. An enhancement of the figures of merit (current efficiency and energy consumption) was also achieved even in the absence of supporting electrolyte.

Figure 1 shows a schematic diagram of a microsystem for the methoxylation of methyl-2-furoate which is currently under development in our lab.

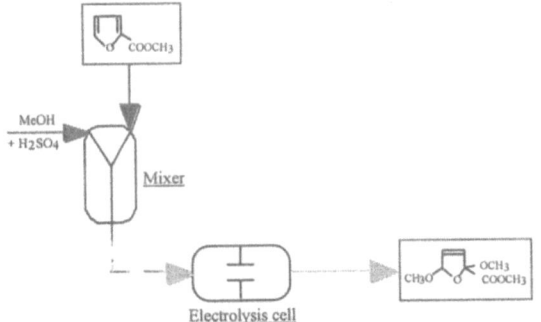

Figure 1: Final microsystem flow sheet, based on a interdigitated band electrode system.

The electrochemical methoxylation was investigated using platinum interdigitated band electrodes (inter-electrode gap of 100 μm) and keeping the same experimental conditions as in the literature (ester in presence of 0.1 M sulfuric acid (97%) in methanol solvent cooled down at 1°C) [8]. The current efficiency was improved from 54 to 96%. The energy consumption was decreased from 2.4 to 1.2 kWh.kg^{-1} for a lower cell potential. In order to scale down this synthesis, an electrochemical microreactor has been designed. The final structure is a sandwich type configuration containing several layers of ceramic sheets (thickness of 250 μm) laminated and sintered (including the interdigitated band platinum electrodes printed on the middle layer).

In order to perform this process on-line, methyl-2-furoate and the solvent employed during the electrolysis should be efficiently mixed prior entering the electrochemical micro-cell. For this purpose we have envisaged the design of a micromixer featuring a ~Y" junction followed by a zigzag channel. The outlet of the channel will eventually feed the microcell. The objective of the present paper is to characterise the mixing behaviour along the zigzag channel as a function of the geometrical parameters and the flow rate.

2 N Theory and numerical description

2.1. Mathematical model

The mixing process involves the momentum and mass transport phenomena. In this model, we consider the laminar mixing of species introduced in a microchannel via two similar fluids having same viscosity and same density. Our approach consists in solving the momentum and mass transport phenomena in two steps. This model is based on:

1- the variations of concentration do not modify the viscosity of the fluid (diluted solutions)

2- channel walls are supposed smooth (roughness and experimental imperfection are neglected)

3- wall capillary forces are neglected

First, the hydrodynamic behaviour of a laminar flow is described by the Navier-Stokes equation:

$$\rho \frac{\partial \mathbf{V}}{\partial t} + \rho \, \mathbf{V}.\nabla \mathbf{V} = -\nabla p + \rho \, g + \mu \, \nabla^2 \mathbf{V} \tag{1}$$

where $\mathbf{V}$ is the velocity vector, ρ is the fluid density, p the pressure, g the gravity and μ the fluid dynamic viscosity.

In this continuous flow configuration, we assume that a steady-state flow is already established at the inlets. The transient term in the above mentioned equation can be neglected. The velocity vectors have been calculated in Cartesian coordinates by solving Eq. (1) coupled with the continuity equation expressed here for an incompressible and non-expandable fluid ($\Box \rho = 0$):

$$\nabla.\mathbf{V} = 0 \tag{2}$$

For the hydrodynamic calculations, a parabolic Poiseuille profile is assumed in the $(x\text{-}y)$ flow direction for both inlet boundaries. In addition, the transversal components are taken as zero along the wall boundaries (no slip conditions).

The concentration distribution is finally obtained by solving the diffusion-convection equation in which is introduced the velocity solution:

$$\frac{\partial c}{\partial t} - D\nabla^2 c + \mathbf{V}.\nabla c = 0 \tag{3}$$

where c is the concentration of the specie and D its diffusion coefficient.

The velocity parameter $\mathbf{V}$ calculated in Eq. (1) and (2) is introduced in Eq. (3) as an input parameter in order to determine the concentration gradient all along the mixer. This approach assumes that the species have no effect on the flow viscosity. As boundary conditions, the parameter c is imposed as zero and as one at the first and at the second inlet respectively.

2.2. Numerical parameters

All the simulations have been carried out using the numerical software Flux Expert® (Simulog, 60 rue Lavoisier, 38 330 Montbonnot, France) based on the finite element method. A non-linear steady-state algorithm is used for hydrodynamic calculations whereas a linear steady-state one is used for the concentration solving. In this work, the equations are solved on two-dimensional Cartesian geometry and in a dimensional form, on a Silicon Graphics Indigo 2 workstation.

The channel geometry studied in this report is schematically shown in Figure 2. The 100 μm wide channel features zigzag angles of 90° and a ~Y" junction in order to connect the two inlets.

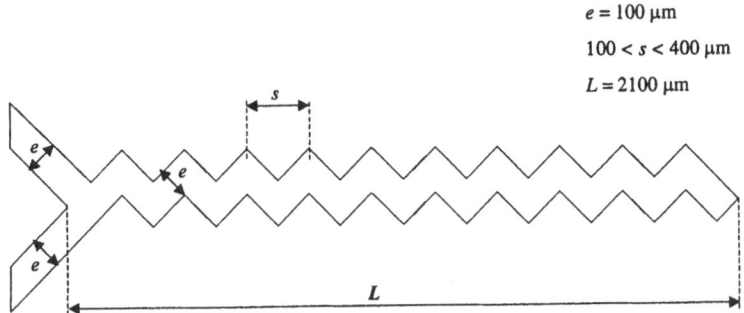

$$e = 100\ \mu m$$
$$100 < s < 400\ \mu m$$
$$L = 2100\ \mu m$$

Figure 2: Zigzag geometry of the mixing channel integrating a ~Y junction ", with e the width of the channel, s the length of the periodic step and L, the total length of mixing.

This work focuses on the effect of the length of the periodic step (parameter s), on the quality of the mixing. So, three zigzag derived geometries are studied corresponding to the cases where s is taken as 100, 200 and 400 μm. In addition, two reference geometries are taken into account in order to illustrate both asymptotic configurations: $s \rightarrow 0$ and $s \rightarrow \infty$. All the simulations have been achieved for flow velocity ranging from 10^{-3} to 1 m. s^{-1}. In order to highlight the effect of the flow rate without changing the chip length, the diffusion coefficient is adapted to maintain a constant length Peclet number, $Pe = \dfrac{\overline{V} L}{D}$. Therefore, D is taken as an artificial value from 10^{-9} to 10^{-6} m^2.s^{-1}. The classical Reynolds number is defined as:

$$Re = \frac{\rho \overline{V}\, Diam}{\mu} \qquad \text{with } \overline{V} = \frac{2}{3} V_{max} \text{ in a rectangular channel}$$

and where $\overline{V}$ is the average velocity and Diam the hydraulic diameter of the rectangular channel.

The values of the other parameters used are the following:

| Fluid density | $\rho = 10^3$ kg.m^{-3} |
| Fluid dynamic viscosity | $\mu = 10^{-3}$ Pa.s |

Two parameters have been defined in order to interpret the calculations. The first one is non-dimensional and represented by the ratio, s/e. The mixing efficiency is here quantified by the ratio C_{min}/C_{max} between the minimal and the maximal concentrations along a cross section of the channel at a given distance from the inlet.

3 N Results and discussion

The effect of the geometry on the mixing process is first investigated. In addition, the effect of the Reynolds number is quantified for a given zigzag channel configuration, showing recirculation phenomena.

3.1. Effect of different geometries

The role of the length of the periodic step, s is first investigated. The three different zigzag geometries studied are presented in Figure 3.

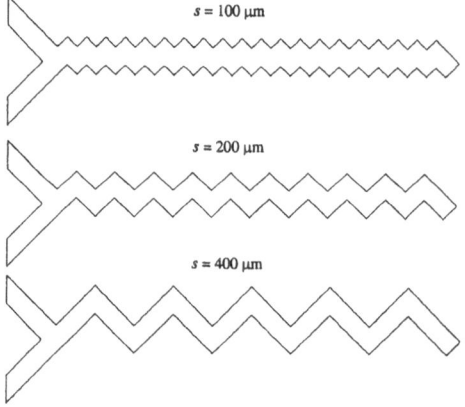

Figure 3: Different zigzag configurations investigated ($e = 100$ µm).

Both asymptotic geometries correspond to a straight channel whose features are reported in Table I.

Reference case	Asymptotic value for s	Channel width / µm	Channel length, L / µm
A	0	141	2100
B	∞	100	2850

Table I: Features of both asymptotic geometries.

Due to the zigzag configuration, for the limiting case A ($s \rightarrow 0$), the effective cross-section of the channel is increased by a factor of $\sqrt{2}$. Concerning the limiting case B ($s \rightarrow \infty$), the channel length is enhanced of the same factor.

For all the geometries, the mixing efficiency is calculated just before the end of the channel (2030 μm except 2760 μm for case B). As illustrated in Figure 4, this parameter is represented versus the geometry ratio for two different velocity values ($Re = 0.67$ and 66.7).

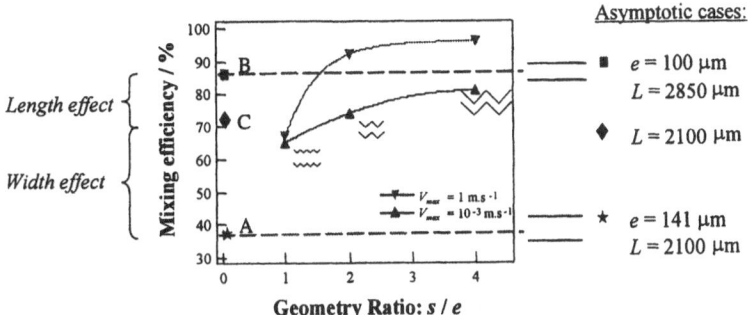

Figure 4: Effect of the geometry ratio on the mixing efficiency for two different velocities ($Re = 0.67$ and 66.7).

For 10^{-3} m.s^{-1}, the mixing efficiency varies from 65 to 81% when s increases from 100 to 400 μm. The most efficient zigzag configuration in our range is then obtained when $s = 400$ μm. This increase of the mixing efficiency is due to two effects:

- the reduction of the channel width decreasing the diffusion path (width effect)
- the increase of the length and thus the increase of the residence time (length effect)

The first effect is illustrated by comparing both cases A and B. A width reduction of 41 μm increases the mixing efficiency by a relative gain of 95%. The second effect is shown by the cases C and B with a relative gain of 13%.

Here are presented both extreme velocity cases ($Re = 0.7$ and 66.7). In a practical case, high flow rates imply long microchannels. In order to increase the velocity without changing the channel length, our simulations take into account different values of the diffusion coefficient. The flow rate parameter seems to have a great mixing effect for s / e value above 2 by improving the efficiency to 96%. Next study will correlate this effect with the existence of recirculation regions. Nevertheless,

3.2. Recirculation phenomenon

The following study is focused on the more efficient zigzag configuration corresponding to $s = 400$ μm. As already established, the mixing efficiency is calculated just before the end of the channel, at 2030 μm. The different cases of Reynolds number are reported in Table II. Numerical calculations are achieved for different diffusion coefficients in order to insure a constant length Peclet number of 1400.

Velocity, V_{max} / m.s^{-1}	Diffusion coefficient / m^2.s^{-1}	Péclet number	Reynolds number	Mixing efficiency / %
10^{-3}	10^{-9}	1400	0.07	81
10^{-2}	10^{-8}	1400	0.67	81
10^{-1}	10^{-7}	1400	6.67	81
3.10^{-1}	3.10^{-7}	1400	20	83
5.10^{-1}	5.10^{-7}	1400	33.33	87.5
1	10^{-6}	1400	66.66	96

Table II : Influence of the zigzag configuration s = 400 μm on the mixing efficiency for different velocities (Pe = 1400).

Mixing efficiency values are plotted versus the Reynolds number, see Figure 4 illustrating two mixing behaviours. For Reynolds numbers below to 10, the mixing efficiency remains constant at the end of the channel. Mixing is only governed by the diffusion length.

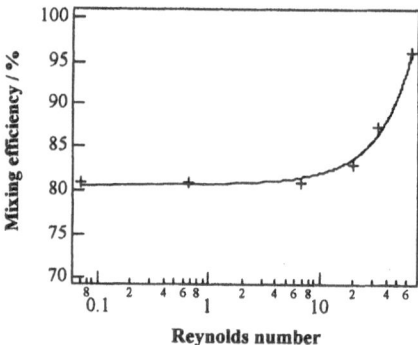

Figure 4: Mixing efficiency evolution with the Reynolds number for the zigzag geometry where s = 400 μm (Pe = 1400).

When the Reynolds number reaches a minimal value of 10 (V_{max} > 1,5.10^{-1}m.s^{-1}), the mixing is greatly improved due to the existence of laminar recirculations, see Figure 5. More precisely, back flows are localised along the walls of the channel after each angle. Therefore, diffusion and recirculation processes finally provide the mixing and improve the efficiency, as shown in Figure 6.

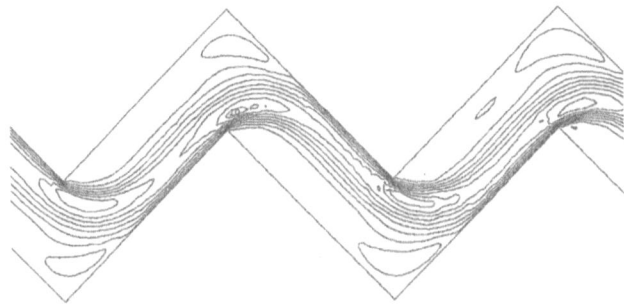

Figure 5: Isovalues of the V_x velocity component, at $V_{max} = 5.10^{-1}$ m.s^{-1} (Re = 33.3) for the zigzag geometry where s = 400 μm (Pe = 66.7).

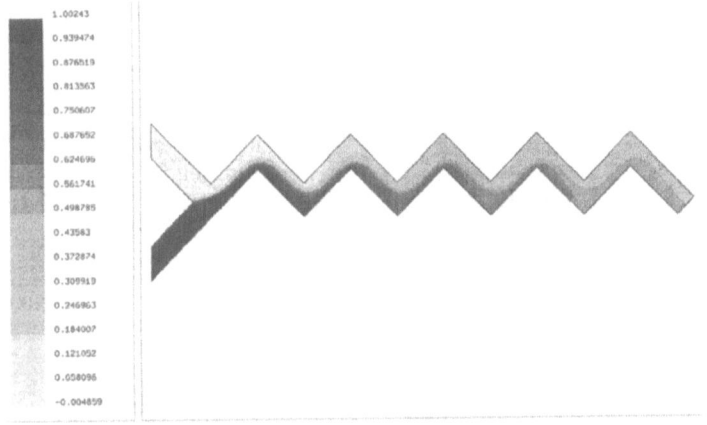

Figure 6: Isovalues of the concentration at $V_{max} = 5.10^{-1}$ m.s^{-1}, for the zigzag geometry where s = 400 μm (Pe = 66.7).

4 N Conclusion

For velocities of the order of 10^{-2}m.s^{-1}, zigzag geometries can provide mixing efficiencies up to 81%, which is slightly larger than for an equivalent straight channel. At higher velocities, laminar recirculation phenomena are observed along the channel walls increasing the mixing efficiency up to 96%. From a practical standpoint (without the artificial diffusion scaling factor used in this work), high velocity of 1 m.s^{-1} requires the use of rather long channels (of the order of 2 m) to insure a minimal residence time. Even if this corresponding geometry is not trivial to achieve on small chip (centimetre scale), it allows working at high throughput (with multi-channel configuration). Another alternative to increase the mixing efficiency at lower Reynolds number could rely on narrowing the angles of the zigzag geometry.

Acknowledgements

The authors would like to thank the Commission pour la Technologie et l'Innovation for the financial support.

References

[1] C. M. Henry, Anal. Chem., **71** (1999) 264A.

[2] A. T. Wooley, K. Q. Lao, A. N. Glazer, R. A. Mathies, Anal. Chem., **70** (1998) 684.

[3] W. Ehrfeld, V.Hessel, H. Lehr, *Microreactors for chemical synthesis and biotechnology - Current developments and future applications in* Topics in Current Chemistry, 194, Berlin, (1998), 233.

[4] K. P. Jäckel, Microtechnology : Application opportunities in the chemical industry, Microsystem technology for chemical and biochemical microreactors : DECHEMA Monograph., (1996), 29.

[5] W. Ehrfeld, K. Golbig, V. Hessel, H. Löwe, T. Richter, Ind.Eng.Chem.Res., **38** (1999) 1075.

[6] H. Löwe, W. Ehrfeld, Electrochim. Acta, **44** (1999) 3679.

[7] R. Ferrigno, V. Reid, H. H. Girault, Microreaction Technology : Third International Conference on Microreaction Technology, proceedings of IMRET 3, (1999), 294.

[8] N. Clauson-Kaas, F. Limborg, K. Glens, Acta Chem. Scand., **6** (1952) 551.

Fuel processor development for a soldier-portable fuel cell system

D. R. Palo, J. D. Holladay, R. T. Rozmiarek, C. E. Guzman-Leong,
Y. Wang, J. Hu, Y.-H. Chin, R. A. Dagle, E. G. Baker

Battelle, Pacific Northwest National Laboratory, Richland, WA, USA

1. Introduction

The remarkable recent advances in wireless and portable communications devices (e.g., laptop computers, cellular phones, portable digital assistants) have fueled a need for high-energy-density portable power sources for consumer use. Similarly, interest in portable power sources has increased in the military and intelligence communities. Currently, portable military electronics are dependent on batteries to supply electrical power for long-duration missions. This poses two major problems which result from the low energy density of current battery systems: excessive weight/bulk, and reduced mission duration.

Compact fuel cell systems that operate on liquid hydrocarbon fuels offer an efficient, lightweight alternative to batteries, thus allowing for greater portability and longer mission lifetime. For instance, the energy densities of diesel fuel and methanol are each at least an order of magnitude greater than that of lithium-ion batteries. Another option, hydrogen storage (compressed or chemical), provides an energy density not much greater than can be found in batteries. As shown in Table 1, a hydrocarbon-based fuel cell system operating at just 5% overall efficiency has a higher energy density than a lithium polymer battery and at least equal energy density as a polymer-electrolyte membrane (PEM) fuel cell system operating on stored hydrogen. Clearly, liquid hydrocarbons would be the preferred energy source for a portable power system if a rugged, reliable, and lightweight fuel processor were available to convert hydrocarbon fuels to hydrogen. An appropriate fuel processor would efficiently produce hydrogen of sufficient quantity and purity to drive a PEM fuel cell, and would do so within a small volume.

Table 1. Comparison of energy densities from various sources

Fuel	LHV (kJ/mol)	Energy Density (kW-hr/kg)	Efficiency Required (to match batteries)
Methanol	639	5.5	5.5 %
n-Butane	2650	12.7	2.4 %
n-Octane	5100	12.4	2.4 %
H_2 storage	242	0.5-1.0	30-60 %
Lithium-ion battery	--	0.3 (projected)	--

Battelle is a leader in the development and demonstration of small-scale, hydrocarbon fuel processors to generate high-purity hydrogen for portable fuel cells. For several years, Battelle has led the development of micro-process technology for various applications and device sizes. These technologies include fuel vaporization,[1-4] gas conversion,[5-7] fuel processing,[2, 6-12] heat transfer,[1, 3, 4, 13, 14] mass transfer,[15, 16] catalytic combustion,[3, 4, 13, 17] and partial oxidation.[12] In each application, the microchannel architecture drastically reduces the heat- and mass-transfer resistances relative to conventional systems.

Many of these advances are built upon Battelle's aggressive effort to develop catalysts for microreactor applications. Battelle has developed highly active steam reforming catalysts for various hydrocarbon fuels. These catalysts were developed for millisecond contact time (GHSV = 10^4-10^5 hr^{-1}) applications for use with many different fuels, including methanol, butane, iso-octane, diesel, and JP-8.

Such micro-process and catalyst technology is, thus, a natural fit for portable power systems, where size and weight must be minimized. For instance, the required catalyst volume for a 15-W steam reformer operating at 100 ms contact time is less than 0.5 cm^3. This translates to correspondingly small device footprints, such that the majority of the weight and bulk of the final system is dictated by the related fuel supply rather than the fuel processor (based on multi-day missions).

2. Objectives

Under this development project, funded by the U.S. Army Communications-Electronics Command, Battelle is developing a 15-W$_e$ fuel processing system suitable for portable power applications. The need arises from a desire to reduce the weight and bulk, and thus increase the mobility and effectiveness of the Army's combat divisions. This portable power source is expected to provide a clean hydrogen stream to a small fuel cell according to the target specifications listed in Table 2.

Table 2. Target specifications for soldier-portable power system

Average Power	15 W
Peak Power	25 W
Volume	< 100 cm3
Weight (excluding fuel)	< 1 kg

For military use, the system will ultimately operate on diesel or some other logistics fuel. However, for commercial use and the first military applications, the fuel of choice is methanol. Not only can methanol be reformed at lower temperatures than other hydrocarbons, but it also is miscible with water, which is a key issue in fuel storage and delivery. The lower reforming temperatures of methanol allow for lower heat losses, require less insulation, and simplify the thermal management of the integrated system.

3. Experimental

Development of the portable fuel processor begins with the design, fabrication, and testing of individual unit operations. Afterward, these unit operations are integrated at the bread-board* level. This occurs simultaneously with catalyst testing and selection for both the combustor and the steam reformer. The next step, which is beyond the scope of this paper, includes the full integration of all unit operations in a single, rugged device suitable for field testing.

All catalyst preparation, device fabrication, system testing, and product analysis were performed on site. The current system is composed of stainless steel process units connected by 0.125-in or 0.250-in stainless steel tubing. Thermocouples and pressure transducers are placed strategically throughout the device, and system data is collected through an on-line data acquisition system.

Catalytic combustion of methanol is used to provide system heat which is necessary for reactant vaporization and preheat as well has for heat of reaction for the endothermic steam reforming of methanol (+50 kJ/mol). As illustrated in Figure 1, methanol (from a syringe pump) and compressed air are fed separately to the vaporizer/preheater, from which the combined vapor stream enters the combustion zone. Hot combustion gases are then used to heat the steam reformer. Downstream of the reformer, the combustion gases then provide heat to the two vaporizers that feed the combustor and the reformer feed stream.

A premixed solution of methanol and water (1:1 ratio by weight) is fed to the system using an HPLC pump. The combined stream first enters the reactant vaporizer, then flows through the steam reformer, where it is converted to H_2, CO_2, and CO. The reformate is chilled, passes through a vapor-liquid separator to

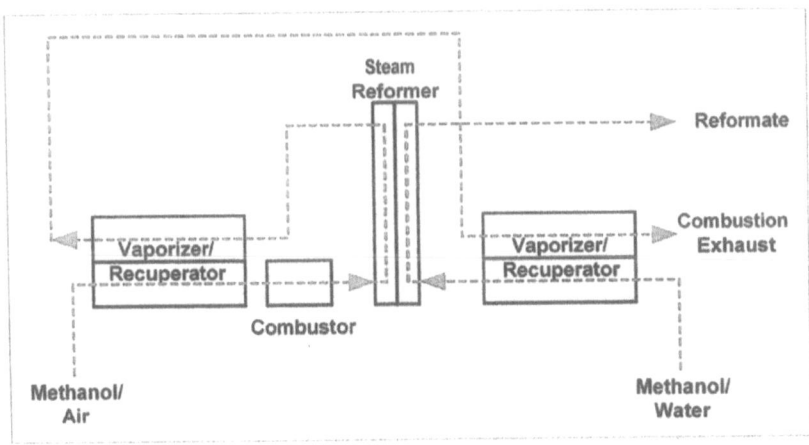

Figure 1. Bread-boarded* fuel processing system for methanol

---

* The term "bread-board" refers to a process train in which all major components are present and connected, but not fully integrated.

remove residual water and methanol, and then flows to the online gas chromatograph for analysis.

The gas chromatograph used in our investigation is an Agilent Technologies Micro-GC capable of detecting gases and hydrocarbons as large as C_8. However, in the methanol-reforming system, detection of compounds up to C_2 is sufficient. All gases other than H_2, CO_2, CO, and CH_4 remain below the detection limit (100 ppm) of the instrument under the system conditions investigated.

4. Results and Discussion

4.1. Steam Reformer

Figure 2 illustrates the performance of the steam reforming reactor over a range of temperatures and contact times. As can be seen from the plot, the reactor can be operated at temperatures as low as 300 °C, but this requires a contact time of 300 ms. Even at very fast throughput (50-ms contact time), the reaction only requires a temperature of 375 °C.

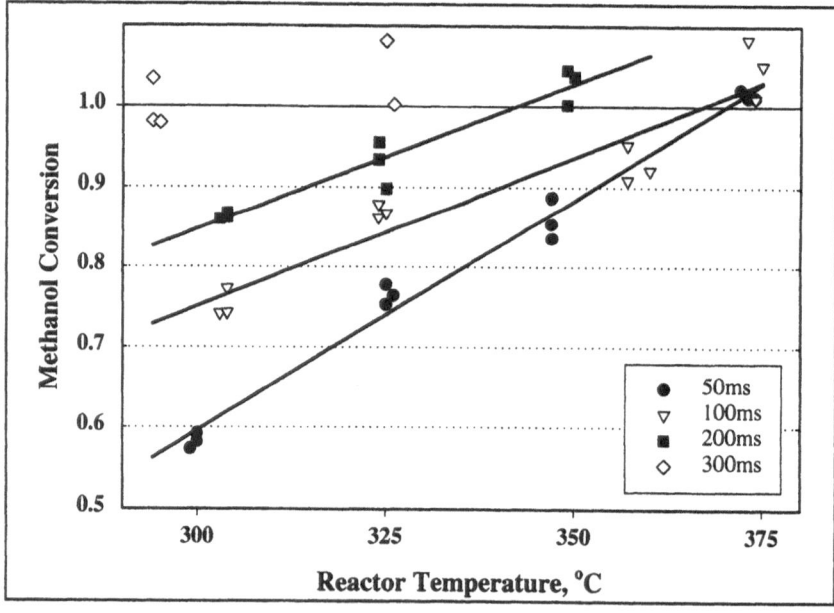

Figure 2. Steam reforming of methanol, reactor performance

Under these conditions, typical dry reformate composition was 73-74% H_2, 25-26% CO_2, and 0.6-1.2% CO. This represents 100% selectivity to H_2 and 95-98% selectivity to CO_2. The low levels of CO (lower than equilibrium) are achievable only with methanol, which does not need to be converted through the water-gas shift mechanism, as described by Takahashi, et al.,[18] and illustrated in Scheme 1, below. Furthermore, the low level of CO leads to two major advantages in this methanol-reforming system. First, it provides for higher hydrogen production. That is, the theoretical maximum hydrogen production for methanol steam reforming is 3 mol H_2 per mol MeOH converted. Under these conditions, this hydrogen production ratio is 2.7 to 2.8, or 90% to 93% of the theoretical maximum yield. The second advantage of low CO concentrations is the simplification of CO cleanup. Since current PEM fuel cells cannot tolerate even these low levels of CO, a CO cleanup step must eventually be incorporated into the process train. However, such low levels of CO eliminate the need for a water-gas shift reactor.

Water-Gas-Shift Pathway	Alternate Pathway
$CH_3OH \rightarrow CO + 2H_2$	$CH_3OH \rightarrow HCHO + H_2$
$CO + H_2O \rightarrow CO_2 + H_2$	$HCHO + H_2O \rightarrow HCOOH + H_2$
	$HCOOH \rightarrow CO_2 + H_2$
Overall Reaction:	Overall Reaction:
$CH_3OH + H_2O \rightarrow CO_2 + 3H_2$	$CH_3OH + H_2O \rightarrow CO_2 + 3H_2$

Scheme 1. Optional pathways for methanol steam reforming

4.2. Bread-board System

The same reactor that produced the results of Figure 2 was integrated into the "bread-board"* system consisting of a combustor, two vaporizers, and a steam reforming reactor. Under self-sufficient operating conditions, this bread-board system yielded hydrogen outputs equivalent to 10 to 64 W_e. The equivalent wattage was calculated assuming a fuel cell with an energy efficiency of 60% and a hydrogen utilization of 80%. Based on these assumptions, the electrical wattage is equal to about half the thermal wattage of the hydrogen produced due to inefficiencies of the fuel cell.

Figure 3 illustrates the operating conditions under which 100% conversion of methanol is attained. As expected, higher contact times (lower throughput) are necessary at lower operating temperatures. Operation at a temperature as low as 325 °C is possible at a contact time of only 270 ms. Conversely, the system can be operated at a very short contact time of 50 ms, but at such high throughput, the

---

* The term bread-boarded refers to a process train in which all major components are present and connected, but not fully integrated.

reactor temperature must be raised to 440 °C. A system analysis will be performed to determine the optimal balance between contact time and temperature. For instance, higher operating temperatures may allow for smaller reactor volumes, but they also increase the need for insulation. At such small device sizes, bulk may ultimately be more important than weight, so there is likely to be a tradeoff between reactor volume and insulation volume.

Efficiency calculations were performed on the bread-board system using a base case. Heat recovery and insulation issues have not yet been fully addressed, so the current process train represents a worst-case scenario. As listed in Table 3, the base-case operating conditions were 350-°C reactor temperature, 140-ms contact time, and a steam to carbon ratio of 1.8:1. Under these conditions, methanol conversion was 100%, and the system produced 27 W_t of hydrogen. Based on the amount of methanol fed to the system, this translates to a fuel processor efficiency of 45%. Based on the previously mentioned fuel cell efficiency assumptions, the overall fuel processor/fuel cell system would produce 13 W_e, an overall efficiency of 22%.

Under these same operating conditions, the dry gas composition was 74% H_2, 25% CO_2, and 0.8% CO. Any methane production during base-case operation was below the detection limit of the gas chromatograph. The very low CO concentration of the reformate eliminates the need for a water-gas shift reactor, significantly simplifying the overall system. In fact, at these low CO concentrations, a water-gas shift reactor would facilitate the reverse-water-gas shift reaction, resulting in higher CO levels in the product gas and consuming some of the product H_2.

The power density of the bread-board system is 720 W-hr/kg, a level that is already several times higher than the best lithium-ion battery. There are several obvious areas which, if addressed, would significantly raise the efficiency and power density of the system. For instance, line losses and exhaust heat were found to account for more than 40% of the combustion heat produced in the system. Eliminating these losses alone would increase the fuel processor efficiency to 60% and the power density to 870 W-hr/kg.

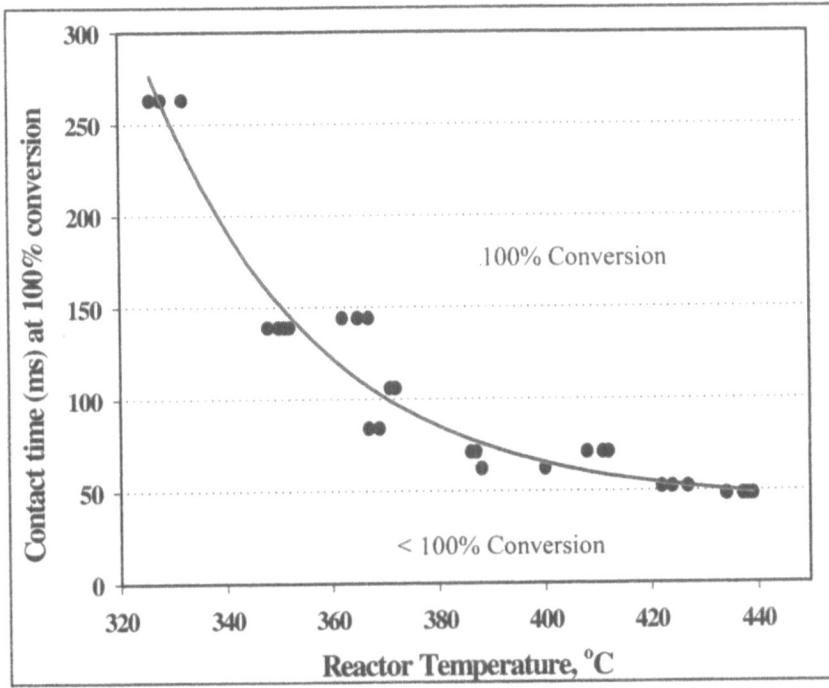

Figure 3. Contact time versus temperature at 100% conversion of methanol

Table 3. Base case demonstration results

Operating Conditions	
Reformer Temperature	350 °C
Contact Time	140 ms
Water/MeOH molar ratio	1.8
Fuel Processor Results	
MeOH conversion	100 %
Gas Composition	
H_2	75 %
CO_2	24 %
CO	0.8 %
Hydrogen Production	27 W_t*
Methanol Fed to System (reformer and combustor)	60 W_t*
Fuel Processor Efficiency	45 %
Fuel Cell Power Estimate[#]	13 W_e*
System Efficiency[#]	22 %
Estimated System Power Density[†]	720 W-hr/kg

*W_t = watts thermal, W_e = watts electric
[#]Assuming fuel cell efficiency of 60% with 80% H_2 utilization
[†]Assuming 14-day mission and 1-kg fuel processor/fuel cell system

5. Conclusions

A bread-boarded 15-W$_e$ methanol fuel processor has been developed for portable power applications. The current reactor train includes a combustor, two vaporizers, and a steam-reforming reactor. The device has been demonstrated under thermally self-sufficient conditions over the range of 10 to 64 W$_e$.

Assuming a 14-day mission life and a 1-kg fuel processor/fuel cell assembly, a base case was chosen to illustrate the expected efficiencies. Operating at 13 W$_e$, the system has a fuel processor efficiency of 45% and an estimated overall efficiency (including fuel cell) of 22%. This translates to an energy density of 720 W-hr/kg, which is several times the energy density of the best lithium-ion batteries. Some immediate areas of improvement in thermal management also have been identified.

6. Acknowledgment

This work was funded by the U.S. Army Communications-Electronics Command, and their support is gratefully acknowledged.

7. References

[1] J. L. Zilka-Marco, A. Y. Tonkovich, M. J. LaMont, S. P. Fitzgerald, D. P. VanderWiel, Y. Wang, R. S. Wegeng, *AIChe 2000 Spring National Meeting* (Atlanta, Georgia), **2000**.

[2] A. Y. Tonkovich, S. Fitzgerald, J. L. Zilka, M. J. Lamont, Y. Wang, D. P. Vanderwiel, R. S. Wegeng, *3rd International Conference on Microreaction Technology* (Frankfurt, Germany), **1999**.

[3] K. P. Brooks, C. J. Call, M. K. Drost, *AIChE 1998 Spring National Meeting* (New Orleans, Louisiana), **1998**.

[4] M. K. Drost, C. J. Call, J. M. Cuta, R. S. Wegeng, *Journal of Microscale Thermophysics Engineering 1* (1997) 321.

[5] D. P. VanderWiel, J. L. Zilka-Marco, Y. Wang, A. Y. Tonkovich, R. S. Wegeng, *AIChE 2000 Spring National Meeting* (Atlanta, Georgia), **2000**.

[6] A. Y. Tonkovich, J. L. Zilka, M. J. LaMont, Y. Wang, R. S. Wegeng, *Chem. Eng. Sci. 54* (1999) 2947.

[7] A. Y. Tonkovich, D. M. Jimenez, J. L. Zilka, M. J. LaMont, Y. Wang, R. S. Wegeng, *Second International Conference of Microreaction Technology* (New Orleans, Louisiana), **1998**.

[8] R. S. Wegeng, L. R. Pedersen, W. E. TeGrotenhuis, G. A. Whyatt, *Fuel Cells Bulletin 28* **2001**, 8.

[9] E. A. Daymo, D. P. VanderWiel, S. P. Fitzgerald, Y. Wang, R. T. Rozmiarek,

M. J. LaMont, A. Y. Tonkovich, *AIChE 2000 Spring National Meeting* (Atlanta, Georgia), **2000**.

[10] S. P. Fitzgerald, R. S. Wegeng, A. Y. Tonkovich, Y. Wang, H. D. Freeman, J. L. Marco, G. L. Roberts, D. P. VanderWiel, *AIChE 2000 Spring National Meeting* (Atlanta, Georgia), **2000**.

[11] W. E. TeGrotenhuis, R. S. Wegeng, D. P. Vanderwiel, G. A. Whyatt, V. V. Viswanathan, S. K. P., *AICHE 2000 Spring National Meeting* (Atlanta, Georgia), **2000**.

[12] A. Y. Tonkovich, J. L. Zilka, M. R. Powell, C. J. Call, *Second International Conference on Microreaction Technology* (New Orleans, Louisiana), **1998**.

[13] M. K. Drost, R. S. Wegeng, P. M. Martin, K. P. Brooks, J. L. Martin, C. Call, *AIChE 2000 Spring National Meeting* (Atlanta, Georgia), **2000**.

[14] A. L. Y. Tonkovich, C. J. Call, D. M. Jimenez, R. S. Wegeng, M. K. Drost, *National Heat Transfer Conference* (Houston, Texas), **1996**.

[15] W. E. TeGrotenhuis, R. Cameron, M. G. Butcher, P. M. Martin, R. S. Wegeng, *AIChE 1998 Spring National Meeting* (New Orleans, Louisiana), **1998**.

[16] W. E. TeGrotenhuis, R. J. Cameron, V. V. Viswanathan, R. S. Wegeng, *3rd International Conference on Microreaction Technology* (Frankfurt, Germany), **1999**.

[17] C. J. Call, M. K. Drost, R. S. Wegeng, *AIChE 1996 Spring National Meeting* (New Orleans, Louisiana), **1996**.

[18] K. Takahashi, H. Kobayashi, N. Takezawa, Chem. Lett., **1985**, 759.

MiRTH-e: Micro Reactor Technology for Hydrogen and electricity

E.R. Delsman, E.V. Rebrov, M.H.J.M. de Croon, J.C. Schouten
Eindhoven University of Technology, the Netherlands

G.J. Kramer*
Shell Global Solutions International B.V., the Netherlands

V. Cominos, Th. Richter
Institut für Mikrotechnik Mainz GmbH, Germany

T.T. Veenstra, A. van den Berg
MESA⁺ Research Institute, Twente University, the Netherlands

P.D. Cobden, F.A. de Bruijn
Netherlands Energy Research Foundation ECN, the Netherlands

C. Ferret, U. d'Ortona, L. Falk
Laboratoire des Sciences du Génie Chimique, CNRS, France

Abstract

Research groups of six companies, institutes and universities have joint forces in a European Community funded project, to develop a miniaturized, low-power fuel processor for the conversion of methanol into clean hydrogen for use in a proton exchange membrane (PEM) fuel cell. The integrated unit will provide a portable power source and is an alternative for battery packs or hydrogen storage in metal hydrides. In the realization of this small-scale fuel processor, microreactor technology will play a key role. Based upon a so-called pinch analysis, a conceptual design of the fuel processor is made, consisting of three combined microreactors/heat exchangers.

Introduction

Fuel cells are regarded as a promising alternative for battery packs, given the much higher energy density of fuel cell systems [1]. The breakthrough of fuel cell technology at the consumer market, however, will largely depend on the development of an easy way to deliver the fuel to the cell. Since the storage of dilute hydrogen is difficult and inefficient, the *in-situ* generation of clean hydrogen from a liquid fuel like methanol offers an attractive alternative [2].

For portable applications, it is clear that the fuel-processing unit should be small and lightweight. In the development of such a unit, microreactor technology will play a key role. With microreactor technology, it is possible to attain high selectivity and productivity as well as efficient energy utilization within a very small unit [3].

In this way the methanol container, fuel processor, and fuel cell will, together, form an efficient and clean power generating system with a high energy density. It will then compete with both battery systems and internal combustion engines for low- and medium-power portable applications. Possible applications will be in laptop computers, portable TV's, camping equipment, and lawn mowers.

Goals and Targets

The objective of the project is to design, micro-fabricate, and test a miniaturized, integrated fuel processor for the conversion of methanol to clean, fuel cell-grade hydrogen for low-power (20-100 W_e) electricity generation. The process scheme in Figure 1, which is analogous to schemes employed in larger-scale applications, will serve as a starting point for the design work.

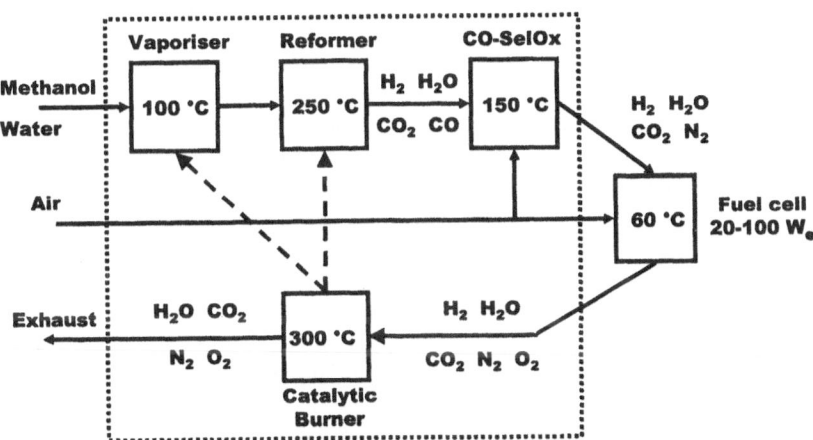

Figure 1 Basic process scheme of the MiRTH-e fuel processor, indicated by the dotted line, in combination with a PEM fuel cell. The heat generated in the catalytic burning of the leftover hydrogen from the fuel cell is used to evaporate and reform the methanol-water mixture.

The technical targets set for the project are summarized in Table 1. The aim is to reach a five to ten-fold improvement of the energy to weight ratio, as compared to battery packs and hydrogen storage in metal hydrides. Since the units have to be used in consumer applications, high standards are set for the allowable emissions of the fuel processor and fuel cell combination.

Table 1 MiRTH-e technical targets, assuming a 500 cc (400 g) methanol tank and weight of the empty fuel tank plus fuel processor of 600 g.

Power range	20 to 100 W(e)
Volume fuel processor	< 500 cc
Efficiency (methanol to electricity)	> 35% (at full load)
System energy density	> 3 MJ/kg (> 800 Wh/kg)
Emissions	low enough to allow for indoor use
Life time	> 40,000 hrs.
Cost of fuel processor unit (mass production)	< 2,000 euro (professional market)
	< 200 euro (consumer market)

Project Organization

The MiRTH-e project started in December 2000 and will run for three years. It is funded by the European Community within the Fifth Framework Programme for Research and Development. In the MiRTH-e project, various partners collaborate, each with a specific expertise in the fields of fuel processing, microtechnology and chemical reactor design. Next to the development of a prototype fuel processor, scientific advances are expected in the fields of metal- and silicon-based micromachining, the integration between metal- and silicon based micro reactors, sensors and actuators, the application of low-cost coated catalysts for microreactor systems, and the development of computer models for the design of microsystems.

In the first phase of the project, individual reactors for the methanol reforming, selective oxidation, and catalytic burning will be fabricated and tested. In this phase, information is gathered about the fabrication and coating techniques used. Furthermore, microreactors will be made in particular for catalyst activity tests, kinetic testing, and the development of appropriate reactor design models. Based upon the information gathered in this first phase of the project, reactor prototypes will be developed in the second phase. In these prototypes, two or more process steps will be integrated. The realization of these prototypes will truly be a scientific and technological step forward.

Work Programme

In the project, the work is divided between the partners according to their specific fields of expertise:

- *Shell Global Solutions International (the Netherlands and Great-Britain)* acts as the project co-ordination and is responsible for the design of a safe methanol container and of a safe coupling between the container and the fuel processor.

- The *Institut für Mikrotechnik Mainz (Germany)* has broad expertise in the design, development, and fabrication of microdevices. Its main activities will be the fabrication of the test reactors and the microreactor prototypes, and the investigation of the issues concerning the use of several different materials in one microsystem.

- The *MESA⁺ Research Institute (the Netherlands)* is a leading institute in the fields of micro electro mechanical systems and silicon micro machining. At MESA⁺ the methanol micro vaporizer will be constructed as well as the sensors and actuators needed for the integrated fuel processor.

- The expertise of the *ECN Energy Research Foundation (the Netherlands)* lies in fuel processing technology, gas cleaning, catalysis and the application of fuel cell technology. ECN will deal with the development of catalytic coatings for the various microreactors and applying these coatings to microstructured platelets.

- The *Laboratory of Chemical Reactor Engineering at Eindhoven University of Technology (the Netherlands)* is specialized in chemical reactor design and catalytic reactor engineering. In Eindhoven, the kinetics of the coated catalytic test reactors and the performance of the microreactor prototypes will be tested. These tests will then serve as input for the design calculations of the integrated fuel-processing device.

- The *Laboratoire des Sciences du Génie Chimique of CNRS (France)* has expertise on both macro- and micro-scale heat exchangers and reactors. CNRS will be involved in the modelling and the design of the methanol-water vaporizer and of the various reactors and heat exchangers of the fuel-processor system.

Process Scheme

As indicated in Figure 1, the fuel processor consists of a vaporizer and three catalytic reactors. In the methanol reformer reactor the endothermic methanol-steam reforming reaction (1), and the reverse water-gas shift reaction (2) take place, over a copper/zinc oxide/alumina catalyst.

$$CH_3OH + H_2O \rightarrow CO_2 + 3H_2 \tag{1}$$

$$CO_2 + H_2 \leftrightarrow CO + H_2O \tag{2}$$

To avoid poisoning of the fuel cell's anode catalyst, it is necessary to bring down the carbon monoxide concentration in the gas mixture to below 10 ppm. For this purpose, a second reactor is needed, where carbon monoxide is selectively oxidized with air (3), using a supported noble metal catalyst.

$$x\,CO + (1-x)\,H_2 + \tfrac{1}{2}O_2 \rightarrow x\,CO_2 + (1-x)\,H_2O \tag{3}$$

In the fuel cell, most hydrogen will be converted into heat and electricity. The leftover hydrogen from the fuel cell will be catalytically combusted in the after-burner reactor, using a platinum/alumina catalyst, to provide the necessary energy for the reforming reaction.

To research the system's feasibility and to point out the energy bottlenecks of the system, a pinch analysis was made of the combined fuel processor, fuel cell system, based upon the general process scheme in Figure 1. In the calculations the electrical power output was set to 100 W. A fuel cell efficiency of 50 % was assumed. An overall process efficiency of 35 % results in a methanol feed rate of 1 ml min^{-1}. To reduce the carbon monoxide formation in the reformer reactor and to avoid dehydration of the fuel cell, an excess of water is used in the process. In the first designs, the water is not recycled, but is fed to the system together with the methanol. Although this dilution of the feed reduces its energy density, it also reduces the system's complexity, since it eliminates the need for a condenser in the system.

In the resulting energy diagram, Figure 2, the amount of energy needed or produced in a specific process step is plotted against the temperature of that specific step, which is a measure for the quality of the energy. The conclusions to be drawn from this diagram are twofold. Most importantly, the system is feasible, i.e. enough energy is available at a high enough temperature to drive all endothermic processes. Secondly, most of the energy losses occur in the fuel cell. Unfortunately, this energy cannot be used in the process, since the fuel cell operates at a low temperature of 60 °C.

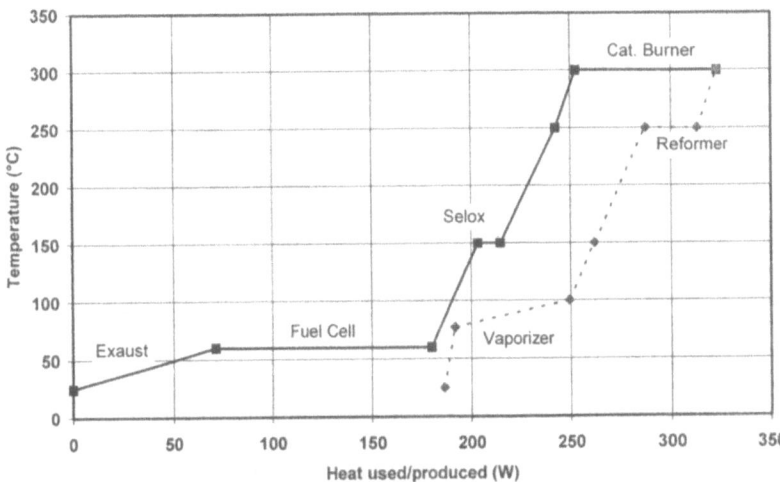

Figure 2 Pinch diagram of an integrated system consisting of a methanol fuel processor and a PEM fuel cell. The dashed line indicates endothermic processes, i.e. the feed vaporization and the reforming as well as several heating steps. The full line indicates exothermic processes: the catalytic burning, the selective oxidation, the fuel cell conversion, and cooling steps. The left-most part of the full line indicates the energy still present in the exhaust gas.

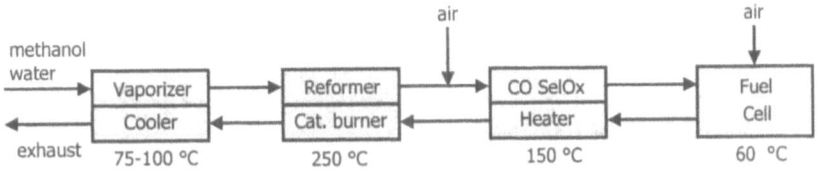

Figure 3 Possible energy integration scheme of the fuel processor. The process is divided into three separate combined reactors/heat exchangers, which all operate at different temperatures.

The pinch diagram, Figure 2, can also be used to show the most efficient way to couple the energy flows in the system, which is coupling the processes of comparable temperature. As a first design, this leads to the division of the fuel processor into three separate units, as shown in Figure 3. In this preliminary scheme, the vaporizer is heated by the hot off-gases of the catalytic burner. The catalytic burner itself is coupled to the methanol reformer, to efficiently deliver the energy needed for the reforming reaction. The heat released from the selective oxidation reaction is recovered by preheating the catalytic burner feed. In this way a simple and elegant design results as a basis for further developments.

Outlook

Based upon the process scheme outlined above, a start will be made with the fabrication and testing of individual reactor components. IMM, MESA$^+$, and ECN will focus on the material issues, as there are the chemical resistance of various possible construction materials, the application of a catalyst coating on micro structured platelets, and the application of different materials within one microreactor. CNRS and TU/e will conduct kinetic and flow tests for the development of kinetic and reactor models, using for instance CFD modelling, suitable to describe the behaviour of microreactors.

Acknowledgement

This work is carried out within the Competitive and Sustainable Growth Programme of the Fifth Framework Programme for Research and Development, funded by the European Community under contract number ENK6-2000-00110. This work, however, does not necessarily represent the opinion of the Community, which therefore shall not be responsible for any use that might be made of data presented here.

We thank M.J.F.M. Verhaak and S.B. van der Molen of ECN, V. Hessel of IMM, J.G.E. Gardeniers of MESA$^+$ and F.J.J.G. Janssen of TU/e for their contributions during the preparation of the project.

Literature

[1] Hebling, C., Heinzel, A., Golombowski, D., Meyer, T., Müller, M., Zedda, M., *Proceedings of the Third International Conference on Microreaction Technology (IMRET 3)*, 1999, 383-393.

[2] Dönitz, W., *Int. J. Hydrogen Energy*, 1998, **23**(7), 661-615.

[3] Ehrfeld, W., Hessel, V., Löwe, H., *Proceedings of the Fourth International Conference on Microreaction Technology (IMRET 4)*, 2000, 3-20.

Part 6

Microreactors as Tools in Chemical Research

.

Microreactors as Tools in Chemical Research

Otto Woerz, BASF Aktiengesellschaft, Ludwigshafen, Ammonia Laboratory

Abstract

Microreactors enable both temperature and velocity gradients to be minimized. That is why only microreactors allow maximum selectivity to be attained when it comes to fast, complex, exothermic or endothermic reactions—multi-phase ones in particular. Several examples demonstrate this. In many cases only microreactors make it possible to measure the kinetics. Another advantage of microreactors is that only very small quantities of reactants are needed—even in continuous process microreactors. It has to be accepted that the large surface-to-volume ratio may have a "detrimental" effect, but by applying different materials such effects can be deliberately used. Examples are given in which industrial production in conventional reactors is impossible (direct fluorination) or extremely expensive (low-temperature reactions), but which can be realized in micro- or milli-reactors. However, the greatest practical use of microreactors is the information they provide in helping us to quickly determine the most suitable conventional reactor to employ.

The outstanding characteristics of a microreactor

In general, a chemical reactor has to perform three tasks: it must make available the necessary reaction time, it must remove or introduce the reaction heat and, in the case of a multiphase system, it must provide an interface between the phases. In each of these tasks, a microreactor offers particular advantages: the microchannel structure allows the streams to be conveyed very precisely, the residence time to be controlled accurately and backmixing to be minimized. In particular, very short residence times are possible. The second task, that of heat transport, is achieved particularly well by the microreactor, since the ratio of surface area to volume increases rapidly with decreasing diameter [1]. When one goes from a 1-litre laboratory reactor to a $30 m^3$ stirred vessel, this ratio drops by a factor of 30. On the other hand, if we scale down to a microchannel having a diameter of 30 µm, the ratio increases by a factor of 3000, i.e. one hundred times greater in the opposite direction. At the same time, the heat transfer coefficient increases by a factor of at least ten [2, 3, 4, 5]. The microreactor's overall effectiveness as a heat exchanger is at least three orders of magnitude higher than that of a conventional apparatus and, therefore, even at very short residence times, the temperature gradients are minimal. The third task, namely the provision of a sufficiently large interfacial area in multiphase systems, can also be achieved extremely well in a microreactor since droplet and bubble sizes are limited by the dimensions of the microchannel. The avoidance of long mass transfer paths as well as hot spots may allow higher selectivities to be achieved in complex reactions. Microreactors are thus instruments by which fast—seconds or minutes in the case of a liquid-phase reaction, milliseconds or less in the case of a gas-phase reaction—exothermic or endothermic, complex and multiphase chemical reactions in particular achieve their maximum selectivity.

The first person to deliberately use microreactors and find significantly higher selectivities was Lerou from Dupont [6]. The first to use microreactors in Germany were Wießmeier and Hoenicke, formerly of "Forschungszentrum Karlsruhe" (FzK). They investigated the hydrogenation of cyclododecatriene [7, 8]. They were interested in maximizing the selectivity of the intermediate cyclododecene (CDE). They compared an ordinary conventionally coated granular catalyst (Fig. 1, Cat A) with anodic oxidized and catalytically activated aluminium wires and foils. The differences in selectivity are impressive. Because the reactant concentration was low, the reaction took place under isothermal conditions, so that differences in selectivity could not be attributed to thermal effects. In the case of anodic oxidation we obtain highly regular pores, and in the case of Cat C in addition we have regular channels. Cat C produces the highest selectivity. That means selectivity increases with increasing structural regularity and decreasing velocity gradients. Similar observations were made in the case of single- and two-phase liquid reactions. If we transport two liquids of different colour in a microchannel we very soon observe a pearl necklace when we reduce the channel diameter and increase the velocity, despite the Reynolds numbers being very small. As a consequence, we measured high interfacial areas and maximum selectivities. Minimal velocity gradients seem to be as important a characteristic of microreactors as minimal temperature gradients.

This view is supported by the following facts. In fixed-bed gas-phase processes, the velocity maximum is close to the reactor wall while a minimum is found at the core of the fixed-bed (Fig. 2). And this minimum is a mean value: in the slipstream of the pellets we have a maximum of residence time and that is why often soot formation occurs. In many fixed-bed production reactors, it is necessary to burn off the soot once a month or even once a week. By contrast, despite months in operation, our microreactors experienced no soot formation and no attendant pressure drop. As the effect of temperature and backmixing on reaction velocity is highly non-linear, a higher uniformity means that reactions can be performed at much lower temperatures. Bruening (IMM) reported recently that steam reforming can be performed in a microreactor at 230°C, as opposed to the 280°C needed in a conventional reactor. We observed in a partial oxidation reaction that the temperature in a microreactor was 80°C lower than in a fixed-bed reactor.

Examples

The microreactors used by Lerou and Wießmeier were not heat exchangers; in other words, they work more or less isothermally, depending on the size of the reactor (heat-transfer through the wall) and the concentration of the reactants. Together with the FzK and the "Institut für Mikrotechnik Mainz" (IMM), we developed microreactors with integrated heat transfer. We started to use microreactors to "come to terms with the past" [9]. In two cases we had been producing chemicals in sub-optimal reactors for many years. The first was a liquid-liquid isomerization reaction catalyzed by concentrated sulfuric acid. Production was initially by means of a semi-batch process with a yield of 70%. After many years a continuous process comprising a mixer and a cooler was developed by an enterprising engineer. The residence times in the mixer and cooler were 0.2 seconds and 4 minutes respectively. The temperature in the mixer increases by 35°C, or half of the adiabatic temperature rise, giving a 50% conversion rate there. The rest of the reaction proceeds in the cooler under isothermal conditions at 50°C. This reactor design increased the yield to 80–85%. To understand this increase in performance at least afterwards, we carried out kinetic experiments in the laboratory. However, we were unable to comprehend these results because of the high reaction rate. Together with the IMM we therefore developed a microreactor, which allowed this reaction to be performed isothermally at residence times in the range of 1 second to several minutes. Soon we could reproduce the results of the production reactor. Because we were able to control the temperature in the microreactor independently of the residence time, we were able to determine reaction conditions that produced yields of 90–95%. We used the same knowledge to modify the production reactor and generate similar yields. We can

Table 1: Systematic overview of used catalysts and fixed beds.

catalyst	description of the catalyst	pore system		cross section of the catalyst	distribution of catal. active component		fixed bed		cross section of the fixed bed
		reg.	irreg.		unif.	non un.	reg.	irreg.	
CAT A	conventionally coated granules	+		⬭	+			+	
CAT B1	pieces of activated Al-wires	+		⬭	+			+	
CAT B2	pieces of activated Al-foils	+			+			+	
CAT C	stack of activated and microstructured Al-foils	+			+		+		

reg. = regular; irreg. = irregular; unif. = uniform; non un. = non uniform

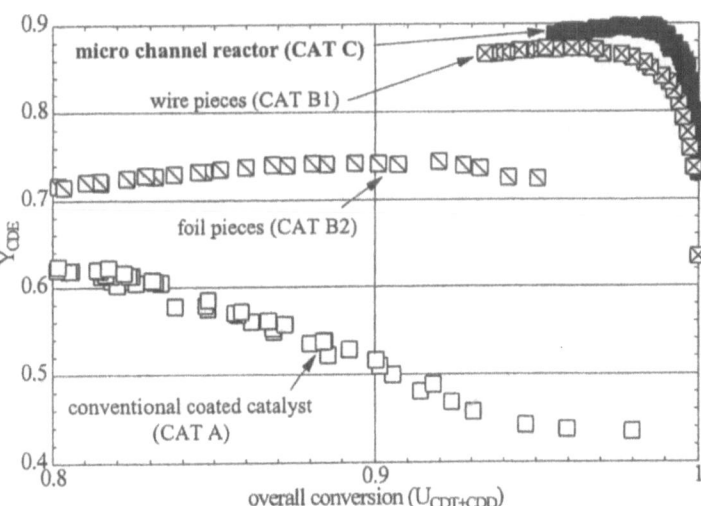

Fig.1 Dependence of CDE-selectivity upon the different catalysts of table 1.

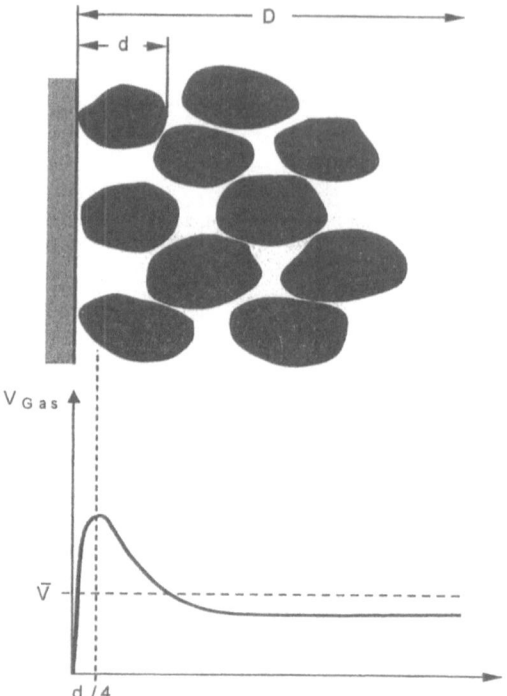

Fig.2 Gas velocity distribution in a fixed-bed.

therefore regard our microreactor as a tuning instrument that enables us to attain the maximum achievable selectivity.

The second example is a heterogeneously catalyzed gas-phase reaction, a new oxidative hydrogenation using a silver catalyst. The most important industrial dehydrogenation is the synthesis of formaldehyde from methanol. Production is carried out in a pan-like reactor in which a 2-cm-thick catalyst layer is placed on a gas-permeable plate. A selectivity of 95% is obtained at nearly complete conversion. These results are achieved regardless of reactor size. Understandably, we decided to adopt this successful technology too. A selectivity of 90% was obtained at a conversion of 50% in a pan-like reactor having a diameter of 5 cm. A production reactor of 3 m diameter was then built. Unfortunately, the selectivity dropped to about 40%. Since this was one step in a multistage process, the result was therefore a real catastrophe. Many theories were formulated to explain this scaling problem. We finally settled on the following. It can be calculated that with the laboratory pan-like reactor roughly 50% of the reaction heat is removed via the wall. By contrast, a larger reactor of this type having a diameter of several meters operates virtually adiabatically. This is not a problem in the case of formaldehyde synthesis since all reactants and products are stable. However, in our case this obviously does not apply. The adiabatic temperature increase in the production reactor leads to a dramatic loss in selectivity. This problem should be solvable by means of a shell-and-tube reactor, which operates isothermally, because the heat of reaction is removed. We therefore returned to the laboratory scale and constructed a single tube of extremely short diameter (12mm) and length (10cm), knowing that the reaction is very quick and very exothermic. We now again obtained a selectivity of 90%. On the laboratory scale the pan-like reactor, however, gave the same selectivity, i.e. our theory could only be proven by constructing a shell-and-tube reactor on the production scale. We did this and achieved success with a selectivity of 85%, nearly the same as in the laboratory. This too appeared to be a suitable example for a microreactor because, in comparison with the short shell-and-tube reactor, it is possible to achieve an even shorter residence time and at the same time even better heat transfer. Together with FzK, we therefore developed a microreactor for this reaction and obtained a maximum selectivity of 96%.

In both cases the use of microreactors quickly gave us information that would not have been obtainable using conventional laboratory techniques. If we had had this information earlier, we would have saved many years of production at low yields. In other words, microreactors can significantly reduce development periods and in many cases they reveal potential for optimization.

Sometimes only the good mixing effects in combination with a short residence time of microreactors are relevant. Merck investigated the reaction of a ketone with a Grignard reagent. Conventionally the reaction was performed in a stirred tank reactor. On the production scale, about 5 hours were needed to mix the reactants and to transfer the heat of the reaction. Using adiabatic micro- or mini-mixers the reaction time could be reduced to less than 10 seconds and the yield increased from 72% to 92%. When a small micromixer was used and imbedded in a cooling bath, the reaction could be performed under near isothermic conditions. Here, a further increase of only 3% was obtained [10]. This means that good mixing within a short residence time is decisive. At Axiva the continuous radical polymerization of acrylates is performed in a pilot plant. The reactor is a tube type device equipped with static mixers. Precipitation occurred in the static mixers, which was presumed to be due to imperfect mixing of the reactants. Therefore micromixers were added as pre-mixers. Although the channels of the micromixers were only 25μm wide, the fouling problems were reduced significantly. Gel permeation chromatography revealed why: high-molecular-weight polymers were no longer formed and that meant a higher product quality too [11].

However, many reactions can only be performed if all the characteristics of a microreactor—maximum heat transfer, minimal backmixing, short residence times—are put to use. The direct fluorination of toluene, for instance, cannot be run on an industrial scale because the reaction is extremely fast, complex and exothermic. A falling film microreactor was constructed and the sulfite oxidation provided an interfacial area of 23 500m^2/m^3. After a few experiments, selectivities of about 40% were attained [12]. We recently investigated the synthesis of a temperature-sensitive precursor for a crop-protection product (Fig. 3). In a conventional laboratory reactor, working semi-batch, we needed a temperature of –70°C to obtain a quantitative yield; at higher temperatures, the yield rapidly dropped to zero. It would be very expensive to produce at a temperature of –70°C. We therefore performed this reaction in a microreactor with a channel width of 60 μm and a residence time of some seconds. A blockage occurred after some minutes despite the reaction mixture being a single-phase liquid. The reason was that the one reactant is not completely soluble in the other. We therefore added a short pre-reactor with a bigger diameter (1 mm). In this pre-reactor some conversion was performed in order to produce a truly single phase. Downstream of that was a narrow-channelled (40 μm) micromixer array, which ensured complete homogenization, followed by a post-reactor to complete the conversion. With this reactor we obtained the same yield at 0°C and residence time of six seconds as we did at –70°C.

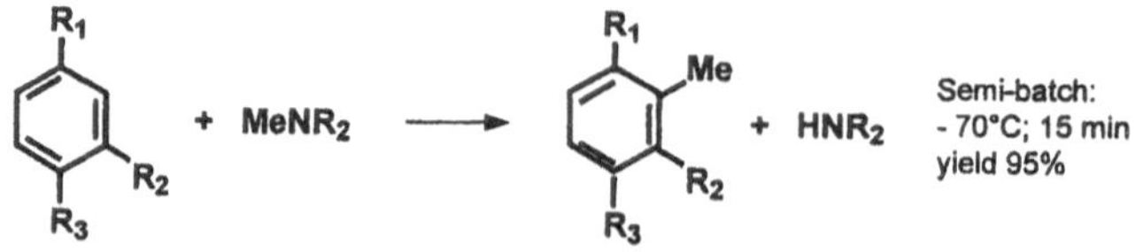

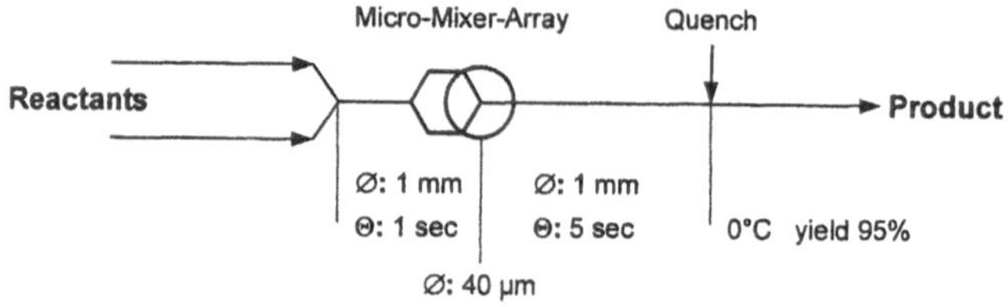

Fig.3 Microreactor-combination: tubular pre-reactor, micro-mixer-array, tubular reactor.

This example reveals a general principle. Kinetically controlled chemical reactions can only be performed in a small temperature range of 30–50°C. At lower temperatures the reaction velocity is too small, while at higher temperatures it is the selectivity that suffers. If one of the reactants is unstable at such temperatures, its concentration can be minimized by using the semi-batch mode. If other reactants or a product are unstable too, this process will not work. In all such cases the highest selectivities can only be obtained in a reactor capable of providing short residence times, high heat transfer and minimal backmixing. Indeed, this is the definition of a microreactor.

Many reactions are so fast and so exothermic that with conventional laboratory techniques kinetic and mechanistic information cannot be determined. As temperature and residence time can be made independent of each other in a microreactor, such information can be obtained anyway. In investigating the reaction of an amine with an aldehyde in a microreactor, we proved the existence of the postulated Schiff base and the kinetics of its formation and decomposition (Fig. 4). Furthermore, we detected an intermediate that was hitherto unknown. In this way, it is possible to understand how some by-products are formed.

A further characteristic of microreactors is that only small quantities of reactants are needed. This makes the parallel screening of reactions for active substances, catalysts and materials possible. Recently we investigated the synthesis of a heterocycle in the liquid phase, and wanted to know whether the reaction could be performed in the gas phase. The problem was that we only had 30 ml of the reactant. Use of the microreactor enabled us to perform 7 experiments, which varied according to temperature, residence time and concentration, as well as the material of the reactor. Unfortunately, selectivities in the gas phase were found to be hopelessly low. On the positive side, however, we found out very quickly that we were completely on the wrong track.

Hydroformylations are high-pressure gas–liquid reactions. We often have the task of finding out with a very small quantity of a new catalyst whether it is better than the one used in

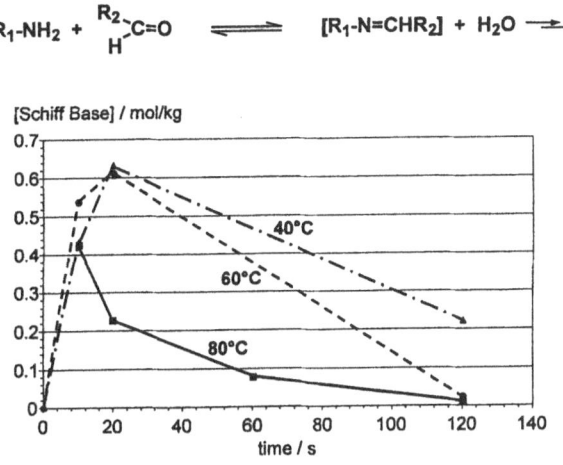

Fig.4 Formation and decomposition of a Schiff-base at different temperatures.

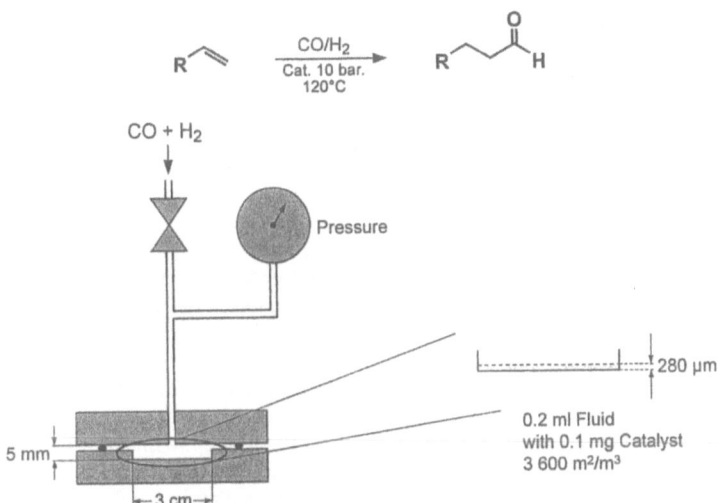

Fig.5 Batch microreactor for a hydroformylation reaction.

production. We constructed a simple batch microreactor (Fig.5). 0.2 ml of solvent containing 0.1 mg of catalyst was spread over a circular area of 3 cm diameter to form a liquid film having a thickness of 280 μm (and thus having an interfacial area of 3600m^2/m^3). Under these conditions, it can be assumed that the quantity of the interfacial area is not rate-limiting. After the reaction, but before lowering the pressure, the whole microreactor was cooled to prevent losses. We succeeded in reproducing the reaction velocity and the selectivity of the industrial reactor.

384

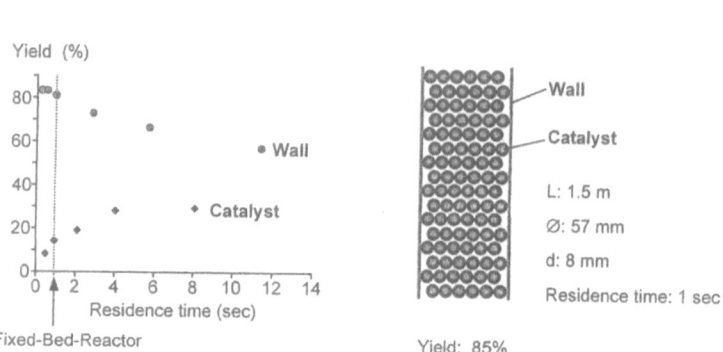

Micro-Reactor Capillary

Fixed-Bed-Reactor

Wall

Catalyst

L: 1.5 m

Ø: 57 mm

d: 8 mm

Residence time: 1 sec

Yield: 85%

Fixed-Bed-Reactor

L: 5 m   Ø: 2 mm

Fig.6 Comparision of a microreactor made of the catalyst respectively the wall material with a fixed-bed reactor.

Microreactors are not really new; every capillary tube is a microreactor. In the past it was often observed that results in a capillary tube did not correspond with the results in bigger reactors. This was attributed to the "detrimental" effect of the relatively large surface area of the reactor wall. But this characteristic can be used to advantage. Last year we were investigating a fixed-bed reaction, but were unsure whether or not a thermal reaction or a catalytic effect of the reactor wall played a role in the production reactor. We therefore performed the reaction in three capillaries, one made of quartz, one made of catalyst material, and one made of reactor-wall material. The dimensions of the capillaries were matched to the ratio of the catalyst's surface to the volume of the fixed-bed reactor. If we assume that the reaction mainly occurs on the surface of the catalyst, we can compare the residence times. In the quartz capillary we had no reaction at all—so no thermal reactions occur. We found that the reactor-wall material produced a greater catalytic effect than the actual catalyst material (Fig.6), which required the temperature to be 100 degrees higher if the same conversion rate was to be attained. From an analysis of the kinetics, we found that the wall material is 70 times more active than the actual catalyst. In the production reactor therefore, the reaction is solely catalyzed by the wall material. In the production reactor, air has to be added to the reactants. Not doing so would result in a rapid decrease in activity. Despite this, it is necessary to burn off carbon deposits once a month. In the capillary reactors adding air has no significant effect and we never were forced to burn off. The explanation is that the soot that forms in the fixed-bed reactor deactivates the catalyst. As we mentioned above, we have no soot formation in the microreactor and therefore the addition of air is not necessary.

Microreactors in production?

The above examples have shown that by using microreactors it is possible to find solutions to numerous problems that would otherwise either remain unsolved or require an inordinate amount of time and money. In most cases the information gained led to optimization of existing reactors or reactor concepts. The result is a more or less conventional reactor. It could be that the improvement potential indicated by microreactors cannot be implemented using conventional reactor engineering. Is it conceivable that microreactors can also be used for production? [13] In principle, yes. However, serious problems would be encountered. One or more streams would have to be divided uniformly over hundreds of thousands or even millions of reaction channels, and then there would naturally be the problem of blockage not only on the product side but also on the side of the heat transfer medium, namely the fouling problem. On the laboratory scale we have these problems under control, not least because we can take our microreactors apart at any time and clean them, or even replace a microcomponent. This is inconceivable for a production reactor. We therefore have to assume that production carried out in microreactors will be the exception rather than the rule. In most cases, it is unnecessary to carry out production in microchannels in order to achieve maximum selectivity. Rather, it would be sufficient for the diameter of the reaction channels to be in the order of millimeters. Use will therefore be made of channels that are as small as is necessary to exploit the optimization potential, but as large as possible to keep the blockage and distribution problems at a manageable level. There are already production reactors with channels of 2 cm diameter. To find the optimum reactor with channels in the millimeter range, we need a tool whose dimensions are one or two orders of magnitude smaller—the microreactor. In our opinion, this could be the most important result.

References

[1] Ehrfeld, W.; Hessel, V.; Möbius, H.; Richter, Th.; Russow, K.; „Potential and Realization of Microreactors" in Microsystem Technology for Chemical and Biological Microreactors, DECHEMA Monographs, Vol. 132, VCH Verlagsgesellschaft, Weinheim 1996, pp. 1-28.

[2] Hagendorf, U.; Janicke, M.; Schüth, F.; Schubert, K.; Fichtner, M.; „A Pt/Al_2O_3 coated microstructured reactor/heat-exchanger for the controlled H_2 /O_2-reaction in the explosion regime", in Ehrfeld, W.; Rinard, I.H.; Wegeng, R.S.(Eds.) Process Miniaturization: 2nd International Conference on Microreaction Technology;IMRET 2; Topical Conference Preprints, pp. 81-87, AICHE, New Orleans, USA, (1998)

[3] Schubert, K.; Bier, W.; Brandner, J.; Fichtner, M.; Franz, C.; Linder, G.; „Realization and Testing of Microstructure Reactors , Micro Heat Exchangers and Micromixers for Industrial Applications in Chemical Engineering", in Ehrfeld, W.; Rinard, I.H.; Wegeng, R.S.; (Eds.) Process Miniaturization: 2nd International Conference on Microreaction Technology, IMRET 2; Topical Conference Preprints, pp. 88-95, AICHE, New Orleans, USA (1998).

[4] Schubert, K.; „Entwicklung von Mikrostrukturapparaten für Anwendungen in der chemischen und thermischen Verfahrenstechnik", KfK Ber.6080, pp. 53-60 (1998).

[5] Adams, T.M.; Abdel-Khalik, S.I.; Jeter, S.M.; Qureshi, Z.H.; „An Experimental Investigation on Single Phase Forced Convection in Microchannels", AICHE Symp.Ser.314, pp.87-94, (1997).

[6] Lerou, J.J.; Harold, M.P.; Ryley, J.; Ashmead, J.; O`Brien, T.C.; Johnson, M.; Perrotto, J.; Blaisdell, C.T.; Rensi, T.A.; Nyquist, J.; „Microfabricated Minichemical Systems: Technical Feasibility", in Microsystem Technology for Chemical and Biological Microreactors, DECHEMA Monographs, Vol. 132, VCH Verlagsgesellschaft, Weinheim 1996, pp. 51-70.

[7] Wießmeier, G.; Schubert, K.; Hönicke, D.; „Monolithic Microreactors Possessing Regular Mesopore Systems for the Successful Performance of Heterogeneously Catalysed Reactions", in Microreaction Technology, Springer, 1998, pp. 20-26.

[8] Wießmeier, G.; Hönicke, D.; „Strategy for the Development of Micro Channel Reactors for Heterogeneously Catalyzed Reactions", in Ehrfeld, W.; Rinard, I.H.; Wegeng, R.S.; (Eds.) Process Miniaturization : 2nd International Conference on Microreaction Technology; IMRET 2; Topical Conference Preprints, pp.24-32, AICHE, New Orleans, USA, (1998).

[9] Wörz, O.; „Microreactors – a New Efficient Tool for Reactor Development", Chemical Engineering & Technology, 2, Vol.24 (2001), pp. 138-142.

[10] Krummradt, H.; Kopp, U.; Stoldt, J.; „Experiences with the Use of Microreactors in Organic Synthesis", in Ehrfeld, W.; (Ed.) Microreaction Technology: 3rd International Conference on Microreaction Technology, Proceedings of IMRET 3, pp. 181-186, Springer Verlag, Berlin, (2000).

[11] Bayer, T.; Pysall, D.; Wachsen, O.; „Micro Mixing Effects in Continuous Radical Polymerization", ibid, pp.165-170.

[12] Hessel, V.; Ehrfeld, W.; Golbig, K.; Haverkamp, V.; Löwe, H.; Storz, M.; Wille, C.; Guber, A.; Jänisch, K.; Baerns, M.; "Gas/liquid Microreactors for Direct Fluorination of Aromatic Compounds using Elemental Fluorine", in Ehrfeld, W. (Ed.) Microreaction Technology: 3rd International Conference on Microreaction Technology, Proceedings of IMRET 3, pp. 526-540, Springer-Verlag, Berlin, (2000).

[13] Ehrfeld, W.; Hessel, V.; Löwe, H.; „Microreactors", pp. 277-283, WILEY-VCH Verlag, Weinheim, (2000).

Microchannel Reactor for the Partial Oxidation of Isoprene

Stefanie Walter, Eric Joannet [1], Monika Schiel, Isabelle Boullet [2], Robert Philipps, Marcel A. Liauw

LTC I, Universität Erlangen-Nuremberg, Egerlandstr. 3, D-91058 Erlangen, Germany
[1] LGRC-EPFL, CH-1015 Lausanne, Switzerland
[2] LSGC-ENSIC, F-54001 Nancy, France

1 Introduction

The partial oxidation of isoprene to citraconic anhydride (CA) is used to test the opportunities of the microchannel reactor system. CA is mainly used in the preparation of synthetic resins but also in the synthesis of pharmaceuticals [1]. Up to now CA is produced in a complex process with itaconic acid as starting material. An easier way of synthesizing CA could be the heterogeneously catalyzed oxidation of isoprene in the vapor phase. As shown in figure (1) the reaction is highly exothermic. In addition there is a high temperature sensitivity of the selectivities to the partially oxidized products.

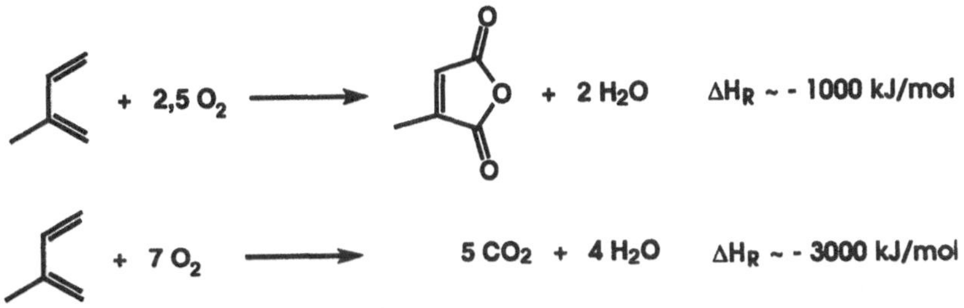

Figure 1: *Reaction equations for the formation of citraconic anhydride (CA) and for the deep oxidation of isoprene.*

The reaction is investigated in a conventional laboratory fixed bed reactor and a microchannel reactor where the channel walls are coated with the catalyst. The catalyst screening was exclusively done in the conventional reactor. For the microchannel reactor a coating method for a conventionally prepared catalyst had to be developed. In both reactor systems the influence of the reaction parameters was examined. The results were compared and evaluated with a focus on the influence of the coating method.

388

2 Experimental

2.1 Preparation of Conventional Catalysts

The conventional catalysts used are prepared by impregnation of the support (TiO$_2$, anatase). The support is added to the vanadium oxide dissolved with oxalic acid in warm water. The solvent is evaporated under stirring. The paste like substance is dried over night at 120 °C and calcined in air at 500 °C for 4 h. The material is ground and tableted without any binders. The catalyst particles used in the fixed bed reactor have a medium diameter of 1 mm.

2.2 Anodic Oxidation Followed by Impregnation

The microstructured aluminum plates are anodically oxidized to obtain a porous oxide layer which is impregnated [2]. To prevent the oxidation of areas outside the channels, the resist used in the etching process is kept and the inlet and outlet chambers are additionally varnished. The anodic oxidation is done in oxalic acid (1.5 wt.-%) at 12 °C with an applied voltage of 50 V for 8 h. After the oxidation the resist and varnish are removed with methanol. The layer is dried over night at 120 °C and calcined in air for 6 h at 450 °C. For the impregnation a solution of 5 g vanadyl acetylacetonate and 5 g titanyl acetylacetonate in 100 ml methanol is prepared. The plates are wetted with the solvent methanol in vacuum prior to the impregnation for 8 h at room temperature. Without washing they are dried at 120 °C and again calcined in air for 6 h at 450 °C. The impregnation with subsequent calcination is repeated twice.

2.3 Application of Catalyst Suspension

2.2 g sodium silicate solution (27 wt.-% SiO$_2$, 10 wt.-% Na) is diluted in 10 ml distilled water. The sodium is removed by ion exchange with Dowex cation exchanger (H$^+$-form). Then the conventionally prepared catalyst (2 g) is ground, sieved (<100 μm) and suspended in the filtered silicate solution. The suspension is stirred for 1 h and then applied into the channels using a fine brush. Within the channels the suspension is distributed by capillary forces. The microstructured plates are dried for 1 h at 120 °C, then a second layer is applied. The plates are again calcined in air for 1 h at 450 °C. Either anodically oxidized aluminum plates or calcined (6 h in air at 450 °C) aluminum plates with a thin oxide layer are used for the coating. The plates are weighed before and after the coating to determine the catalyst mass. The surplus is also dried and calcined. It is used for experiments in the fixed bed reactor and for catalyst characterization.

2.4 Experimental Setup

The microchannel reactor used has been developed within the joint project "Periodic Processing in Microchannel Reactors" [3] and constructed by the IMM

(Institut für Mikrotechnik Mainz) [4]. The reactor consists of a stack of 6 or 10 microstructured plates (40 x 40 x 0.5 mm), respectively, made from aluminum with each having 34 microchannels (hydraulic diameter: 230 - 280 μm, length: 20 mm). The microchannels are fabricated by isotropic wet chemical etching of the metal plates. The plates are either glued, stacked together with graphite sealing or fixed within a closed stainless steel housing (fig. 1). For the reactor with glued plates a countercurrent heat exchanger is integrated but not used in the catalytic experiments. The microchannel reactor is heated electrically with one heating cartridge in each housing plate. It is isolated with glass wool. During the experiments the temperature of the reactor is controlled.

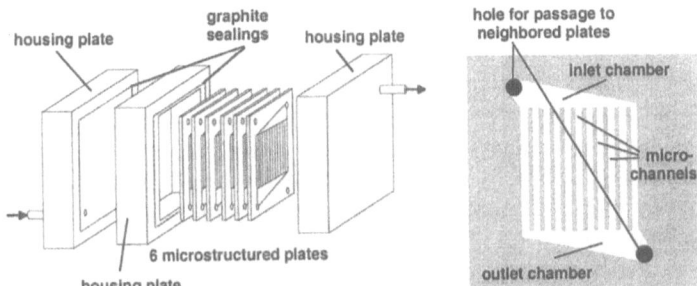

Figure 2: Schematic of microchannel reactor system with six interchangeable microstructured and three housing plates (left). Instead of using the ring plate in the middle, the microstructured plates can also be assembled with graphite sealings or with high temperature resistant glue. Schematic of microstructured plate (right).

The conventional fixed bed reactor is made of a ceramic tube (α-Al$_2$O$_3$, inner diameter: 10 mm, length: 300 mm) with a central stainless steel capillary to measure the axial temperature profile. 200 mg of the catalyst is diluted with inert material (Al$_2$O$_3$/SiO$_2$) to a volume of 3 ml. The tubular reactor is heated electrically with three independently controlled heating bands.

The experiments are performed in a continuous flow system. The gases (N$_2$, O$_2$, CH$_4$ as internal standard) are dosed by mass flow controllers whereas the liquid isoprene is dosed by a liqui-flow system and subsequent evaporation in nitrogen. The whole system is heated to prevent condensation of reactants or products. The bypassed reactant and product mixture are analyzed online by means of gas chromatography. A GC equipped with three columns (HP-5, PoraPlotQ, molecular sieve 5 Å) and two detectors (FID, TCD) is used to analyze the organic components and the permanent gases. In the experiments, the temperature is varied for two different total flows. The concentrations of isoprene and oxygen in nitrogen are kept constant. The same mass of catalyst (200 mg) is used in both reactor systems.

3 Results and Discussion

3.1 Comparison of Coating Methods

Two different coating methods for the microchannel reactor are tested. The first method is the anodic oxidation of microstructured aluminum plates with subsequent impregnation [2]. In the following this method is abbreviated as *anodic oxidation*. The second method is reminiscent of the well know application of porous washcoats in monolithic reactors. Here powdered catalyst is directly fixed on the microchannel walls with silica. This second method is named *suspension method*. The recipe is taken from the suspension used for the spray drying of e.g. VPO catalysts [5]. The spray drying is done to improve the attrition resistance by encapsulating the catalyst in a highly porous silica shell.

The catalyst screening using a conventional fixed bed reactor showed that vanadium oxide supported on titanium oxide is among the best catalysts for the partial oxidation of isoprene. For the titania supported vanadium oxides the molar vanadium content (V/V+Ti) is varied between 5% and 75%. The maximum selectivities of 27% - 30% are almost independent on the composition whereas for a vanadium content of < 1%, the maximum selectivities are fairly lower (5%).

Using both coating methods three different catalyst coatings based on vanadium and titanium oxide were prepared. In the case of the preparation by anodic oxidation, vanadium and titanium oxide with a molar vanadium content of 50% are deposited on the porous alumina produced during the anodic oxidation ($V_{50}Ti_{50}O_x-Al_2O_3$). For the suspension method conventionally prepared vanadium oxide on titanium oxide with molar vanadium content of 30% ($V_{30}Ti_{70}O_x-SiO_2$) and pure vanadium oxide (VO_x-SiO_2) are used on microstructured aluminum plates. The catalyst coatings are characterized by means of scanning electron microscopy (SEM), elementary analysis with inductively coupled plasma optical emission spectroscopy (ICP-OES), N_2-adsorption and tested for the isoprene oxidation.

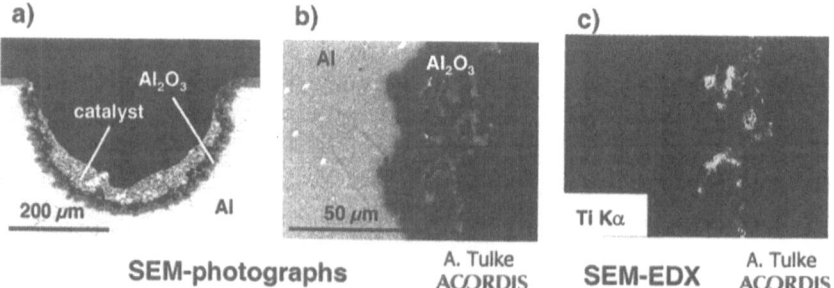

Figure 3: a) SEM photo of etched microchannel on anodically oxidized aluminum plate coated with VO_x by the suspension method. b) SEM photo of anodically oxidized and impregnated alumina layer in etched microchannel. c) SEM-EDX analysis of the distribution of titanium in the impregnated alumina layer shown in b).

The SEM photograph shows the semicircle like geometry of the etched microchannels (fig. 3a). The thickness of the catalyst layer prepared by the suspension method with pure vanadium oxide on an anodically oxidized plate is 10 to 40 μm. As expected for the manual application the layer is irregular. The cross section of the microchannel is reduced to about 70%. In the enlarged detail of the anodically oxidized alumina layer (fig. 3b) the rough surface is displayed. This is due to the etching during the microstructuring process and confirmed by SEM of a not oxidized aluminum plate. The SEM-EDX analysis reveals the dispersion of the impregnated titanium (fig. 3c): the titanium is preferably located in the holes of the rough surface. This is also the case for the vanadium (not shown).

The ICP-OES analysis is done to determine the elementary composition of the catalyst coatings. This is representatively done for the anodic oxidation with aluminum wires treated like the aluminum plates and for the suspension method with the surplus of the suspension. In the case of the anodic oxidation the molar ratio of vanadium and titanium corresponded to that in the impregnation solution 1:1. For the suspension method the ICP-OES analysis revealed that the original composition of the catalyst does not change. The content of the additional silica and remaining sodium is determined (table 1). The sodium is not totally removed by the ion exchange. 2 respectively 7 mole-% of residual sodium are detected.

Table 1: Molar contents of elements in catalyst coatings determined by ICP-OES according to the equation e.g. for V molar content = $n_V / (n_V + n_{Ti} + n_{Si} + n_{Na})$.

Nominal composition	V [mole-%]	Ti [mole-%]	Si [mole-%]	Na [mole-%]
$V_{50}Ti_{50}O_x$-Al_2O_3	49	51	-	-
$V_{30}Ti_{70}O_x$-SiO_2	20	51	27	2
VO_x-SiO_2	78	-	15	7

N_2 adsorption with anodically oxidized aluminum wires resulted in BET surfaces below 5 m²/g. The BET surface of the conventionally prepared catalyst is for the VO_x 6 m²/g and the $V_{30}Ti_{70}O_x$ 15 m²/g. The addition of silica for the suspension method increases these surfaces to 35 m²/g and 110 m²/g, respectively.

Concerning the mechanical stability no problems are expected for the anodic oxidation, but the adhesion of the catalyst layer for the suspension method is tested. During two months of operation no catalyst loss is observed. In addition more severe conditions are simulated by blowing air with a high speed onto the plate for 120 s (fig. 4). No loss of catalyst is observed. Here the surface roughness of the etched channels seems favorable for a good adhesion of the catalyst layer to the channel walls.

392

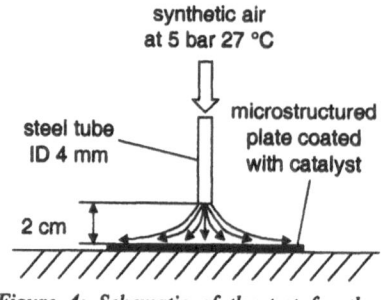

synthetic air
at 5 bar 27 °C

steel tube
ID 4 mm

microstructured
plate coated
with catalyst

2 cm

Figure 4: Schematic of the test for the mechanical stability of the catalyst coating.

All three catalyst coatings were tested in the isoprene oxidation. The temperature dependence of the isoprene conversion and CA selectivity for the three coatings is depicted in fig. 5. While all curves exhibit similar shape, there are quanitative differences. The maximum CA selectivities for the catalysts prepared with the suspension method are with 25% and 22% significantly higher than the 12% reached with the anodic oxidation. For the latter the conversion and selectivity curves are also shifted to higher temperatures. It should be noted that the experiments were carried out in two different microchannel reactors. To eliminate the influence of the different numbers of plates (for the anodic oxidation 10 plates and the suspension method 6 plates) the flow per channel was kept constant. Other differences e.g. the hydrodynamic residence time due to the reduction of the cross sections for the suspension method, the absolute vanadium content or the BET surfaces could not be avoided. Despite these difficulties it can be stated that so far the catalytic behavior of the suspension method seems superior.

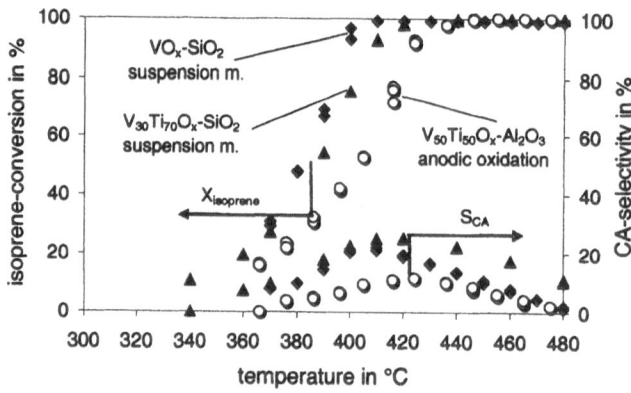

Figure 5: Comparison of three different catalyst coatings for the partial oxidation of isoprene: conversion of isoprene and CA-selectivity as a function of temperature ($V_{per\ channel} = 1.5$ Nml/min, $c_{isoprene} = 0.6$ vol.-%, $c_{O2} = 20$ vol.-%, $p = 1.2$ bar, suspension method: 6 microstructured aluminum plates with 34 channels: $d_H = 230$ µm, $l = 20$ mm, catalyst: 120 mg VO_x-SiO_2, 200 mg $V_{30}Ti_{70}O_x$-SiO_2, anodic oxidation: 10 microstructured aluminum plates with 34 channels: $d_H = 280$ µm, $l = 20$ mm, catalyst: $V_{50}Ti_{50}O_x$-Al_2O_3).

The evaluation of the two different coating methods reveals other advantages of the suspension method. First the method is easier in preparation and less time consuming. There is a high flexibility in choosing the wall material which is completely restricted to aluminum for the anodic oxidation. Conventionally optimized catalyst can be applied. The quantification is easily done by weighing and for the characterization of the catalyst the surplus is used. Questions about the long term stability of the layer, the application of a layer with a constant thickness remain to be solved. The influence of the additional silica and sodium on the catalytic behavior is shown in the next chapter.

3.2 Comparison of Reactors

(a)

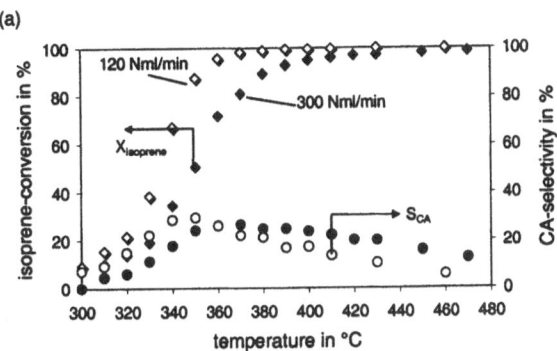

(b)

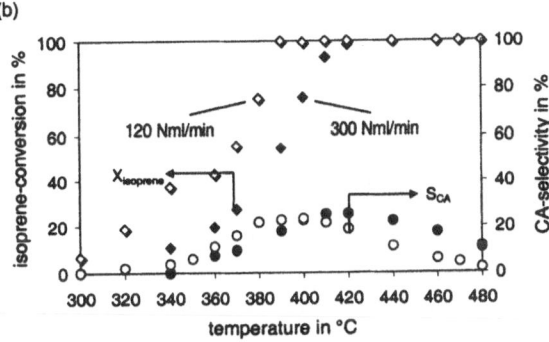

Figure 6: a) Partial oxidation of isoprene in the fixed bed reactor and b) the microchannel reactor for two different total flows: conversion of isoprene and CA-selectivity as a function of temperature ($c_{isoprene}$ = 0.6 vol.-%, c_{O2} = 20 vol.-%, p = 1.2 bar, fixed bed reactor: 200 mg $V_{30}Ti_{70}O_x$, microchannel reactor: 200 mg $V_{30}Ti_{70}O_x$-SiO_2)

First the influence of operating parameters on conversion, selectivity and yield, respectively was investigated in both reactors. The effect of an increase of the total flow velocity is shown in figure 6. The flow rate in the tubular reactor was increased from 120 to 300 Nml/min, corresponding to a decrease of the mean residence time by 60%. This leads to a shift of both conversion and selectivity curve to higher temperatures by about 15 – 25 K (fig. 6a). The maximum selectivity of about 28% is hardly affected. The corresponding experiment in the microchannel reactor yields the same result, albeit at temperatures that are up to 40 K higher with a slight decrease in the selectivity (fig. 6b). Similarly matching responses are also observed with respect to a variation of isoprene and oxygen feed concentration, respectively.

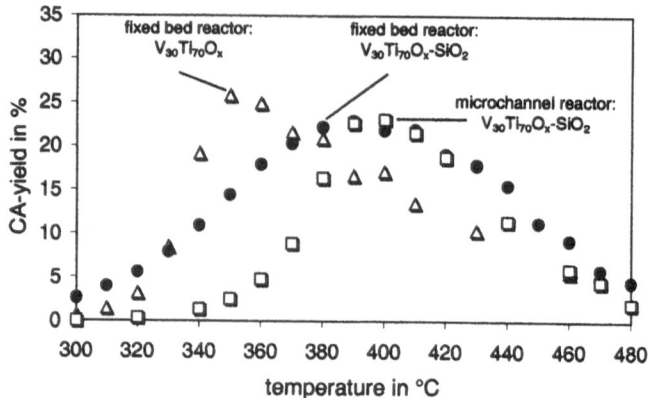

Figure 7: Comparison of partial oxidation of isoprene in the fixed bed reactor and the microchannel reactor: CA-yield as a function of temperature ($V_{tot} = 120\ Nml/min$, $c_{isoprene} = 0.6\ vol.-\%$, $c_{O2} = 20\ vol.-\%$, $p = 1.2\ bar$, fixed bed reactor: 200 mg $V_{30}Ti_{70}O_x$ resp. 200 mg $V_{30}Ti_{70}O_x$-SiO_2, microchannel reactor: 200 mg $V_{30}Ti_{70}O_x$-SiO_2).

To explain this different behavior in the two reactor systems it is important to investigate the influence of the coating method on the catalyst. Therefore the conventionally prepared catalyst and its suspension method counterpart were tested in the fixed bed reactor and compared to the microchannel reactor (fig. 7). With about 26%, the maximum yield reached in the fixed bed reactor with the conventionally prepared catalyst is higher than that reached with the suspension catalyst either in the microchannel reactor or in the fixed bed reactor (23%; fig. 7). In order to explain this apparently detrimental influence of the suspension method, the differing composition as compared to the conventional catalyst must be borne in mind (e.g. residual sodium, see table 1). During the catalyst screening in the conventional fixed bed reactor alkali metals have been found to decrease the catalytic performance of VTiO for this reaction.

A further difference between the two reactor systems is the wall material: ceramic for the fixed bed reactor and aluminum/stainless steel for the microchannel reactor. Measurements of the catalytic activity for the oxidation of isoprene and CA with different materials of the microstructured plates and in the fixed bed reactor without catalyst were performed. For isoprene the activity in the microchannel reactor is independent of the material (aluminum, stainless steel, titanium) higher than within the fixed bed reactor. The experiments with CA show that CA is very unstable almost independently of the wall material.

A major intrinsic difference between the two reactors is their (non)isothermicity. Conventional laboratory-scale fixed-bed reactors tend to support hot spot formation when an exothermic reaction takes place (fig. 8). Note that the temperature decreases downstream. Such an axial temperature gradient may turn out to be acting like a quench, thereby preventing undesirable consecutive reactions like deep oxidation which is beneficial for the selectivity to a thermally unstable intermediate. In contrast, the entire microchannel reactor (fig. 2) is heated. Due to the high heat conductivity the whole reactor including inlet and outlet will be at reaction temperature. The target product may therefore undergo deep oxidation before leaving the reactor. This situation is aggravated by the high surface-to-volume ratio and the fact that the reactor material (aluminum, stainless steel) exhibits a significant activity for the deep oxidation of citraconic anhydride. In order to assess the impact of these temperature-related phenomena, perturbation-free spatially resolved temperature measurements in the microchannels will prove helpful (see e.g.[6]).

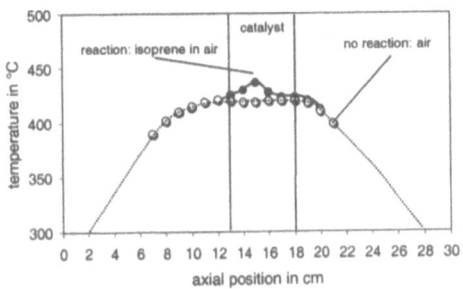

Figure 8: Axial temperature profile in the fixed bed reactor for a non reacting air flow and the partial oxidation of isoprene (V_{tot} = 300 Nml/min, $c_{isoprene}$ = 0.6 vol.-%, c_{O2} = 20 vol.-%, p = 1.2 bar, fixed bed reactor: 200 mg $V_{30}Ti_{70}O_x$)

There is yet another problem in comparing the differently-scaled systems. It is obviously not possible to keep all relevant parameters constant upon moving from macroscale to microscale. Here the parameters kept constant are the catalyst mass, the total flow and the concentrations. However, the linear flow rate and the mean residence time are different in the two reactors. A proper comparison hence requires that all these different parameters be considered and evaluated for their influence.

4 Summary

It is possible to conduct the partial oxidation of isoprene to citraconic anhydride in a microchannel reactor. The conversion/selectivity behavior almost reaches that of a conventional fixed bed reactor. Of the two coating methods tested, the well known anodic oxidation has the advantage of uniform layer thickness and no problems with the mechanical stability whereas the suspension method presented here is less laborious and more versatile. The suspension method yields catalytically activated microchannel plates that exhibit catalytic behavior comparable to a conventional fixed bed. A quantitative comparison reveals a slight decrease in catalytic performance by the suspension method. One objective of current studies is therefore to prevent this catalyst degradation. Another focus is the implementation of reliable temperature measurements in the microchannels to study isothermicity.

Literature

[1] G. L. Castiglioni, F. Cavani, C. Fumagalli, S. Ligi, F. Trifirò: Synthesis of Methylmaleic Anhydride (Citraconic Anhydride) by Heterogeneous Selective Oxidation of Isoprene with V/Ti/O Catalysts, Catalysis of Organic Reactions, edited by M. E. Ford, Marcel Dekker, Inc., New York · Basel (2001) p. 371-377

[2] D. Hönicke: Partial Oxidation of 1,3-Butadiene on Vanadia/Alumina/Aluminum-Coated Catalysts: Products and Reaction Routes, J. Catal. 105 (1987) p. 10-18

[3] J.-M. Commenge, L. Falk, J.-P. Corriou, M. Matlosz: Optimal Design for Flow Uniformity in Microchannel Reactors, Proceedings of 4th International Conference on Microreaction Technology, Atlanta, 5 - 9 March 2000, p. 23-30

[4] R. Schenk, K. Gebauer, W. Ehrfeld, V. Hessel, H. Löwe, Th. Richter: Konzipierung und Herstellung der Mikroreaktionssysteme für die periodische Prozessführung, Schriftliche Projektpräsentation auf der 2. Projektgruppensitzung "Mikroreaktorsysteme in der chemischen Technik", Frankfurt/Main, 11 May 2000

[5] K. Uihlein: Butanoxidation an VPO Wirbelschichtkatalysatoren, Dissertation, Universität Karlsruhe (TH), Karlsruhe, Germany (1993)

[6] I.-M. Hsing, R. Srinivasan, M. P. Harold, K. F. Jensen, M. A. Schmidt: Simulation of micromachined chemical reactors for heterogeneous partial oxidation reactions, Chem. Eng. Sci. 55 (2000) p. 3-13

Acknowledgements

We wish to thank Dr. Broucek and Dr. Lebens (Akzo Nobel), Prof. Renken (EPFL Lausanne), Prof. Matlosz (ENSIC Nancy) and Dr. Schenk (IMM Mainz) for a fruitful cooperation and Dr. Tulke (Acordis) for the SEM-EDX analysis. Financial support by the BMBF and by the DFG (Li 669/2-1) is gratefully acknowledged.

Selective Oxidation of 1-Butene to Maleic Anhydride– comparison of the Performance between Microchannel Reactors and a Fixed Bed Reactor

Stefan Kah, Dieter Hönicke

Technische Universität Chemnitz, Lehrstuhl für Technische Chemie
D-09107 Chemnitz/ Germany

Abstract

Two types of microchannel reactors made of packed wafer stacks were used for the selective partial oxidation of 1-butene to maleic anhydride with hydrocarbon concentrations to some extent within the explosion range. The microstructured wafers were catalytically activated by anodic oxidation of aluminum wafers followed by an impregnation process with $V_2O_5/P_2O_5/TiO_2$. Hereby, channel diameters varied between 80 and 400 μm. Catalytic results in the different microchannel reactors were compared with those of identically prepared catalysts used in fixed bed reactors.

1 Introduction

The selective oxidation of C_4-hydrocarbons in air to maleic anhydride, an important industrial fine chemical, is a highly exothermic gas phase reaction and normally carried out in fixed bed reactors at high conversion degrees and low hydrocarbon concentrations (approximately 1.5 %). Contrary to the partial oxidation of n-butane the maximum achieved selectivity to maleic anhydride is significantly lower in case of n-butene as educt [1-5]. Nowadays, vanadium phosphorous oxide-based catalysts are used without exception whereas product selectivity can be increased to approximately 60% by promoters or additions in the feed, e.g. steam. The reaction proceeds via butadiene and other intermediates like furane or crotonaldehyde to the main product or respectively towards carbon oxides which can be summarized in the following simplified reaction scheme:

$\Delta_r H^0_{400} = -1315$ kJ/mol

$CO_2 + H_2O$ $\Delta_r H^0_{400} = -2542$ kJ/mol

Due to the high exothermicity hot spots in conventional fixed bed reactors lower drastically the selectivity towards the main product in favour of parallel reactions with higher activation energies. Improvements should be made by faster heat removal out of the reaction zone. With their superior mass and heat transfer abilities microchannel reactors are therefore an alternative reactor type for this kind of reaction. Additional, in contrast to conventional fixed bed reactors it is possible to investigate the partial oxidation at hydrocarbon concentrations within the explosion range.

2 Experimental

The aim of this study was to show suitability of microchannel reactors for the partial oxidation of 1-butene in comparison to fixed bed reactors with identical catalytic active surfaces. For the different reactor types, microstructured aluminum wafers (AlMg3) made by the "Forschungszentrum Karlsruhe GmbH" and cut aluminum wires (Al 99.95%), respectively, were used as initial materials. Both substrates were catalytically activated in exactly the same manner by anodic oxidation to obtain a porous oxide layer and subsequent impregnation/calcinations steps. Hereby, the limited stability of aluminum oxide in acid media prevented the use of common procedures for catalyst preparation. Thus, precursors like Titaniumisopropoxide, Vanadyl-(V)-acetylacetonate, and Triethylphosphate were solved in toluene or chloroform in absence of any promoters. The resulting catalysts were investigated in microchannel reactors (see below) and in the fixed bed reactor made of quartz, respectively.

Catalytic experiments were performed in a continuous flow apparatus consisting of gas supply including feed dosage at constant pressure, the reactor itself, traps to condense the liquid products as well as analytical devices. The composition of the feed and the organic products were analysed on-line using a gaschromatograph equipped with a SIL-8 column and a flame ionisation detector. Additionally, the concentrations of carbon oxides were monitored on-line by an IR photometer.

3 Microchannel Reactors

Each microchannel reactor consisted of a stack of microstructured wafers with quadratic microchannels manufactured by mechanical micromachining. In order to investigate influences on catalytic behaviour the channel diameters were varied, ranging from 80 to 400 µm. Two types of microchannel reactors were used in the present study. The first type (fig. 1) was assembled by the "Forschungszentrum

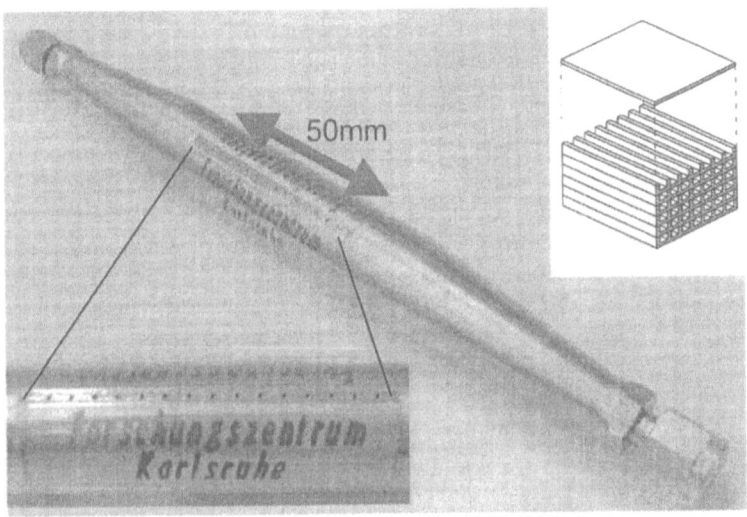

Figure 1: Example of a microchannel reactor (diffusors attached) containing a waferstack divided by a steel plate which contains inserts for thermocouples

Karlsruhe GmbH" by stacking the wafers and a subsequent sealing process. For a reasonable comparison the geometric surface areas as well as the outer dimensions of all waferstacks were approximately identical which is summarized in Table 1.

Table 1: Engineering figures of microchannel reactors used for the partial oxidation of 1-butene; wafer (AlMg3-alloy): 50 x 10 mm^2; Al$_2$O$_3$-layer thickness: 40 µm, metal loading (weight-%): V$_2$O$_5$: 1.7, TiO$_2$: 1.87, PO$_4^{3-}$: 0.17

Reactor	Number of wafers	Number of Channels (total)	Total flow cross section [mm^2]	Channel width [mm]	Total channel volume [cm^3]	Total geom. Surf. area [cm^2]
MCR1	15	255	35,5	0,4	1,77	190,4
MCR2	25	550	16,5	0,2	0,826	190,7
MCR3	37	1165	3,31	0,08	0,237	177,6

400

Diffusors for the inlet and outlet of the fluid complete the microchannel reactor. In order to prevent further reaction of products at the stainless steel surface the diffusor at the reactor outlet was coated with aluminum oxide via a sol gel method. The corresponding fixed bed catalyst consisted of 12.5 g cut aluminium wires with an total geometric surface area of 191.4 cm^2, which were catalytically activated like the wafers of the microchannel reactors.

The second type microchannel reactor, developed at the Technische Universität Chemnitz [6] was of a modular construction providing flexible stack dimensions and a fast exchange of wafers. The wafers including inlays to fill up possible void space were stacked into an inner housing with a cover plate on top. Diffusors were attached to both sides of the housing to provide accurate gas flow. The system was incorporated into an outer housing including flange, copper sealing and necessary connectors. In contrast to the first type reactors metallic reactor surfaces after the reaction zone were uncovered. Optionally, the metallic diffusor at the reactor outlet could be exchanged by an inert glas insert. The same kind of wafermaterial as for the first type microchannel reactor was used. However, with the modular microchannel reactor, different wafer lengths were also tested at lower wafer numbers (tab.2). The aim was to investigate the influence of mass transport in the microchannels due to the very different linear flow velocities in the first type microchannel reactors. To maintain nomenclature M was added to the specific reactor name indicating the modular construction. Wafers used, e.g. in MCR1, exhibited identical catalytic surfaces like those of MMCR1.

Table 2: Engineering figures of modular microchannel reactors used for the partial oxidation of 1-butene; Al$_2$O$_3$-layer thickness: 40 μm, metal loading (weight-%): V$_2$O$_5$: 1.7, TiO$_2$: 1.87, PO$_4^{3-}$: 0.17

Reactor	Number of wafers	Wafer length [mm]	Total flow cross section [mm^2]	Channel width [mm]
MMCR1	4	12.5, 37.5, 50	9,5	0,4
MMCR2	8	12.5, 37.5, 50	5,3	0,2
MMCR3	8	12.5, 37.5, 50	0,7	0,08

4 Results

4.1 Partial Oxidation at low Hydrocarbon Concentrations

Catalytic runs for fixed bed catalysts and first type microchannel reactors were carried out by varying residence times. At a low partial pressure of 0.1 MPa (fig. 2) and conversion degrees between 75 to 80%, maximum selectivites of 33% to maleic anhydride were obtained in the microchannel reactor MCR3 and the fixed bed reactor. For the calculation of the conversion degrees the formed isomeric n-butenes were summed up. By reason of better comparison the analogous curve for MCR1, which lies between those of MCR2 and MCR3, was neglected. The influence of the channel diameters on the selectivities was small. Although the diffusors of the microchannel reactors were coated, the postcatalytic activity could not be excluded at all. In addition, selectivity to the sum of organic products shows

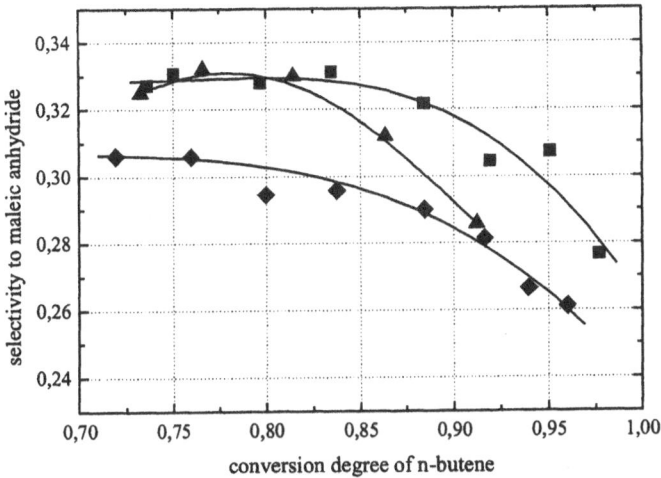

Figure 2: Selectivity to maleic anhydride vs. conversion degree of n-butene in microchannel reactors MCR2 (♦),MCR3 (▲) and a fixed bed reactor (■). Reaction conditions: T=400°C, $c_{1\text{-butene}}$=0.4% in air, p=0.1 MPa

a similar behaviour like those of maleic anhydride. As organic byproducts, e.g. acetaldehyde, phthalic anhydride, acroleine, furane and acetic acid were observed. Differences in catalytic behaviour may also originate in the different alloys used for the catalysts of the microchannel reactors and the fixed bed reactor. By reducing channel widths in microchannel reactors the conversion degrees increased at constant residence times (fig.3). This is likely the influence of the increased mass transport in the smaller channels. The space time yield of maleic anhydride was related to the volume of an waferstack. It is noteworthy that the space time yield is approximately proportional to the total channel cross section. For example, the ratio of the space time yield of maleic anhydride in MCR1 to MCR3 is 2.3 whereas the ratio of the total cross section is 2.2. Residence times to yield comparable conversion degrees were in all microchannel reactors up to max. one order of magnitude lower than in an equivalent fixed bed reactor. To reach a conversion degree of 90%, a contact time of 750 ms in the fixed bed reactor was necessary while 25 ms were sufficient in microchannel reactor MCR3. Due to the resulting higher flowrates at constant conversion degrees the space time yields of maleic anhydride could be increased in microchannel reactors by a factor of up to 5 in

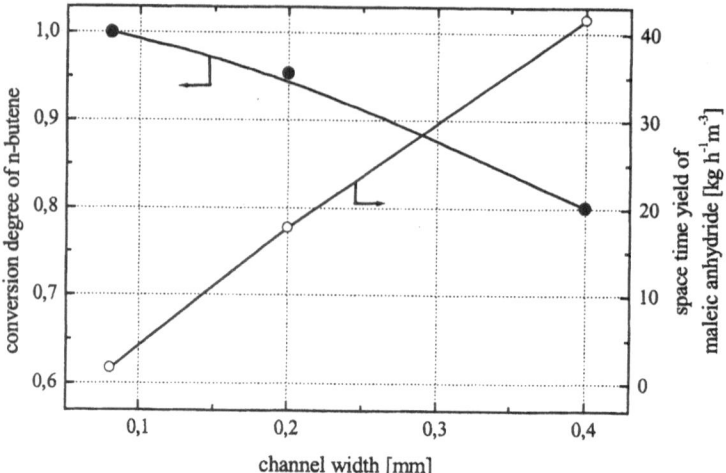

Figure 3: Conversion degree (•) of n-butene and space time yield of maleic anhydride (o) vs. channel widths in microchannel reactors MCR1-MCR3. Reaction conditions: T=400°C, $c_{1-butene}$=0.4% in air, p=0.1 MPa, t=120 ms (STP)

comparison to the fixed bed reactor. The maximum respectively minimum space time yields in the microchannel reactors are represented in fig. 4. The space time yield was related to the volume of the waferstack (microchannel reactor) and the volume of the fixed bed, respectively. With decreasing flow conversion degrees

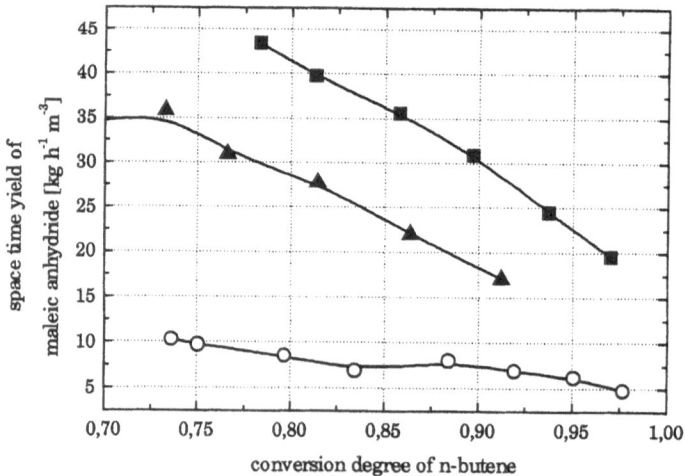

Figure 4: Space time yield of maleic anhydride vs. conversion degree in microchannel reactors MCR1 (■), MCR3(▲) and a fixed bed reactor (○). Reaction conditions: T=400°C, $c_{1-butene}$=0.4% in air, p=0.1 MPa

increased to a lesser extent leading in all reactors to lower space time yields. This is especially obvious for the microchannel reactors, so that the differences to the fixed bed reactor are diminished. At 91% conversion a space time yield of maleic anhydride in microchannel reactors solely more than 2.4 times higher than those obtained in the fixed bed reactor can be found.

Catalytic runs in the modular microchannel reactors were performed in order to investigate mass transport for the different channel widths. Several circumstances have to be kept in mind with regard to possible catalytic acticvity at the reactor walls: due to the different foil lenghts, the void space and free surfaces in the postcatalytic zone vary. Equal contact times in waferstacks differing only in waferlengths are equivalent to different residence times in the postcatalytic zone and vice versa. Therefore, by shortening the waferlength from 37.5 mm to 12.5 mm the residence time in the region from the waferend to the begin of the diffusor at the reactor outlet increased by factor 9. One method to scrutinize if any postcatalytic activity occured was to exchange the metallic diffusor at the outlet with an outlet made of glass which was known to be inert. While at a constant volume flow of 110 ml/min (STP) and wafer lengths of 37.5 mm the conversion degree diminished from 72% to 70%, selectivity to maleic anhydride increased from 22% to 28%. It must be concluded that onto the surface of the metallic diffusor approx. 21% of the formed maleic anhydride reacted further, probably to carbon oxides because the selectivity to organic products changed like the selectivity to the main product. Furthermore, catalytic activity of the metallic diffusor towards 1-butene was proved with the formation of carbon dioxide and water as only products. As a result, catalytic experiments concerning mass transport were all performed with the glasoutlet instead of the metallic diffusor. It is known that diffusors distribute the incoming stream equally among the different channels of the waferstack. For that reason the diffusor in front of the waferstack was further used.

4.2 Partial Oxidation within the Explosion Range

In order to demonstrate the behaviour of microchannel reactors at high thermal powers, catalytic runs with butene concentrations up to ten times higher than the lower explosion limit were performed. If necessary oxygen was used instead of air. During operation a slight deactivation of the catalysts was observed (5% butene in air), which was not the case for concentrations ten times less. Hereby, conversion degrees stabilize shortly with time on stream. This effect may be due to adsorption processes onto active centers. The catalysts could be regenerated by reoxidation. To demonstrate the inherent safety in microchannel reactors an ignition-experiment (fig. 5) was performed with a butene concentration of 5% in air by successively raising temperature at a constant volume flow. As a result, in the observed temperature range from 320°C up to 400°C no instability occurred. The conversion degree linearly increased up to 375°C and then slowly converged from 85% conversion onwards. The liberated power due to reaction ranged up to 6.5 W with an adiabatic temperature rise of more than 2000°C. It is noteworthy that with regard to the safe operation of the microchannel reactors, the prevention of pluggings due to the high product concentrations in the product stream was the main obstacle

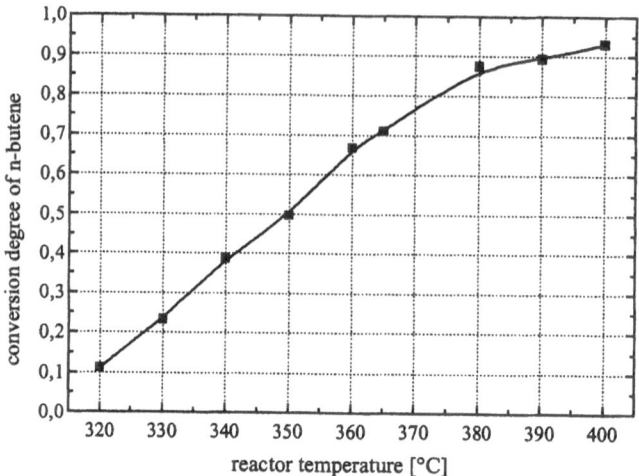

Figure 5: Conversion degree of n-butene vs. reactor temperature in microchannel reactor MCR2. Reaction conditions: $c_{1\text{-butene}}$=5 % in air, p=0.1 MPa, Flow=100 ml/min (STP)

during operation. Additional experiments were performed to proof the uncritical behaviour of microchannel reactors during operation by evoking strong thermal changes inside the reactors. This should be caused by applying strong changes in conversion degrees. Hereby, the microchannel reactors were flushed with air to reach stationarity followed by adding 5% butene to the feed. The feedstream, therefore, was held constant and the time dependent response of the system was monitored along the reactor axis. The resulting temperature difference curves for two points of measurement, one near the entrance and outlet of the reactor, respectively, depicts figure 6. Both curves elapse parallel in time whereas 110 seconds after the butene switch maxima occur. Because the reaction rate at the entrance is much higher than at the outlet, the maximum temperature difference is nearly three times higher (6K). The temperature jump due to the onset of reaction is with increasing time accompanied by the reduction of external heating power. This leads to different stationary temperatures lying relatively under (wafer outlet) or above (wafer inlet) the initial values. As a result, the reaction imposed an axial temperature gradient of 3 K in the microchannel reactor. By performing the experiment at higher thermal power (ca. 22 W) which is proportional to the total flow, the higher temperature maxima is located then near the outlet of the waferstack. The resulting temperature gradient due to reaction along the reactor axis decreases although the thermal power is several times higher than that at lower contact time (ca. 6 W). Figure 7 represents a snapshot for several conversion degrees from 40 to 93% at the time when maxima temperature differences will be achieved. Clearly visible is that with decreasing conversion degree the maxima wanders through the reactor towards the reactor outlet. However, volume flows

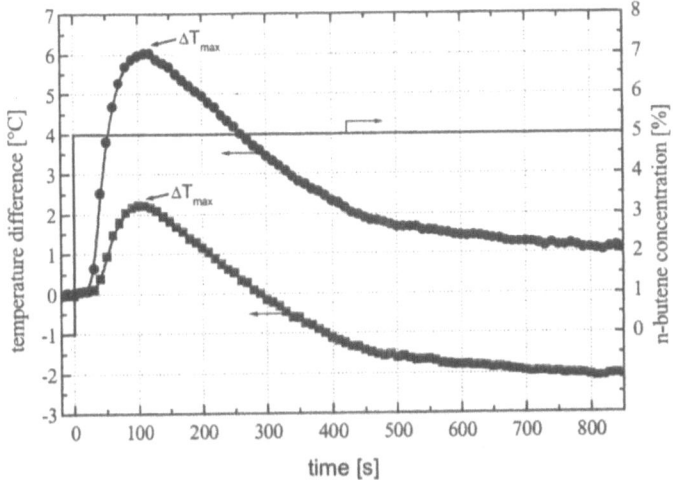

Figure 6: Temperature difference (•: wafer inlet, ■: wafer outlet) in microchannel reactor MCR2 after the 1-butene flow was switched on. Reaction conditions: T=400°C, p=0.1 MPa , Flow = 100 ml/min (STP), P=6 W

where a linear increase of the temperature difference along the reactor axis should be obtained, could not be applied due to limitations of the mass flow controller. As time elapses each curve will shift horizontally towards the stationary values which

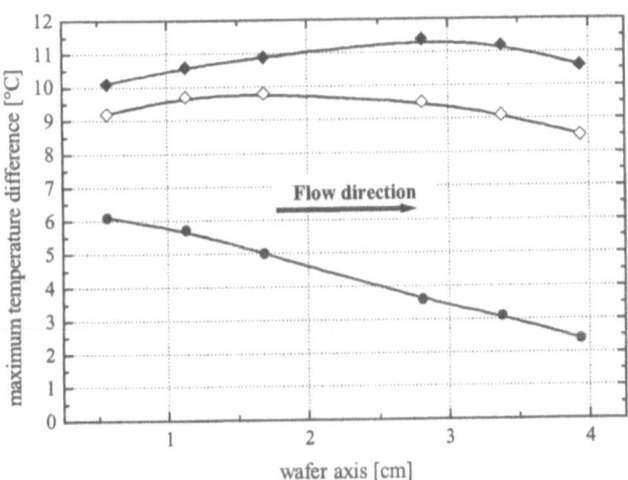

Figure 7: Maximum temperature difference along the reactor axis for different conversion degrees (X=0.4 ♦,X=0.58 ◊, X=0.93 ●) in microchannel reactor MCR2. Reaction conditions: T=400°C, $c_{1\text{-butene}}$=5 % in air, p=0.1 MPa

will be achieved in the above case after nearly an quarter of an hour. It should be kept in mind that in industrial fixed bed reactors a temperature difference of 1 degree centigrade can cause a thermal runoff during operation. Additionally, the partial oxidation at concentrations of 5% butene in air lead in the fixed bed reactor with identical reaction parameters to hotspot formation followed by a thermal runoff. Contrary, even at hydrocarbon concentrations of 15% butene in oxygen microchannel reactors operated without instabilities.

5 Conclusions

The partial oxidation of 1-butene to maleic anhydride was successfully performed in microchannel reactors with different channel widths at high degrees of conversion and hydrocarbon concentrations to some extent within the explosion range. The catalyst components V_2O_5, TiO_2 and P_2O_5 were immobilized into a porous aluminum oxide layer without any promoters. Catalytic investigations on two different types of microchannel reactors and a fixed bed reactor with an identical catalytic surface were performed. While the influence of the channel diameter on the product selectivity was small, the space time yield to maleic anhydride could be increased in all microchannel reactors up to factor 5 in comparison to the fixed bed reactor. With the reduction of the channel size mass transport improved. However, at butene concentrations within the explosion range selectivitiy to maleic anhydride could not be increased. Hereby microchannel reactors run stable without any ignition. The partial oxidation caused remarkably low temperature gradients even within the explosion regime at high thermal powers. Nevertheless, a shift of the zone of highest reaction rate along the reactor axis could be observed.

Acknowledgements

This study was supported by the "Arbeitsgemeinschaft industrieller Forschungsvereinigungen" (AiF) under grant number 11146B/1 with financial support from the German Federal Ministry of Economics and Technology (BMWi). Three microchannel reactors were made in cooperation with Dr. Schubert and Dr. Fichtner by the "Forschungszentrum Karlsruhe GmbH". The Al_2O_3 coating of the diffusors was performed by Dr. Haas-Santo from the "Forschungszentrum Karlsruhe GmbH".

References

[1] F.Cavani, F. Trifirò; Chemtech **24** (4) 1994, p.18-25

[2] V.V. Guliants, J.B. Benziger, S. Sundaresan; Stud. Surf. Sci. Catal., part B
     101 (1995), p. 991-1000

[3] F. Trifirò; Catal.Today **16** (1993), p. 91-98

[4] K. Mori, A. Miyamoto, Y. Murakami; J. Chem. Soc., Faraday Trans. **1** (1986),
     p. 13-34

[5] M. Ai, J. Catal; **100** (1986), p. 336-344

[6] A. Kursawe, R. Pilz, H. Dürr, D. Hönicke; Proc. 4th Int. Conf. on
     Microreaction Technology, (AIChE 2000), p. 227-235

Application of a Micromixer for the High Troughput Screening of Fluid-Liquid Molecular Catalysis

C. de Bellefon*, S. Caravieilhes, P. Grenouillet

Laboratoire Génie des Procédés Catalytiques
CNRS/ESCPE Lyon, F-69100 Villeurbanne.
Fax : +33 4 72 43 16 73   Email : cdb@cpe.fr;

High throughput synthesis methodologies, such as combinatorial techniques, are now applied to the discovery of molecular catalysts.[1] Libraries of ligands which can be turned into catalysts by complexation to transition metals, are accessible. The effectiveness of this approach has been demonstrated for restricted libraries of compounds and in the case of single liquid phase catalysis.[2] However, numerous reactions of interest such as olefin oligomerisation, hydrogenation, carbonylation, hydroformylation, etc. are operated in gas-liquid or liquid-liquid systems. Inadequate control of phase and catalyst presentation, resulting from non-optimised agitation, may effectively have dramatic consequences in the estimation of selectivity and reactivity. For example, enantio- and regioselective catalysed reactions susceptible to mass transport effects, are known. Thus, a major challenge is to develop special reactors for rapid catalyst screening that would ensure good mass and heat transport, in a small volume. Application of such apparatus for intrinsic kinetic determination would also be of high interest. In this paper, a new concept to achieve High Throughput Screening (HTS) of polyphasic fluid reactions is proposed. As test reactions, a gas-liquid asymmetric hydrogenation and a liquid-liquid isomerisation have been chosen to validate our approach of HTS experiments. A preliminary work with the liquid-liquid system has been published recently.[2]

Concept

To our knowledge,[3] commercial apparatus for High Throughput Screening of multiphase catalytic reactions are all based on the very simple concept (Figure 1): several batch type tank reactors are downscaled (step a) and placed in a rack (step b). Ideally, one reaction/catalyst or substrate can thus be screened in each of the millireactors (step c).

That parallel/batchwise approach is just derived from traditional glassware and pressure apparatus adding small scale parallelisation and ca. a tenfold downscaling of reaction volumes. Most of the commercial apparatus are equipped with agitation/heating devices and the consumption of gaseous reagents can also be monitored. Because of their batchwise concept, their familiar design and the absence of other commercial HTS multiphase apparatus, these equipment's are enjoying a good commercial success.

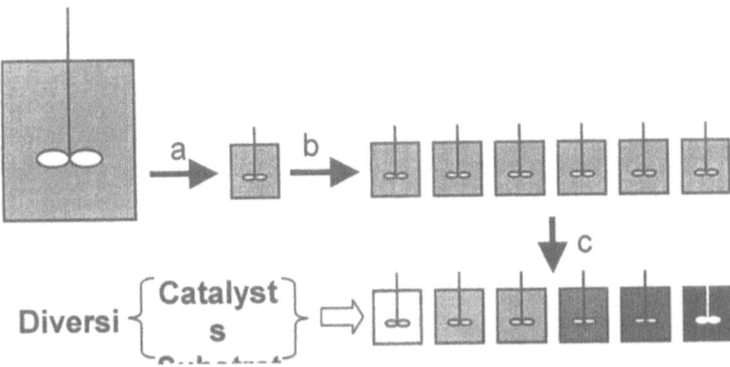

Figure 1: Traditional concept for speeding-up testing of multiphase (pressure) catalysis via downscaling and parallelisation of batch reactors.

However, many drawbacks may drive to disappointments when using these apparatus for actual HTS experiments: i) mixing and multiphase mass transfer may be a issue in the case where agitation is insured by mean of a magnetic stirring rod, ii) quite a small number of samples can be screened per day (typically 8 to 24), iii) monitoring of the reaction selectivity, a key consideration in fine chemistry, is seldom proposed, iv) when such a monitoring can be performed, it relies on sampling with syringes/septum thus limited to low pressure chemistries and involving complex electromechanical moving parts, v) for the same reasons, the pressure and temperature range is limited. Finally, the parallel reactor racks are themselves quite expensive and required time and manpower consuming reconditioning steps (decontaminating/cleaning/drying) that cannot be easily automated and which is often the rate limiting step in HTS experiments.

Rather than this parallel/batchwise approach, we propose a new concept for high throughput experiments that is based on dynamic sequential operations with a combination of pulse injections and micromachined elements.[3] Some expected advantages over traditional batch parallel operations are a lower inventory of sample (down to μg), a larger range of operating conditions (pressure, temperature), simpler and fewer electro-mechanical moving parts. A schematic representation of the concept for gas-liquid catalysis is given in figure 2.

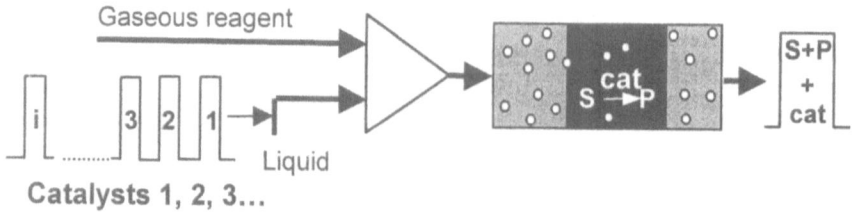

Figure 2 : Principle used for high throughput sequential fluid-liquid catalyst screening.

During operations, the micromixer is continuously fed with the gaseous reagent and the liquid phase. The later does not contains the catalyst (**cat**) to be screened. The composition of the liquid carrier is chosen to ensure the formation of a gas-liquid foam at the exit of the micromixer.[4] At periodic time intervals, a pulse consisting of a mixture of the catalyst to be screened and the substrate (**S**) to be hydrogenated is injected. After being mixed with the gas reagent, the reacting segment containing the catalyst, the substrate and the product (**P**) will travel along a tube whose length determine the residence time. At the outlet of the system, the pulse is analysed to check for activity and selectivity. Since there is no need to wait for the first reacting segment to come out from the reactor, sequential injections of other catalysts/substrates may be performed in a short time. The limiting time will be that required to avoid overlapping of the reacting segments all along the tube reactor.

Results and discussion

The experimental set up for the is described in figure 3. The micromixer used in this work was provided by the Institut for Micromechanics at Mainz (IMM), Germany. The micromixer has been described previously.[4] For the gas-liquid hydrogenation reaction, the liquid is composed of water/ethyleneglycol (40/60 wt %) and a surface active agent sodium dodecyl sulfate (SDS). That composition is required to ensure a stable foam (no coalescence was noticed for residence times up to 6 min at 60°C) with small gas bubbles (ca. 200 µm average diameter) with a liquid hold-up of 20 %.[4]

A pulse containing the substrate Z-acetamidocinamic methyl ester (MAC) (0.05 to 0.1 M) a Rh/chiral ligand catalyst (the chiral ligands are S,S-CBDTS and S,S-BDPPTS) dissolved in the water/ethyleneglycol/SDS mixture is injected. Using this procedure, we have found that the rate of reaction is proportional to the catalyst concentration (figure 4a). Also, knowing the rate of reaction being independent of the substrate concentration within the range of concentrations used in this study,ref an activation energy of 67 kJ.mol^{-1} was determined (figure 4b).

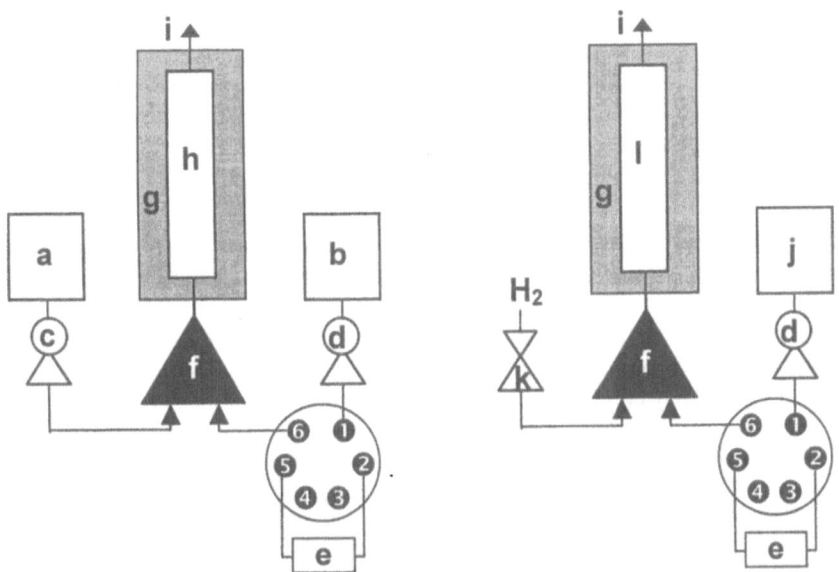

Figure 3: Experimental set up for liquid-liquid (left) and gas-liquid (right) experiments : **a** reservoir for the substrate in *n*-heptane (0.1 M), **b** water reservoir, **c** and **d** high pressure liquid pumps, **e** catalyst injection valve equipped with a 200 µl loop, both valves are HPLC type valves, **f** micromixer, **g** heating mantel, **h** tubular stainlesssteel reactor (0.4 cm i.d., 80 cm length), **i** outlet to analytics, **j** water/ethyleneglycol/SDS reservoir, **k** hydrogen supply (50 bar) with mass flow controler, **l** tubular glass reactor (0.28 cm i.d., 150 cm length).

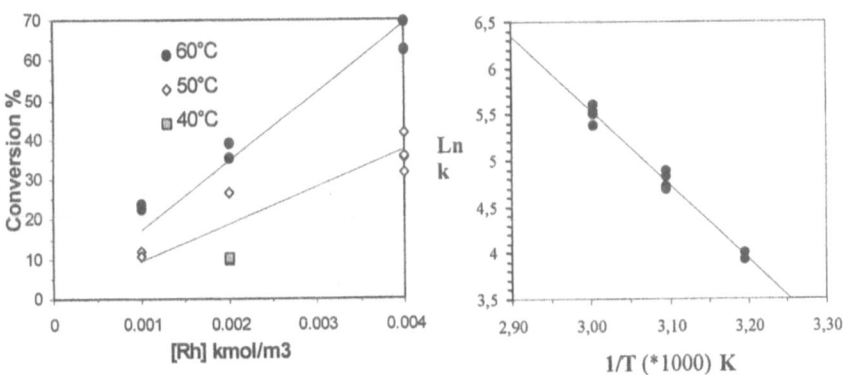

Figure 4: Gas-liquid asymmetric hydrogenation of MAC. (a) Conversion at the microtest outlet as a function of catalyst loading, (b) Arrhenius plot. Condition: C_{MAC} 0.05 kmol.m^{-3}, 1 bar, H_2 flow rate 10^{-4} m^3/h (NPT), liquid flow rate 1 cm^3.min^{-1}, residence time 225 s.

These results strongly support that the microtest is working under chemical regime. A further argument comes from the good comparison of the enantioselectivities obtained in the microreactor and those previously reported for these catalysts under similar conditions. Using this set-up, more than 20 tests per hours can be performed. Further work is in progress to use this microtest for intrinsic kinetic determination.

For liquid-liquid operations, no specific composition of the liquid carriers is required. In this work, we have found that the chosen solvents for the reaction, i.e. n-heptane and water do lead to a very nice emulsion comparable to the oil-water mixture investigated by Hessel et al, with droplet size less than 30 μm.[4] The model reaction for liquid-liquid catalysis is the isomerisation of allylic alcohol's into carbonyles (eq 1).

$$R' \diagdown \diagup \diagdown_{R}^{OH} \xrightarrow[\text{water}/n\text{-heptane}]{\text{Cat. 40-80 °C}} R' \diagdown \diagup \diagdown_{R}^{O} \qquad (1)$$

In this reaction, the substrates and products are mainly located in the organic layer whereas the catalyst is soluble only in the water layer. The thermodynamic and kinetic aspects of this reaction were previously thoroughly investigated.[5] We also have shown in a previous paper that qualitative screening of catalysts and substrates can be achieved using the set-up depicted in figure 3a.[2]

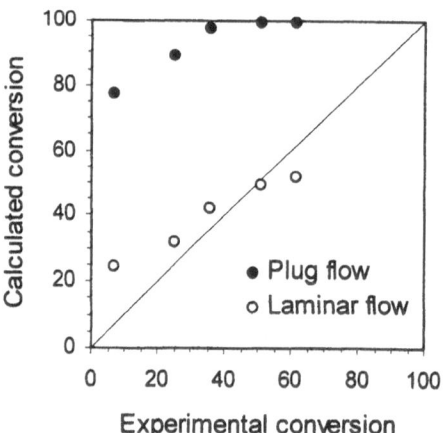

Figure 5 : Parity plot showing the comparison of the conversion with a batch and laminar flow model for five different substrates. The calculations have been performed using the same rate law for all substrates. The observed increase in conversion comes from the increasing solubility's of the substrates, on going from C_8 to C_4 alcohol's, in the catalytic aqueous phase.

However, kinetic studies required some more appropriate reactor description. Simple modelling using plug flow assumption is not valid. The computed conversion are indeed higher than those actually measured. Visualisation of the water layer indicates that laminar flow prevails in the reactor tube. Considering the emulsion as a pseudo homogeneous liquid phase indeed leads to the estimation of low Renoyld numbers in the range . Correction of the batch conversion with laminar flow leads to a better fit (figure 5).

The High Throughput Test proposed in this work relies on a static micromixer. For qualitative screening of liquid-liquid catalysis under various operating conditions, the test reveals very useful and reproducible. However, some drawbacks were noticed. For gas-liquid catalysis, the required composition of the liquid phase makes the scope of the test restricted. Similarly, accurate intrinsic kinetic determination requires detailed modelling of the reactor. For the latter purpose, the perfect control of the flow regime and the interfacial area must be achieved. Microdevices with such properties will be evaluated in future work.

References

1. a) B. Jandeleit, D.J. Schaefer, T.S. Powers, H.W. Turner, W.H. Weinberg *Angew. Chem. Int. Ed.* **1999**, *38*, 2494 ; b) R.H. Crabtree *Chemtech* **1999**, 21.
2. C. de Bellefon, N. Tanchoux , S. Caravieilhes, P. Grenouillet, V. Hessel *Angew. Chem. Int. Ed.* **2000**, *39*, 3442.
3. a) Automated Synthetic Methods for Speciality Chemicals W. Hoyle Ed., The Royal Society of Chemistry, Cambridge, UK, 1999 ISBN 0-85404-825-1 b) M. Harre, U. Tilstam, H. Weinmann *Org. Process Res. Dev.* **1999**, *3*, 304.
4. a) W. Ehrfeld, K. Golbig, V. Hessel, H. Löwe, Th. Richter *Ind. Eng. Chem. Res.* **1999**, *38*, 1075 ; b) H. Löwe, W. Ehrfeld, V. Hessel, T. Richter, J. Schiewe Proceedings of " IMRET 4 ", March 2000, Atlanta, USA.
5. Isomérisation des alcools allyliques : Catalyse moléculaire biphasique et cinétique en réacteur transitoire chromatographique. S. Caravieilhes, PhD Thesis, Université Claude Bernard de Lyon, december 2000.

A Novel Cross-Flow Microreactor for Kinetic Studies of Catalytic Processes

Sameer K. Ajmera, Cyril Delattre, Martin A. Schmidt[‡], and Klavs F. Jensen

MicroChemical Systems Technology Center
Department of Chemical Engineering, [‡]Department of Electrical and Computer
Engineering, Massachusetts Institute of Technology, Cambridge, MA 02139 USA

Abstract. A silicon differential microchemical reactor that utilizes a novel cross-flow geometry has been developed for the testing of porous supported catalysts. The cross-flow design reduces pressure drop by using a shallow but wide catalyst bed. Modeling, flow visualization, and pressure drop studies confirm even flow distribution across the reactor. Kinetic experiments with the oxidation of carbon monoxide as a model chemistry demonstrate the cross-flow microreactor as a practical laboratory tool for catalyst testing.

Keywords. catalyst testing, fixed-bed, packed-bed, kinetics, catalysis, screening, microchemical

1. Introduction

In standard bench-scale tube reactors used for heterogeneous catalyst testing, flow is parallel to the long axis of the tube (axial flow). A microfabricated axial flow packed-bed reactor has been previously demonstrated for determining the chemical kinetics of phosgene formation [1]. Although the small dimensions of the microreactor reduce thermal and mass gradients, the small particle sizes dramatically increase the pressure drop in the reactor. For a given bed length, a packed-bed with 60 µm catalyst particles has a pressure drop ~275x larger than a bed with 1 mm particles, and ~27500x larger than a bed with 1 cm particles. In studies with the moderately fast phosgene reaction in the axial flow microreactor, gas flow rates of only 4.5 sccm with 53-71 µm catalyst yielded a large pressure drop of 0.4 atm. In general, a large quantity of catalyst is desirable for catalyst testing with microreactors to average out variances between catalyst particles, yield practical quantities for handling, and increase flow rates to reduce lag time. However, shortening the length of the catalyst bed is necessary to reduce pressure drop.

In this work, we present a silicon differential cross-flow microreactor for catalyst testing that uses catalyst particles similar to those used in industry instead of thin-films or coatings. This way, results are relevant to catalyst preparation techniques, catalyst-support interactions, and other issues in industrial catalysis. The cross-flow geometry enables the use of practical flow rates and catalyst quantities while minimizing pressure drop. The microreactor design, fabrication, and characterization are presented. Kinetics data for a model chemical reaction are experimentally determined to demonstrate the utility of the cross-flow microreactor for catalyst testing.

2. The Cross-Flow Microreactor Design

Multiple, short packed-beds would increase the effective catalyst area and increase throughput while reducing pressure drop. The cross-flow design integrates short parallel beds into a continuous wide packed-bed (Fig. 1). For a fixed bed volume and flow rate, the cross-flow design yields the same residence time as an axial flow reactor, but with a smaller pressure drop. This geometry is conducive to differential operation (low conversions) as reactants have a small contact time with the catalyst at high flow rates, but without the large pressure drop. Differential operation without recycle is desirable for studying chemical kinetics. Figure 2 shows the silicon cross-flow reactor. The reactor has one inlet and

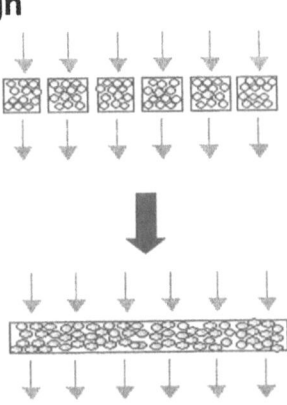

Fig 2. Integrating short beds into a cross-flow packed-bed

one outlet for gas flow. The inlet channels (350 μm wide, 370 μm deep) bifurcate into 64 parallel channels that are integrated into a 2.555 cm wide catalyst bed. The catalyst bed is 500 μm deep and 400 μm long, and is approximately 5.1 μL in

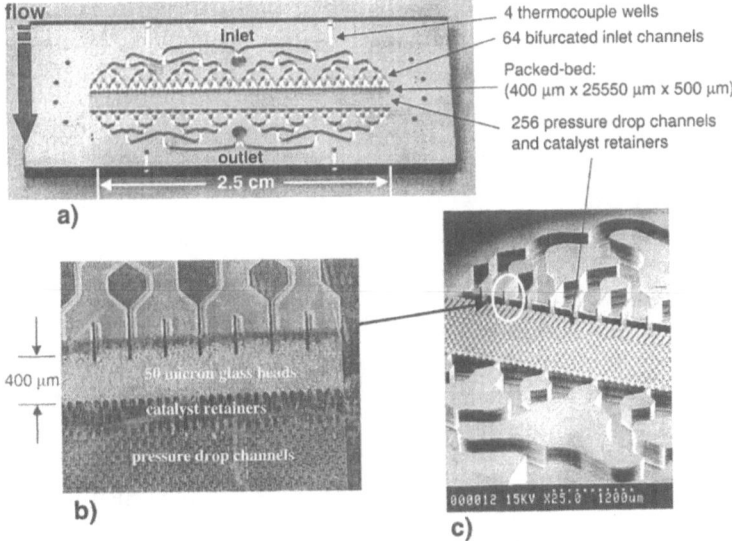

Fig 1. a) Photograph of the silicon cross-flow microreactor chip. b) Photograph of microreactor packed with 50 μm glass beads. c) SEM of the microreactor.

total volume. An array of retainer posts (50 μm wide) acts as a frit to hold the catalyst particles in place. The heads of the retainer posts expand out to leave a 35-40 μm gap between posts, setting the lower limit for particle size.

To achieve even flow distribution, a fixed pressure drop region consisting of 256 shallow channels is incorporated into the microreactor (Fig. 3). The channels are only 40 μm wide and 20-25 μm deep, and meander for ~2.2 mm (net center-line length).

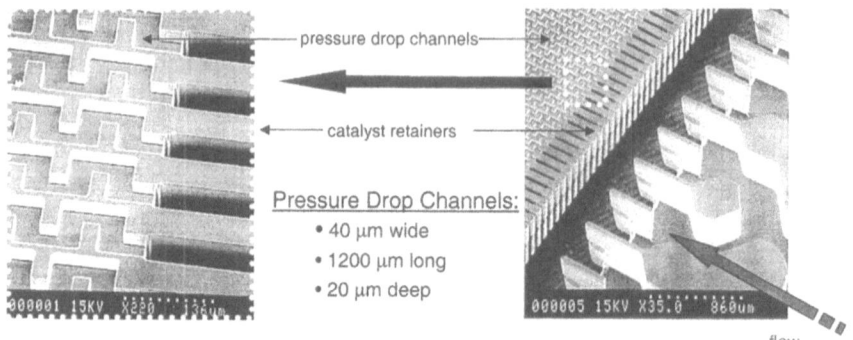

Fig 3. Scanning electron micrographs of the catalyst retainers and pressure drop channels

The pressure drop through these shallow channels is designed to be much greater than the pressure drop through the rest of the reactor, including the packed-bed. Thus, variances in catalyst packing density have minimal effect on the overall flow distribution in the reactor. Along the perimeter of the microreactor are 4 side-wells for thermocouples. The large thermal conductivity of single crystal silicon and high surface to volume ratio helps the packed-bed quickly achieve thermal equilibrium with the reactor bulk.

3. Fabrication

The microreactor is fabricated from single crystal silicon (100 mm diameter wafer, 500 μm thick) with standard microfabrication processes. The features are etched using anisotropic deep reactive ion etching (DRIE) with a time-multiplexed inductively coupled plasma etcher [2]. To avoid the possibility of non-uniform catalyst bed depth, the packed-bed region is etched all the way through the wafer. The region before the packed-bed and the channels between the catalyst retainer posts are also etched through to ensure even flow of gases at the entrance and exit of the catalyst bed. The reactor contains 3 different channel depths. The pressure drop channels are ~20 μm deep, while the inlet and exit bifurcated flow channels are etched 370 μm deep. The packed-bed region is 500 μm deep (a single silicon wafer thickness). Both front-side and back-side etches are used to fabricate the three different channel depths.

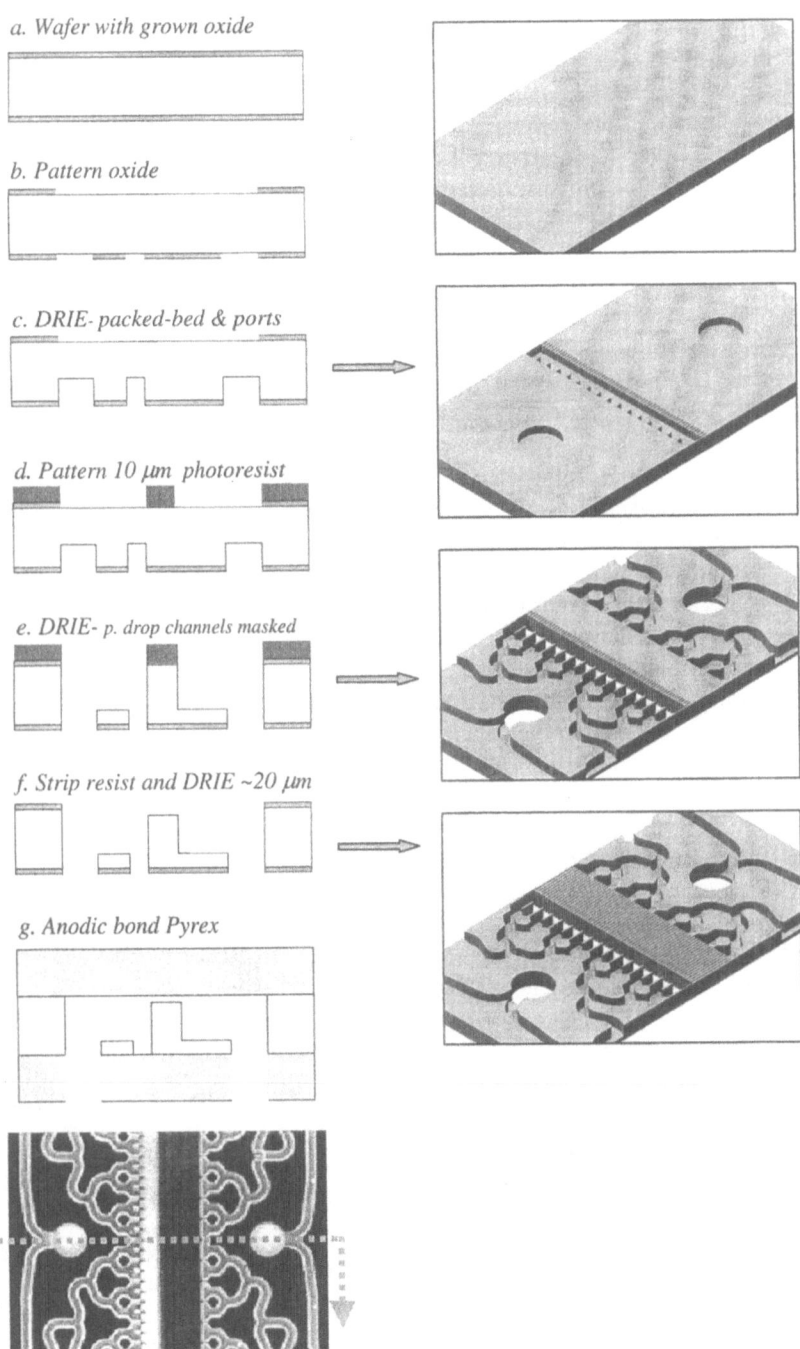

a. Wafer with grown oxide

b. Pattern oxide

c. DRIE- packed-bed & ports

d. Pattern 10 μm photoresist

e. DRIE- p. drop channels masked

f. Strip resist and DRIE ~20 μm

g. Anodic bond Pyrex

Fig 4. The microreactor fabrication sequence.

In the first fabrication step, 0.5-1.5 µm of oxide is grown on both sides of the starting wafer (Fig. 4a). The oxide protects the bonding surfaces during processing and serves as a mask for etching the pressure drop channels on the front-side. Next, the entire reactor geometry is patterned into the oxide on the front-side of the wafer. This mask includes the inlet/outlet ports, flow channels, packed-bed region, pressure drop channels, and thermocouple wells. On the back-side, the features to be etched through the wafer (packed-bed region and inlet/outlet ports) are patterned (Fig. 4b). The back-side features are etched with timed DRIE ~250 µm deep (Fig. 4c). Using a thick photoresist, the entire reactor geometry *except* the pressure drop channels is aligned to the oxide and patterned on the wafer front-side (Fig. 4d). A second DRIE is performed approximately 350 µm deep (Fig. 4e). This step opens up the features etched on the back-side and creates the flow channels. Removal of the photoresist reveals the previously patterned oxide layer. A timed etch is performed to etch all the features ~20-25 µm, creating the pressure drop channels (Fig. 4f). The oxide layer is thick enough to serve as a resist against the shallow etch. The flow channels and thermocouple wells end up ~370 µm deep. In the final step, the channels are capped on the top and bottom with Pyrex wafers *(Corning 7740)* by anodic bonding [3] (Fig. 4g). Mechanically drilled holes (2 mm) in the bottom Pyrex wafer were aligned to the inlet/outlet ports during the bonding process. The bonded wafer stack is diced to obtain eight individual reactor chips (15 mm × 40 mm × 1.5 mm). The optically transparent Pyrex enables visualization of the reactor, and sets the maximum operating temperature of the device at ~550°C. If higher temperatures are desired, silicon capping wafers can be used, raising the operating limit above 1000°C.

4. Packaging and Catalyst Loading

The reactor is compressed with a metal cover plate against a thin elastomer gasket (0.8 mm thick Kalrez™ or Viton™) with punched through-holes to form fluidic connections to a stainless-steel base in a fashion similar to Losey et al. [4]. External fluidic connections are made directly to the metal base (Fig. 5). Catalyst between 50-70 µm diameter is loaded into the reactor through the inlet port using a vacuum placed on the outlet. The flow of gas generated by the vacuum fluidizes the small catalyst particles and draws them into the reactor. By applying a high pressure to the

Fig 5. Microreactor assembly with fluidic connections and heaters.

outlet, catalyst can be blown out of the inlet port and the reactor can be reused. The compression plate and the base contain cartridge heaters and thermocouples are threaded into the side wells of the reactor for temperature control.

5. Microreactor Characterization

5.1 Pressure Drop Channels

Pressure drop experiments with nitrogen gas flow were performed for both an empty reactor and one packed with 60 μm glass beads (Fig. 6). The pressure drop across the empty reactor was approximately 0.075 atmospheres for a flow of 100 sccm at room temperature. The addition of beads minimally increased the pressure drop, and contributed only ~7% to the total drop across the reactor. In normal operation with evenly distributed packing, the flow is considered independent of packing voids. Further, the pressure drop across the catalyst bed is negligible with a drop of only 0.006 atm for 100 sccm flow yielding an isobaric catalyst bed. The benefit of the cross-flow design is seen as the pressure drop in an axial flow reactor of same volume packed with the same quantity of glass spheres would be ~1200-2000x larger at similar gas flow rates.

Fig 6. Pressure drop measurements: The packing has only a minor affect on the total pressure drop.

5.2. Flow Distribution- Visualization and Modeling

The ability of the pressure drop channels to ensure uniform flow distribution was studied by visualization experiments using liquid flow through an empty reactor. 1 μL plugs of 20 mM phenol red in ethanol were sent through the reactor. The Reynolds number based on the inlet channel width was approximately 2.6, well

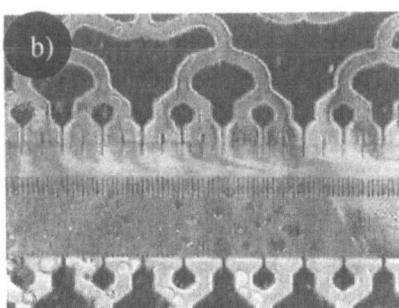

Fig 7. Still frames of 1 μL plugs of phenol red: a) typical microreactor with even flow distribution, b) reactor with extra shallow pressure drop channels. The flow curves sharply towards regions of lower pressure drop.

420

into the laminar flow regime. Fig. 7 (a) shows a still frame of a phenol red plug flowing through a typical microreactor where the pressure drop channels are the normal 20-25 µm deep. The flow is well distributed across the width of the reactor. Fig. 7 (b) shows still frame of the flow through a different reactor with an extra shallow pressure-drop section. The discolorations in the pressure drop channels are areas that were etched only 5-10 µm, significantly increasing the pressure drop in this region of the reactor. The flow curves sharply towards regions of lower pressure drop giving uneven flow across the reactor, illustrating the importance of having a well defined, uniform pressure drop region.

To understand the flow characteristics across the depth of the reactor, 2-D CFD calculations (CFD-ACE, *CFD Research Corp, USA*) were performed using the reactor cross-section shown in Fig. 8. The simulations used air at 373K flowing at an average velocity of 1 m/s for both an empty reactor and one with 2 layers of 50 µm spherical particles. In both cases, the convective velocity fields were calculated for the reactor cross-section. Fig. 8 shows the velocity profiles at particular cross-sections.

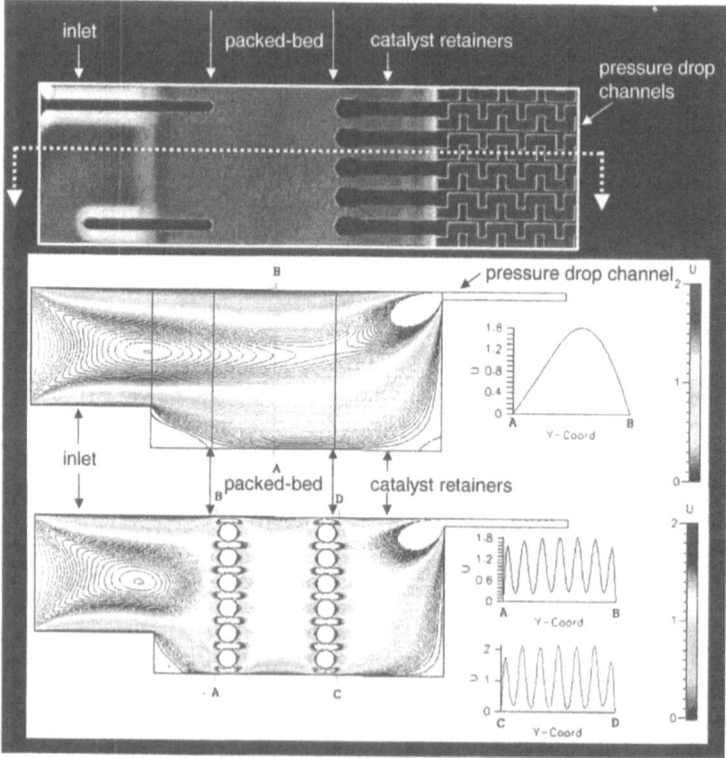

Fig 8. 2-D CFD model of the velocity profiles through the cross-section indicated by the dashed line.

The flow through the empty reactor shows a laminar parabolic profile with non-symmetric distribution across the reactor depth due to the placement of the pressure drop channels (section A-B, Fig. 8). Recirculation zones at the beginning and end of the packed-bed region extend close to the bed region. When 2 layers of particles are placed in the reactor, the previously uneven velocity profile (section C-D, Fig. 8) flattens out across the depth of the reactor. The recirculation zones are considerably smaller (approximately 50-100 μm) and are away from the packed-bed region. At this length scale (~50 μm), diffusion is fast and counters the trapping effects of a stagnant volume. Therefore, the microreactor packed completely with catalyst particles will have an even distribution of flow across the entire reactor depth with minimal effects from recirculation zones, rendering a device favorable for kinetics experiments.

6. Determination of Kinetic Parameters

The rate of reaction and observed activation energy for the oxidation of carbon monoxide to carbon dioxide over supported metal catalysts were determined experimentally to validate the cross-flow microreactor as a tool for obtaining chemical kinetics. The experiments were performed with metal catalyst (Rh, Pt, or Pd) supported on alumina particles (~3 mg) sieved between 53-71 μm and loaded with a vacuum in air. A reactant feed of $1\%CO/1\%O_2$ in helium was fed into the reactor at flow rates ranging between 20 and 600 sccm and differential kinetics were measured for conversions of 10% or less at steady-state. Temperatures up to 530K were used and the reactor assembly was heated without external cooling. Figure 9 shows the experimental rate of reaction given as the turnover frequency (TOF-molecules of CO formed per catalyst site per second) for 0.3% Pd/Al_2O_3 (% by weight) and a comparison to previously published values by Rainer et al. [5].

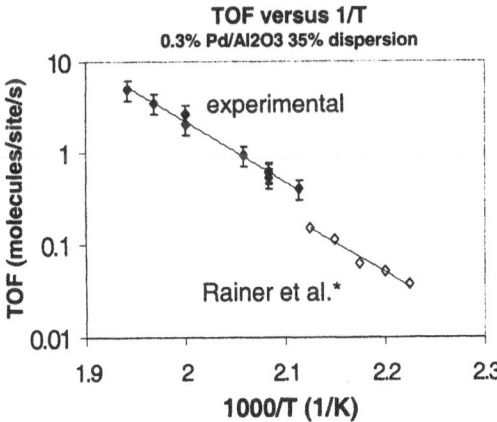

Fig 9. Experimental TOF versus literature for CO oxidation over Pd/Al_2O_3. (*Rainer et al. [5])

Both the TOF and the apparent activation energy (30 kcal/mol) compare favorably to literature values. Activation energies obtained from oxidation experiments for Pt/Al_2O_3 (22.8 kcal/mol) and Rh/Al_2O_3 (27.6 kcal/mol) shown in Table 1 also compare well with literature values [5-8].

Table 1. Experimental apparent activation energy for CO oxidation over different catlaysts compared to literature values

Catalyst	Microreactor E_{act} (kcal/mol)	Literature E_{act} (kcal.mol^{-1})	Literature Reference
Pd/Al_2O_3	30.1	30.0	[5]
Pt/Al_2O_3	22.8	21.6	[8]
Rh/Al_2O_3	27.6	24-31	[6,7]

The consistency of the data with different catalysts is important for demonstrating the use of the cross-flow microreactor for catalyst testing. The good agreement between the experimental and literature values indicates that the cross-flow microreactor, with a low pressure drop and even flow distribution, can be a useful laboratory tool for catalyst testing.

7. Conclusions

A silicon cross-flow microreactor has been fabricated for heterogeneous gas phase catalyst testing. The microreactor uses catalyst particles relevant to macro-scale industrial processes instead of thin films or unsupported coatings. Unlike axial flow designs that are susceptible to large pressure drops, characterization of the flow and pressure drop shows that cross-flow yields an isobaric catalyst bed with even flow distribution. By eliminating gradients within the reactor and minimizing the pressure drop, the cross-flow microreactor provides a useful tool for catalyst testing and determination of kinetic parameters. The microfabrication approach opens the possibility for using several of these reactors in parallel to accelerate catalyst discovery and optimization.

423

Acknowledgements

The authors thank Dr. James Ashmead, Dr. Aleks J. Franz, Justin T. McCue, Dr. Shinji Isogai, and Felice Frankel for their help during the course of this work. The expertise of the personnel at the Microsystems Technology Laboratory (MIT) is also gratefully acknowledged. C. Delattre would like to thank the Institut Français du Pétrole (IFP) for financial support. S.K. Ajmera would like to thank the National Science Foundation Graduate Fellowship Program. The authors also thank IFP for catalyst samples. The DARPA Micro Flumes Program (F30602-97-2-0100) is gratefully acknowledged for financial support.

Citations

1. S.K. Ajmera, M.W. Losey, K.F. Jensen, M.A. Schmidt. "Microfabricated Packed-bed Reactor for Distributed Chemical Synthesis: The Heterogeneous Gas Phase Production of Phosgene as a Model Chemistry," Submitted to *AICHE JOURNAL* (2001).
2. A.A. Ayon, R. Braff, C.C. Lin, H.H. Sawin, M.A. Schmidt. "Characterization of a Time Multiplexed Inductively Coupled Plasma Etcher," *J. Electrochem. Soc.* **146** (1999) pp. 339-349.
3. M.A. Schmidt. "Wafer-to-Wafer Bonding for Microstructure Formation," *Proceedings of the IEEE* **86** (1998) pp. 1575-1585.
4. M.W. Losey, M.A. Schmidt, K.F. Jensen. "Microfabricated Multiphase Packed-Bed Reactors: Characterization of Mass Transfer and Reactions," to appear *Ind. Eng. Chem. Res.* (2001).
5. D.R. Rainer, M. Koranne, S.M. Vesecky, D.W. Goodman. "CO + O_2 and CO + NO Reactions over Pd/Al_2O_3 Catalysts," *J. Phys. Chem. B* **101** (1997) pp. 10769-10774.
6. S.H. Oh, G.B. Fisher, J.E. Carpenter, D.W. Goodman. "Comparative Kinetic Study of CO-O_2 and CO-NO Reactions over Single Crystals and Supported Rhodium Catalysts," *J. Catal.* **100** (1986) pp. 360-376.
7. S.H. Oh and C.C. Eickel. "Influence of Metal Particle Size and Support on the Catalytic Properties of Supported Rhodium: CO-O_2 and CO-NO Reactions," *J. Catal.* **128** (1991) pp. 526-536.
8. R.H. Venderbosch, W. Prins, W.P.M. van Swaaij. "Platinum catalyzed oxidation of carbon monoxide as a model reaction in mass transfer measurements," *Chem. Eng. Sci.* **53** (1998) pp. 3355-3366.

A New Microreactor for the Solution-Phase Synthesis of Potential Drugs

Ulrich Kunz*, Andreas Kirschning**

*Institut für Chemische Verfahrenstechnik, Leibnizstr. 17

Technische Universität Clausthal, D-38678 Clausthal-Zellerfeld

email: kunz@icvt.tu-clausthal.de

**Institut für Organische Chemie, Am Schneiderberg 1B,

Universität Hannover, D-30167 Hannover

email: andreas.kirschning@oci.uni-hannover.de

Abstract

For optimizing lead-structures that evolve from high-throughput parallel synthesis (combinatorial chemistry) in increasing numbers of tools for automated synthesis in millimolar scale becomes necessary. For this purpose, the combination of solution phase and solid phase synthesis would be ideal. This can be achieved by immobilizing reagents and catalysts on a polymer support. By fixing one of the reactants on such a support the starting material and product stay in solution. Applied to a flow through system, the product leaves the reactor in solution. For this purpose we developed a new microreactor.

Our approach employs an inorganic carrier material with irregularly shaped channels which is readily available at low price. The polymer is introduced by a precipitation-polymerization process, resulting in a fixed polymeric phase inside the channels of the microreactor. The average size of these interconnected polymeric particles was determined to be about one micrometer. In between the polymeric particles channels in the tenth to hundredth micrometer range exist. This morphology ensures high accessibility to the active sites of the polymer, which are for example quaternary ammonium cations. These functional sites can be loaded with various anions, including organic anions, which act as reactants. As a model reaction the reduction of ketones to alcohols as well as a set of selected synthetic examples are presented.

Keywords

Polymer assisted solution phase synthesis (PASS-flow), millimolar scale drug synthesis, pharmaceuticals, polymers, polymer/glass microreactor, kinetics

Introduction

The recent introduction of high-throughput synthesis in pharmaceutical and agrochemical research for the discovery of lead-structures resulted in a dramatic increase of libraries with huge numbers of components. The use of microreactors in the shape of high density well plates will further enlarge the number of substances synthesized. This encloses the use of solid supports as reactants or catalysts too [1]. If a hit is found the researcher needs a larger quantity of this substance on a millimolar scale for evaluation. Up to date conventional laboratory equipment is used to prepare industrial scale production (figure 1).

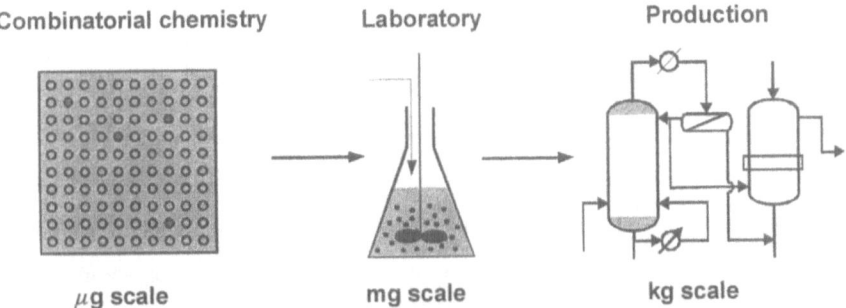

Combinatorial chemistry	Laboratory	Production
µg scale	mg scale	kg scale

Figure 1: Conventional development of pharmaceuticals production

The combination of solution phase and solid phase synthesis is an emerging tool for high throughput synthesis [2]. One of the reactants is fixed on a solid support, in most cases a polymer. The other reactant is applied as a solution. During the course of the chemical reaction a soluble product is formed. In this context, the use of polymer beads with a size of 50 to 500 µm as support is state of the art. The polymer is loaded with the reactant, filtered and the appropriate amount of the resin is filled into a stirred vessel where a solution of the second reactant is added.

426

When the reaction is completed the product in solution is separated by filtration and is purified by additional procedures according to figure. 2.

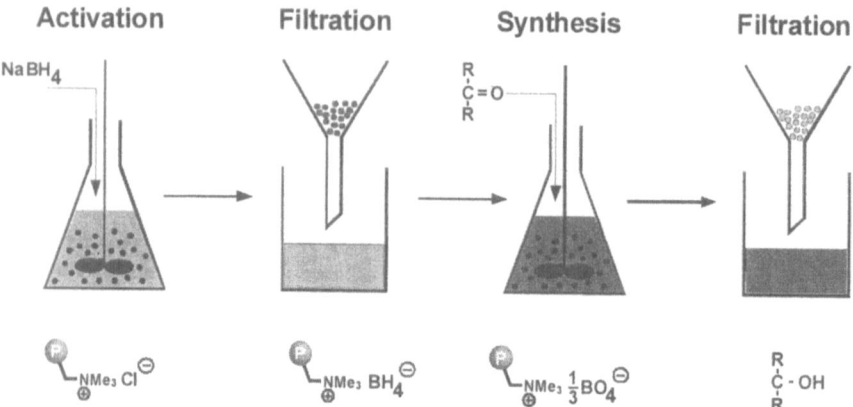

Figure 2: State of the art in polymer assisted solution phase synthesis. As an example the reduction of a ketone to an alcohol with immobilized borohydride is given.

This approach is time consuming and the productivity of the laboratory is limited. Automation with the help of traditional laboratory equipment is therefore difficult to achieve. As a result of the conventional approach a gap evolves which needs to be filled with new methods suitable for mg scale drug synthesis. Microreactors can play a part in this growing field [3,4]. Single microreactors are especially useful for small scale synthesis and production. But so far microreactors for the production of potential drugs on a millimolar scale using solid phase supports are not readily available.

Microreactor concept

Besides other reasons there are two main objectives to use microreactors:
- Enhanced heat transfer
- Enhanced mass transfer

In organic synthesis most reactions proceed rather slowly. Thus, the enhanced heat transfer is of less importance than the improved mass transfer. Especially in polymer supported synthesis the reaction rates are slow because, the reactants having a relatively high molecular mass have to diffuse to the active sites inside the solid polymer particles. State of the art is to use crosslinked resins, frequently based on styrene/divinylbenzene. The degree of crosslinking is kept at a low level to allow the resins to swell in the organic solvents. This makes the handling of the solid support uncomfortable because the swelling dramatically changes the volume of the resin bed. In packed bed reactors swelling forces can destroy the polymer particles or reduce the volumetric flow rate. So most of the organic synthesis are done in stirred tank reactors with suspended resin beads. High stirrer speed can damage the particles, fines have to be removed by filtration.

During the last years much effort has been conducted to improve the reactivity of polymer supports. In a recent review the situation is described in detail [5,6]. The field of polymer assisted solution phase synthesis is growing steadily and becomes a standard method for the organic chemist [2]. Although many polymer modifications were investigated, monolithic polymer/carrier material combinations are not in use, although the benefits of monoliths are known for a long time [7]. The implementation of monolithic polymer/carrier composite components allow to prepare microreactors for resin based synthesis.

The limitations of classical polymer assisted solution phase synthesis are:
- High expenditure of dosing work
- Filtration necessary
- Mass transport limited reactions
- Complicated flow through technique

The requirements for microreactors suitable for organic synthesis are:
- Production on a millimolar scale
- Integration of synthesis, purification and analysis in a flow system
- High accessibility of active sites
- Low pressure drop
- Simple dosing and separation of solid reactants
- Automation by available technique

In order to construct a microreactor for the polymer assisted solution phase synthesis it is necessary to fill the channels of the reactor with polymer particles. This is not an easy task, since the channels are small in diameter and filling these channels by conventional techniques is complicated and time consuming. In addition the polymer particles can leave the reactor during operation, because the polymer is not anchored to the reactor walls. Up to now most microreactors are fabricated by expensive microstructuring processes resulting in parallel channels.

Our approach to incorporate polymer particles inside the channels is to start with the monomers and polymerize the particles inside the channels. To ensure that the polymer does not leave the channels open connections between the channels to wedge the polymer particles is required.

The requirements for a microreactor material with respect to the polymerization process are:

- Large volume of the channels
- Channel diameter 10 – 500 µm
- Interconnected channels
- High mechanical strength
- Free of substances influencing the polymerization process
- Chemical stability during all preparation steps

Porous glass with irregular shaped channels fulfills these requirements best, so it was chosen as the material of choice.

The requirements for the polymerization process of the polymer particles inside the microreactor channels are:

- Homogeneous polymer distribution inside the whole microreactor
- Formation of small interconnected polymer particles
- Possibility to control the particle size
- Control of the polymer load
- Suitable for the processing of different microreactor materials

- Possibility to extend the procedure to a wide variety of monomers
- Removeability of additives which are used during the polymerization process
- Easy and cost effective preparation process

After evaluation of several polymerization processes precipitation polymerization was chosen to prepare finely divided interconnected polymer particles inside the microreactor channels. For the preparation of microreactors a porous glass rod with a length of 110 mm and a diameter of 5.3 mm was used as a carrier material for the polymer.

Then the glass rod was encapsulated with a pressure resistant fiber reinforced housing. Standard fittings were used for easy connection with conventional laboratory equipment.

Results and discussion

Polymerization and characterization

The polymerization procedure was described elsewhere [8]. The polymer particles have the shape of spheres which are connected by polymer bridges. The result is a polymer phase inside the microreactor channels which consists of one piece. The polymer phase can not leave the microreactor structure even under convective flow conditions because the irregular shaped channels are fixing the whole polymer piece. There are still large pores in between the particles with diameters of several micrometers. This morphology ensures a good accessibility to the active sites of the polymer. The morphology of the polymer particles is depicted in figure 3. In figure 4 the bridges between the polymer particles can be seen. Figure 5 shows a comparison of the precipitated polymer particles inside the microreactor channels with commercial resin beads. The black areas in the pictures are the microreactor channels, the white dots in the black areas are the interconnected polymer particles, the white area is the glass carrier material. The polymer particles do not form a dense packing, but a loose structure with large channels between the particles. Typical dimensions are given in figure 6.

430

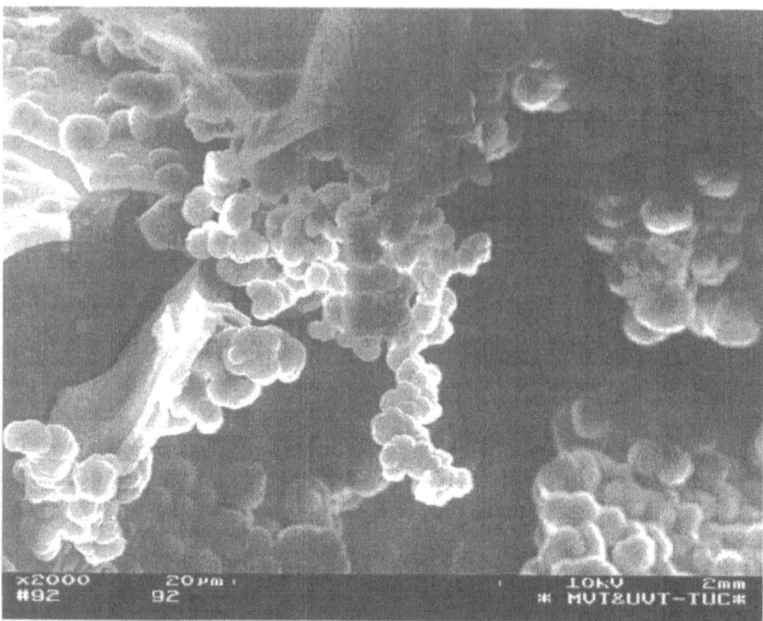

Figure 3: Morphology of the resin particles inside the microreactor channels

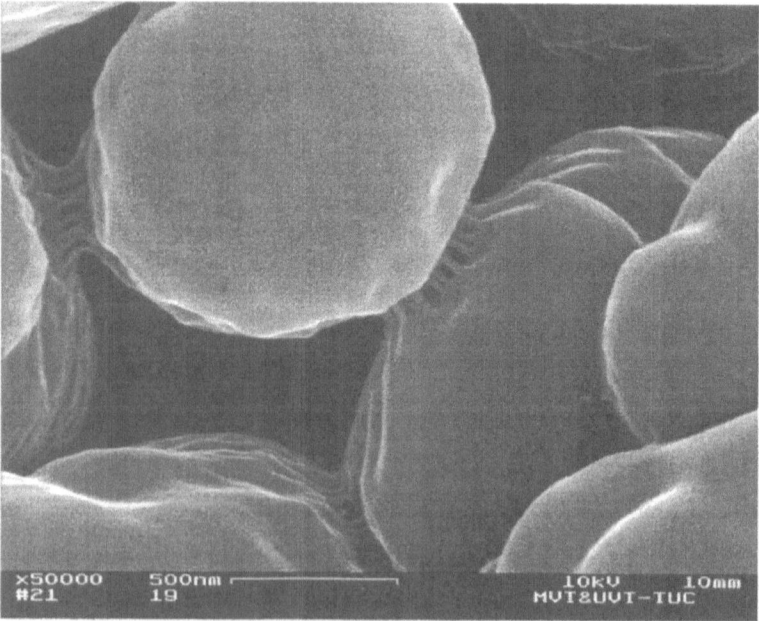

Figure 4: Connections of the polymer particles inside the microreactor channels

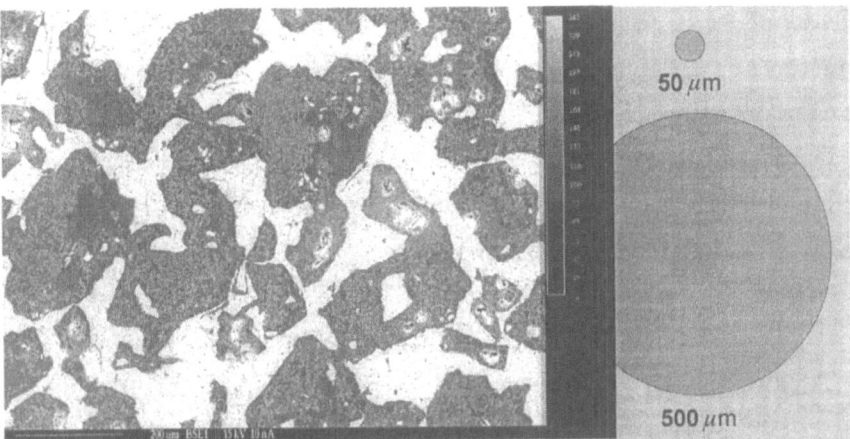

Figure 5: Polymer particles inside the microreactor channels in comparison to commercial resin beads (black areas are the microreactor channels, white dots in the channels are the interconnected polymer particles, white areas are the glass carrier)

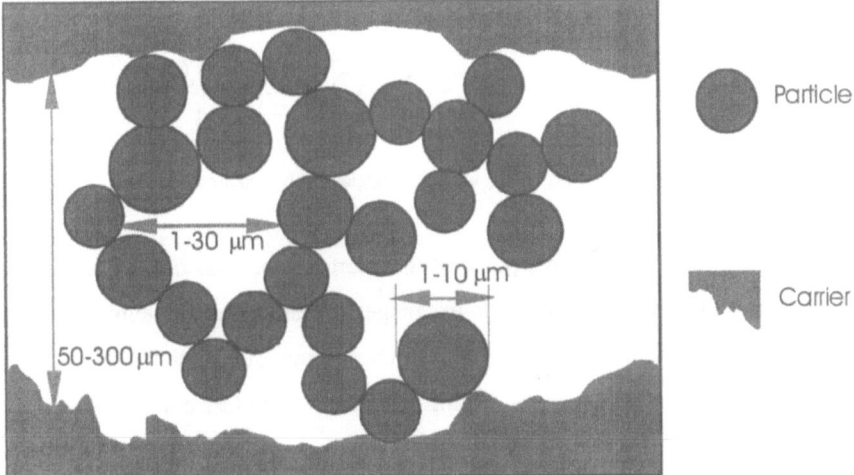

Figure 6: Dimensions of polymer particles and channels inside the microreactor

Functionalization of the polymer with organic reactants

There are several possibilities to attach organic reactants to the polymer particles. One approach is to prepare a basic ion exchange resin and attach anions as reactive groups. For this purpose we prepared a crosslinked polyvinylbenzylchloride from the monomers. Then the benzylchloride group was converted to a quarternary amino group by reaction with anhydrous trimethylamine. The resulting polymer is a cation which allows anchoring of many organic or inorganic anions. Some reactive groups can be introduced in the polymer phase directly by polymerization of the appropriate monomers. Examples are polyvinylaniline and polyvinylpyridine. A selection of some functionalized polymers inside microreactors prepared by us is given in figure 7.

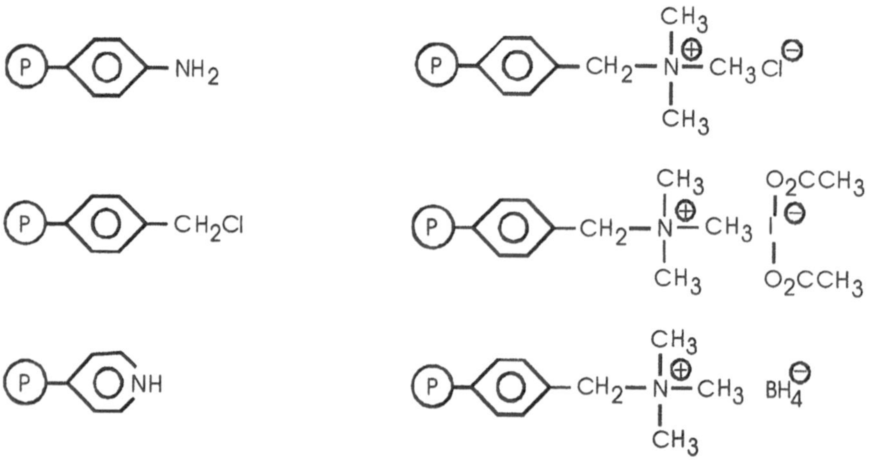

Figure 7: Examples for reactive groups inside polymer/glass microreactor channels

Characterization of the polymer phase of the microreactors was done by elemental analysis and by titration. Typical ion exchange capacities of a single microreactor are about 0.1 – 1 mmol, depending on the chosen polymer load. The size of the polymer/glass composite material of this microreactors is 110 mm length and 5.3 mm diameter.

Casing of the microreactor

In order to apply the polymer/glass material as a microreactor in a flow through application it is necessary to install a pressure resistant housing. The organic reactants which are attached to the polymer must not be heated to high temperatures, so only a low temperature process is allowed for the casing preparation. In organic chemistry a wide variety of solvents is used, that means the casing has to withstand these solvents. For solving this problem the glass rod was lined with a PTFE hose which was applied as a shrinking hose. This results in chemically inert reactor walls. The ends of the hose were left longer than the reactor length to fasten standard connectors made of PEEK. The connectors were glued with epoxy resin, the Teflon lined glass rod was wrapped with glass fiber clothing and put into a polymer tube. The gap between the tube walls and the glass rod was filled with epoxy resin and cured under vacuum at 80°C. Figure 7 depicts a cross sectional drawing of the microreactor.

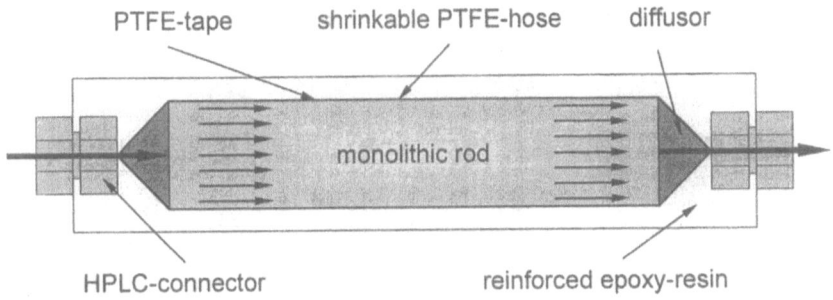

Figure 7: Sectional drawing of a polymer/glass microreactor

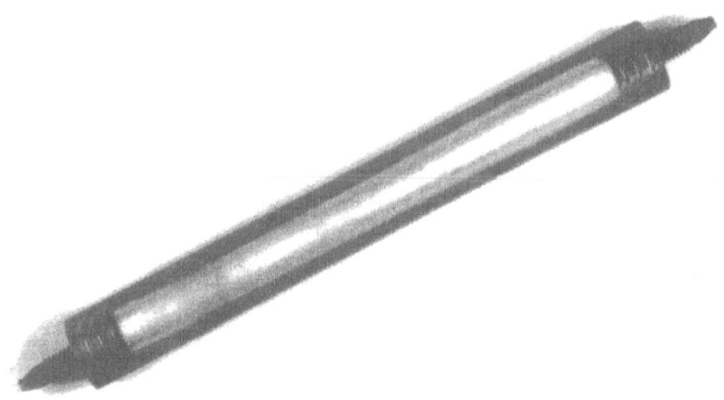

Figure 9: Photo of the microreactor

In figure 9 a photo of one of our microreactors can be seen. The casing made by this method has an outer diameter of 14 mm and can withstand pressure up to 100 bar. In figure 10 the flow dependant pressure drop across the reactor length is given.

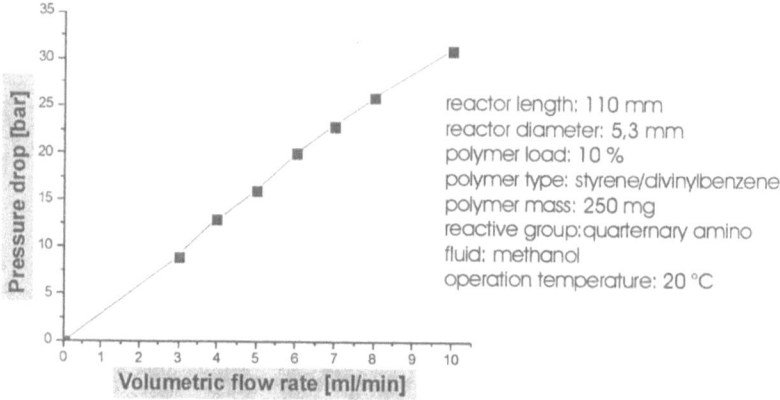

Figure 10: Technical data of a polymer/glass microreactor

The flow behavior inside the microreactor was observed by dynamic NMR-microscopy. The method is described elsewhere [9, 10]. By pumping water through the glass rod the flow patterns shown in figure 11 were obtained.

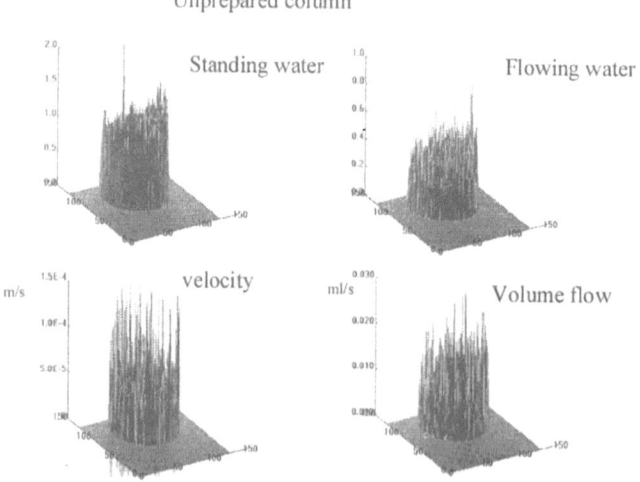

Figure 11: Flow profile across a polymer/glass microreactor

Compared to packed beds of polymer particles of the same size the pressure drop in this microreactor is much lower. The reason for this is, that the interconnected polymer particles do not form a dense packing like single spheres would do. Between the bonded particles are channels which allow convective flow through the microreactor. This ensures good accessibility of the small polymer particles. The flow pattern reveals that we do not have a plug flow behavior. But this was not expected, because we used a random pore material as the carrier. The channels in the glass rod were made by a dissolving procedure. A more defined flow pattern can be reached by the use of sintered glass rods. But sintered rods are more expensive. For most organic synthesis the observed flow pattern is not a disadvantage. Reaction rates are slow and in a recycle reactor operation with high flow rates the reactor operates as a differential reactor. Of much more importance is the fact, that the polymer particles are by more than an order of magnitude smaller than commercial resin beads. The diffusion path length is much shorter allowing higher reaction rates.

Application of the microreactor in organic synthesis

We started our investigations with the reduction of ketones to alcohols. As a model substance we chose acetophenone which was reduced to 1-phenylethylalcohol. The reaction equation is given in figure 12.

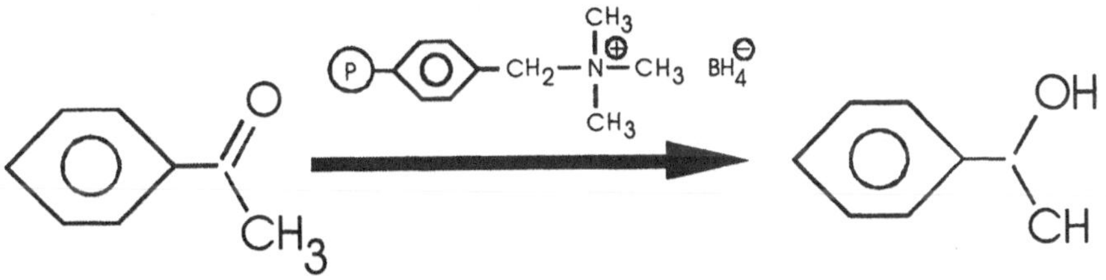

Figure 12: Reduction of acetophenone with borohydride

For this purpose we used a microreactor which contained borohydride anions. The first aim of the experiments was to compare the observed reaction rate in the microreactor with the rate observed in the classical procedure. For this purpose we operated the microreactor batch wise in a recycle mode. The flow sheet of the equipment is given in figure 13.

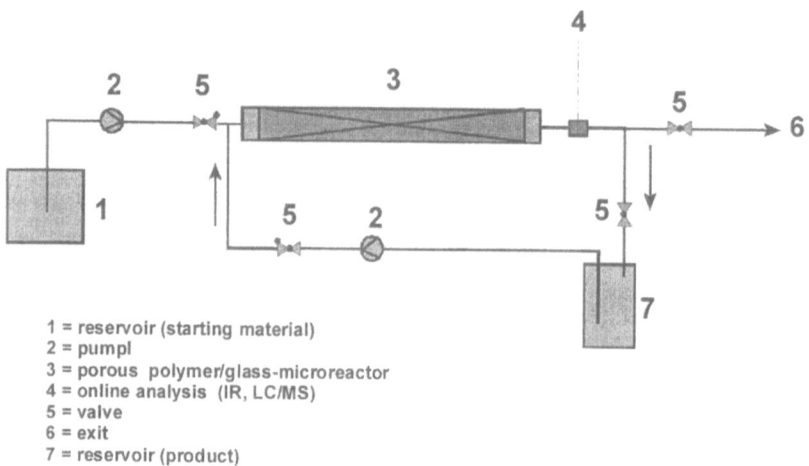

1 = reservoir (starting material)
2 = pumpl
3 = porous polymer/glass-microreactor
4 = online analysis (IR, LC/MS)
5 = valve
6 = exit
7 = reservoir (product)

Figure 13: Polymer/glass microreactor synthesis station for one step organic reactions

With this equipment at hand the activation of the resin particles was also investigated. A solution of sodium borohydride in water was pumped through the microreactor. Chloride anions were exchanged by borohydride anions by this procedure. Then the reactor was rinsed with methanol, the recycle loop operation was started and acetophenone was added with a syringe. Samples over the reaction time were analyzed by chromatography.

In a second experiment polymer/glass rods were activated by sodium borohydride and test in conventionally way in a stirred tank reactor. One rod was crushed to a fine powder, an other rod was broken into several pieces. With comparable reaction parameters the reduction of acetophenone was monitored. The experimental set up for the evaluation of mass transport effects in polymer/glass microreactors is depicted in figure 14. The results of these experiments are summarized in figure 15. As expected the observed reaction rates are much faster with the powdered material than with the coarse pieces of the rod. The fastest reaction was obtained with the microreactor.

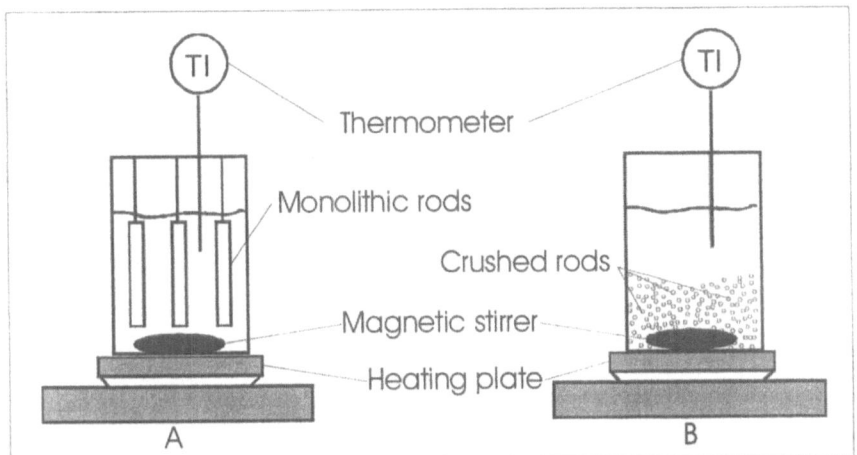

Figure 14: Experimental set up for the evaluation of mass transport effects in polymer/glass microreactors

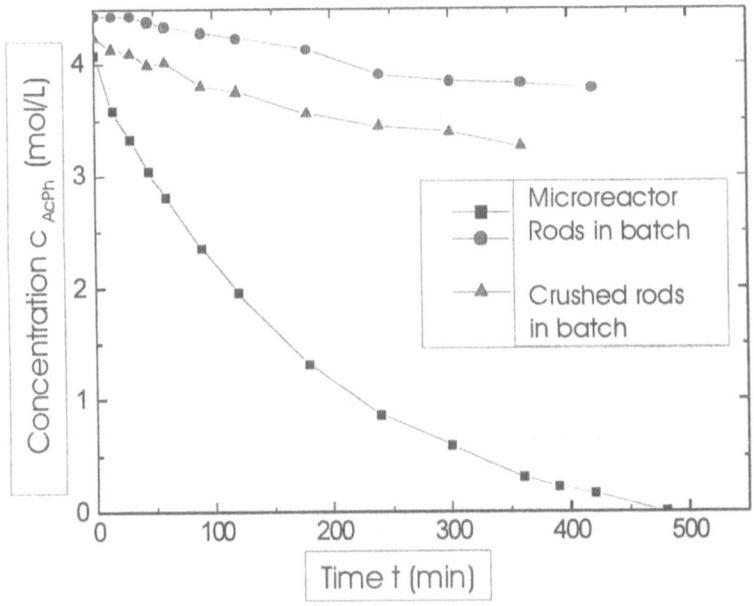

Figure 15: Comparison of the microreactor with rods and powdered rods in a batch experiment.

Kinetic investigations

For the reduction of acetophenone to 1-phenylethanol a more detailed kinetic study was done. It was assumed, that the hydrogen transfer occurs step wise. For a detailed kinetic approach all steps should be taken in consideration. In principal borohydride ions are able to transfer 4 hydrogen atoms. The reactivity of the stepwise dehydrogenated hydride declines with the loss of hydrogen. Our investigations reveal, that only the first two steps are important for the development of a rate equation. It was further assumed that the reaction mixture does not change the volume, which is justified by the fact that the concentration of the acetophenone was kept at a low level (about 1 mmol in 50 g of methanol). The microreactor was considered to work as a differential reactor, so an integral approach over the whole reactor length is suitable. For three temperatures (40°C, 60°C, 80°C) and three ratios of acetophenone to the solid reactant borohydride (0.9, 0.45, 0.2) kinetic experiments were performed. To describe the rate of the chemical reaction the following set of equations was used:

$$\frac{dc_{AcPh}}{dt} = -r_1 - r_2 = -k_1 \cdot c_{AcPh} \cdot c_{(BH_4)^-} - k_2 \cdot c_{AcPh} \cdot c_{(BRH_3)^-}$$

$$\frac{dc_{(BH_4)^-}}{dt} = -r_1 = -k_1 \cdot c_{AcPh} \cdot c_{(BH_4)^-}$$

$$\frac{dc_{(BRH_3)^-}}{dt} = +r_1 - r_2 = +k_1 \cdot c_{AcPh} \cdot c_{(BH_4)^-} - k_2 \cdot c_{AcPh} \cdot c_{(BRH_3)^-}$$

$$\frac{dc_{1-PhE}}{dt} = r_1 + r_2 = +k_1 \cdot c_{AcPh} \cdot c_{(BH_4)^-} + k_2 \cdot c_{AcPh} \cdot c_{(BRH_3)^-}$$

Only two rate constants (k_1, k_2) have to be determined. The result of the parameter evaluation is given in table 1. The values of the rate constants for two temperatures were averaged. As can be seen the average values $\overline{k_1}$ and $\overline{k_2}$ are very close to the single values. In figure 16 a comparison of calculated curves and measured concentration profiles are depicted. It can be seen that the mathematical model describes the measurements very well. These investigations reveal that the new microreactor is a convenient tool for kinetic studies in organic synthesis as a first step for a reactor scale up. Scale up can be done easily with this microreactor concept. One approach is to use a larger number of reactors and to operate them in parallel whereas another possibility is to enlarge the dimensions of the polymer/glass material. Sizes up to 20 mm in diameter and a length of 500 mm are possible.

temperature ϑ [° C]	molar ratio acetophenone/ borohydride	k_1 [L/mol*s]	k_2 [L/mol*s]	$\overline{k_1}$ [L/mol*s]	$\overline{k_2}$ [L/mol*s]
40	0,90	0,091	0,0001	0,092	0,0001
40	0,45	0,093	0,0001		
60	0,90	0,20	0,001	0,21	0,001
60	0,45	0,22	0,001		
80	0,90	0,35	0,009	0,355	0,0095
80	0,45	0,36	0,01		

Table 1: Reaction rate constants for different temperatures and varied molar reactant ratios

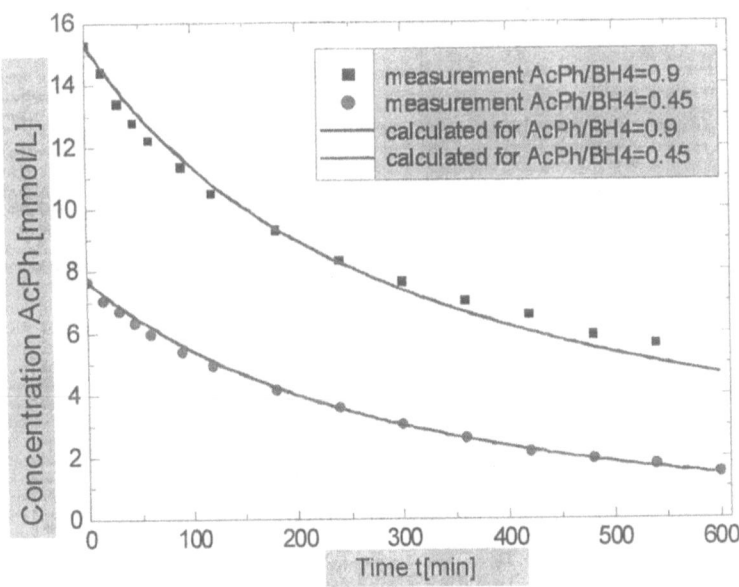

Figure 16: Comparison of calculated and measured concentration/time profiles for the reduction of acetophenone with borohydride in a polymer/glass microreactor

For the reactant ratios 0.9 and 0.45 the temperature dependency was evaluated. Using the well known Arrhenius equation in the linear logarithmic form figure 17 gives calculated and measured results. A linear relation can be seen which represents the measured values very well.

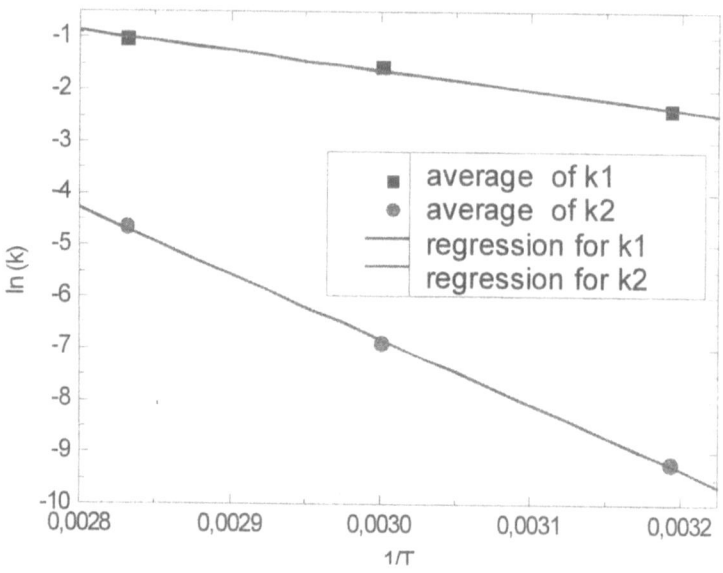

Figure 17: Temperature dependency of the rate constants for the reduction of acetophenone with borohydride in a polymer/glass microreactor

The temperature dependencies for the rate constants are:

$$\ln(k_1) = 9{,}6088 - 3745{,}4 \cdot \frac{1}{T}$$

$$\ln(k_2) = 30{,}9208 - 12577{,}9 \cdot \frac{1}{T}$$

Also the frequency factors and the activation energies were determined:

$$k_{1,\infty} = 14895{,}285 \, \frac{L}{mol \cdot s}$$

$$k_{2,\,\infty} = 2{,}684 \cdot 10^{13}\ \frac{L}{mol \cdot s}$$

$$E_{1,\,A} = 31139{,}256\ \frac{J}{mol} = 7437{,}5\ \frac{cal}{mol}$$

$$E_{2,\,A} = 104572{,}66\ \frac{J}{mol} = 24976{,}8\ \frac{cal}{mol}$$

The obtained values are in agreement with values common for chemical reactions. This result confirms the developed mathematical model.

Organic synthesis

In a second step the application of the new microreactor was extended to a number of other organic reactions. In table 2 the reactants, the functionalized polymers, the reaction conditions, the products and yields are compiled. All isolated products were pure (>95%) as judged by NMR-spectroscopy.

Reactant	Functionalization	Conditions	Produkt	Yield (%)
O Ph Me	$\oplus$ NMe$_3$ $\ominus$ BH$_4$	MeOH, 60°C, 6h	OH Ph Me	>99%
OH (cyclohexanol)	$\oplus$ NMe$_3$ $\ominus$ Br(OAc)$_2$ cat. TEMPO	CH$_2$Cl$_2$, rt, 6h	O (cyclohexanone)	>99%
Ph Br	$\oplus$ NMe$_3$ $\ominus$ Nu	Nu= N$_3$ C$_6$H$_6$, 70°C, 12h	Ph Nu	>99%
		Nu= CN C$_6$H$_6$, 70°C, 12h		>99%
		Nu=SCN C$_6$H$_6$, 70°C, 12h		>99%

Table 2: Compilation of selected organic synthesis performed in polymer/glass microreactors

442

The reactions chosen cover the fields of reduction, oxidation and nuceophilic substitution. In all these syntheses the microreactor has proven that it is well suited as a fast and convenient tool in organic chemists´ laboratories. In all these applications a high selectivity towards the products was obtained. It was possible to use the same microreactor for different type of reactions (e.g. reduction, oxidation) due to the fact that the reactor can easily be regenerated by state of the art ion exchange procedures. This is an important feature, enabling microreactors to contribute to "green chemistry".

In organic synthesis reaction sequences are very common. First experiments to use microreactors in a multi step synthesis with different functionalization are under development in our laboratories. The results will be published elsewhere [11]. In these studies different microreactors are in use with different functional groups in such a way, that a one step laboratory apparatus was used and the microreactors were changed from step to step. For a more convenient work it would be desirable to have a cascade of microreactors with different functionality and to use them in the same equipment. A sketch of such a cascade is given in figure 18.

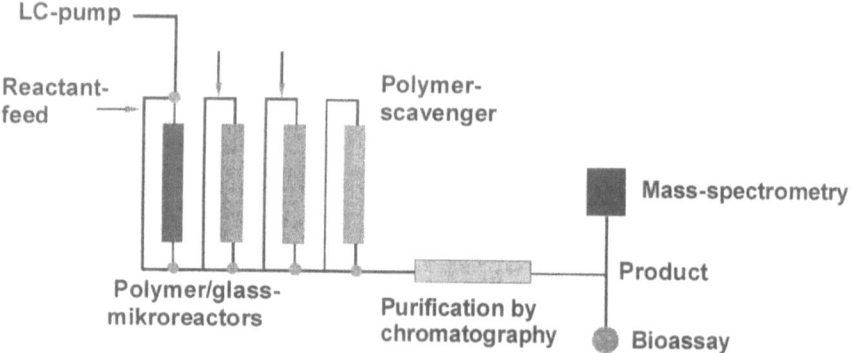

Figure 18: Assembly of microreactors for organic reaction sequences

In figure 19 the state of our development and future perspectives are given. The shaded symbols are the fields that we already developed, the blank fields are under current development. As can be seen a wide field of applications can be covered by the new microreactors we presented here.

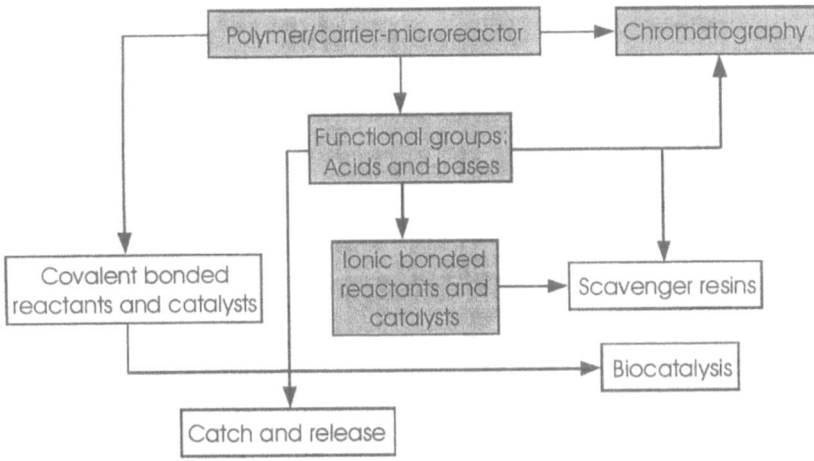

Figure 19: State of the art and perspectives of polymer/glass microreactors

Conclusions

A new microreactor was developed. The carrier material consists of a glass rod, in which micrometer sized channels hold polymer particles. This rod with a typical length of about 110 mm and a diameter of 5.3 mm is lined by PTFE and equipped with a pressure resistant fiber reinforced epoxy resin casing including standard fittings for easily connecting to existing laboratory equipment.

The microreactor was applied in the synthesis of organic compounds. As a model reaction the reduction of ketones to alcohols was studied. It was found that the forced convective flow inside this reactor ensures high productivity compared to classically stirred tank reactor designs. The microreactor is a versatile tool for kinetic studies and the evaluation of reaction rate equations, which is a first step for scale up. The handling of solid phase reactants during the synthesis is extremely convenient as tedious workup (filtration, aqueous hydrolysis etc.) becomes irrelevant. A wide range of reactants can be fixed on this type of polymeric backbone which allows to perform a large set of organic transformations.

The advantages of the new microreactor for the synthesis of potential pharmaceuticals are:

- active polymer phase and carrier are a monolithic composite
- small polymeric particles ensure good accessibility of the dissolved reactants
- large channel diameters result in low pressure drop
- simple regeneration by well known classical chemical methods can be performed
- all materials of the reactor which are in contact with the reaction mixture are either glass or the functionalized polymeric phase, PTFE and PEEK, allowing for operation with most solvents
- low cost fiber reinforced epoxy resin casing with standard fittings enables easy integration in standard laboratory equipment

In summary, we developed a new tool for polymer-assisted solution-phase synthesis which allows for the rapid multistep synthesis of potential drug candidates without workup problems that are typically associated with classical solution phase chemistry. Automation of this technique is feasible by merging it with automated HPLC-technology.

Acknowledgement

Thanks to Dr. G. Dräger, J. Harders, N. Hoffmann and Dr. W. Solodenko from the Institut für Organische Chemie, Universität Hannover for their work on the organic synthesis using microreactors and to H. Langer, H. Schönfeld and Dr. C. Altwicker from the Institut für Chemische Verfahrenstechnik, Technische Universität Clausthal for the preparation of polymers, kinetic studies and analysis. Thanks also to the Wageningen NMR Centre for the dynamic NMR-microscopy measurements and to the European Union for the financial support for this part of our investigations.

Literature

[[1] H. Becker, T. Klotzbücher: Polymer nanowell plates with variable well slope angles, Microreaction Technology – Industrial Prospects, Springer, (2000), ISBN 3 540 66964 7, 102-112

[2] A. Kirschning, H. Monenschein, R. Wittenberg: Functionalized polymers – Emerging versatile tools for solution-phase chemistry and automated parallel synthesis, Angewandte Chemie Int. Ed. (2001) 40, 650-679

[3] W. Ehrfeld: Microreaction Technology, Springer, (1998), ISBN 3 540 63883 0

[4] W. Ehrfeld: Microreaction Technology – Industrial Prospects, Springer, (2000), ISBN 3 540 66964 7, 102-112

[5] D. Hudson, Journal of Combinatorial Chemistry Vol. 1, No. 5, (1999), part I, 333-360

[6] D. Hudson, Journal of Combinatorial Chemistry Vol. 1, No. 6, (1999), part II, 403-457

[7] A. Cybulski, J. A. Moulijn: Structured catalysts and reactors, Marcel Dekker, (1998), ISBN 0 8247-9921-6

[8] U. Kunz, A. Kirschning, U. Hoffmann: PCT patent application PCT/DE01/01092, March 15th, (2001)

[9] U. Tallarek et al.: Dynamic NMR-microscopy of chromatographic columns, AIChE Journal, 44, (1998), 9, 1962-1975

[10] U. Tallarek et al.: Electroosmotic and pressure-driven flow in open and packed capillaries: Velocity distribution and fluid dispersion, Analytical Chemistry, 72, (2000), 2292-2301

[11] A. Kirschning, U. Kunz et al.: Funktionalisierte Glas/Polymer-Komposite in Mikroreaktoren: Herzstücke für eine automatisierte Synthese in Lösung, in preparation

Investigation, Analysis and Optimization of Exothermic Nitrations in Microreactor Processes

J. Antes, T. Türcke, J. Kerth, E. Marioth, F. Schnürer,
H.H. Krause, S. Löbbecke*

Fraunhofer-Institut Chemische Technologie ICT
P.O. Box 12 40, 76318 Pfinztal, Germany

Introduction

Nitration reactions are among the basic reactions used in chemical synthesis. Products of nitrations find many applications. Nitrocompounds are, for example, used as precursors or final products for pharmaceuticals, agricultural and pest control chemicals, pigments, polymers or explosives.

The majority of nitrations give off considerable amounts of heat. The highly exothermic nature of the reactions - sometimes with explosive potential - along with the acidic corrosivity of the nitrating agent, makes nitration processes potentially very hazardous.

Furthermore, strong exothermicities in combination with short reaction times cause large numbers of secondary, consecutive and decomposition reactions due to hot spots and accumulated reaction heat. The occasional result is a wide spectrum of products with correspondingly low selectivity, yield and purity.

Application of Microreactors

One way to greatly reduce the potential hazard of highly exothermic or explosive nitrations is to miniaturize the synthesis apparatus and auxiliary equipment by employing microreactors [1 - 3]. Furthermore, the high surface to volume ratio in microreactors provides much better temperature control than that in macroscopic batch or flow-through reactors. Microreactors thus not only permit safe operation, but also suppression of secondary and consecutive reactions due to isothermal processing, short retention times (continuous processing) and intensive mixing of the reactants.

Nitration of Ureas in Microreactors

In this work nitration reactions were carried out in different types of microreactors made of silicon and glass, each of them well characterized with respect to their mixing quality (by applying modified Villermaux/Dushman reactions [3]), their fluidic dynamics (by applying CFD calculations) and their heat transfer characteristics [3, 4]. CFD simulations were also used to optimize process conditions (e.g. educt input, flow rates, temperature).

The paper focuses on the exothermic nitration of N,N'-dialkyl substituted ureas which is the most difficult reaction step in the synthesis of a new energetic plastiziser, the so-called DNDA (fig.1).

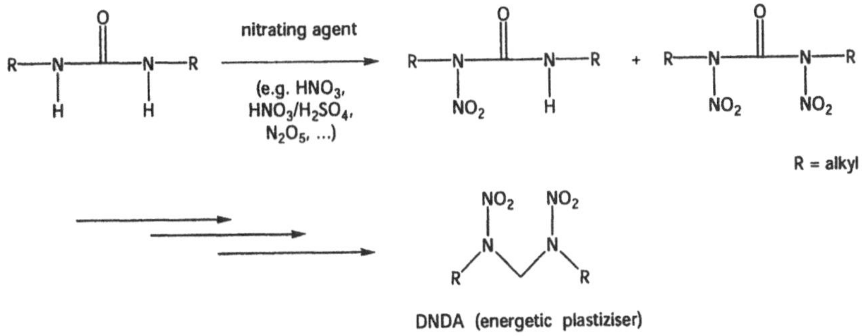

DNDA (energetic plastiziser)

Fig. 1: Nitration of N,N'-dialkyl substituted ureas for the synthesis of DNDA

From macroscopic batch reactions it is known that always both the mono- and dinitro-substituted products are obtained. With respect to a further processing of the nitroureas to the final DNDA plastiziser a quantitative synthesis of N,N'-dialkyl-N,N'-dinitro-urea is preferably wanted due to better properties of the final product. However, until now a deliberate synthesis of mononitro resp. dinitro substituted ureas has not been achieved, so that in conventional DNDA synthesis both derivatives are immediately processed to the final product.

Detailed information about a deliberate control of the nitration degree (number of introduced nitro groups), selectivity and yield of urea nitration products are not available so far. Hence, to improve yield and product quality of the entire DNDA synthesis the nitration step has to be investigated and analyzed in detail. For this reason microreactors were applied.

Nitration in microreactors were performed by systematic variation of the process conditions (e.g. temperature, flow rate resp. retention time, nitrating agent, stoichiometry, etc.). Since educts like diethyl urea are very expensive precursors for the DNDA synthesis alternative educts like the corresponding thiourea (diethyl thiourea is 35 times cheaper) were also investigated in microreaction processes.

Furthermore, the specific features of microreactors allow also to carry out nitrations under new process conditions, e.g. at room temperature (in contrast to minus 20°C in macroscopic batch reactors) due to the good heat and mass transfer performance of the microreactors combined with the short retention times of the reactants.

Experimental

Two different types of microreactors were used for the nitration experiments.

A commercially available silicon micromixer based on a microfluidic split-and-recombine structure (MiMoCo GmbH, Ilmenau/Germany) and a microreactor array made of glass containing 20 reaction channels in parallel with integrated educt mixing zones and cooling structure (development in cooperation with mgt mikroglas technik AG, Mainz/Germany) [2, 3].

Since nitrating agents are highly corrosive media only resistant materials can be applied for such reactions. Silicon, glass and titanium were proved to be suitable for nitration processes. Microstructured devices made of stainless steels did not withstand such corrosion processes.

Different nitrating agents were used for the microreaction experiments. Besides concentrated and fuming nitric acid (65% resp. conc. HNO_3), nitrating acid ("mixed acid": $HNO_3/H_2SO_4 = 1:1 - 6:1$) and dinitrogen pentoxide (N_2O_5) as a less acid nitrating agent were used. N_2O_5 was either dissolved in CH_2Cl_2 or used as a gaseous reactant (in-situ production: $N_2O_4 + O_3$). Nitrating agents were used in up to 6-fold excess.

Ureas (dissolved in dichloromethane) and nitrating agents (liquid or dissolved) were supplied by pulsation-free pumps, mixed inside the microreactors (resp. micromixers) and then passed through reaction capillaries (e.g. PTFE tubes) of different lengths representing different retention times (0.6 s - 82 s).

Nitrations were carried out in a continuous mode at defined temperatures between 0°C and 20°C. The reaction mixture eluting out of the microreaction system was quenched in ice water and/or cold dichloromethane, extracted and passed to NMR, MS, IR, HPLC or GC-MS for analysis.

For a more effective investigation and optimization of nitrations analytical devices and sensors for process control (like temperature monitoring, flow control, etc.) were adapted to the microreaction processes.

FTIR microscopy was applied for the online monitoring of nitrations in silicon microreactors. Due to a high spatial resolution of this analytical method ($\geq 10 \mu m$) educts and products of nitrations can be identified and distinguished at different positions inside the microreactor [5]. Both intermediates and final products of nitrations can be IR spectroscopically detected by focusing the IR beam consecutively on different reaction zones. Hence, the progress of nitrations can be monitored online in microreactors.

Besides IR spectroscopy HPLC analysis was directly adopted to the microreaction process. Computer-controlled sampling devices allows a regular transfer of an aliquot of the reaction mixture to the HPLC chromatograph where separation, identification and quantification takes place.

Both analytical methods are very helpful tools for a better understanding of the individual chemical processes taking place in microreactors and thus for a better understanding of the entire nitration process which is investigated.

Exemplary results

The nitration of N,N'-diethyl-urea in macroscopic batch reactors led to mixtures of mono- and dinitro derivatives, which were chromatographically separated and clearly identified by NMR, MS and IR spectroscopy (fig. 2 and 3). Yields of up to 75% could be achieved for N,N'-diethyl-N,N'-dinitro-urea.

Experiments in microreactors carried out so far did not show significant differences in the product spectrum compared with macroscopic experiments. Despite of ensuring short retention times and isothermal conditions to a great content yields and selectivities for the dinitro resp. mononitro derivative could not be dramatically enhanced, although a tendency for higher yields of N,N'-diethyl-N-nitro-urea could be observed.

Furthermore, there were no convincing evidences to clear up whether the mononitro derivative is a key intermediate for the synthesis of the dinitro substituted product.

A new approach to a quantitative synthesis of N,N'-diethyl-N,N'-dinitro-urea was investigated by using the much cheaper N,N'-diethyl-thiourea as starting material. In macroscopic batch reactors four products were obtained, two - at that time - unidentified products of 90% yield as well as small amounts of N,N'-diethyl-N-nitro-urea and the target compound N,N'-diethyl-N,N'-dinitro-urea (10%). Surprisingly, no C=S functionality was found in the products anymore.

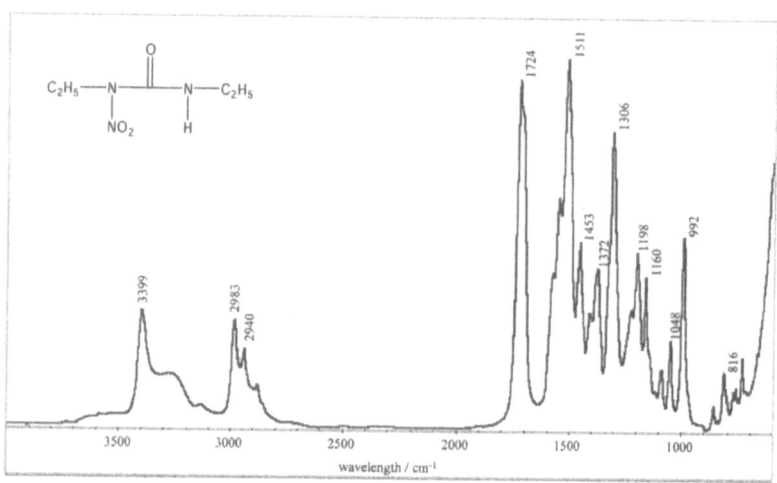

Fig. 2: IR spectrum of N,N'-diethyl-N-nitro-urea

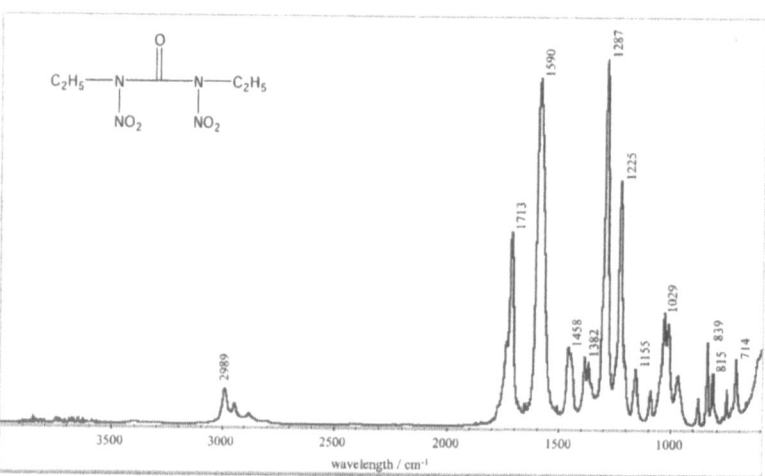

Fig. 3: IR spectrum of N,N'-diethyl-N,N'-dinitro-urea

The nitration of N,N'-diethyl-thiourea was repeated by applying microreactors under systematic variation of process parameters and conditions.

Even under mild conditions (e.g. weak nitrating agents) only one product with approx. 100% selectivity was obtained in all microreaction experiments, which could be easily isolated and identified by MS (fig. 4) and IR (fig. 5) as N,N'-diethyl-N-nitroso-urea:

$$C_2H_5-N(H)-C(=S)-N(H)-C_2H_5 \xrightarrow[\text{(e.g. HNO}_3, \text{ HNO}_3/H_2SO_4, N_2O_5, ...)]{\text{nitrating agent}} C_2H_5-N(H)-C(=O)-N(NO)-C_2H_5$$

On the basis of these experiments the main products of the macroscopic experiments could be also identified as N,N'-diethyl-N-nitroso-urea and probably N,N'-diethyl-N,N'-dinitroso-urea.

It could be shown that microreactors turned out to be suitable tools for the safe and highly selective synthesis of N,N'-diethyl-N-nitroso-urea. A selective synthesis and isolation of this stable product made also a consecutive nitration step possible. For this reason N,N'-diethyl-N-nitroso-urea was nitrated again in microreactors under similar conditions as described above:

$$C_2H_5-N(H)-C(=O)-N(NO)-C_2H_5 \xrightarrow[\text{(e.g. HNO}_3, \text{ HNO}_3/H_2SO_4, N_2O_5, ...)]{\text{nitrating agent}} C_2H_5-N(NO_2)-C(=O)-N(NO_2)-C_2H_5$$

The product obtained with approx.100% selectivity could be identified as N,N'-diethyl-N,N'-dinitro-urea, the target product for the further DNDA synthesis.

This result, obtained by performing the nitration of N,N'-diethyl-thiourea in microstructured reactors under systematic and controlled process variation, are of significant importance for a better understanding and subsequent optimization of the entire DNDA synthesis. A promising new synthetic route was discovered based on the nitrosation of N,N'-diethyl-thiourea and subsequent nitration of the formed nitroso derivative.

Furthermore, for the first time a successful "step-by-step" nitration process was carried out by applying microreaction technology.

452

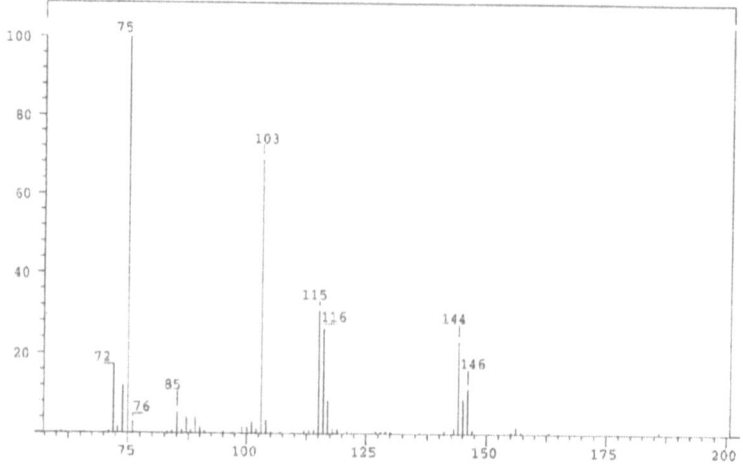

Fig. 4: MS spectrum of N,N'-diethyl-N-nitroso-urea (MH⁺: m/z = 146)

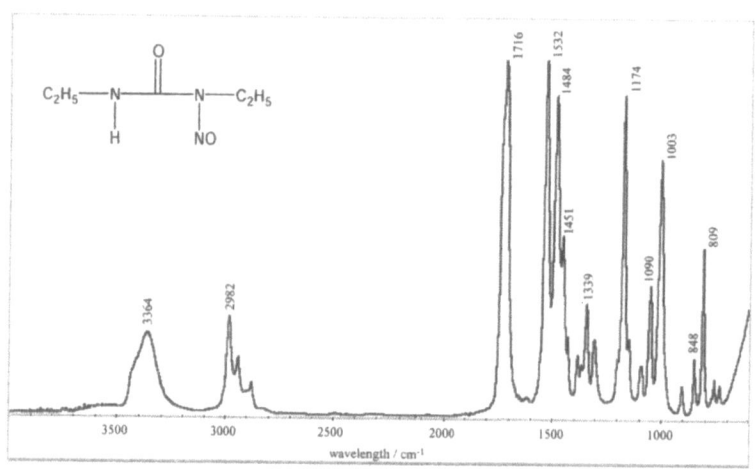

Fig. 5: IR spectrum of N,N'-diethyl-N-nitroso-urea

From the mechanistic point of view the reactions could be explained as follows:
The sulphur of the thiourea functionality is oxidized by the nitrating agent (e.g. HNO_3) forming simultaneously amounts of nitrous acid (HNO_2) which acts as an nitrosation agent. Furthermore, in an excess of HNO_3, HNO_2 exists essentially as N_2O_4, which is nearly completely ionized as $NO^+ + NO_3^-$ [6].
Such nitroso compounds are excellent precursors for the introduction of nitro groups as evidenced by the second nitration step.

Very similar results were obtained for the nitration of N,N'-dimethyl-urea and N,N'-dimethyl-thiourea, respectively, confirming the described reaction pathway via the corresponding nitroso derivatives.

Hence, these experiments show the great potential of microreaction technology as a suitable tool for process development, process analysis and process engineering. The results obtained by applying microreactors can now be transfered to the synthesis of N,N'-diethyl-N,N'-dinitro-urea on a macroscopic scale. First experiments have already shown that the macroscopic nitrosation of the dialkyl substituted thioureas can be also achieved by conventional nitrosation agents like $NaNO_2/H_2SO_4$. However, the excellent heat transport properties of microreactors makes it more easier to achieve high selectivities, since it has to be considered that the oxidation resp. nitrosation step is the most exothermic step of the entire process.

References

[1]   J.R. Burns, C. Ramshaw, *A Microreactor for the Nitration of Benzene and Toluene*, 4th Int. Conference on Microreaction Technology (IMRET 4), 5-9 March 2000, Atlanta, GA, USA

[2]   J. Antes, T. Tuercke, E. Marioth, K. Schmid, H. Krause, S. Loebbecke, *Use of Microreactors for Nitration Processes*, 4th Int. Conference on Microreaction Technology (IMRET 4), 5-9 March 2000, Atlanta, GA, USA

[3]   S. Loebbecke, J. Antes, T. Tuercke, E. Marioth, K. Schmid, H. Krause, *The Potential of Microreactors for the Synthesis of Energetic Materials*, 31st Int. Annu. Conf. ICT: Energetic Materials - Analysis, Diagnostics and Testing, 33, 27 - 30 June 2000, Karlsruhe, Germany

[4]   E. Marioth, S. Loebbecke, M. Scholz, F. Schnuerer, T. Tuercke, J. Antes, H.H. Krause, *Investigation of microfluidics and heat transferability inside a microreactor array made of glass*, 5th Int. Conference on Microreaction Technology (IMRET 5), 27-30 May 2001, Strasbourg, France

[5]   T. Tuercke, W. Schweikert, F. Lechner, J. Antes, H.H. Krause, S. Loebbecke, *Monitoring of Chemical Processes in Microreactors by Adapting FTIR Microscopy and HPLC*, 5th Int. Conference on Microreaction Technology (IMRET 5), 27-30 May 2001, Strasbourg, France

[6]   G.A. Olah, R. Malhotra, S.C. Narang, *Nitration*, VCH New York, 1989, p. 129

Development of a Microreactor for Solid Phase Synthesis

Elke Bremus-Köbberling, Arnold Gillner, Martin Wehner, Ulrich Russek,
Fraunhofer Institute for Lasertechnology, Aachen, Germany
Johannes Köbberling, Dieter Enders, Institute for Organic Chemistry, RWTH
Aachen, Germany
Siegfried Brandtner, Innolabtec GmbH, Stolberg/ Rhld., Germany

Abstract:
Laser micromanufacturing is well suited for the fabrication of microreaction
systems. The use of high performance polymers in combination with
microstructuration and microjoining technologies allows us to develop sealed
microreaction systems. These are suited for organic synthesis even with
aggressive reagents and under inert atmospheres.
The microreaction system presented in this paper will enable us to perform
combinatorial solid phase chemistry very efficiently. The most striking novelty of
our microreactor matrix is the mixing of the reaction vessels by means of an
oscillating membrane. That generates turbulence in the above suspension of a
polymer resin. The addition of reagents and the work up / washing of the resin will
be extremely fast and gentle.

Keywords: laser microstructuration, microjoining, polymers, combinatorial solid
phase synthesis, mixing, microreaction system

1 Introduction

The trends of miniaturization, parallelization and automation are continuously
gaining influence on modern fields of chemistry and life sciences.

Chemical and biological micro total analysis systems (μTAS) are a subject of
steady development since more than 10 years.[1,2]

In the last few years advantages of miniaturization for chemical synthesis such
as use of the high surface to volume ratio, good heat conduction and mass
transport have drawn the attention to developing microreactors for single
reactions, often as tools for optimization purposes. [3] The many applications of
microreactors are reviewed by Ehrfeld et al.[4]

1.1 Miniaturization of Combinatorial Chemistry

Combinatorial chemistry is a topic of great interest for microreaction technology due to high economic importance and the fact that single micro-reactors can easily produce sufficient amounts of substances for HTS purposes.

Modern ultra High Throughput Screening systems (uHTS) can test up to 100.000 compounds per day, which is only accessible as a result of the miniaturization of all components in screening systems and a high degree of parallelization and automation.

Within the last 5 years the standard microtiter plate used for screening has been redesigned twice. Starting from 96 well plates followed by 384 well plates, these are today replaced by state of the art 1536 well plates. The increased screening capacities lead to an exponential increase in the demand for new substances for testing while the needed amounts are decreasing at the same ratio.

The increase in test wells per plate has led to a decrease in volume by a factor of 16. Following the decrease in the needed amounts of test substances, modern combinatorial / parallel synthesis does not need to produce several (10-20) milligrams of new compounds any more, less than one milligram will be satisfactory in the future to be used in several dozens of essays.

This goal can be accomplished using the modern method of solid phase organic chemistry (SPOS) on less than 30 mg of resin. Today's technologies in SPOS do not offer any satisfactory solutions for this demand - more compounds in smaller amounts – in a fast and reliable automated synthesis yet.

1.2 Solid Phase Organic Chemistry

One of today's most effective and reliable methods for the preparation of large libraries is based on the use of solid phase organic synthesis due to the ease of its automation and the good purities of the final products.

This technology – the build up of a molecule on a template which is covalently bonded via a linker to a polymer support, followed by cleavage of the bond to the linker – has been developed to a very reliable tool in parallel synthesis [5].

The main advantages are:
• The ease of driving a reaction to completion by employing excess reagent which can be washed off after the reaction.
• The avoidance of difficult purification steps such as chromatography, distillation or crystallization.

SPOS needs only few repetitive unit operations which have to be realized under inert conditions:
• Addition of liquid reagents / solvents
• Agitation
• Filtration
• Heating / cooling

Today's solutions to this approach mainly consist of two models: First, tubes with a frit at the bottom and a septum layer on top which are arranged in an array (e.g. 6*8) and which can be filled by a pipetting robot with reagents. These robots will eventually be too slow to fulfill the increasing demand for new compounds. The rack of vessels is agitated by means of an orbital shaker which causes vibrations that may interfere with other lab equipment.

The second approach consists of macroscopic amounts of resin which are portioned in a permeable housing (tea bags, cans) and which are reacted in standard glassware. Up to now this approach utilizes only little automation.

Both approaches have some disadvantages, which shall be circumvented by the microreactor design presented below.

2 Fabrication Technologies

The production of microfluidic devices and microreaction systems requires manufacturing methods which meet the demands on accuracy, material flexibility and processing speed. Laser processing offers appropriate solutions for this purpose as well as for the packaging and interconnection technologies. Also the tool manufacturing for succeeding mass production can be done using laser technology.

2.1 Microstructuring

The laser beam becomes an important tool for microstructuring due to smaller geometries and a great variety of materials which are processable. In comparison to standard microfabrication techniques like etching or LIGA [6] laser technology has some advantages regarding flexibility, material variety, structure size, processing speed and ease of integration into existing fabrication plants.

High resolutions due to the extreme focussibility of laser radiation and only little thermal stress induced in the material qualifies the laser process ideally for microfabrication especially for prototyping and small series production.

Polymers and other thermally sensitive materials are microstructured with excimer laser radiation due to the good absorption of UV-radiation.[7] A mask projection mode is applied, where the laser beam is shaped and homogenized to illuminate a mask which is imaged onto the substrate by a projection lens in reduced size. Therefore the pattern on the substrate is defined by the pattern of the mask. In our experimental set-up a mask is imaged onto the substrate in reduced size with a demagnification ratio of 4:1 to 25:1. For the structuring of bigger areas e.g. 70 x 20 mm^2 a conformal mode is applied. Thus the mask and substrate are moved simultaneously resulting in smooth extensions of the different structural elements.

Regarding to the short wavelength of ArF and KrF excimer lasers (λ = 193 / 248 nm) and high pulse energies this procedure is excellently suited for applications that afford high accuracies and in which the bulk material should not

458

be damaged or where no residues from processing are allowed to remain on the device. These demands apply to microfluidic devices where channel structures with high surface qualities have to be realized. Many different polymers e.g. PC, PMMA, PI, PEEK, PFA etc. have been microstructured using UV-laser radiation (see Fig. 1).

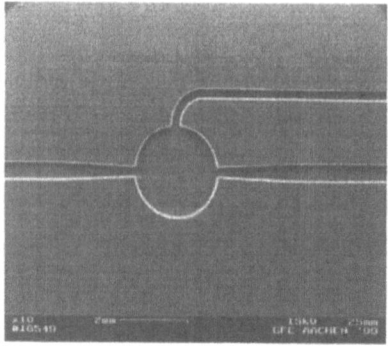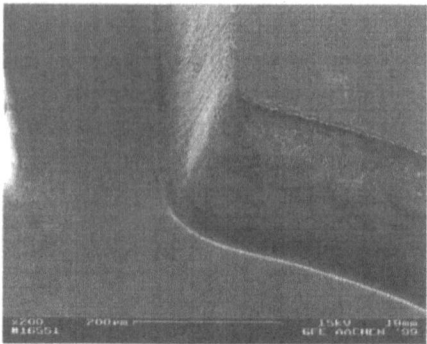

Fig. 1: Laser (248 nm) ablated microfluidic device in PC; right: detailed view of the inner walls.

The availability of DUV-Lasers, namely F_2-Lasers with 157 nm wavelength, enlarges the material spectrum to Quartz glass and PTFE. Perfluorinated polymers like PTFE are of high interest in chemistry due to their exceptional properties like chemical inertness and heat resistance. It can be processed mechanically, but only with limited spacial resolution. Most IR and UV lasers show little effect. Only at high fluences they lead to thermal decomposition of the material.

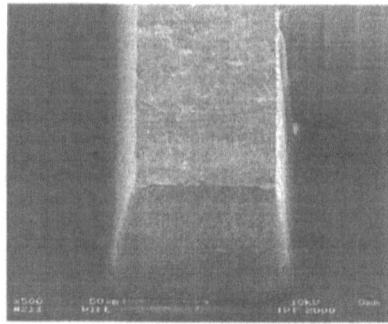

Fig. 2: PTFE microstructures processed with 157 nm laser radiation

A geometrically resolved ablation occurs when F_2-Laser radiation is used. Their short wavelength of 157 nm is absorbed in all polymers and almost all other materials within a few micrometers and leads to instant vaporization of the material. In Fig. 2 a microfluidic device made from PTFE is shown. Therefore laser ablation with F_2-Lasers is the only technology for microstructuring this kind of material for microfluidic devices.

2.2 Laser polymer welding

The described laser microstructuring is ideally accompanied by laser joining techniques. Laser joining [8] and laser polymer welding [9] are processes which offer appropriate solutions for packaging and interconnection of various devices such as μ-TAS or microreactors using processes with high precision and long term stability of the joint.

The specific advantages of the laser joining technologies are:
- minimum of energy input without heating the whole device,
- possibility of selective joining,
- no additional materials like adhesives required,
- joining geometries $\geq 20\ \mu$m.

The laser welding of polymers uses the different absorption characteristics of the two polymer parts that shall be joint.

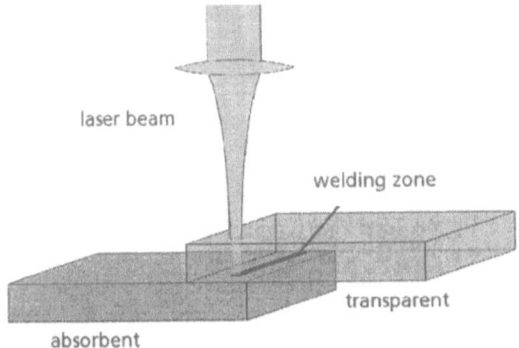

Fig. 3: Scheme of laser polymer welding.

The upper material is transparent for the laser beam (often diode laser radiation in the infrared range) but the material underneath absorbs the energy, melts and by thermal diffusion the upper polymer is also molten. The two parts are bonded together with an inner welding zone (see Fig. 3). For this process it is important that no decomposition of the polymers occurs near the melting point and the two components have to be miscible.

This process allows us to interconnect polymer parts under mild conditions and to achieve a strong, fluid sealed bond. Even modified perfluorinated polymers can be welded with a PFA sheet to yield capillaries in these chemically inert materials. Additionally the process is suited for fixation of the oscillation membrane in our solid phase microreactor described in the following section.

3 Microreactor for Solid Phase Chemistry

3.1 Design of our new microreactor concept

Several objectives for a new design of an effective microreactor matrix were defined:

- Omitting the need of pipetting into individual reactor cells with a robot arm by accessing rows or columns of cells within a reactor matrix simultaneously through a system of capillaries.
- Replacing the use of orbital shakers (or magnetic stirring) for mixing the resin by a new approach enabling the design of a completely capsuled compact reaction block containing internal mixing.
- Enabling efficient washing of the resin by pumping solvent through the resin while it is being agitated.
- Extending the accessible temperature range for SPOS due to highly efficient heat transfer into the capsuled reaction block.

For this purpose, a novel microreactor [10] for the combinatorial solid phase chemistry has been developed (see Fig. 4.; patent pending).

To achieve the goals stated above, we designed a microreactor matrix with 96 reactors in standard MTP size. Each cylindrical reactor can be filled with up to 30 mg resin. On top and bottom of each reactor there is a polymer-sieve allowing reagents and solvents to enter or exit. Below the reactor there is a plate containing capillaries allowing the simultaneous addition of reagents to either columns or rows of reactors. The design ensures the dispensing of uniform amounts to every vessel which has been calculated by CFD simulation of eight vessels belonging to one fluid line. During the reaction the inlets are closed by valves.

Mixing in the reactors is ensured by pumping reagents and solvents in and out of the reaction chamber by means of actuating a fluoropolymer membrane pneumatically.

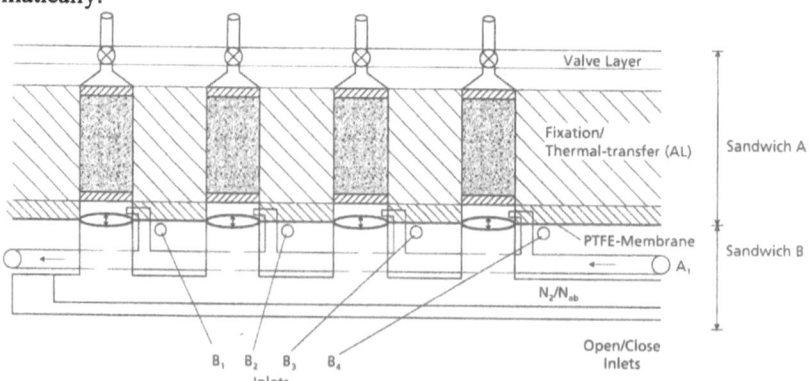

Fig. 4: Scheme of the functional principle: An oscillating membrane agitates the above suspension of polymer resin which results in gentle but highly efficient mixing of the polymer beads.

Up to now a single cell functional model and an eight cell prototype have been built. The single cell functional model consists of a PTFE-membrane underneath of which either vacuum or air pressure is applied to excite it. On the side is an inlet for reagents and solvents. The reaction chamber is a small syringe (5 ml volume) with a teflon frit on top and bottom to contain the resin inside. This model has been used for testing the agitation principle in resin washing and product cleavage.

Fig. 5: Single cell functional model.

The eight cell prototype (Fig. 6) is built of the following chemically resistant fluoropolymers: PTFE, Zedex^{TM}, TFM^{TM} and PFA. The membrane underneath the reaction cavity (volume 1.5 ml) is fixed by laser welding process (contour weld) which also facilitates the sealing of the microchannels for the solvent / reagent distribution system.

The housing is made of steel, so that the sandwich layers are sealed up by pressure. The middle layer can be shifted from the agitation position into the filling position. This operates as a valve by simultaneous closing of the eight connecting fluid channels of a row through this shift mechanism. The Zedex^{TM} material is ideally suited to this approach since it has been developed for applications in sleeve bearings.

Fig. 6: Prototype of an eight cell membrane agitated solid phase microreactor; left: complete system, right: details of the set-up.

3.2 Application tests

First tests have proven extremely efficient washing of resins. Merrifield® resin was coloured with Sudan Red in Dichloromethane (DCM). Washing was done by pumping DCM and afterwards Methanol through the resin while it was agitated. After two cycles the resin was completely colourless, and a collected third washing cycle did not contain any Sudan Red after evaporation. The whole procedure lasts only two minutes. A conventional bottom fritted syringe on a vortex shaker needed at least four cycles of adding solvent, shaking, and draining with both solvents i.e. in total more than 20 min to achieve comparable results. The washing of contained resin portions (cans) was not complete after eight runs.

A first example of synthetic chemistry was performed on 100 mg of T2-triazene resin [11]. Acylating cleavage with Acetyl Chloride (AcCl) in absolute Tetrahydrofuran (THF) and agitation for 4h yielded substituted Piperazine in excellent purity:

purity 98% (GC)

4 Conclusion

Microstructuring and packaging of high performance polymers like PEEK, PTFE or PFA will allow numerous new applications in microreaction technology.

With our presented microreactor design for solid phase chemisty we have shown the beneficial use of lasers in microtechnology for the development of synthesis equipment in the field of combinatorial chemistry.

We believe that the numbering up of microreactors with similar but not identical products in every microreactor will be of at least equal economical importance compared to the production for macroscopic amounts of a single compound in microreactor arrays.

5 References

[1] Manz, A.; Becker, H.: Microsystem Technologies in Chemistry and Life Sciences, Topics in Current Chemistry Vol. 194, Springer 1998.

[2] Micro Total Analysis Systems '98, Proceedings of the μ-TAS'98 Workshop, Banff, Canada 13-16 October 1998 ; Harrison, D.J. ; van den Berg, A.(Eds.), Kluwer Academic Publishers 1998.

[3] Microreaction Technology : Industrial Prospects, Proceedings of the 3rd International Conference on Microreaction Technology, Ehrfeld, W. (Ed.), Springer Verlag Berlin 2000.

[4] Ehrfeld, W. ; Hessel, V.; Löwe, H.: Microreactors – New Technology for Modern Chemistry, Wiley-VCH Verlag GmbH, Weinheim, 2000.

[5] Dörwald, F. Z.: Organic Synthesis on Solid Phase: Supports, Linkers, Reactions, 2000, John Wiley & Sons, Weinheim.

[6] Aachener Werkzeugmaschinen-Kolloquium (Hg), Wettbewerbsfaktor Produktionstechnik, VDI-Verlag Düsseldorf 1996.

[7] Srinivasan, R. : Ablation of Polymers and Biological Tissue by Ultraviolet Lasers, Science 234 (1986), pp. 559.

[8] Wild, M.J.; Legewie, F.; Gillner, A. ; Poprawe, R.: Laser Joining of Dissimilar Materials, Micro System Technology 98, H. Reichl, E. Obermeier (Editors), 6th International Conference on Micro Electro, Opto, Mecanical Systems and Components, Potsdam, December 1-3, 1998, VDE-Verlag GmbH, Berlin, pp.59-16.

[9] Puetz, H.; Haensch, D.; Treusch, H.-G.; Pflueger, S.: Laser welding offers array of assembly advantages, Modern Plastics, Sept. 1997, pp. 121-124.

[10] E. Bremus-Köbberling, A. Gillner, J. Köbberling, D. Enders, Deutsche Patentanmeldung 100 57 827.6

[11] S. Bräse, J. Köbberling, D. Enders, R. Lazny, M. Wang, S. Brandtner, THL 1999, 40, 2105-2108.

CATALYSTS AND REACTION CONDITIONS SCREENING BY MICROREACTOR DEVICES FOR EXHAUST GAS PURIFICATION

Irina RodicaTamas, Rodica Ganea and Grigore Pop
S.C. ZECASIN S.A. Splaiul Independentei 202, PO – BOX 12-304, Bucharest / RO

ABSTRACT

It is well established that noble metal catalysts for exhaust gases purification are not enough active in oxidant environments resulting from car engines with fuel injection. Better results can be obtained with Cu – modified zeolites. There are very numerous factors to be take into consideration for obtain a good, practical usable catalyst.

It is proposed a combinatorial statistical model with seven level for optimization of the catalyst composition and reaction parameters.

The experiments were made in a pulse microreactor coupled to a chromatographic assembly which analyses simultaneous CO_2, NO_x, hydrocarbons, O_2, N_2 and CO_2.

Hydrocarbons and NO_x are easy converted over Cu – HZSM – 5 zeolites but for reduction more than 50% CO are necessary high Cu concentration in catalyst or more than 2 vol. % O_2 in exhaust gases.

INTRODUCTION

In the car engines with fuel injection air / fuel ratio is very well controlled and an encess of oxygen is always present. The catalytical control of CO emission in exhaust gas by "three – ways catalyst" is not satisfactory due to the oxidation of the metallic Pt active centers [1].

Selective reduction of NO_x by hydrocarbons and CO in the presence of oxygen is feasible on copper ion – exchanged ZSM – 5 zeolite [2], [3]. In the last years this topic has been extensively studied all over the world and now many comprehensive reviews were published [4], [5], Cu ion – exchanged zeolites including ZSM – 5, ferierite, mordenite, Y, L – zeolite were studied. Cu – ZSM 5 showed the highes activity but Cu – Y the lowest under the same conditions. Recently was published studies which show some advantages of Cu – or Ce – Cu – MCM – 41 catalysts [6]. So, it seems that the zeolite framework composition plays an important role as base support for Cu active center in the very complex process of exhaust gases decontamination by NO_x reduction and CO and hydrocarbons oxidation upon the same catalyst and in a single step.

Both ZSM – 5 and MCM – 41 molecular sieves can be obtained with very different framework SiO_2 / Al_2O_3 mol. ratio from 2 – 10 till 100 SiO_2.

MCM – 41 has more open porosity and higher specific surface area than ZSM – 5 but the last has better ion – exchanged properties.

Much other variables are important for catalysts activity in exhaust gases purification like temperature, space velocity, ion – exchange degree. Is also interesting to clarify whether these new catalysts can be applied to the emission from diesel engines.

It is evident that the manufacture of a good catalyst based upon Cu – zeolites systems, usable in practice for exhaust gases decontamination, is a very complex problem.

After our knoledges till now there are no practical application of these catalyst as catalytic converter of the vehicle engine exhaust gases.

Such a solution should have also economic consequence in comparison with three – ways catalysts, based on noble metal as Pt, Pd and Rh.

In this work a microreactor coupled with original chromatographic arrangement is used to screen the variation limits of the catalyst composition and reaction conditions for oxygen containing automative exhaust gases cleaning.

A combinatorial statistical model is proposed for catalyst and process optimization based upon Cu – HZSM – 5 system.

EXPERIMENTAL

A. Catalysts preparation.

Zeolites with MFI structure were obtained by hydrothermally treatment of silica – alumina gel in the presence of di-n-propylamine as "template". By our procedure ZSM – 5 zeolite with large variation limits of SiO_2 / Al_2O_3 ratio, can be obtained.

Silica-alumina gel with desired SiO_2 / Al_2O_3 ratios is prepared by precipitation from Na_2SiO_3 and $Al(NO_3)_3$, filtration and drying at 120^0C, 4 yours.

The gel is suspended in demineralized water in concentration of 20 wt %, and under continous steering di-n-propylamine is added. The template / Al_2O_3 mol ratio reaction mixture is 1.2. The crystallization step is realized by hydrothermally treatment at 180^0C, 24 to 100 hours at pH = 12.2 – 12.3. When modification with copper is made in crystallization step the copper salt is added in gel suspension in desired Cu / Al ratio. The crystallization time depends upon SiO_2 / Al_2O_3 gel ratio and Cu / Al ratio in reaction mixture. Crystalline zeolite separated by centrifugation is dried and calcined 4 hour at 550^0C. XRD purities of the zeolites obtained are more than 95% H – form of the zeolite is obtained by ion exchange with 2M NH_4NO_3, 5 hours at 95^0C. NH_4 ZSM – 5, after separation and drying, by calcination at 550^0C is converted in H – ZSM – 5.

For copper zeolite modification three method were used:
- ion exchange of H – ZSM – 5 with $Cu(NO_3)_2$
- ion exchange of H – ZSM – 5 with $Cu(CH_3COO)_2$
- introduction of Cu – salt in the crystallization step of the amorphous silica – alumina gel.

Ion exchange was made with Cu – salt solutions at different ratio between Cu^{2+} in solution and Al in zeolite at 105^0C and 4 to 6 hours. Samples with different concentrations of copper in zeolite were obtained. XRD measurements show that copper is distributed between Cu into the zeolite chanel probably onto cationic sites and CuO on the external surface. The surface CuO increase with total Cu concentration. Cu – modified zeolites were dried 4 hours at 120^0C, formulated with 30% alumina matrix, extruded in cylindrical shape with 1.5 mm diameter and finally calcined 4 hours at 550^0C. The catalyst was ground and the 0.1 mm dimension was selected for the test reaction.

Copper in the zeolites was analyzed by atomic adsorption method.

Some experiments were made with MCM – 41 mesoporous materials [7].

B. Reaction test

Exhaust gas purification tests were made in a microreactor filled with 0.1 – 0.2 g catalyst.

Space velocity was 10.000 h^{-1} and temperatures between 350 – 450^0C. Exhaust gas collected from a car type DACIA 1300 was used as feed.

The microreactor is coupled with a special chromatographic assembly [8] which comprises two chromatographic columns, a FID detector and a TCD detector. The two detectors are separated by a hydrogenation reactor which converts at 270^0C, CO and CO_2 in CH_4 to increase the detection sensibility of these components. The feed and carrier gas are directed towards a six way valves and an eight way valves to the columns and detectors circuits. The first chromatographic column is located out side and the second inside of the chromatograph chamber, thermostated at 100^0C. Hydrogen was used as carrier gas at 0.5 at pressure.

NO_x, N_2 and O_2 are detected by TCD and CO, CO_2(as CH_4) and hydrocarbons by FID. The feed is injected as pulses in the chromatographic assembly two times, first by – passing the microreactor with decontamination catalyst and second through the reactor maintained at programmed reaction temperature.

An example of the results obtained is presented in Figure 1. The catalyst performance in decontamination of each component is evaluated by % modification of the two chromatogram peak surfaces and indicates the relative modifications of the component concentrations.

As is expected when a catalyst works N_2 and CO_2 peaks increase over 100% by oxidation of NO_x, hydrocarbons and CO while O_2, NO_x, hydrocarbons and CO peaks decrease under 100% till zero, when the components are totally converted.

REZULTATS AND DISCUSSIONS

The main purpose of this work being the elaboration of combinatorial model for the catalyst manufacture and reaction conditions optimization, the variables were fuzzy modified in the limits acceptable by practical reasons.

In Table 1 are shown results obtained over catalysts based upon ZSM – 5 zeolites modified with copper in the crystallization step. Cu^{2+} ions existing in this phase strongly reduce the rate of crystallization and also zeolite crystallographic phase purity. Only under 1% Cu can by introduced in the zeolite composition by this method. The experiments were made with 60, 80 and 100 SiO_2 / Al_2O_3 mol ratio in zeolites. When Cu / Al ratio in the crystallization mixture was 1.000, maximum copper found in zeolite composition was 0.33 wt %.

Samples with 0.04 – 0.33 copper concentrations tested at 400^0C show good activity in hydrocarbons but low and very low in NO_x and CO conversion. Maximum 17.3 % CO conversion (82.7% peak deerease after decontamination reaction) was obtained over H – ZSM – 5 zeolite with high SiO_2 / Al_2O_3 ratio, equal with 100 and 0.33 wt % Cu concentration in zeolite.

Table 2 resumes some results obtained with Cu – modified H - ZSM – 5 zeolite by ion exchange with $Cu(NO_3)_2$ solution.

In a reasonable ion exchange time max 0.87% Cu was exchanged. The decontamination grade obtained over these catalysts is unsatisfactory for all components, especially for CO, which is maximum 13.4% converted.

Table 3 resumes the results obtained over catalysts obtained by Cu ion exchange with $Cu(CH_3COO)_2$. With this copper salt high Cu concentration in zeolite are obtained.

Table 3 data show also the temperature dependence of the catalytic activity for medium Cu content. The best results are obtained at 400^0C but CO conversion remains small, under 12%.

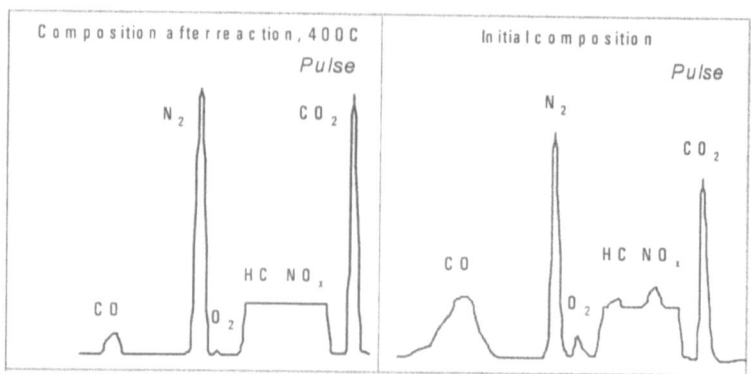

Fig.1. Gas composition before and after catalytic treatment.

As Table 4 shows very good results were obtained with catalyst containing more than 3 wt % Cu. NO_x and hydrocarbons are completely converted while more than 50% CO is transformed.

The data of the Table 5 demonstrate the influence of oxygen content in exhaust gases. Over medium Cu – exchanged catalyst No 27, all the three contaminates NO_x, hydrocarbons and CO are totally converted when feed is diluted with 15% air. This means about 2 vol. % oxygen in exhaust gas.

As our results show the variables which must by considered for combinatorial model and their limits are the following:

1. SiO_2 / Al_2O_3 mol ratio: 30 – 150
2. Cu concentration in crystallization step: 0 – 0.5 wt.%
3. Cu by ion exchange with $Cu(NO_3)_2$: 0 – 2 wt.%
4. Cu by ion exchanged with $Cu(CH_3COO)_2$: 0 – 4 wt.%
5. Reaction temperature: 300 – 450 ^{0}C
6. Space velocity: 10.000 – 100.000 h^{-1}
7. Oxygen content in exhaust gas: 0.5 – 3 vol %

A Box – Willson model indicates a minimum experiment number of

$$N = 2^n + 1$$

Our model having n = 7 levels, results 129 experiments.

This combinatorial model will by applied for the following zeolites: ZSM – 5, MCM – 22 and MCM – 41.

CONCLUSIONS

With Cu – HZSM – 5 catalysts complete decontamination of the vehicle engine exhaust gases can be obtained.

Combinatorial statistical model for optimization of the catalyst composition and reaction conditions needs minimum 129 experiments.

Pulse microreactor coupled with a chromatographic system can solve this combinatorial experimental program in reasonable time.

REFERENCES

1. B.K.Cho, J.Catal., 142 (1993), 418
2. H.Hamada, Y.Kintaichi, M.Sasaki, T.Ito and M.Tabata, Appl. Catal., 64 (1990), L1
3. M.Iwamoto, H.Yahiro and N.Mizuno, Proc. 9 – th Int Zeolite Conference, Montreal, 1992, vol II, p.397
4. M.Shelef, Chem. Rev., 95 (1995), p.209
5. A.Fritz and V.Pitchon, Appl. Catal., B, 13 (1997), p.1
6. R.K.Long and R.T.Yang, Ind. Eng. Chem. Res., 38 (1999), p.873
7. Gr.Pop, R.Ganea, R.Barjega, I.R.Tamas, M.Lupascu, Clairea Marichal – Westrich and H.Kessler, Prog. Catal., 8,2 (1999), p.37
8. Gr.Pop and I.R.Tamas, 2000 A.I. Ch.E Spring Meeting, IMRET 4, Marth 5 – 9, 2000, Atlanta, Georgia, SUA, TE 012

Table 1

Results obtained over catalysts obtained by Cu – modification in crystallization step, of ZSM - 5 zeolite
(T = 400⁰C)

Nr.	$\frac{SiO_2}{Al_2O_3}$	$\frac{Cu}{Al}$	%Cu in zeolite	% ratio of the exhaust gas component after and before reaction					
				CO_2	NO_x	Hydrocarbons	O_2	N_2	CO
1	60	-	-	101.3	100	92.1	97.5	100	100
2	60	0.025	0.04	100.2	77.8	90.3	91.2	100.2	100
3	60	0.050	0.12	134.7	57.9	0	73.3	102.4	93.3
4	60	0.100	0.24		64.3	0	85.3	101.3	98.0
5	80	-	-	100	95.3	100	100	100.4	100
6	80	0.125	0.05	101.3	78.3	0	95.3	115.9	100
7	80	0.250	0.25	102.4	61.9	0	92.0	108.3	100
8	80	0.500	0.30	112.2	5.3	0	94.4	103.6	97.6
9	100	-	-	100.3	80.0	94.3	98.4	100.2	100
10	100	0.125	0.06	100.4	78.1	0	97.3	112.3	95.7
11	100	0.250	0.12	101.6	46.4	0	97.1	103.9	94.7
12	100	0.500	0.18	101.8	0	0	96.7	105.1	99.1
13	100	1.000	0.33	102.3	0	0	90.1	106.3	82.7

Table 2.

H – ZSM – 5 modified by Cu (NO₃)₂ ion exchange
(T = 400°C)

Nr.	$\frac{SiO_2}{Al_2O_3}$	% Cu in zeolite	% ratio of the exhaust gas component after and before reaction					
			CO₂	NOₓ	Hydrocarbons	O₂	N₂	CO
14	50	0.08	100.3	86.0	79.0	98.3	100.3	98.4
15	50	0.87	102.4	54.3	20.4	89.4	101.4	86.5
16	100	0.06	100.1	91.0	85.2	98.7	100.2	98.5
17	100	0.75	101.8	63.2	35.6	89.9	101.3	86.6

Table 3

Experiments at different temperatures over H - ZSM – 5 modified by Cu(CH₃COO)₂ ion exchange

Component Reaction temperature °C			CO₂			NOₓ			Hydrocarbons			O₂			N₂			CO		
Nr.	$\frac{SiO_2}{Al_2O_3}$	% Cu in zeolite	350	400	450	350	400	450	350	400	450	350	400	450	350	400	450	350	400	450
			% ratio of the exhaust gas component after and before reaction																	
18	50	-	102.5	103.6	101.1	66.7	87.8	100.0	0	0	0	100	92.3	100.0	100	100.0	100.0	98.6	96.1	100.0
19	50	1.8	122.9	119.8	134.4	0	80.0	91.9	0	0	0	30.8	30.8	27.4	102.5	102.5	100.3	89.1	88.8	91.5
20	52	1.9	102.9	105.0	106.3	0	70.3	90.7	0	73.5	80.0	36.4	27.3	28.0	102.5	100.6	100.6	86.6	86.2	89.0
21	100	-	107.4	101.9	113.0	83.3	94.4	40.0	0	0	0	100	81.5	71.3	100	101.3	104.9	90.6	95.1	97.6
22	100	1.8	121.9	108.8	119.7	38.9	44.4	50.0	0	0	0	62.5	36.1	55.6	102.1	101.8	100.3	95.3	87.6	95.3

Table 4.

Experiments over H-ZSM – 5 zeolite modified with Cu(CH₃COO)₂ at high Cu concentrations in the catalyst (T = 400°C)

Nr.	$\frac{SiO_2}{Al_2O_3}$	% Cu in zeolite	% ratio of the exhaust gas component after and before reaction					
			CO_2	NO_x	Hydrocarbons	O_2	N_2	CO
23	50	1.90	105.4	0	0	27.0	103.1	86.0
24	50	2.84	106.2	0	0	20.1	101.2	37.0
25	50	3.94	107.1	0	0	48.0	103.1	54.4
26	100	1.80	104.1	73.5	0	74.6	100.3	91.2
27	100	2.21	105.2	0	0	46.2	104.8	77.3
28	100	2.63	106.4	0	0	42.6	108.1	73.6

Table 5.

Experiences over catalyst No. 27 with different oxygen content in feed

Nr.	Air added %	T°C	% ratio of the exhaust gas component after and before reaction																
			CO_2			NO_x			Hydrocarbons			O_2			N_2			CO		
			350	400	450	350	400	450	350	400	450	350	400	450	350	400	450	350	400	450
1	0		102.1	104.0	105.3	30.2	71.4	91.6	20.1	74.6	82.4	36.6	27.8	30.1	101.0	101.8	100.2	88.4	89.6	90.3
2	5		102.4	104.1	105.8	15.3	65.2	84.3	5.6	55.4	71.6	48.3	44.1	40.6	101.2	101.9	100.8	80.2	81.1	88.6
3	10		110.4	111.3	112.1	0	20.4	63.3	0	31.2	44.7	63.1	77.2	78.9	101.8	102.3	100.9	49.2	61.3	71.8
4	15		112.1	113.4	110.1	0	0	0	0	0	0	88.4	89.5	90.2	102.4	103.0	101.4	0	10.1	50.4

Initial composition of exhaust gas; % vol
9.7 CO_2, 980 ppm NO_x, 750 ppm Hydrocarbons, 1.1 O_2, 81.9 N_2, 7.3 CO

Liquid phase hydrogenation of p-nitrotoluene in microchannel reactors

Ringo Födisch[1], Dieter Hönicke[1], Yugong Xu[2], Bernd Platzer[2]

1 Lehrstuhl für Technische Chemie, 2 Lehrstuhl für Chemische Verfahrenstechnik, Technische Universität Chemnitz, D-09107 Chemnitz/Germany

Abstract

The hydrogenation of p-nitrotoluol to p-toluidin was done both in a fixed bed reactor and in a microchannel reactor made of a stack of microstructured aluminum wafers. Two procedures for the deposition of the catalytic active component palladium on the microstructured aluminum wafers are described. The hydrogenation results are discussed. Advantages and drawbacks of both reactor types are shown.

1. Introduction

Nitro aromatics are of great importance for chemical synthesis. They are the major intermediates for the production of anilines by hydrogenation of the corresponding nitro compounds. These reactions are highly exothermic having a reaction enthalpy of 500 to 550 kJ per mol hydrogenated nitro group. Due to limitations in the heat removal from the reaction mixture, the reaction rate, usually controlled by the hydrogen supply, is limited. Thus, using microchannel reactors seems profitable because of their excellent heat transfer properties. However, microchannel reactors possess some drawbacks. One important is the very low specific surface area in comparison to that of catalysts in conventional fixed bed reactors, e.g. Pd/C, used for the hydrogenation of nitro aromatics [1]. Therefore, the microstructured aluminum wafers used to build the microchannel reactor need a special pretreatment in order to increase the specific surface area. For this reason, anodization was chosen in the present study, which yields an oxide layer coat, having a regular system of nearly parallel mesopores. The anodization is followed by impregnation with palladium, because it is known, that palladium based catalysts are suitable for the hydrogenation of nitro aromatics to the corresponding amines [2] and that they have an excellent catalytic activity [3]. To provide the reaction system with sufficient amount of hydrogen several options can be applied, e. g. using hydrogen donors, soluble in the reactand mixture [4-6] or applying a loop reactor. More sophisticated is the detailed monitoring [7] and modelling of the three phase flow [8] with the aim to precisely control the parallel flow of liquid and gas inside each channel of a microchannel reactor as described by Hessel et. al. [9]. In the present study, a loop reactor was applied.

2. Experimental

2.1. Catalyst preparation

2.1.1. Catalysts for microstructured aluminum wafers and aluminum wires

Anodization procedure
   Aluminum wires having diameters of 1 mm and additionally aluminum wafers having a geometry as depicted in figure 1 with 14 channels per wafer were

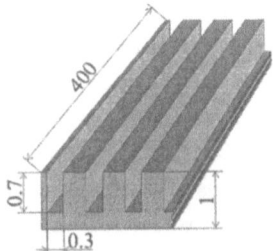

Fig. 1: Geometry of a microstructured aluminum wafer, not true to scale

degreased using tetrachlorethylene (Merck) and afterwards stained for 1 minute in 10% sodium hydroxide (Merck) and 5 seconds in 10% nitric acid (Merck). The aluminum specimens were then rinsed with distilled water and placed as anode in 1.5% aqueous oxalic acid (Acros). As cathode an aluminum plate was used. Anodization was done at 25°C and 50 V dc for 2 hours. After anodization the specimens were placed in methanol (Merck) for 2 hours at room temperature for water removing and afterwards calcined using the following temperature program: with 60 K/h to 423 K; 6 hours at 423 K; with 120 K/h to 903 K; 6 hours at 903 K.

Electrochemical deposition

For the electrochemical deposition of palladium the described anodization procedure was modified as follows: At the end of the anodization process, voltage was decreased slowly with 10 V/min to 0 and the dehydratisation in methanol and calcination were omitted. Immediately after anodization the microstructured aluminum wafers were placed in the palladium electrolyte, which had the following composition: 4 g boric acid (Merck), 5 g citric acid (Merck) and 1,5 g $PdSO_4$ (Alfa Aesar) in 250 g distilled water. Electrochemical deposition was done at 298 K and 7.5 V ac, 50 Hz for 3 minutes. Finally, the catalyst was calcined: with 60 K/h to 423 K; 6 hours at 423 K; with 120 K/h to 773 K; 6 hours at T_{Calc}=773 K. This catalyst is referred to as MEl.

Chemical deposition

Aluminum wafers and aluminum wire were anodically oxidized as described. The aluminum wire was then cut into extrudate like peaces of 3 to 5 mm in length. Afterwards, the specimens were placed in a solution of 10% formalin (Acros) in distilled water. The gas phase above this mixture was evacuated to eliminate air inside the mesopores of the oxide layer. Subsequently, the specimens were placed in a fresh solution of 10% formalin in distilled water for 1 hour and afterwards transferred in a solution of 200 mg $PdCl_2$ (Alfa Aesar) in 40 ml distilled water for 5 hours, whereby the $PdCl_2$ was reduced to metallic palladium by the formalin still present inside the mesopores. No longer impregnation time is possible because $PdCl_2$ in water is stable only for a few hours. Finally, the specimens were calcined using the following temperature program: with 60 K/h to 423 K; 6 hours at 423 K; with 120 K/h to T_{Calc}=773 K or T_{Calc}=903 K (table 1) for 6 hours. This impregnation procedure except the anodization step was repeated once in order to increase the palladium amount (m_{Pd}). The microstructured wafers are referred to as MCh and the aluminum wires as WCh respectively.

2.1.2. Conventional fixed bed catalyst

The prepared catalyst was of sol-gel-type starting from an alkoxide suspension as described by Yoldas [10]. 10.2 g aluminumtriisopropylate (Merck) in 80 ml distilled water were stirred for 1 hour at 353 K. Then, 1.21 g concentrated nitric acid were added. In a second flask 46 mg $PdCl_2$ were given to 100 ml 0.1 n EDTA (Titriplex, Merck) at 323 K. The resulting mixtures were stirred over night at 353 K and 323 K, respectively. The two mixtures were combined and stirred for 5 minutes at 353 K and afterwards poured in a petri dish, where the water was evaporated at 353 K. The resulting catalyst was dried at 373 K for 24h. Calcination was done using the following temperature program: with 50 K/h to 773 K; 6 hours at T_{Calc}=773 K. This fixed bed catalyst is referred to as FB.

2.2. Hydrogenation experiments

Hydrogenation experiments were done in a flow apparatus as depicted in figure 2 at 370 K and a pressure of 2 MPa. Liquid reactand was a solution of 10%

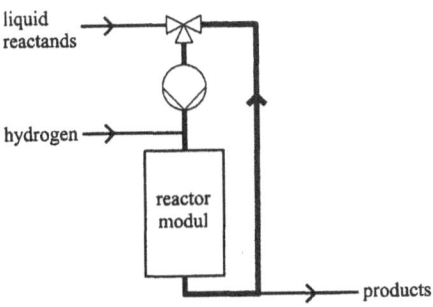

Fig. 2: Flow apparatus used for the hydrogenation of p-nitrotoluene to p-toluidine with hydrogen

p-nitrotoluene (Acros) in 2-propanol (August Hedinger GmbH). Hydrogen was of purity 5.0 (Messer Griesheim). The reactor was of modular construction and could be operated as microchannel reactor as well as as fixed bed reactor as shown in

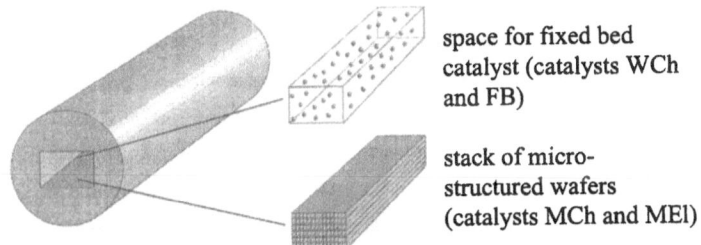

space for fixed bed catalyst (catalysts WCh and FB)

stack of micro-structured wafers (catalysts MCh and MEl)

Fig. 3: Reactor module equipped with a fixed bed catalyst or a stack of microstructured wafers

figure 3. In case of a microchannel reactor a stack of 6 microstructured wafers was used. The amount of aluminum wires peaces was chosen to have the same geometric surface area (A_{geom}) as the wafer stack, which equals 4.4 g. For the experiment using a conventional fixed bed catalyst, 1 g of catalyst FB was used. Analysis was done using a DSMS mass spectrometer by Hiden Analytical. Because of too many overlapping signals in the mass spectra of educt and product, evaluation of the product composition had to be done using a modified linear regression instead by selecting one characteristic signal for each component.

3. Results and Discussion

The hydrogenation results are listed in table 1. From that, hydrogenation of p-nitrotoluene to p-toluidine is practically free of byproducts, i. e. selectivity to p-toluidine is 100%. Thus, the conversion degree equals the yield of p-toluidine.

Tab. 1: Results of the hydrogenation of p-nitrotoluene to p-toluidine using micro-channel reactors, aluminum wires and a conventional fixed bed catalyst; $\dot{m}_{RNO_2}$: p-nitrotoluene flow, τ: residence time, rr: recycle ratio

experiment no.	catalyst	m_{Pd} [mg]	T_{calc} [K]	$\dot{m}_{RNO_2}/A_{geom}$ [g/(h·cm²)]	τ [s]	rr	conversion degree [%]
1	MEl	23	773	0.013	280	43	58
2	MCh	24	773	0.013	280	43	96
3	MCh	21	903	0.013	280	43	98
4	MCh	21	903	0.045	85	21	58
5	WCh	20	903	0.045	260	21	89
6	FB	10	773	$1.7 \cdot 10^{-6}$	90	21	85

Electrochemical deposition of palladium (catalyst MEl) as well as chemical deposition using formalin (catalyst MCh) are easy reproducible procedures. Major difference is the resulting palladium distribution inside the oxide layer as depicted in figure 4. To understand its importance, CFD-calculations were performed. The objective of these calculations was the concentration profile inside one single mesopore of the oxide layer of the anodically oxidized aluminum specimen. Assuming a uniform palladium distribution inside the mesopores, the resulting concentration profile is as depicted in figure 5. Despite, the actual palladium distribution as depicted in figure 4 was different than assumed, one important fact remains. Most of the hydrogenation takes place at the first 4 µm of a pore. Thus, palladium deposited at the pore base is not available to chemical reaction. In case of electrochemical deposition, about one third of the palladium was deposited at

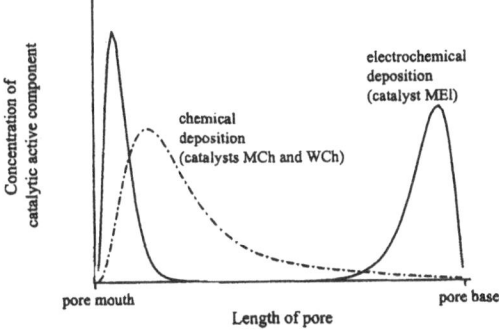

Fig. 4: Typical distribution of the catalytic active component palladium inside the oxide layer of an anodically oxidized aluminum wafer

the pore base, whereas when using chemical deposition, most of the palladium was deposited at the pore mouth. This is the major drawback of electrochemical palladium deposition and concurrently one major advantage of chemical

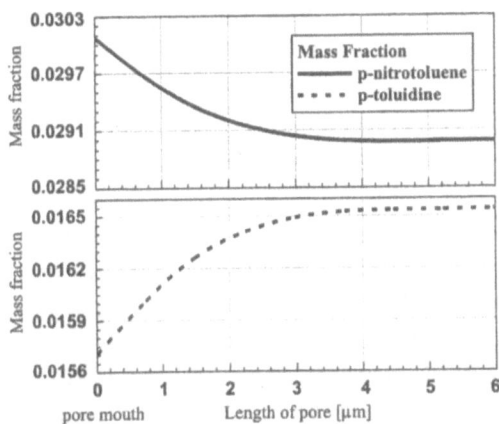

Fig. 5: Calculated distribution of reactand and product inside of a pore which is located at the channel inlet and having a uniform palladium distribution

deposition. Experimental results reflect this. The degree of p-nitrotoluene conversion was much lower when using catalyst MEl (experiment 1) instead of catalyst MCh (experiment 2). Disadvantageous of chemical deposition was the relatively long impregnation time of several hours in comparison to only a few minutes in case of electrochemical palladium deposition.

Hydrogenation of nitro aromatics to the corresponding anilines has water as a stoichiometric byproduct. Considering the applied reaction conditions, 370 K and 2 MPa, partial hydrothermal dissolution of the oxide layer and thus, catalyst deactivation seemed possible. To test this assumption, calcination temperature of the catalyst MCh was increased by 130 K from 773 K to 903 K. This higher calcination temperature should stabilize the oxide layer against the attack of water and should lead to an increase in catalyst activity. Indeed, the corresponding experiments 2 and 3 showed a slight increase in p-nitrotoluene conversion from 96% to 98%.

The studied liquid phase hydrogenation in microchannel reactors is difficult because of the presence of three phases, viz. liquid educt solution, gaseous hydrogen and solid catalyst. Unless special precautions are taken [9], there is a substantial risk of uncontrolled separation of these phases. As test experiment, hydrogenation was done using the microstructured catalyst MCh calcined at 903 K, and palladium impregnated aluminum wires, catalyst WCh, experiments 4 and 5. The impregnation procedure for both catalysts was identical. The amount of aluminum wires was chosen, that both catalysts have the same geometric surface areas. Thus, any differences in catalyst activity should be due to their different geometry, a well defined channel system against a irregular fixed bed. As can be seen in table 1, using catalyst MCh a conversion degree of 58% was achieved, which is much lower than when using catalyst WCh, showing a p-nitrotoluene conversion of 89%. This implicates, that separation of the phases occured. Some channels of the microchannel reactor could have been filled with liquid, others with gas. Because the channels are separated from each other, there would have been no mass transfer between these phases and thus, conversion degree would be limited. In a irregular fixed bed, separation of liquid phase and gas phase might occure. However, mass transfer between both of these phases would still be possible, resulting in a higher conversion degree.

On the other hand, a fixed bed of aluminum wires has a less compact geometry than a stack of microstructured wafers. There is three times the space between the various wire peaces resulting in three times the residence time. This difference in residence time is significant. Therefore, the differences in conversion degree might as well be simply attributed to the different residence times. On the basis of our results, it is not possible to distinguish between these two cases. One possibility would be, to use hydrogen donors instead of gaseous hydrogen. Thus, eliminating the gas phase, the first case would be excluded.

The final experiment in this study was to compare the activity of the catalyst MCh based on microstructured wafers (experiment 4) with that of the conventional fixed bed catalyst FB (experiment 6). The conversion degree of 85% using catalyst FB was much higher than when using catalyst MCh at the same residence time even though the sol gel catalyst FB exposed only half the palladium amount. Most of this high activity is attributed to the high BET surface area of the sol-gel-type catalyst compared to that of microstructured wafers. Thus, using microstructured wafers as catalysts seems disadvantageous. But, the crucial point in the use of the fixed bed catalyst is, that in industrial processes reaction rate must be limited because of the high reaction enthalpy of the studied hydrogenation reaction. Considering this, the activity of the described microstructured wafers is still sufficient.

4. Conclusions

Hydrogenation of p-nitrotoluene to p-toluidine in microchannel reactors is a favourable option. Microstructured wafers have a sufficient catalytic activity. Because of the aluminum core, they have an excellent thermal conductivity. Thus, limiting reaction rate is not necessary. However, one difficulty is to provide a sufficient amount of hydrogen. A possible solution is to dissolve hydrogen in the liquid phase. But, because of the poor solubility of hydrogen in most solvents, this requires a sophisticated reactor design or at least the use of a loop reactor with recycle ratios above 20. Major drawback of the latter is the very wide residence time distribution, similiar to that of a continuously stirred tank reactor. Thus, it is nearly impossible to achieve a conversion degree well above 99%. However, if the advantages, viz. very high reaction rate without heat transport difficulties and small, modular and easy modifiable reactor design are more important than the high conversion degree, then the described reactor type with microstructured aluminum wafers as catalyst is a suitable alternative.

478

5. Acknowledgements

We thank the Sächsisches Staatsministerium für Wissenschaft und Kunst and the Fonds der Chemischen Industrie for financial support. We thank Dr. Hulzer (Asta Medica Dresden) and Prof. Reschetilowski and Mr. Leuteritz (TU Dresden) for stimulating discussions.

6. References

[1]   V. R. Choudhary, M. G. Sane, S. S. Tambe; Ind. Eng. Chem. Res. **37**, pp. 3879-3887 (1998)

[2]   B. Yang, Z. K. Yu, Y. Xu, S. J. Liao, D. R. Yu; Chin. Chem. Lett. **7**, pp. 663-664 (1996)

[3]   G. Cordier, J.-M. Grosselin, R.-M. Ferrero; Ind. Chem. Libr. **8**, pp. 336-342 (1996)

[4]   A. Leuteritz, W. Reschetilowski, R. Födisch, D. Hönicke; in: Tagunsgband XXXIII. Jahrestreffen deutscher Katalytiker, March 22nd-24th 2000, Weimar, Germany, pp. 197-200

[5]   A. Leuteritz, W. Reschetilowski; in: Tagunsgband XXXIV. Jahrestreffen deutscher Katalytiker, March 21st-23rd 2001, Weimar, Germany, pp. 278-279

[6]   M. Burli, T. Bühlmann, B. Troxler; Spec. Chem. **13(6)**, pp. 346-348 (1993)

[7]   J. Chaouki, F. Larachi, M. P. Dudukovic; Ind. Eng. Chem. Res. **36**, pp. 4476-4503 (1997)

[8]   F. Stüber, A.M. Wilhelm, H. Delmas; Chem. Eng. Sci. **51**, pp. 2161-2167 (1996)

[9]   V. Hessel, W. Ehrfeld, K. Golbig, V. Haverkamp, H. Löwe, M. Storz, Ch. Wille, A. E. Guber, K. Jähnisch, M. Baerns; in: Proc. of 3rd International Conference on Microreaction Technology, W. Ehrfeld, eds., Topical Conference Preprints, April 18th-21st 1999, Frankfurt (Main), Germany, pp. 526-540

[10]   B. E. Yoldas; Am. Ceram. Soc. Bull. **54(3)**, pp. 289-290 (1975)

Monitoring of Chemical Processes in Microreactors by Adapting FTIR Microscopy and HPLC

T. Türcke, W. Schweikert, F. Lechner, J. Antes,
H.H. Krause, S. Löbbecke[*]

Fraunhofer-Institut Chemische Technologie ICT
P.O. Box 12 40, 76318 Pfinztal, Germany

Introduction

In recent years microreactors have received an increasing interest as useful devices for chemical reactions and processes. In comparison with conventional scales the application of microfluidic devices give the opportunity to change significantly product distribution and thus yield and selectivity for specific products.

Up to now, the analysis of products synthezised in microreactors is usually carried out offline by chromatographic and/or spectroscopic techniques. As a consequence, both the evaluation of experimental data and the assessment of the reactor performance is relatively time-consuming.

To improve the application of microfluidic devices for chemical processes fast analytical techniques are required allowing a simultaneous process monitoring while reactions are taking place. Such monitoring comprises both analysis of chemical reactions and control of process parameters and performance (mixing, flow distribution, blockages, etc.); hence it will give an access to an increased through-put and automation of microreaction processes.

Application of FTIR Microscopy

Since many microreactors are based on fluidic structures made of silicon, infrared spectroscopy is a suitable method to analyse chemical processes within these structures due to the transparency of silicon in the mid-IR spectral range.

Guber et. al., for example, applied IR spectroscopy for the analysis of chlorine-fluorine exchange reactions [1]. Jackmann et. al. describe the IR analysis of the hydrolysis of propionyl chloride in a silicon microreactor [2]. In both cases silicon cells with rectangular structures were developed and applied for these analyses.

In own experiments infrared microscopy was applied to investigate nitration reactions and hydrolysis of acid chlorides in commercially available silicon micromixers (MiMoCo GmbH, Ilmenau/Germany) [3, 4].

The key advantage of applying microscopic techniques is the high spatial resolution ($\geq 10\ \mu m$) allowing IR spectroscopic analysis within three-dimensional, complex fluidic structures as they are required for chemical processes, e.g. to enable high mixing qualities.

Furthermore, high spatial resolution ensures also that educts and products of chemical reactions can be identified and quantified at different positions inside the microreactor. By focusing the IR beam consecutively on different reaction zones, both final products and intermediates of chemical reactions can be IR spectroscopically detected. Hence, the progress of a reaction can be monitored.

Experimental

The setup of the FTIR spectroscopic microprocess monitoring is schematically shown in fig. 1. The flat silicon microreactor is mounted on a positionable stage and connected with the reactants supply (e.g. syringe pumps) A video camera on the top of the FTIR microscope (Nicolet Magna-IR 750/NIC-Plan) allows a computer controlled viewing and positioning of the reactor. The IR beam can thus be focused on different parts of the microreactor. Furthermore, the video-controlled computer allows a pre-programmed collection of IR spectra along a designated line (fig. 2) or even a complete IR spectroscopic mapping of the entire microreaction process. A time resolution of one spectrum per second can be achieved.

Exemplary results

In complex microfluidic structures inhomogeneous flow distributions may occur due to blockages of individual microchannels influencing significantly the overall reactor performance. Such blockages are often caused by gas bubbles, fouling effects or precipitated solids which can easily be identified and located by IR microscopy in sense of an on-line process control.

Fig. 3 shows as an example the spatially resolved finding of gas bubbles and solid blockages in different microchannels of a silicon micromixer which is continously supplied with methylene chloride by syring pumps. Although the microfluidic structure of the mixer is quite complex due to the "split-and-recombine" mechanism ensuring high mixing quality (fig. 4) those areas of bubbles ands blockages can be located.

Furthermore, by filling the whole microdevice with IR absorbing fluids a mapping of the internal structures is possible.

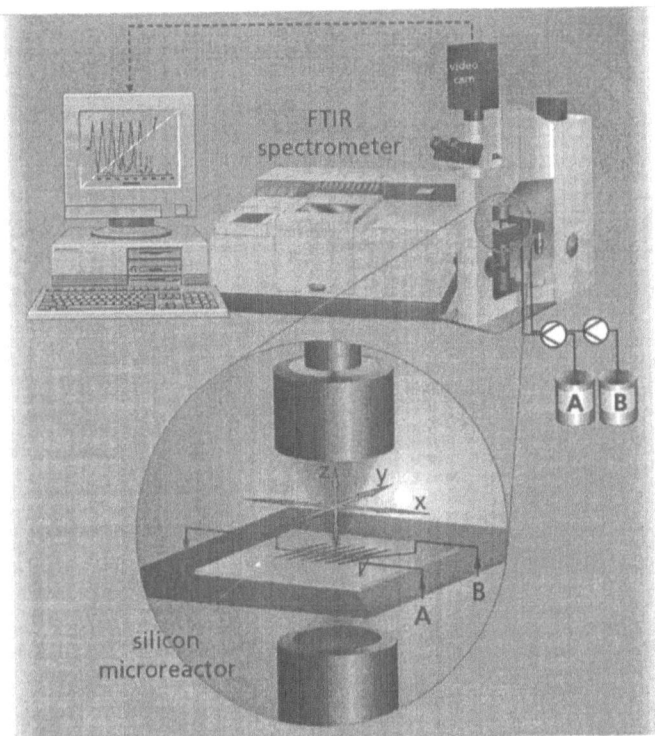

Fig. 1: Scheme of experimental setup for FTIR microscopic monitoring of chemical processes in microreactors

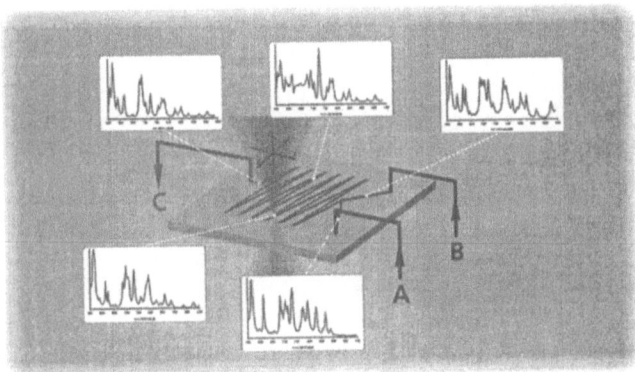

Fig. 2: Scheme of pre-programmed collection of IR spectra at different positions within a microreactor

482

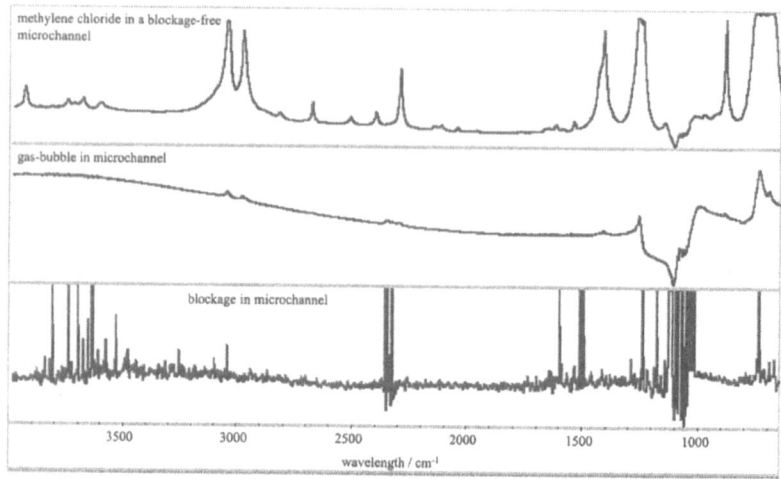

Fig. 3: Finding of gas bubbles and solid blockages in individual microchannels by FTIR microscopy

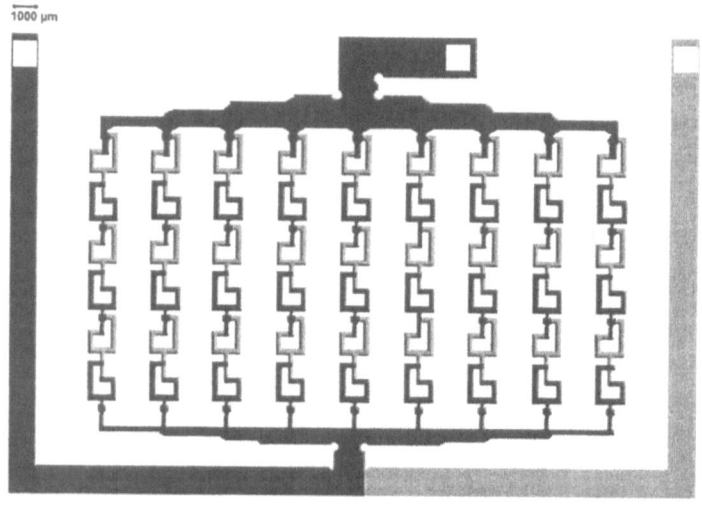

Fig.4: Simplified scheme of a complex "split-and-recombine" fluidic structure of a silicon micromixer (Mimoco GmbH, Ilmenau/Germany)

An example for the IR spectroscopic monitoring of a chemical reaction is shown in fig. 5.

The exothermic hydrolysis of benzoyl chloride to benzoic acid was investigated in a silicon microreactor. Since it is a fast reaction of two immiscible fluids the reaction rate is controled by mass transfer:

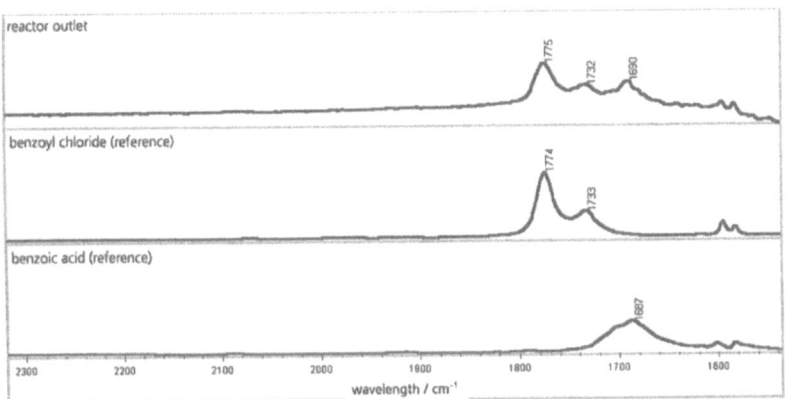

The IR-spectrum detected at the outlet of the microreactor indicates clearly the formation of the benzoic acid at 1690 cm^{-1}. Furthermore, the spectrum shows also that conversion was not completed at the reactor outlet (IR absorption bands of benzoyl chloride at 1775 cm^{-1} and 1732 cm^{-1}) due to the immiscibility of the educts.

Hence, FTIR microscopy does not only allow monitoring of the chemical conversion but also of the mixing resp. emulsion process.

Fig. 5: IR spectroscopic on-line detection of benzoic acid as a hydrolysis product of benzoyl chloride in a silicon microreactor (absorption spectra)

The time-resolved monitoring of the emulsion formed by benzoyl chloride and water in a silicon microemulsifier (MiMoCo GmbH, Ilmenau/Germany) is shown in fig. 6. The spectra were measured every 21 seconds at a fixed microchannel position near the inlet of the mixer. Uniform pulses of water and benzoyl chlorid were detected demonstrating a homogeneous emulsification.

484

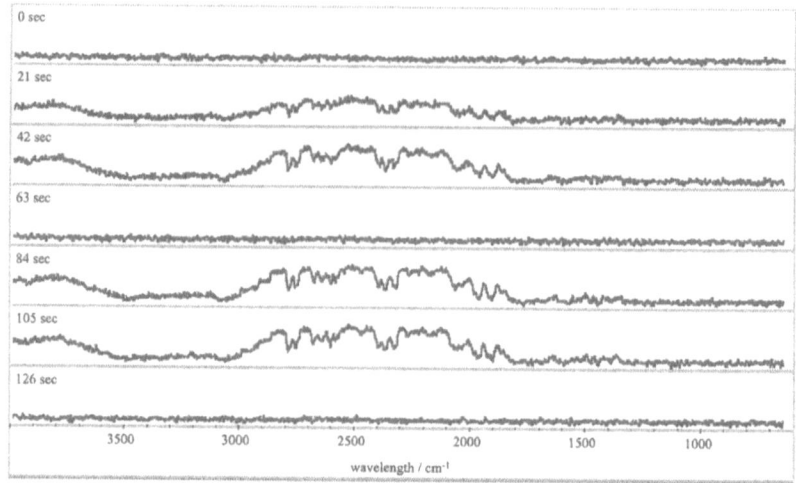

Fig. 6: Time-resolved monitoring of the emulsification of water and benzoic chloride in a silicon microchannel (transmission spectra)

Another example for the IR spectroscopic analysis of a highly exothermic reaction in a silicon microreactor is the nitration of N,N' dialkylurea (fig. 7).

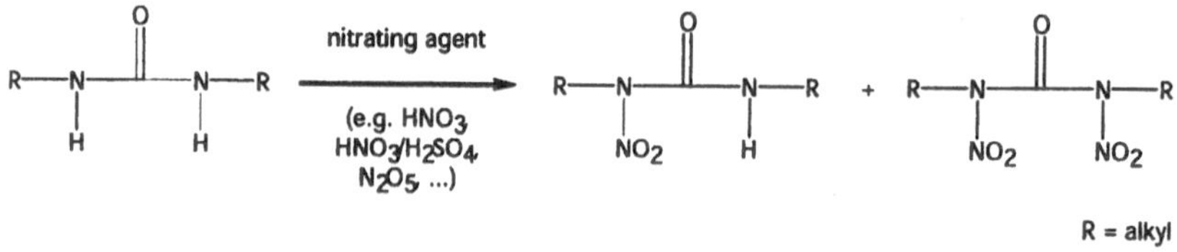

Fig. 7: Nitration of N,N'-dialkylurea

Although solvent, nitrating agent, educt and product(s) are all IR active substances causing a large number of, partly overlapping, absorption bands in the IR spectral range a characteristic product absorption band could be indentified at 1522 cm^{-1} (fig. 8).

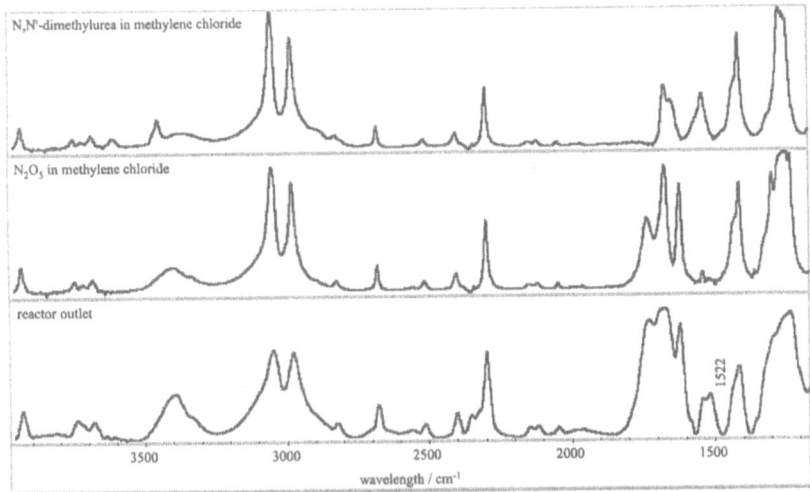

Fig. 8:   IR spectroscopic detection of the nitration product of N,N'-dialkyl urea
          inside of the silicon microreactor

Application of HPLC

The application of spectroscopic techniques for the simultaneous on-line
identification and quantification of educts and products in a microreaction process
is limited by the number of the different substances which have to be analyzed.
Overlapping absorption bands, for example, make it difficult to distinguish
between various analytes, and especially to quantify them. For this reason
chromatographic techniques like HPLC are required allowing a separation and
subsequent quantification of each component.

Experimental

Commercially available HPLC analysis was adapted to the above described
microreaction devices. Continous sampling was realized either by sample loops
of different volumes or an automated microsyringe system taking samples from
septum-sealed flow-through vials.

The sampling devices were connected to the outlet of the microreactor via a T-
type mixer allowing quenching of the reactant stream by cold solvents (here:
methylene chloride or ice water); fig. 9.

The nitration of toluene was chosen as a typical strong exothermic test reaction
with relatively broad product spectrum (mono- and dinitro toluenes, oxidation

and decomposition products, cresols, etc.). Microreactors were applied to achieve a more selective synthesis of o-, m- and p-mononitro toluene (fig. 10).

UV detectors and optical cells of different sensitivity were used depending on the educt concentrations which were applied and the product concentrations which were expected.
Sample loops with volumes of less than 10 µL turned out to be not suitable for transfering samples from the microreaction process to the HPLC analysis due to significant pressure drops.

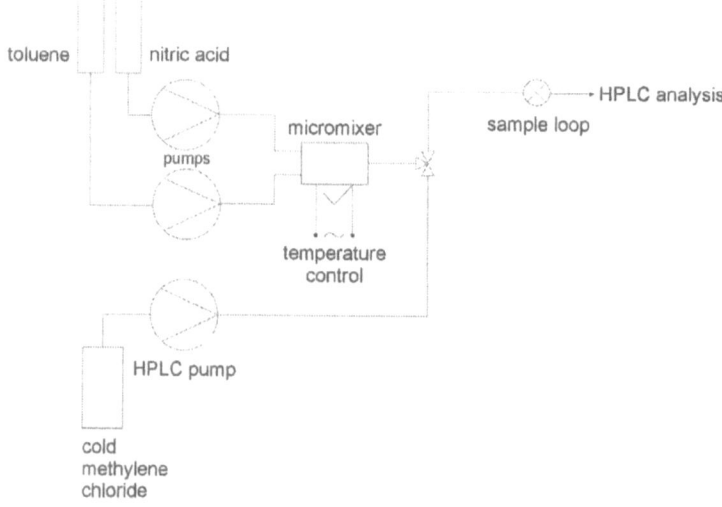

Fig. 9:   Scheme of experimental set-up for the adaptation of HPLC analysis to a microreaction process for the nitation of toluene

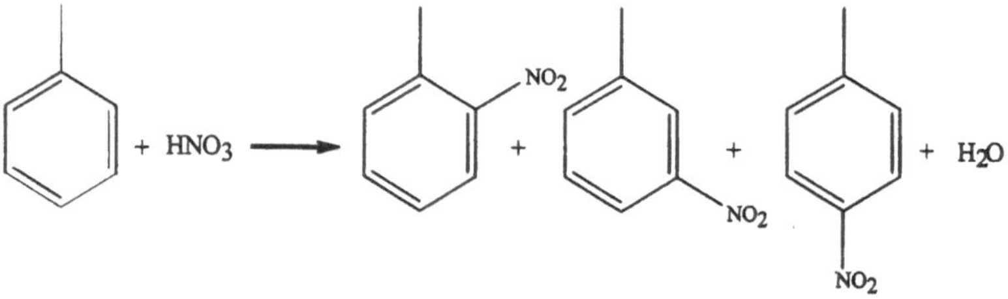

Fig. 10: Nitration of toluene to mononitro toluenes (isomeric ratio in industrial processes: 63% ortho, 4% meta, 33% para [5])

Exemplary results

HPLC allows the separation, identification and quantification of the different nitro toluenes synthesized in the microreaction process under systematic variation of the process conditions.

By applying HPLC-coupled microreactors the influence of quenching on the product distribution could be systematically investigated. Fig. 11 shows as an example how increasing volume flow of the quenching solvent (here: cold methylene chloride) increases yield and selectivity of mononitro toluenes while the degree of conversion is kept constant.

Due to the retention time of the analytes chromatography is no on-line technique in the true sense. Nevertheless, since the chromatogram is obtained in direct connection to the microreaction process it contains not only valuable information about the composition of the reaction mixture but also about the progress and influencing parameters of the entire reaction. Hence, monitoring and regulation of the running reaction process is possible.

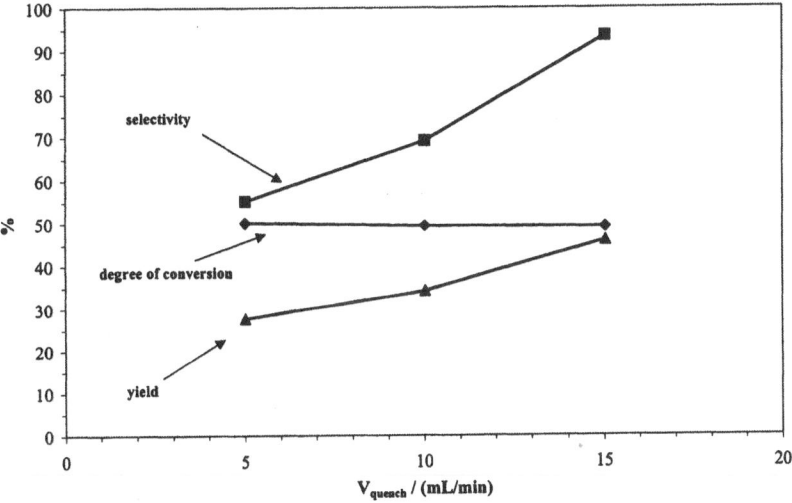

Fig. 11: The influence of quenching on yield and selectivity of mononitro toluenes (sum parameter) analyzed by HPLC-coupled microreactors

488

References

[1] A.E. Guber, W. Bier, K. Schubert, *IR Spectroscopic Studies of a Chemical Reaction in Various Micromixer Designs,* 2nd Int. Conference on Microreaction Technology (IMRET 2), 1997, New Orleans, USA

[2] R. J. Jackman, T. M. Floyd, M. A. Schmidt, K. F. Jensen, *Development of Methos for on-line chemical detection with Liquid Phase microchemical reactors using conventional and unconventional techniques,* Micro Total Analysis systems 2000 (μ-TAS), 14-18 May 2000 Enschede, The Netherlands

[3] W. Schweikert, T. Tuercke, S. Loebbecke, *Monitoring of Nitration Reactions in Microreactors with FTIR Microscopy,* 31st Int. Annu. Conf. ICT: Energetic Materials - Analysis, Diagnostics and Testing, 98, 27 - 30 June 2000, Karlsruhe, Germany

[4] S. Loebbecke, W. Schweikert, T. Tuercke, J. Antes, E. Marioth, H. Krause, *Application of FTIR Microscopy For Process Monitoring in Silicon Microreactors,* MICRO.tec 2000 - VDE World Microtechnologies Congress, 43, 25 - 27 September 2000, Hannover, Germany

[5] K. Weissermehl, H.J. Arpe, *Industrielle Organische Chemie,* VCH-Verlag, 4th edition (1994) 409

Suitability of Various Types of Micromixers for the Forced Precipitation of Calcium Carbonate

R. Schenk, M. Donnet*, V. Hessel, Ch. Hofmann, N. Jongen*, H. Löwe
Institut für Mikrotechnik Mainz GmbH, Carl-Zeiss-Str. 18-20, D-55129 Mainz, Germany
* Swiss Federal Institute of Technology (EPFL), Materials Science Department, Powder Technology Laboratory, CH-1015 Lausanne, Switzerland

1. Introduction

Inorganic compounds having uniform morphology are highly valuable materials for applications as diverse as catalysis, electronics, medicine, or paint, just to name a few. A typical synthesis procedure of these materials are forced precipitation experiments. An innovative approach for this type of processing is the continuous operation in a new plug-flow reactor (Fig. 1), the segmented flow tubular reactor (SFTR) [1]. The main units of this reactor are a mixer, a segmenter, and a tubular section (termed bubble tube).

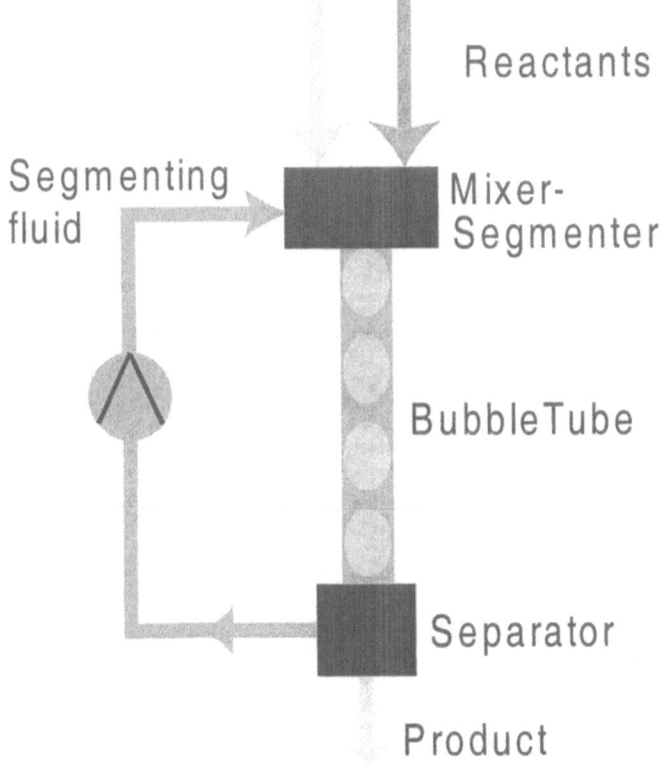

Fig. 1:   Scheme of the Segmented Flow Tubular Reactor (SFTR)

In the SFTR system, various inorganic materials with distinctive crystalline morphology were synthesized, exceeding the performance of samples gained by batch-type processing [1]. The success of these exploratory laboratory experiments motivated to undergo a project focusing on the industrial implementation of SFTR systems [2]. In this context, scale-up procedures are currently being developed, taking into account especially the minimization of fouling in the mixing section, e.g. in the case of $CaCO_3$ formation, first generation mixer, termed Vortex-type, is blocked after an operation time of 40 minutes. For this purpose, microstructured mixers were evaluated offering new means for chemical processes to benefit from micromixing physics [3].

2. Microfabrication

Several designs of so-called interdigital micromixers [4,5] were discussed with respect to avoid fouling and to facilitate fabrication technologies. Ideally, such a design optimization should be based on a theoretical understanding of all mixing and reaction processes. However, a simulation of the nucleation and crystal growth of the forced precipitation process is not possible yet. Accordingly, existing designs of interdigital mixers were optimized to yield an optimal fluid flow without eddies, small fluid layers for short mixing times, a straight fluid routing to prevent fouling, and a precise fitting to the segmenter of the SFTR. Taking all these design guidelines into account, it was found that an interdigital mixer with a triangular mixing chamber (see Figure 2) is the best suited device, currently available.

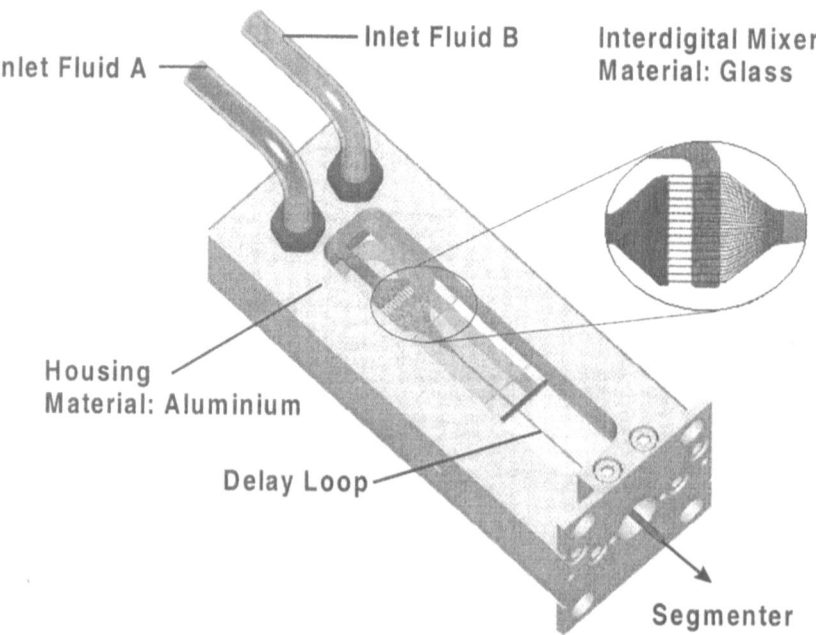

Fig. 2:   Assembled interdigital mixer

The housing of the interdigital mixer was made of aluminum and contains the fluid inlets as well as the outlet. The latter was formed so that it fits to the segmenter of the SFTR. For direct observation of mixing and fouling, the housing comprises an inspection window. The mixing device was realized in glass which is shown in Figure 2. The channels have a width of 60 µm and a depth of 500 µm, the walls being 50 µm wide.

The fluids A and B are divided into several substreams. At the beginning of the triangular chamber, fluid lamellae of a width of 110 µm are formed. In the triangular chamber this width is reduced to a lamellae thickness of about 10 µm by means of geometric focusing, i.e. a reduction of lamellae thickness by means of geometry constraints. Such thin lamellae are necessary to achieve a reasonably short mixing time, typically in the range of several tens of milliseconds. Furthermore, the cover plate can be removed partially for cleaning purposes.

Existing designs of so-called caterpillar mixers [6], comprising an 3D mixing structure, were adapted for similar reasons (Fig. 3). In particular, a delay loop for residence time prolongation, a replaceable inlay for material screening, an inspection window for observation of possible fouling, and a voltage supply for fouling prevention were integrated. The housing of the mixer was adapted to the needs of the SFTR. The 3D mixing structure was formed by two different techniques, namely micro spark erosion and micro 3D milling. For the gaskets, graphite and rubber were chosen as material and structured by laser cutting.

a)   b)

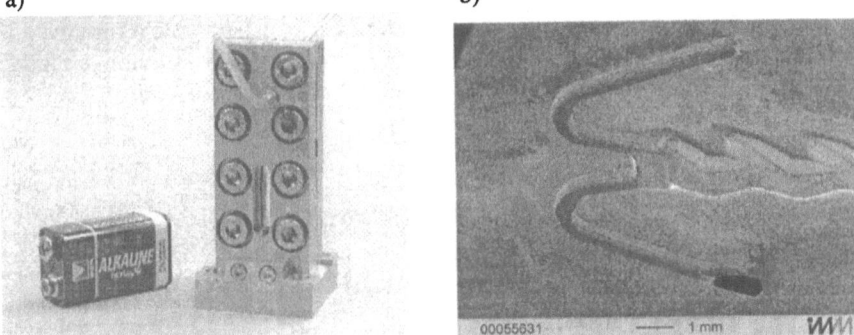

Fig. 3:   Caterpillar mixer with inspection window and delay loop: a) assembled micromixer and b) SEM image of the fluid inlet and the mixing structure.

For reactor material testing with regard to their fouling properties, the delay loops were made of different materials, including stainless steel (1.4301), PTFE and PMMA as well as surface-modified PMMA according to the following procedures:

- Exposure to oxygen plasma yielding hydrophilic surfaces
- Plasma deposition yielding hydrophobic surfaces
- Plasma deposition generating surfaces with an amino-group containing layer.

3. Results and Discussion

3.1. Characterization of micromixers

The fluid flow within the transparent interdigital mixer was observed using a water stream and a stream of an aqueous solution of a blue color with flow rates of 200 ml/h each. Figure 4 shows the thinning of the fluid streams by means of geometric focusing in the triangularly shaped mixing chamber. Furthermore, due to the residence time and the diffusion rate constants of the reactants it is evident that mixing is not completed inside the triangular chamber of the mixer.

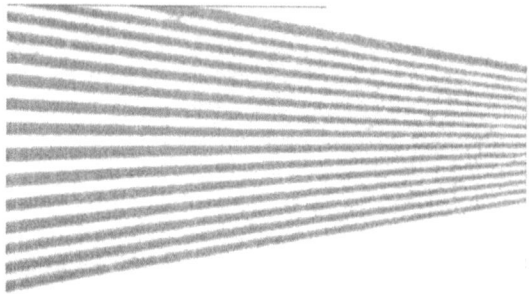

Fig. 4:   Photo of the fluid flow in the interdigital mixer.

Before undertaking the forced precipitation experiments, all micromixers were characterized with respect to their mixing performance, using integral information gained by a method based on competing reactions [5] with colored products, originally developed by Villermaux [7]. The UV absorption of such mixtures is reciprocally proportional to the mixing quality.

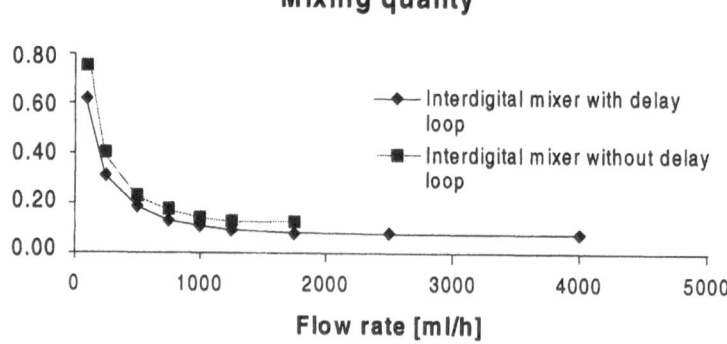

Fig. 5:   Comparison of mixing quality of the interdigital mixer with and without a delay loop. Absorption is inversely proportional to mixing quality.

The characterization of the mixing quality of the interdigital mixer (Fig. 5) shows that a delay loop serves to enhance mixing quality.

Figure 6 shows the UV absorption as function of the total volume flow rate for the different caterpillar mixers. The mixing quality of the milled mixer is slightly better than the mixing quality of the mixer made by spark erosion. In the case of using the milled caterpillar mixer, optimal mixing is obtained at a total flow rate higher than 2 l/h. Both curves display a qualitatively similar behavior as already found for interdigital micromixers (see Figure 5). In the case of applying a flow rate less then 1.5 l/h, the mixing quality is better when using the interdigital mixer compared to using caterpillar mixers. In contrast to this result, the mixing quality at high flow rates (> 3 l/h) is better in the case of using caterpillar mixers. According to this measure, interdigital mixers with delay loops proved to be the superior devices at flow rates below 2 l/h which are standard in the SFTR.

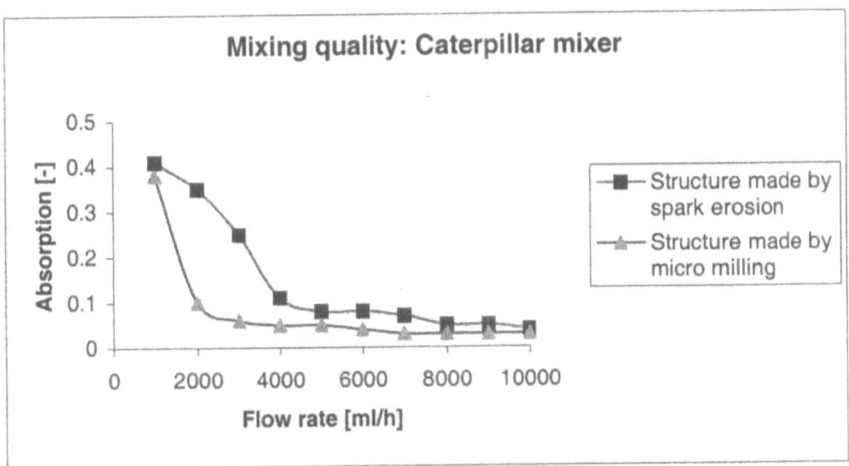

Fig. 6: Mixing quality of the caterpillar mixers

One explanation for the difference between the two caterpillar mixers is that small deviations from the ideal 3D geometry, due to the fabrication process, may strongly affect the splitting and recombination of the fluid lamellae. Accordingly, the microstructures of the different caterpillar mixers were characterized by SEM and optical profilometric measurements.

Figure 7 shows the profile of microchannel fabricated by micro spark erosion and micromilling, respectively. The channel of the mixer made by micro spark erosion is 'V'-shaped, has a depth of 255 μm, and the walls are smooth. In contrast to this result, the mixer made by micromilling has rough walls, the channel is 'U'-shaped, and has a depth of only 170 μm.

This result is – on a first sight – surprising since it stands to reason to expect that the more precise microstructure will give a better-defined laminated fluid system. However, parasitic mixing effects may be more important than originally

494

supposed. In this context, it may be possible that the rough surfaces of the mixer made by micro milling induce eddies in the fluid resulting in a higher mixing quality compared to a mixer with a smooth surface. Furthermore, fluid wakes in the small and narrow grooves of the eroded channel may enhance the average mixing time and may reduce the mixing quality. However, deeper insight in the hydrodynamics of fluid flow in the respective caterpillar structures can only be gained by visual observation in a transparent structure, an important R & D future task.

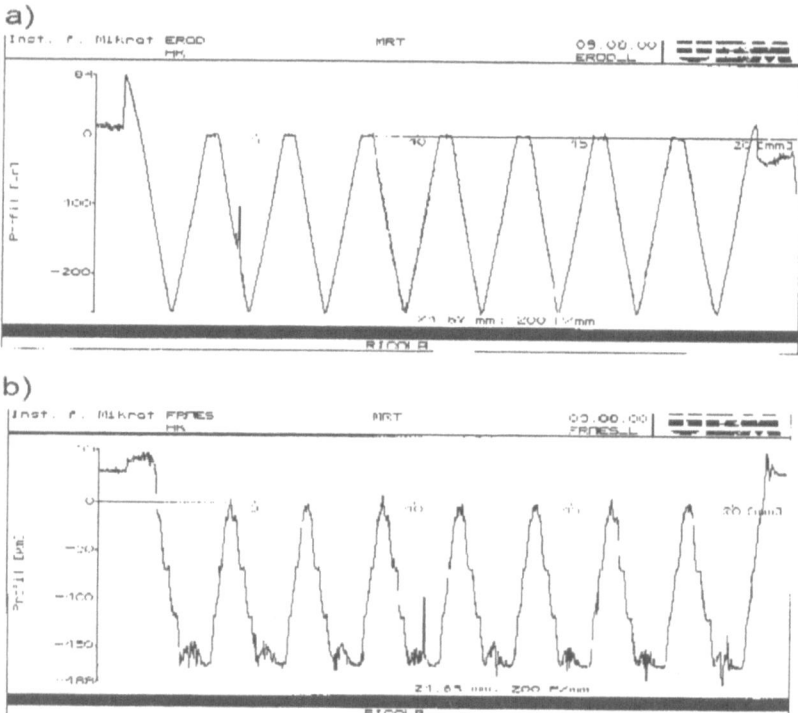

Fig.7:   Profiles of channels, measured along the channel, i.e. the fluid flow axis of a caterpillar mixer which are fabricated by a) micro spark erosion and b) micromilling.

3.2. Fouling investigations

In a next step, the forced precipitation of calcium carbonate, being an interesting material for medical applications, was carried out. At the beginning of the investigations, a stable operation was achieved only for a few minutes, being accompanied with plugging of the mixer outlet by crystalline calcium carbonate particles. Changing the composition of the solutions and other process parameters resulted in a longer processing time, up to ten minutes. For another forced-precipitation system, namely the formation of copper oxalate, this time turned out to be sufficient to take samples by SFTR operation and demonstrated the feasibility of the micromixers as laboratory tools [8].

Nevertheless, long-term operation still was hindered by fouling, in particular, along the streamlines in the triangular chamber (Fig. 8). $CaCO_3$ precipitates immediately when the $Ca(NO_3)_2$ (c = 40 mmol) and K_2CO_3 (c = 40 mmol) aqueous feeding solutions are contacted. The precipitation occurs mainly at the outlet of the interdigital structure. Blocking of the mixer happens after about 3 minutes at a total flow rate of 2000 ml/h. Applying ultrasound (frequency = 35 kHz) during the operation led to a about twofold increase of operation time until blocking occurs.

a)

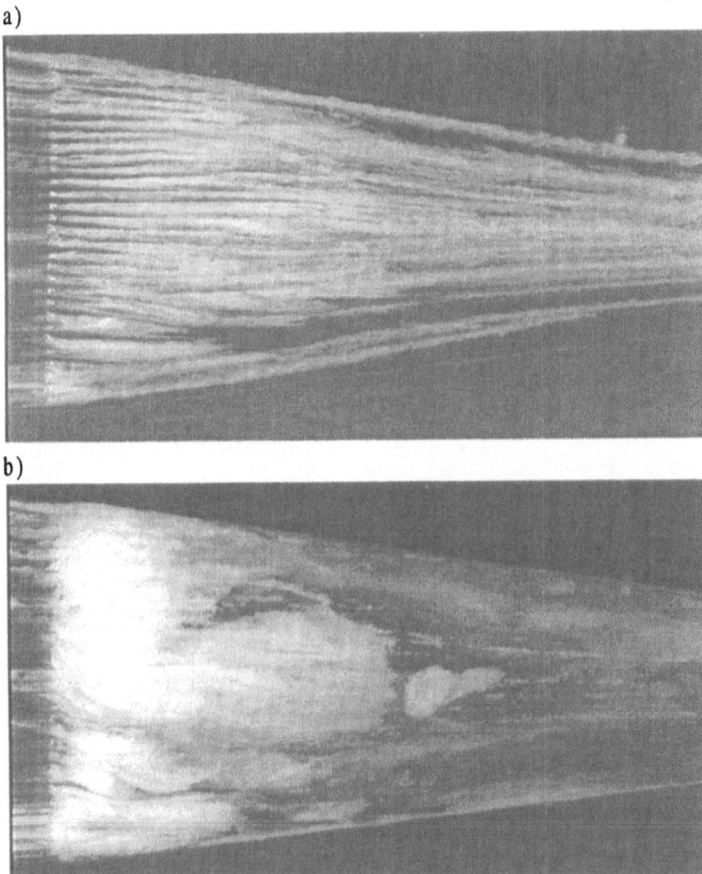

b)

Fig. 8: Deposition of calcium carbonate along the streamlines of reactant solution flow in a tri-angular-shaped interdigital glass micromixer: after (a) 30 s and (b) 180 s.

This type of self-clogging was attributed to the intimate mixing characteristics of these devices. At standard flow conditions mixing occurs through diffusion. Additional induced convection by ultrasound leads to enhanced mixing and to the mobilization of particles. Overall, the maximum process time does not exceed 10 minutes. As a consequence, caterpillar mixers were tested which do not rely on feeding fluid streams through very small openings. Indeed, it was achieved, by re-designing the mixing chamber, that fouling was excluded for several hours from the major part of the mixing chamber (see below).

First exploratory investigations of fouling were performed with a caterpillar mixer made of Poly(methyl methacrylate) (PMMA). This mixer allows the visible observation of fouling. Figure 9a shows the precipitation of $CaCO_3$ after an operation time of 3 min at different process conditions. In the case of processing at standard conditions (room temperature, without ultrasound), the mixer was blocked after this time. Applying ultrasound reduced the deposition of $CaCO_3$ (Fig. 9b) significantly. An additional enhancement of temperature to 50 °C led to a further reduction of the deposition of $CaCO_3$ (Fig. 9c) due to the increased solubility of $CaCO_3$. Despite this technical progress, the possible process time was limited to 10 minutes.

a)

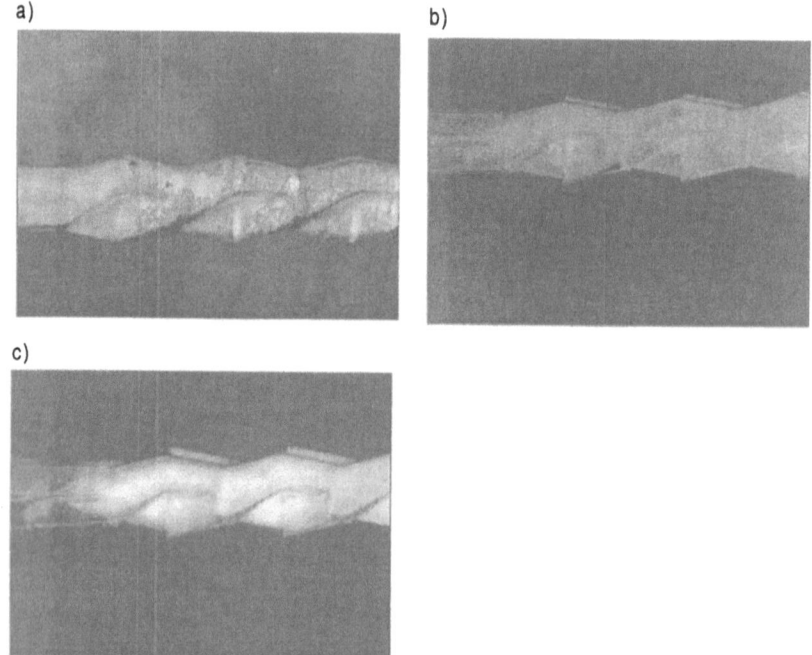

b)

c)

Fig. 9:   Precipitation of CaCO3 in a caterpillar mixer made of PMMA after a process time of 3 min a) standard conditions, b) room temperature + ultrasound, and c) T = 50 °C + ultrasound.

The following fouling investigations were performed with a caterpillar mixer of stainless steel. This device has mixing channels larger than those of the mixer made of PMMA. This increase in characteristic width should enhance the process time. To monitor the extent of fouling, the pressure drop was measured as a function of time. It was expected that the pressure drop rises continuously with the amount of precipitated $CaCO_3$ in the mixer. Figure 10 shows several of such plots at different process conditions with regard to $CaCO_3$ precipitation. In contrast to the expectation, the pressure drop stays at a low level for a long time and increases in short period before blocking dramatically. In all cases, the precipitation of $CaCO_3$ occurs mainly in the mixing structure, and not in the delay loop, which leads to blocking. All variations of conditions like the change of temperature, applying ultrasound, and current lead to a reduction of the maximal process time compared to the best result under standard conditions.

It turned out that stainless steel mixers were least impacted by fouling, whereas polymeric mixers exhibited a higher degree of particle deposition, including PTFE devices. Figure 10 reveals that using steel mixers operation times up to 5 hours were realized, which is significantly longer as compared to the state-of-the-art mixer at the beginning of the project, namely the Vortex mixer (40 minutes). Applying ultrasound and electric current obviously improved mixing, thereby increasing the contents of precipitates.

A detailed analysis of the morphology and composition of the calcium carbonate particles made by the caterpillar mixers revealed that during long-term operation the precipitated product is influenced by the deposited layer inside the mixer. This is not desired since the otherwise favorable SFTR operation has no impact on product quality in this case. Hence current investigations aim to minimize this effect further.

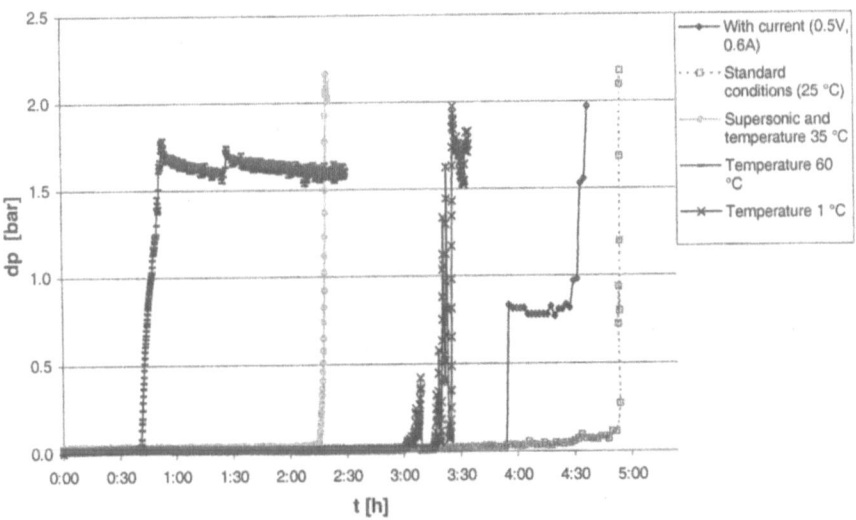

Fig. 10: Pressure drop versus operation time for forced precipitation of calcium carbonate in a stainless steel caterpillar mixer. An increase in pressure drop indicates precipitation, i.e. fouling.

4. Conclusion

The presented first results of the integration of micromixers in the case of a $CaCO_3$ powder production process enhances the process time from 40 minutes, which is obtained with a conventional mixer, to 5 hours. Nevertheless, there remains a need for further enhancement of process times. So other types of mixers are currently under investigation. There is first experimental evidence that these microstructured devices show reduced fouling characteristics as well.

The present results show that the aim of micromixing is not exclusively dedicated to achieve fast mixing. A further application field, as shown here, is the controlled and precise dosing of reactants aiming at a "smooth" mixing process, here to avoid particle clogging. Hence ranking of the devices described according to their mixing quality and processing reliability is opposite in this special case. This nicely supports the need for developing different, individually adapted microdevices. Thereby, even precipitation processes on an industrial scale are going to become accessible in near future to reactors equipped with tiny microstrucures [2], in contrast to the simple suggestion of the irreconcilable nature of both features.

Further information concerning the Segmented Flow Tubular Reactor are given in the internet on the page www.bubbletube.com.

5. Acknowledgement

For financial support from the European Union (Contract Number: G5RD-CT1999-00123) is gratefully acknowledged..

6. References

[1]   Jongen, N., Lemaitre, J., Bowen, P., Hofmann, H.; *Proceedings of the "5th World Congress of Chemical Engineering"*, 1996; pp. 31-36; San Diego, CA

[2]   EU Project G5RD-CT-1999-00123

[3]   Löwe, H., Ehrfeld, W., Hessel, V., Richter, T., Schiewe, J.; *Proceedings of the 4th International Conference on Microreaction Technology, IMRET 4*, Atlanta, USA, pp. 43-44 (2000)

[4]   Herweck, T., Hardt, S., Hessel, V., Löwe, H., Hofmann, C., Weise, F., Dietrich, T.R., Freitag, A.; *Proceedings of the 5th International Conference on Microreaction Technology*, Strasbourg, France, in press

[5]   Ehrfeld, W., Golbig, K., Hessel, V., Löwe, H., Richter, T.; Ind. Eng. Chem. Res. **38**, 3 (1999) 1075-1082

[6]   Ehrfeld, W., Hessel, V., Löwe, H.; *Microreactors*, VCH-Wiley, pp. 64-73, pp. 164-166 (2000)

[7]   Villermaux, J., Falk, L., Fournier, M.-C., Detrez, C.; AIChE Symp. Ser. 286 (1999) 6

[8]   Hessel, V., Ehrfeld, W., Löwe, H., Schiewe, J., Donnet, M.; "Nano- and Microscale Materials in Microreactors", *Proceedings of the MicroMat 2000*, Berlin, to be published

Microdevice-Based System for Rapid Catalyst Development

R.S. Besser, X. Ouyang, J. Fort, H. Surangalikar
Chemical Engineering and Institute for Micromanufacturing
Louisiana Tech University
Ruston, LA 71272
(318)257-5134
(240)255-4028 (fax)
rbesser@coes.latech.edu

Abstract

The following describes the implementation of a silicon microreactor as a tool for characterizing materials for use as heterogeneous gas-solid catalysts. We developed a complete system that can replace typical laboratory reactors that are known to have higher infrastructure requirements and to operate more sluggishly than microreactor systems. Using a model hydrogenation reaction, the microreactor system proved to generate catalytic data equivalent to literature reports in a timely, cost-effective manner. These results show the viability of microreactors for laboratory reaction systems for rapid and efficient catalyst development.

Keywords

catalysis, catalyst, heterogeneous, microreactor, microsystem, microfluidic

Introduction

Catalytic processes are largely responsible for the successful expansion of the international chemical industry. Because most chemical products are derived by catalytic processes, the impact of catalysis has been to make available a diverse array of low-cost versatile products in essentially every industry including food, plastics, textiles, pharmaceuticals, agriculture, etc.

As a result of this impact, new catalysts and improved versions of existing catalysts are continually sought. New catalyst discoveries will lead to the development of chemical processes with even greater efficiency in producing critical products. Moreover, chemicals now considered scarce because they are too expensive to produce in volume may become more widely available at reasonable cost.

The economic impact of catalyst technology is evident from the leverage that catalysts exert on the chemical process industry. The total annual investment in commercial catalysts by chemical producers results in revenues from catalytic processes far exceeding the investment. This factor was recently estimated in the U.S. to be sixty or so [1]. Because of this leverage, even incremental improvements in catalyst efficiency will result in substantial increases in profitability as a given catalyst investment will produce greater revenues through increased yield, less waste, and lower energy requirements.

Catalyst science has moved from a purely empirical to a more theory-based discipline, especially over the past three decades or so [2]. Now, models based on molecular-dynamics and other computational methods are aiding prediction and pre-selection of candidate catalysts [3]. Combinatorial methods have also been adapted in an attempt to accelerate catalyst screening, however, efforts to date have mostly focused on schemes that assess catalytic activity without directly measuring conversion [4, 5]. The most heavily relied upon method of catalyst assessment today still consists of measuring reactant conversion under conditions closely matching those expected in the ultimate industrial process system.

Traditionally, catalyst screening has been performed in small laboratory reactors. For gas-solid catalysis, these reactors typically consist of a tube packed with a catalyst sample and operated as a plug flow reactor. The reactor is interfaced to suitable equipment for control of temperature, pressure, and flow. Conversion and selectivity are monitored with analytical instruments. These experimental setups typically require moderate laboratory space and require support personnel for their setup and monitoring during an experiment. In addition to time for setup, adequate time is needed for stabilization whenever reaction conditions are modified. It is not uncommon for this stabilization time period to last several hours, and often only a single set of experimental conditions can be explored per day. As a result, the experimental evaluation of a single catalyst may require a week or more.

The desired characteristics of laboratory reactors were defined in a frequently cited paper by Weekman over twenty-five years ago [6]. These characteristics relate to the quality and applicability of the data obtained by using the reactor, as well as the relative convenience and cost of operation. It was a goal of this work to construct a laboratory reaction system that would possess these attributes as applied to the screening of candidate catalysts and for taking kinetic data useful for the large-scale reactor design. We determined to incorporate a chemical microdevice for the reactor in this system in order to exploit the many advantages of microscale systems.

The advantages of chemical microreactors have been well documented in the literature [7]. Included in these advantages and of particular interest to this project are the following:

1. High surface-to-volume ratio that suppresses heterogeneous, free-radical chain reactions in favor of the heterogeneous reactions under study. The large surface area relative to volume also facilitates thermal management.

2. Precise control over geometry of the fluidic structures on the device resulting in a well-defined and narrow distribution of residence time.

3. Fast equilibration due to minuscule heat and mass transfer resistances that derive from microscale geometry.

4. Materials of construction that are robust mechanically and chemically under harsh conditions of elevated temperature and pressure. These materials can be manipulated and built into practical devices with a well-known and versatile set of fabrication processes. We chose to adopt conventional silicon micromachining procedures for these reasons.

5. Compatibility with a variety of catalyst formation processes including physical and chemical vapor deposition, sol-gel deposition, solution impregnation, and nanoparticle layered assembly.

6. Low consumption of reactants.

7. Inherent safety of operation due to minimal likelihood of explosion and reduced hazard of toxic leaks.

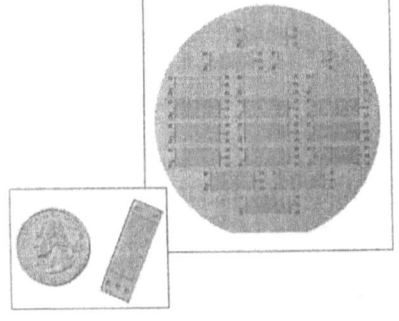

Figure 1: Microreactor device and 100-mm microreactor wafer.

Experimental

Several system design parameters were defined at the outset in order to be successful in fabricating a microreactor for gas-solid catalytic process evaluation. Maximum temperature of operation was set at 520 K as the goal for an initial design, with 770 K as the ultimate goal. Likewise, the initial pressure specification was taken as 5 bar, with an ultimate goal of 30 bar. Other factors affecting the design were the need for catalysts to be flexibly and rapidly deposited to facilitate quick experiment turnaround. A further consideration was for the microreactor to be based on a standard chip footprint to allow future designs to take advantage of the peripheral equipment setup constructed for the initial system.

The outcome of the design process was a simple microreactor module consisting of a bulk micromachined silicon chip with a glass cover (Figure 1). The chip is 1 x 3 cm^2 in size and has a reaction zone consisting of a number of parallel microchannels coated with catalyst.

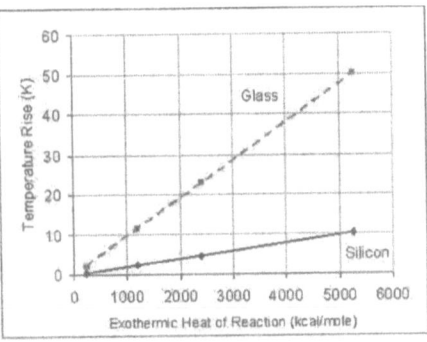

Figure 2: Results of solving 2-D steady-state heat equation for reaction zone cross-section.

Devices with 100-μm and 5-μm channel widths have been fabricated. Gas-phase reactant chemicals are introduced from the bottom side of the chip, passing upward through vias that connect to access channels leading to the reaction zone. Similar vias take the reaction products out the bottom side of the exit end of the

502

chip where they can be analyzed. The layout can accommodate up to three separate feeds and two separate product streams.

The heat transfer characteristics of the microreactor were assessed with a simple thermal model. A two-dimensional finite difference approach was taken to obtain a solution to the steady-state heat equation with suitable boundary conditions. The effect of enthalpy of reaction was investigated by assuming that the heat generated was dissipated uniformly around the perimeter of the microchannel cross-section and ignoring advection. We found that the reactor can manage very large heats of reaction without appreciable temperature increase. The graph in Figure 2 shows that even exothermic heats as high as 5000 kcal/mole result in a manageable temperature of only tens of degrees or less above the temperature of the heat sink supporting the device.

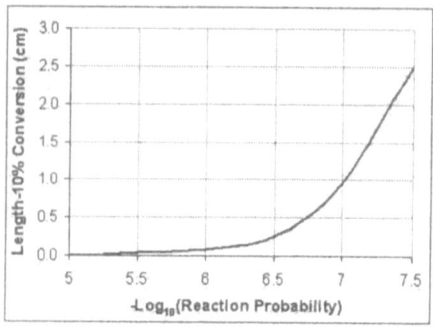

Figure 3: Results of conversion model. The 1.9-cm reaction zone will support 10% conversion at RP as small as $10^{-7.3}$.

The microreactor was sized to accommodate a range of reaction probability (RP). The RP is defined as the probability that an impinging reactant molecule will undergo transformation on any given collision with the catalyst surface and varies between 10^{-5} and 10^{-10} or so for many classes of hydrocarbon reactions. RP values for reactions and catalysts of importance are documented in many literature sources [2]. A simple model for a continuous flow reactor taking into account geometry, temperature, pressure, and flow rate was used to determine conversion as a function of RP. The result of this analysis for typical flow conditions and reactor geometry is shown in Figure 3. We took 10% as an arbitrary level of conversion at which catalysts can be reasonably compared. From the figure, we see that the 1.9 cm reaction zone length supports 10% conversion for reactions with RP as small as $10^{-7.3}$ and larger.

The requirement that the microreactor be fabricated with few steps and high yield led to the selection of well-known bulk micromachining processes [8,9]. The

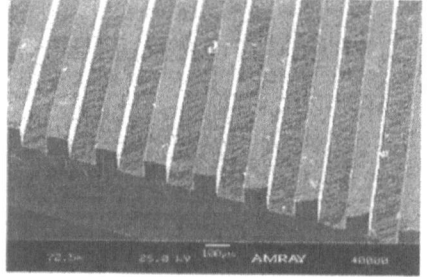

Figure 4: SEM image of microchannel reaction zone cross-section. Bar is 100μm.

process steps, described previously [9], are based on simple photolithography and etching of a silicon dioxide layer that acts as a mask during subsequent KOH etching of silicon [10]. Four-inch silicon wafers of (110) orientation were used in order to produce vertical microchannel sidewalls as illustrated in Figure 4. The device photomask set incorporates several options to allow for flexibility in fabrication. As few as two masking steps can be used to complete the silicon process, however, greater control over the depth of individual fluidic structures on the device can be gained by using up to four separate masking steps.

Catalyst films are deposited in the microreactors after the fluidic passages are formed in the silicon. This step may be completed on all devices on the wafer simultaneously (eighteen microreactors per wafer), or on individual chips after

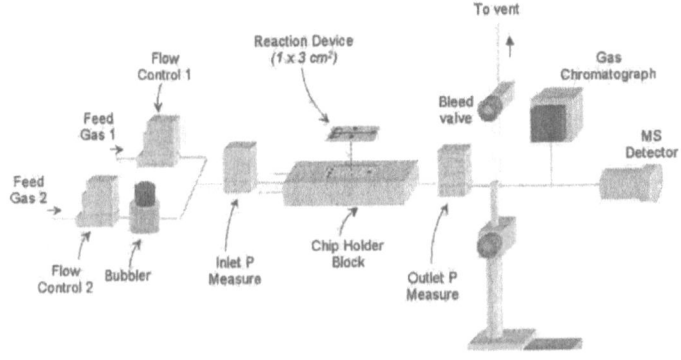

Figure 5: Schematic of catalyst characterization experiment.

sawing the wafer. For the model reaction study described later in this paper, a Pt film 20 nm thick was deposited by DC magnetron sputtering. The final step of fabrication is sealing the chip with a glass cover. Anodic bonding was used to bond the Pyrex glass cover to the silicon. The resulting seal was found to be hermetic under all conditions tested (temperature and pressure as high as 570 K and 7.9 bar).

The microreactor was interfaced to an experimental setup for controlling temperature, introducing and removing reactants and products, and analyzing the composition of the gas streams. A diagram of the setup is shown in Figure 5. The block supporting the microreactor was custom fabricated. It is fitted with cartridge heaters and thermocouples for temperature control. The gas passages in the block seal directly to the bottom side of the silicon chip with 2-mm silicone o-rings. The flow controllers and pressure gauges are commercially available components. Chemical analysis was performed by an Extrel C-50 research grade mass spectrometer housed in a custom-built vacuum chamber. A Varian Saturn gas chromatograph has been connected but was not used in the experiments described in this paper.

We conducted conversion experiments in order to assess the usefulness of the system for determining catalytic activity. The hydrogenation of cyclohexene (C_6H_{10}) over platinum catalyst was selected as a model reaction for these experiments. This reaction is representative of the important class of hydrocarbon hydrogenation reactions ubiquitous in the chemical and petroleum processing industries.

In the experiment described here, C_6H_{10} vapor was introduced into the reactor by bubbling argon through the liquid hydrocarbon. The reaction conditions were the following: 0.1 sccm of hydrogen, 0.1 sccm of argon saturated with C_6H_{10}, 1.01 bar pressure, and temperature settings applied in the order

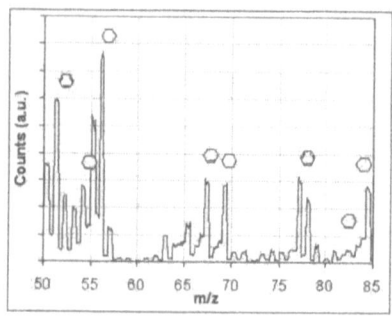

Figure 6: Typical mass spectrum with signature peaks labeled.

473 K, 523 K, and 423 K. Temperature was held constant for a period of six hours before adjusting to a new setting.

A typical mass spectrum taken during the experiment is shown in Figure 6. The spectrum indicates the presence of the cyclohexene reactant, and cyclohexane (C_6H_{12}), the hydrogenation product. Also visible in the spectrum is evidence for the dehydrogenation product, benzene (C_6H_6).

Table I shows the measured conversion during the temperature cycles. We observe a monotonic decrease in conversion with increasing time on stream, irrespective of temperature. It was suspected that this decrease is due to the dominance of catalyst deactivation on conversion.

Table I: Measured reactant conversion during the progress of the cyclohexene hydrogenation experiment

Cycle	Temperature (K)	Time (h)	Conversion (%)
1	473	6	91
2	523	6	74
3	423	6	51

To examine whether surface chemical changes led to deactivation of the catalyst, we removed the Pyrex cover and performed x-ray photoelectron spectroscopy (XPS) analysis on the reaction zone area. XPS analysis combined with sputter depth profiling revealed that the platinum signal is absent upon initial survey of the surface. A high carbon signal in the near-surface depth range indicated the presence of a carbonaceous overlayer. The formation of this layer is consistent with a decrease in catalytic activity due to the elimination of active catalytic surface available to participate in the reaction. This phenomenon (known as "coking") is well known in catalytic hydrocarbon reactions [11].

The measured conversion data permit the determination of catalytic parameters that can be compared to literature values. We consider the conversion for the 473 K case since the reaction was as yet relatively unaffected by catalyst deactivation. The RP responsible for this conversion level was determined to be $1 \cdot 10^{-6.2}$. This result was then used in conjunction with the density of active catalytic sites to calculate turnover frequency (TOF). The site density was estimated for the sputtered platinum layer by assuming a perfectly planar morphology and taking every surface atom as an active site. The TOF calculated thereby was 1.7 molecules/site/s. These values of RP and TOF compare well with literature values for hydrogenation reactions [12].

Discussion

The above results demonstrate the potential of the experimental system as a tool for rapid catalyst development. An important aspect of the demonstration is that data were obtained under conditions approximating an industrial reaction. Data quality was ensured by adequate levels of single-pass conversion and throughput. The catalytic reaction parameters obtained (RP, TOF) allow catalyst candidates to be confidently compared with literature values, or with one another in screening experiments.

A second important attribute is the ability to conduct experiments rapidly. We used six-hour cycles in this experiments to make certain we had achieved steady-state operation. The data taken in the experiment made it clear that the cycle length could be reduced without loss of quality. Because of the fast heat and mass transfer kinetics, the limiting factor in turnaround time is the ability to fill the system volume with chemical streams at low flow rates. We are presently working to eliminate excess volume in the system to verify that turnaround times can be reduced to only minutes per data point.

Ultra-low consumption of reactants was verified in the above experiment. In the 18 hour time period of the experiment, a total of less then 32 mg of cyclohexene was consumed. This low usage means that the cost of reactants is not a limitation in these kinds of experiments. Studies on expensive fine or pharmaceutical reactants will be significantly more economical than with conventional methods. Moreover, costly isotopically marked chemicals can be used routinely. The low volume of chemical usage also implies the inherent safety of the system from the standpoint of explosion or toxicity hazard. Potential negative impact on the environment is similarly minimal.

Another useful attribute of the system is the ability to rapidly perform surface analysis on the catalyst for correlation with reaction results. The XPS data above support the conclusion that deactivation was due to the formation of a carbonaceous overlayer that had deposited on the catalyst. The rapidity of obtaining the spectroscopic result was due to the ability to easily remove the microreactor from the experimental system and to quickly remove the cover. The initial surface spectroscopic information was available within minutes of completing the reaction experiment. This aspect of rapid post-reaction characterization applies to other analyses as well, for example BET surface area measurement.

Conclusions

In summary, we have demonstrated a flexible, microsystem-based test platform for heterogeneous catalysis development. The design, modeling, and fabrication approaches used for the silicon micromachined reactor chip led to a useful module that was integrated with analysis and flow control equipment to form the complete system. The model hydrogenation reaction that we characterized yielded catalytic reaction parameters that compare well with literature values. The system was also useful for gaining understanding of the catalyst surface after the reaction as the entire reactor was able to be quickly loaded into a surface analysis system for spectroscopy. The experimental reaction system was shown to have faster turnaround, less chemical usage, and greater safety than conventional laboratory systems.

Acknowledgements

The authors gratefully acknowledge the assistance of M. Prevot, W. Yang, S. Zhao, S. Williams, S. Moncrief, F. Jones and M. Vasile. This work was performed under the support of the Louisiana Board of Regents, LEQSF-RD-A-20.

References

1.   J.D. Hewes, L.P. Herring, M.A. Schen, B. Cuthill, and R. Sienkiewicz, "Combinatorial Chemistry-The Discovery of Catalysts and New Materials," *ATP Position Paper, Advanced Technology Program*, National Institute of Standards and Technology, Gaithersburg, MD, 1998.

2.   G.A. Somorjai, *Introduction to Surface Chemistry and Catalysis*, (Wiley, New York, 1994), 1st ed., Chap. 7, pp.444-5.

3.   L. Deng, T. Ziegler, and L. Fan, "Computer Design of Living Olefin Polymerization Catalysts: A Combined Density Functional Theory and Molecular Mechanics Study," *Organometallics*, 17 (15), 3240 (1998).

4.   F.C. Moates, M. Somani, J. Annamalai, J.T. Richardson, D. Luss, and R.C. Willson, "Infrared Thermographic Screening of Combinatorial Libraries of Heterogeneous Catalysts," *Ind. Eng. Chem. Res.*, 35 (12), 4801-4803 (1996).

5.   S.M. Senkan and S. Ozturk, "Discovery and Optimization of Heterogeneous Catalysts Using Combinatorial Chemistry," *Angew. Chem. Int. Ed.*, 38, 791 (1999).

6.   V.W. Weekman, "Laboratory Reactors and Their Limitations," *AIChE Journal*, 20, 833 (1074).

7.   For an exhaustive coverage of the subject, see W. Ehrfeld, V. Hessel, and H. Lowe, *Microreactors: New Technology for Modern Chemistry*, Wiley, New York, 2000, and references contained therein.

8.   M. Prevot and R.S. Besser, "Fabrication Process for a Microreaction Device," *Abstracts of the Materials Research Society Spring Meeting*, Materials Research Society, p. 491, 2000.

9.   R.S. Besser and M. Prevot, "Linear Scale-Up of Micro-Reaction Systems," *Topical Conference Proceedings of the Fourth International Conference on Microreaction Technology*, American Institute of Chemical Engineers, pp. 278-283, 2000.

10. For a review on the subject, see M. Madou, "Wet Bulk Micromachining," *Fundamentals of Microfabrication*, (CRC Press, New York, 1997), 1st ed., Chap. 4, and references contained therein.

11. Reference [2], p. 451.

12. G. A. Somorjai and J. Carrazz, "Structure Sensitivity of Catalytic Reactions," *Industrial Engineering Chemistry Fundamentals*, **25**, 63 (1986).

The Synthesis of Peptides using Micro Reactors

Paul Watts,[a] Charlotte Wiles,[a] Stephen J. Haswell,[a] Esteban Pombo-Villar[b] and Peter Styring[c]

[a] Department of Chemistry, The University of Hull, Cottingham Road, Hull, UK, HU6 7RX.
[b] Nervous System Research, WSJ-386.07.15, Novartis Pharma Ltd., CH4002, Basel, Switzerland.
[c] Department of Chemical and Process Engineering, University of Sheffield, Mappin Street, Sheffield, UK, S1 3JD.

Abstract

The synthesis of peptides has been successfully performed using a borosilicate glass micro reactor, in which a network of channels has been produced using a photolithographic and wet etching method. The reagents were mobilised by electroosmotic flow (EOF) and electrophoretic mobility. In addition, microporous silica frits were used in the micro reactor channels, these not only aided the flow mechanism but also removed the problem of hydrodynamic pressure within the reagent reservoirs.

The micro reactor was evaluated using a carbodiimide coupling reaction of Fmoc-β-alanine and Dmab protected β-alanine to form the dipeptide. However, the methodology has been extended such that the peptides can be produced *via* the pentafluorophenyl ester derivatives of the amino acids. It was found that performing these reactions in the micro reactor for 20 minutes, in order to acquire sufficient volume (*ca.* 10 μl) for detection, resulted in a two-fold increase in the reaction efficiency over the batch method.

It has also been demonstrated that selective deprotection of the resultant dipeptides can be achieved and further peptide bond forming reactions may be performed, resulting in the synthesis of longer chain peptides.

Introduction

During the past ten years, there has been a rapid growth in the development of micro-Total Analytical Systems (μ-TAS)[1-5] which exploit electroosmotic flow (EOF).[6-9] The development of micro reactor devices for chemical synthesis based on complimentary technology is less common. However, recent research has shown that Suzuki[10] and Wittig[11] reactions may be performed using micro reactor systems.[12] A micro reactor is generally defined as a series of interconnecting channels (10 to 300 microns in diameter), formed in a planar surface, in which small quantities of reagents are manipulated. The reagents can be brought together in a specific sequence, mixed and allowed to react for a specified time in a controlled region of the reactor.

Since the discovery by Merrifield in 1963,[13] peptides have been commonly prepared *via* solid supported techniques. Solid phase peptide synthesis (SPPS) is based on the addition of a protected amino acid residue to an insoluble polymeric support. The acid-labile Boc group[14] and base-labile Fmoc group[15] have been commonly used for *N*-protection. After removal of the protecting group the next protected amino acid may be added using either a coupling reagent or a pre-activated amino acid derivative. If this dipeptide is the desired product, it may be cleaved from the polymer support using various reagents, one of the more common methods being treatment with HF.[16] If a longer peptide is required additional amino acids can be added by repeating further coupling reactions.

Solid phase peptide synthesis has the disadvantage that a fairly expensive polymer support is required. In addition, extra steps are added to the synthesis as a result of initially linking the amino acid to the support and finally having to remove the peptide from the polymer. In this paper, a micro reactor has been used to prepare peptides using solution phase chemistry in an attempt to overcome some of the current problems associated with such syntheses.

We specifically demonstrate the multi-step synthesis of peptides derived from β-amino acids for two reasons, firstly, the simplest β-amino acids lack chiral centres. Hence we avoid potential problems with racemisation and generation of diastereomeric mixtures which would complicate the analysis of the products in our initial studies. In addition, β-Peptides have attracted much interest due to their structural[17,18] and biological properties.[19] In particular, their stability to degradation by peptidases[20,21] makes them potentially superior to the drugs derived from α-amino acids. Both solution and solid phase synthetic methods for these have been developed which would serve to benchmark the micro reactor synthesis.[22,23]

Fabrication of micro reactors

A number of materials such as silicon, quartz, glass, metals and polymers can be used to construct micro reactors. The micro reactor devices used in this work were prepared from borosilicate glass using standard procedures developed at Hull, as illustrated in Figure 1.[12,24] Microporous silica frits[25] were placed in the channels to prevent hydrodynamic flow occurring.

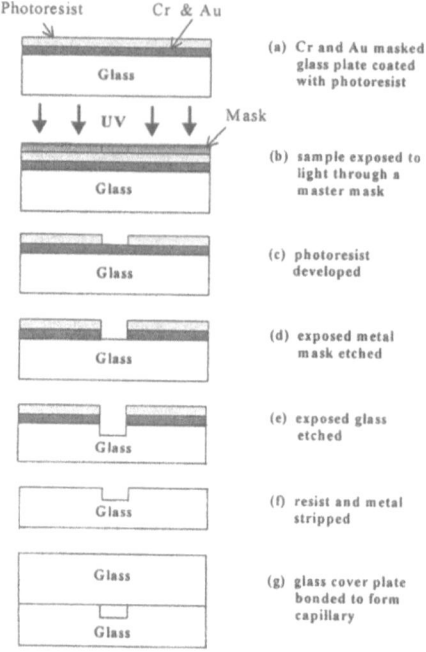

Figure 1. Fabrication of a micro reactor

Electroosmotic flow (EOF) was used to move the reagents and solvents around the system. To illustrate the principles of EOF, one can consider a micro channel fabricated from a material having negatively charged functional groups on the surface. If a liquid, displaying some degree of dissociation, is brought into contact with the surface, positive counter ions will form a double layer such that the positively charged ions are attracted to the negatively charged surface. Application of an electric field causes this layer to move towards the negative electrode, thus causing the bulk liquid to move within the channel (Figure 2).

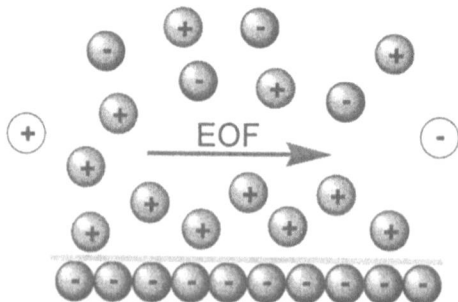

Figure 2. Principle of electroosmotic flow.

Results and Discussion

In the first instance, a one-step reaction was considered in which an *N*-protected β-amino acid was reacted with an *O*-protected β-amino acid, to prepare the protected β-dipeptide. To enable the methodology to be applicable to the synthesis of more complex peptides, the use of orthogonal protecting groups was clearly required. After careful consideration, the base-labile Fmoc protecting group[15] was selected for *N*-protection while the Dmab ester[26] was chosen for protection of the carboxylic acid. Importantly, both protecting groups may be removed under mild conditions, since electroosmotic flow is retarded if the pH of the reaction is outside the range 3-10.

Commercially available Boc-β-alanine **1** was protected as the Dmab ester using an EDCI/DMAP coupling reaction, to give the ester **2** in 92 % yield in a bulk reaction (Scheme 1). Treatment of **2** with trifluoroacetic acid furnished the desired amine **3** in 61 % yield, which was subsequently reacted with Fmoc-β-alanine **4** *via* a carbodiimide coupling reaction, to give a synthetic sample of dipeptide **5**.

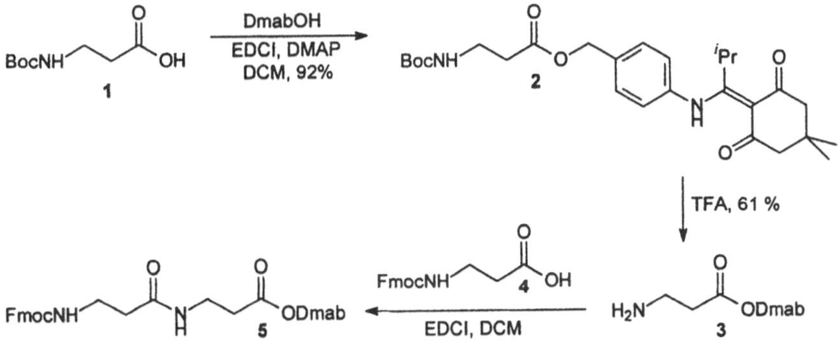

Scheme 1. Synthesis of standard dipeptide derivative.

Having prepared dipeptide **5**, it represented a synthetic target for preparation using the micro reactor. Prior to synthesis, the micro reactor channels were primed with anhydrous *N,N*-dimethylformamide (DMF) to remove any air and moisture from the channels and the microporous silica frits. A standard solution of Fmoc-β-alanine **4** (50 µl, 0.1 *M*) in anhydrous DMF was added to reservoir A, a solution of EDCI (50 µl, 0.1 *M*) was placed in reservoir B and a solution of amine **3** (50 µl, 0.1 *M*) was placed in reservoir C (Figure 2). Anhydrous DMF (40 µl) was placed in reservoir D, which was used to collect the products of the reaction. Platinum electrodes were placed in each of the reservoirs (A, B and C positive, D ground) and an external voltage was applied to the channels inducing electroosmotic flow of the reagents. The reactions were conducted at room temperature for a period of 20 minutes, in order to acquire sufficient volume of product to determine the yield of the reaction. Analysis was achieved by high performance liquid chromatography (Jupiter C_{18} 10 μm, 4.6 x 250 mm, mobile phase composition: 0.1 % trifluoroacetic acid in water and 0.1 % trifluoroacetic acid in acetonitrile, using a gradient system of 30 % aqueous to 70 % aqueous over 20 minutes, with a flow rate of 2.5 ml min^{-1} at room temperature).

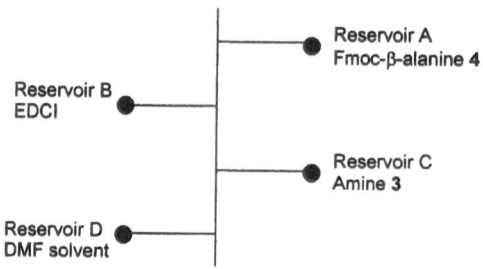

Figure 2. Schematic of the micro reactor used in EDCI coupling reactions.

When stoichiometric quantities of the reagents were used only *ca.* 10 % conversion to peptide **5** was achieved when a voltage of 700 V was applied to the reagents (A, B and C). By lowering the applied voltage to 500 V and hence increasing the residence time of the reaction, no significant increase in yield was observed. However, by using two equivalents of EDCI (0.2 *M* solution) the yield of the reaction was increased to *ca.* 20 %. By applying a stopped flow technique (2.5 sec injection length with stopped flow for 10 sec) the yield of the reaction was further increased to 50 %. Since the yield of reaction appeared to greatly depend on the number of equivalents of EDCI used, we wished to further investigate the effect of carbodiimide concentration on the reaction, however we found that EDCI was insoluble in DMF above 0.2 *M* concentrations. In further experiments dicyclohexylcarbodiimide (DCC) was used as the coupling reagent as it was considerably more soluble in DMF. Using 5 equivalents of DCC (0.5 *M* solution in reservoir B) a 93 % yield of dipeptide **5** was obtained using the optimised conditions described above.

Another common method utilised in peptide bond formation involves the reaction of a pre-activated amino acid derivative, such as a pentafluorophenyl ester, with an amine.[27,28] Fmoc-β-alanine **4** was activated as the pentafluorophenyl ester **6** *via* an EDCI coupling reaction (Scheme 2). The pentafluorophenyl ester **6** was stable and could be stored indefinitely in the freezer. The ester **6** was subsequently reacted in bulk with amine **3** to produce dipeptide **5**. It is however important to note that even after stirring the reagents for 24 hours at room temperature only about 40-50 % conversion to the dipeptide was observed.

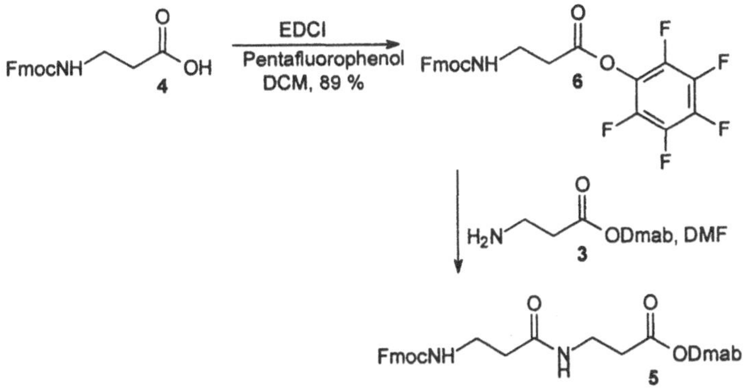

Scheme 2. Preparation and reaction of pentafluorophenyl esters.

Having prepared dipeptide **5** *via* the alternative pre-activated strategy, we wished to investigate if the reaction could be performed in a micro reactor. A standard solution of the pentafluorophenyl ester of Fmoc-β-alanine **6** (50 µl, 0.1 *M*) in anhydrous DMF was added to reservoir A, a solution of amine **3** (50 µl, 0.1 *M*) was placed in reservoir B and anhydrous DMF (40 µl) was placed in reservoir D, which was used to collect the products of the reaction. It was found that using continuous flow of both reagents, where the ester **6** was maintained at 700 V and the amine **3** was maintained at 600V, dipeptide **5** was produced in quantitative yield in just 20 minutes. This represented a significant increase in yield compared with the traditional batch synthesis.

Similarly, the reaction between the pentafluorophenyl ester **7** of Boc-β-alanine and amine **3** was also investigated in the micro reactor (Scheme 3). In this case, when the reagents were mixed using continuous flow, with both reagents maintained at 700V, again a quantitative yield of peptide **8** was observed.

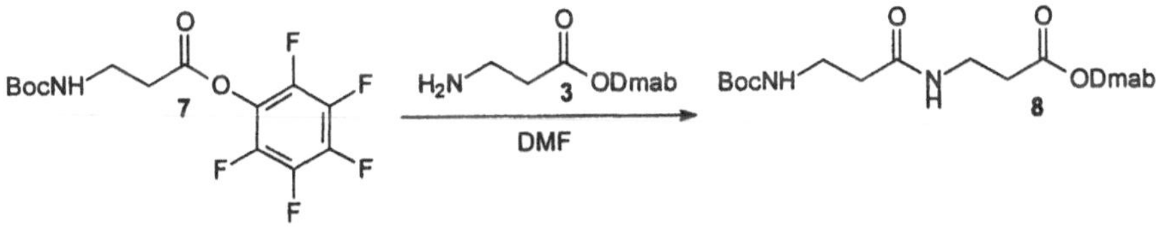

Scheme 3. Reaction of pentafluorophenyl esters of Boc-β-alanine.

The fact that pentafluorophenyl ester **7** needed to be maintained at a higher potential, in order to obtain a quantitative yield, demonstrates that the different pentafluorophenyl esters have different electroosmotic mobilities. Importantly, this result demonstrates that both Boc and Fmoc protecting groups are suitable for use in the preparation of peptides using micro reactors.

Having successfully demonstrated that peptide bonds could be formed in micro reactors, using two common methods, we wished to show that we could extend the methodology to the preparation of longer chain peptides. Consequently, we needed to be able to conduct deprotection reactions in the micro reactor and subsequently perform further peptide bond forming reactions. Fmoc-β-alanine **4** was converted into the Dmab ester **9**, in a bulk reaction, using standard conditions (Scheme 4). It was proposed to convert ester **9** into amine **3** by deprotection of the Fmoc group in the micro reactor and subsequently react the amine *'in situ'* with pentafluorophenyl ester **7**, to give the dipeptide **8**.

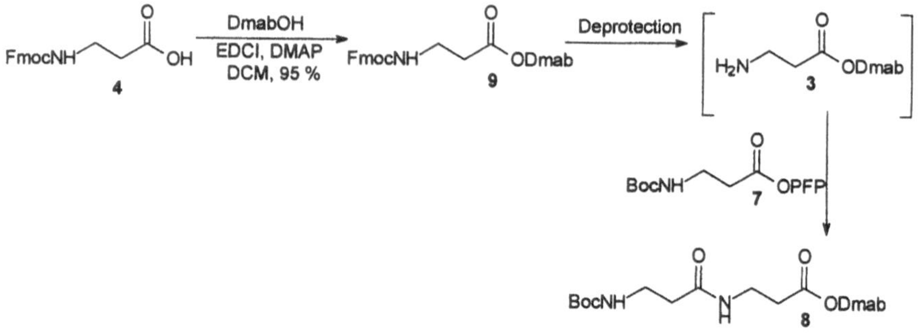

Scheme 4. Multi-step peptide synthesis.

Several methods for the deprotection of the Fmoc group are reported in the literature, the most common method being treatment with 20% piperidine in DMF.[29,30] Treatment of **9**, with 10 equivalents of piperidine in DMF using the micro reactor, resulted in 60-70 % deprotection over a 20 minute period, to give amine **3**.

Subsequently, a standard solution of the Dmab ester of Fmoc-β-alanine **9** (50 µl, 0.1 *M*) in anhydrous DMF was added to reservoir A, a solution of piperidine (50 µl, 1.0 *M*, 10 equivalents) was placed in reservoir B and a solution of pentafluorophenyl ester **7** (50 µl, 0.1 *M*) was placed in reservoir C, in an attempt to prepare dipeptide **8** using this multi-step approach. Anhydrous DMF (40 µl) was placed in reservoir D, which was used to collect the products of the reaction. The HPLC of the reaction mixture showed that Fmoc deprotection had occurred however no peptide was evident. It was however found that the excess piperidine used in the reaction was reacting with the pentafluorophenyl ester **7** to give amide **10** (Scheme 5).

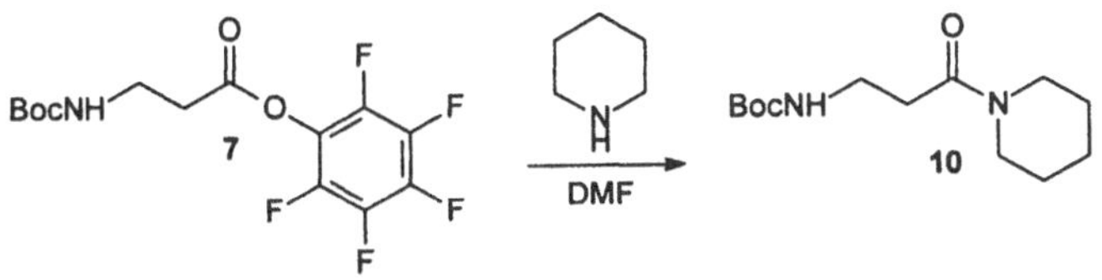

Scheme 5. Reaction of piperidine with pentafluorophenyl esters.

As a result, an alternative method of Fmoc deprotection was required that would not cause the aforementioned problem. Using the micro reactor, the Dmab ester of Fmoc-β-alanine 9 was reacted with one equivalent of DBU to give the free amine 3 which was then reacted with the pentafluorophenylester of Boc-beta-alanine 7, in an attempt to form the dipeptide 8.

In this case, when the reagents were mixed using continuous flow, with the reagents maintained at 700V, product 8 was observed in typically 25 % yield. Surprisingly, when the reagents were mobilised using a stopped flow technique, the yield of product was actually reduced. By comparing the flows of each reagent at this stage we were able to optimise the reaction. The Dmab ester of Fmoc-β-alanine 9 was maintained at 750V while reacted with DBU at 800V. The deprotected amine was then reacted, using continous flow, with the pentafluorophenyl ester of Boc-β-alanine 7 to give a conversion of 96 %, based on the amount of Dmab ester 9 present at the end of the reaction.

Having shown that more complex peptides could be produced by removal of the N-protecting group we wished to determine if we could remove the Dmab protecting group using hydrazine. Hence, a solution of the Dmab ester of Fmoc-β-alanine 9 (50 μl, 0.1 M) in anhydrous DMF was added to reservoir A and a solution of hydrazine (50 μl, 0.1 M) was placed in reservoir B. Anhydrous DMF (40 μl) was placed in reservoir D, which was used to collect the products of the reaction. Using continuous flow of both reagents, maintained at 700 V, quantitative deprotection was observed to give carboxylic acid 4.

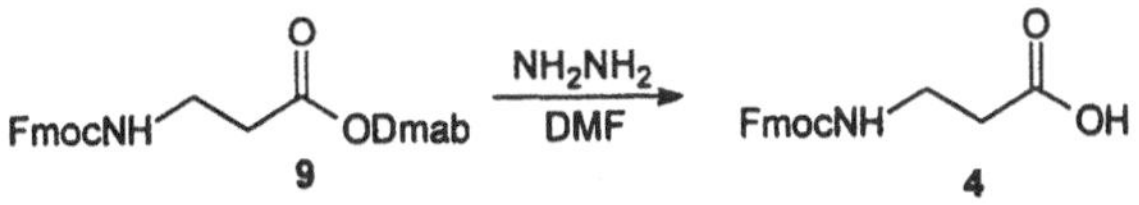

Scheme 6. Removal of the Dmab protecting group.

516

Conclusion

We have demonstrated the potential application of micro reactors to perform multi-step synthesis, allowing high throughput screening of biologically active peptides. Further studies to prepare more complex peptides are currently underway within our laboratories. In addition, new devices are being prepared which would allow 'combinatorial' approaches to be applied to the devices.

Acknowledgements

We wish to thank Novartis Pharmaceuticals (PW and CW) for financial support. We are grateful to Dr. Tom McCreedy (University of Hull) for help in fabricating the micro reactor devices.

References

1.   S. J. Haswell, *Analyst*, 1997, **112**, 1R.
2.   A. Manz, D. J. Harrison, E. Verpoorte and H. M. Widmer, *Adv. Chromatogr.*, 1993, **33**, 1.
3.   D. J. Harrison and A. van den Berg (eds.), *Proceedings of the Micro Total Analytical Systems 98' Workshop*, Kluwer Academic Press, Dordrecht, 1998.
4.   D. J. Harrison, K. Fluri, K. Seiler, Z. H. Fan, C. S. Effenhauser and A. Manz, *Science*, 1993, **261**, 895.
5.   S. C. Jacobson, R. Hergenroder, L. B. Koutny and J. M. Ramsey, *Anal. Chem.*, 1994, **66**, 1114.
6.   D. M. Spence and S. R. Crouch, *Anal. Chem.*, 1998, **358**, 95.
7.   B. H. Schoot, S. Jeanneret and A. Berg, *Anal. Methods Instrum.*, 1993, **1**, 38.
8.   P. K. Das Gupta and S. Lui, *Anal. Chem.*, 1994, **66**, 1792.
9.   P. D. I. Fletcher, S. J. Haswell, and V. N. Paunov, *Analyst*, 1999, **124**, 1273-1282.
10.   G. M. Greenway, S. J. Haswell, D. O. Morgan, V. Skelton and P. Styring, *Sensors & Actuators B*, 2000, **63**, 153.
11.   V. Skelton, G. M. Greenway, S. J. Haswell, P. Styring, D. O. Morgan, B. Warrington and S. Y. F. Wong, *Analyst*, 2001, **126**, 7.
12.   S. J. Haswell, R. J. Middleton, B. O'Sullivan, V. Skelton, P. Watts and P. Styring, *Chem. Commun.*, 2001, 391.
13.   R. B. Merrifield, *J. Am. Chem. Soc.*, 1963, **85**, 2149.
14.   P. Munster and W. Steglich, *Synthesis*, 1987, 223.
15.   B. Penke and J. Rivier, *J. Org. Chem.*, 1987, **52**, 1197.
16.   D. B. Whitney, J. P. Tam and R. B. Merrifield, *Tetrahedron*, 1984, **40**, 4237.

17. X. Daura, K. Gademann, H. Schaefer, B. Jaun, D. Seebach and W. F. van Gunsteren, *J. Am. Chem. Soc.*, 2001, **123**, 2393.

18. D. Seebach, J. V. Schreiber, S. Abele, X. Daura and W. F. van Gunsteren, *Helv. Chim. Acta*, 2000, **83**, 34.

19. M. J. Loeb, H. Jaffe, D. B. Gelman and R. S. Hakim, *Arch. Insect Biochem.*, 1999, **40**, 129.

20. K. Gademann, M. Ernst, D. Seebach and D. Hoyer, *Helv. Chim. Acta*, 2000, **83**, 16.

21. M. Werder, H. Hauser, S. Abele and D. Seebach, *Helv. Chim. Acta*, 1999, **82**, 1774.

22. P. I. Arvidsson, M. Rueping and D. Seebach, *Chem. Commun.*, 2001, 649.

23. J. V. Schreiber and D. Seebach, *Helv. Chim. Acta*, 2000, **83**, 3139.

24. T. McCreedy, *Anal. Chim. Acta.*, 2001, **427**, 39.

25. P. D. Christensen, S. W. P. Johnson, T. McCreedy, V. Skelton and N. G. Wilson, *Anal. Commun.*, 1998, **35**, 341.

26. W. C. Chan, B. W. Bycroft, D. J. Evans and P. D. White, *J. Chem. Soc., Chem. Commun.*, 1995, 2209.

27. L. Kisfaludy and I. Schon, *Synthesis*, 1983, 325.

28. E. Atherton, L. R. Cameron and R. C. Sheppard, *Tetrahedron*, 1988, **44**, 843.

29. L. A. Carpino and G. Y. Han, *J. Org. Chem.*, 1972, **37**, 3404.

30. L. A. Carpino, B. J. Cohen, K. E. Stephens, S. Y. Sadat-Aalaee, J-H Tien and D. E. Langridge, *J. Org. Chem.*, 1986, **51**, 3732.

Part 7

Analytical and Biological Applications

BIOCHIPS: *Lab-on-a-chip and DNA chips,*
new tools for high throughput genetic analysis and diagnostics.

Marc CUZIN, Alexandra FUCHS, Patrice CAILLAT, Yves FOUILLET
CEA-Grenoble
17 rue des Martyrs
38054 GRENOBLE cedex 9
marc.cuzin@cea.fr

DNA chips, also known as DNA arrays, microarrays or oligochips, are miniaturized microsystems based on the ability of DNA to spontaneously find and bind its complementary sequence in a highly specific and reversible manner, known as hybridization. Thus, labeled DNA molecules in a sample are analyzed by DNA probes tethered at distinct sites on a solid support. The composition of the DNA sample is then deduced by analyzing the signal generated by labels present at each probe site.

1. Introduction

Applications of DNA chip analysis are widespread and gaining ground: fundamental research, human genetics, infectious desease diagnosis, genotyping, gene expression monitoring, pharmacogenomics, environment control... *Recent articles describe the identification and detection of mutations in genes responsible for cancers, or describe how DNA chip analysis of individual polymorphisms may guide the doctor towards the most efficient treatment for his patient's particular form of disease. In the environmental and agro-industrial fields, DNA chips show great promise in rapidly and extensively testing microorganism content, contamination and pathogenicity. Access to DNA chip technology has become a technical and economical priority for academic and industrial institutions, reaching a 650 million dollar market in the next 5 years.*

"Lab-on-a-chip" are also miniaturized microsystems but carry out a higher number of analysis steps on a biological sample. They include channels, fluidics and thermal zones to achieve various enzymatic reactions in order to

amplify, purify, and label biological samples. Lab-on-a-chip and DNA chips are complementary parts of a miniaturized analysis and are currently mainly used in conjunction with fluorescent labeling. A laser-based read-out system is commonly used to determine and eventually quantify the hybridized sequences.

Looking back, hybridization techniques for analyzing DNA have been known to molecular biologists for decades. The explosion of interest for these new technologies have stemmed from a number of key innovations: 1) the use of a planar solid support such as glass or silicon, facilitating **miniaturization** and **high density analysis** 2) powerful high resolution detection offered by fluorescence 3) the development of high density spatial synthesis of short DNA probes (**oligonucleotides**) by photolithography or ink jet printer technology. DNA chips dimensions offer hybridization sites in the 50-200 micron range, producing arrays from 100 to over 10,000 or even 400,000 different probes on cm^2 areas.

2. Manufacturing DNA Chips

DNA chip manufacturing is achieved either by «on chip» technologies where oligonucleotide probes are chemically synthesized in situ, or by «off chip» technologies where DNA probes are previously synthesized, purified and controlled before being grafted onto the substrate using a number of technologies such as mechanical, ink jet or electrochemical deposition [1]. Initially, this type of technique was conducted on nylon or nitrocellulose membranes [2] with radioactive detection. Then, in order to increase the number and density of fixed probes and to be compatible with fluorescent detection, other supports have emerged such as glass [3], polypropylene sheets, polyacrylamide gel pads [4] and silicon [5]. The precise location of probes can be reached by photochemistry [6] or through micromechanical devices. The complexity of the array is adapted to the application requirements (cf. Figure1).

Affymetrix, founded in 1992, uses photolithographic masking borrowed from the semiconductor industry and is clearly the current leader in chip manufacturing business. This company has signed collaborations with most big pharmaceutical groups such as Glaxo Wellcome, Pfizer, Aventis, Novartis, bioMérieux, etc..

Figure 1: Different DNA Chips for different applications

• Low complexity: from 100 to 1000 probes
Dedicated analyses, diagnostic tools
➡ "Active" chips
  Nanogen, MICAM™ (CEA, CIS bio international)

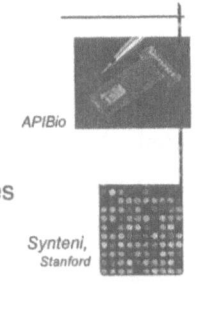

APIBio

• Mean complexity: from 1000 to 10 000 probes
global gene expression analysis
➡ MicroArrays ("off chip" technology)

Synteni, Stanford

• High complexity: 100 000 probes +
global polymorphism analysis
➡ In situ synthesis
  (oligonucleotide probes)

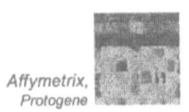

Affymetrix, Protogene

This leadership is challenged today by a growing number of companies: HySeq, Incyte, Nanogen, Caliper, etc.. With a strong foothold in microsystem development, Motorola has also joined the race. Among them, only a few european companies are present (BioRobotics in the UK, Clondiag, BioChip Technologies in Germany). In France, the French Atomic Energy Commission (CEA) has officially launched its «BioChip project» based on its proprietary electrochemical deposition technology developed early in 1993 [7] and is contributing to the foundation of a new company committed to chip manufacturing. These new actors are confident to play a significant role in the coming years as the market needs more flexible and cheaper technologies.

As in situ synthesis remains an attractive method for manufacturing high complexity chips with hundreds of thousands of probes (without having to

524

worry about handling and storing each probe), other companies or research labs have entered the competition by proposing lower cost or higher yield alternatives (without photomasks). Protogene uses piezoelectric pipetting to deposit successively the 4 base precursors to build the DNA molecule in situ on a hydrophilic/hydrophobic 2D structured surface. The University of Wisconsin proposes a Maskless Array Synthesizer where photolithography is not carried out by using masks but uses Digital Light Processor technology from Texas Instruments. The technology revolves around 480,000 tiny aluminum mirrors arranged on a computer chip.

«Off chip» technologies have drawn attention as probes are synthesized beforehand (with better yields), quantified and purified before being grafted on the substrate. Some technologies use contactless techniques (cf. figure 2) The result is a better control over probe quality and quantity, a great versatility in choice of supports and a higher confidence in the chip readout. Changing single bases or whole probes on an array involves little expense other than synthesizing a new DNA molecule and allows more flexible update of products.

In "off Chip" technologies, the nature of the probe is more flexible as well.

Figure 2: Piezoelectric Technology

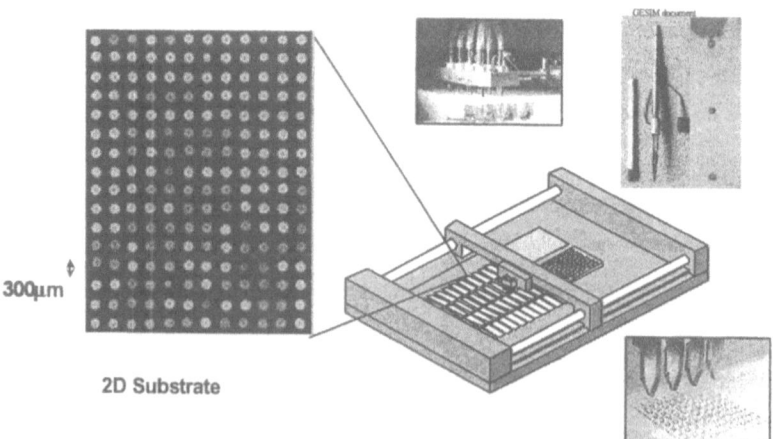

300μm

2D Substrate

Depending on the application, longer probes (typically over a hundred bases long), PCR products, or PNAs may be grafted to the substrate. In future developments aiming the protein world, some technologies have proven to accommodate peptides, antibodies or even enzymes.

Other alternative technologies include using silicon substrates instead of glass (Nanogen, CEA). Microelectronic components are integrated into the silicon base such as multiplexed circuits for active addressing of oligonucleotide probes to electrodes (CEA) or producing electrical fields which enhance the on-chip hybridization and stringency dynamics (Nanogen). These substrates may also integrate thermal control elements, microvalves, and microfluidic devices.

3. MICAM™ chip technology

At CEA, first generation DNA chips have 128 gold sites on a multiplexed silicon substrate on which oligonucleotides are fixed using pyrroles linkers in an electrochemical process. The process is named MICAM™ (MICrosystem for Analysis in Medicine). The construction of MICAM™ DNA chip is based on the

Figure 3: **The MICAM˙Technology**

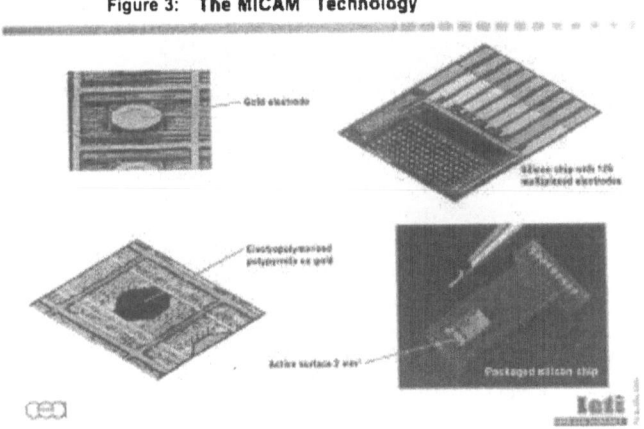

fixation of 5'pyrrole-labeled oligonucleotides. This technology has been previously described and is based on the electrosynthesis of a conducting polymer film bearing oligonucleotides [8-9]. The sites are very well defined (50µm) and the quantity of probes is electrically controlled. In such conditions, a fast read-out system allow quantitative measurement (see figure 3).

Among other potential applications, we are especially investigating the clinical diagnosis of gene mutations in cancer and the control of adequate treatment. A rapid and large scale method to detect K-ras gene mutations in tumor samples was adapted [12]

The K-ras proto-oncogene is altered by point mutations on codon 12, 13 or 61 in a wide variety of tumors (12) [detection of K-ras mutations enables the understanding of cancer biology and pathogenesis]. These alterations have a clinical relevance by providing information for early diagnosis and prognosis. According to the clinical implication of the K-ras gene in human tumorigenesis and its potential role as a target for novel therapeutic approaches, reliable methods are needed for the analysis of K-ras sequence in clinical samples.

The technical development and the use of this format assay for the detection of

Figure 4: Homogeneity of the MICAM™ technology

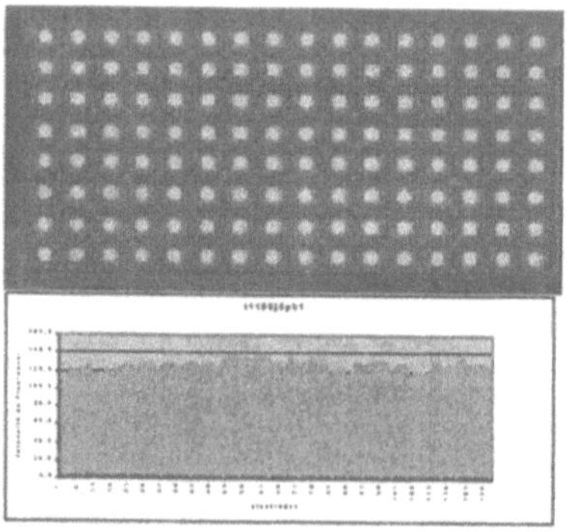

CV # 3%

chip with 128 probes
K-Ras CT, 26 mers, phycoerythrine marker

K-ras mutations in DNA coming from human colorectal carcinoma is described in [10].

The MICAM™ chip has a 4 mm² active multiplexed device. The 50 µm wide microelectrodes are arranged in a rectangular matrix of 8 rows and 16 columns. Each electrode can be selected individually. The silicon chip is integrated in a package compatible with both the electrocopolymerization step and the biological analysis. The copolymerisation process allows the formation of an homogeneous film (20 nm thick). The amount of hybridizable oligonucleotides grafted onto each electrode was measur at 200 fmol/mm². The functionalization of the gold electrodes by electrocopolymerization allows the formation of a controlled thin film ODN-pyrrole which is very stable. The homogeneity of this technology is very good and will be easily improved when developed at the industrial level (see figure 4).

A second generation DNA chips is built on a 3D silicon structured substrate,

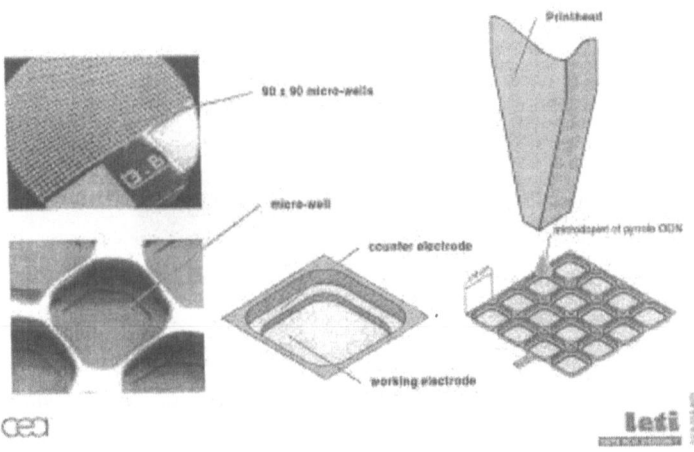

using dispensing addressing and electrochemical grafting in 300 picoliter wells (figure 5). Advantages of this approach are homogeneous probe density at the

528

electrode's surface, highly defined contour of each spot and easy readout of the chip as each well is in a predefined position and size. DNA chips bearing 8100 wells are currently manufactured in CEA (Grenoble).

The development of such chips require fast, adjustable, large dynamic readout systems. CEA/Leti is also engaged in the development of such elements. Figure 6 present an example of new ideas to develop adequate readers. There is no mechanical parts in this system and in such conditions, the readout time is expected to be very fast. Other readers are developed on the basis of a CD

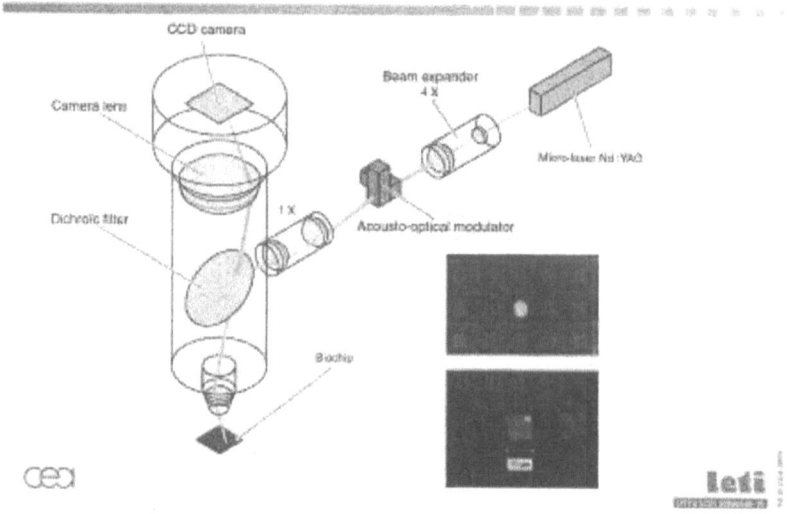

reader and are very compact. They integrate miniaturized laser sources and are glass substrate oriented. The main advantages of this technique are to improve precision, speed of the reading and cost of the system.

4. Lab on a chip

The preparation of the biological sample is also critical if we want to improve the performances of the analysis (reproducibility, time, volume, automatisation). For the miniaturization of the sample preparation of the sample, CEA/LETI has also developed microsystems using microtechnology in silicon. Silicon 3D structured devices allow mixing, heating and parallel sample preparation. The use of very good thermal coefficient material such as silicon allows very fast heating and cooling zones. Surfaces can be adjusted to requirements and to increase the thermal cycling for polymerase amplification, the adequate zones have been thinned. An external heating machine controls

Figure 7: microPCR device for HT-Genotyping

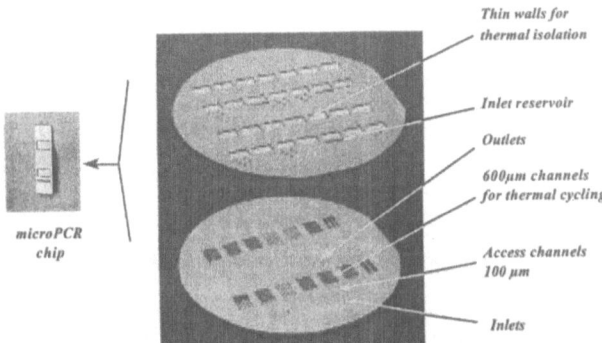

the temperature cycles.

The dimension of the channels is in the range of hundreds of micrometers. Detection with fluorescent markers is carried out off-chip. This solution for reading minimize all sources of noise from substrate and is well adapted for a binary genotyping application. The requirement is to reach hundreds of genotyping per hour at very low cost. Parallelisation and collective manufacturing on standard silicon microsystem foundry is one of the key points to reach such a goal. Figure 7 & 8 represent the main assembling principles.

530

Figure 8 : Disposable HG : Main structure

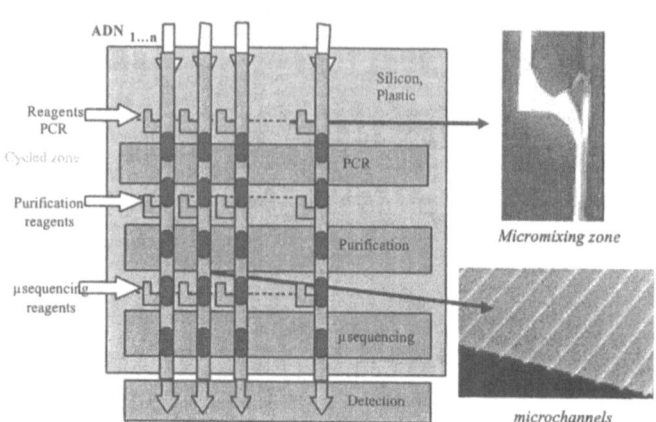

Micromixing zone

microchannels

5. Conclusion

Quantification of hybridized array images remains tedious (due essentially to difficulties in spot detection and inter and intra-spot variability), optimized clustering algorithms have hardly kept pace with the flux of data poured from the chips, and integration of the results with knowledge stored in biological databases is yet a challenge, particularly in the case of mammalian studies.

DNA chip technology has come a long way but it will continue to develop as new ideas, concepts, disciplines join this multidisciplinary field dedicated to better understanding nature and the living being.

In order to be able to provide quantitative analysis in the field of genomics, we must consider all the steps of such analysis. Lab-on-a-chip is consequently a very important device which include various functions (mixing, heating, cooling, detection). Its development is complex and will certainly take few years before a large penetration of market, but actors become numerous and various as the potential market is hudge.

6. Bibliography

1. E. Crapez, Nature Genetics vol. 21 n° 1 - 1999

2. Saiki RK, Walsh PS, Levenson CH, Erlich HA. Genetic analysis of amplified DNA with immobilized sequence-specific oligonucleotide probes. Proc Natl Acad Sci U S A 1989; 86:6230-4.

2. Guo Z, Guilfoyle RA, Thiel AJ, Wang R, Smith LM. Direct fluorescence analysis of genetic polymorphisms by hybridization with oligonucleotide arrays on glass supports. Nucleic Acids Res 1994;22:5456-5465.

4. Khrapko KA, Lysos Yu, Khorlin A, Shick V, Florentiev V, Mirzabekov A. Hybridization of oligonucleotides as a method of DNA sequencing. FEBS Lett 1989;256:118-22

6. Livache T, Fouque B, Roget A, Marchand J, Bidan G, Teoule R, Mathis G. Polypyrrole DNA chip on a silicon device: example of hepatitis C virus genotyping. Anal Bioch. 1998;15:188-94.

6. Pease AC, & al.. Light-generated oligonucleotide arrays for rapid DNA sequence analysis. Proc Natl Acad Sci U S A 1994;24:5022-6.

7. Patent 9303732

8. Livache T, Roget A, Dejean E, Barthet C, Bidan G, Teoule R. Preparation of a DNA matrix via an electrochemically directed copolymerization of pyrrole and oligonucleotides bearing a pyrrole group. Nucleic Acids Res 1994;11:2915-21.

9. Livache T, Bazin H, Mathis G. Conducting polymers on microelectronic devices as tools for biological analyses. Clin Chim Acta 1998;278:171-6.

10. Livache T, Bazin H, Caillat P, Roget A. Electroconducting polymers for the construction of DNA or peptide arrays on silicon chips. Biosens Bioelectron 1998;15:629-34.

11. Schimanski CC, Linnemann U, Berger MR. Sensitive detection of K-ras mutations augments diagnosis of colorectal cancer metastases in the liver. Cancer Res 1999;59:5169-75.

12.   K-ras mutation detection by hybridization to a polypyrrole DNA chip, Evelyne Lopez-Crapez, Thierry Livache , Joseph Marchand, Jean Grenier;

Clinical Chemistry (2001) 47 : 189-194

Fabrication of Disposable Plastic Microplates with 96 Integrated Microfluidic Capillary Electrophoresis (CE) Structures for High Throughput Screening (HTS)

A. E. Guber[1], M. Heckele[1], D. Herrmann[1], A. Muslija[1], Th. Schaller[2],
A. Gerlach[3], G. Knebel[3]

1) Forschungszentrum Karlsruhe GmbH, Institut für Mikrostrukturtechnik
   Postfach 3640, D-76021 Karlsruhe, Germany
   E-mail: andreas.guber@imt.fzk.de
2) Forschungszentrum Karlsruhe GmbH, Hauptabteilung Versuchstechnik
   Postfach 3640, D-76021 Karlsruhe, Germany
3) Greiner Bio-One GmbH
   Maybachstr. 2, D-72636 Frickenhausen, Germany

Abstract

A novel microfluidic platform based on a standard microtiter plate has been developed by the Greiner Bio-One company and the Forschungszentrum Karlsruhe in strategic cooperation. Instead of 96 wells, this novel platform accommodates 96 identical microfluidic CE structures on the same surface area (approx. 128 mm x 85 mm). The metal mold insert generated by micromilling possesses 96 inversely shaped CE structures that are molded into positive structures of PMMA or COC by vacuum hot embossing. In a further fabrication step, the obtained filigree microchannel structures are covered by an adapted, large-area cover plate to obtain real microcapillary structures. First tests demonstrated that these microcapillary structures have a good filling behavior. It may therefore be concluded that the channel geometry is accurate and homogeneous.

Keywords: HTS and UHTS screening, microtiter plates, microfluidic CE structures made of plastic, mechanical micromachining, vacuum hot embossing, bonding technologies

1.   Introduction

Highly efficient substance screening in today´s pharmaceutical industry represents the major application of microtiter plates. Microtiter plates are large-area test platforms made of plastic, which accommodate several hundred or thousand individual, miniaturized reaction chambers or depressions, so-called wells, on a standardized surface area of approx. 128 mm x 85 mm. Conventional standard microtiter plates are available with 96, 384 or 1536 wells and usually made of polystyrene or polypropylene by injection molding [1]. In HTS or UHTS screening, for instance, between 100000 and 200000 substances are tested per test campaign.

This requires comprehensive plate management which includes e.g. filling of the individual wells, specific metering of substances into the wells, taking of samples from the wells, and fluorescence spectroscopy of all wells. All these activities take place largely automatically by means of adapted laboratory automates (pipetting and liquid handling robots or reader systems). Still, certain partial steps, such as capillary electrophoresis (CE) of a substance mixture from a well, can only be performed sequentially; i.e. all wells of a microtiter plate have to be processed successively. Therefore, cycle times are considerably extended, at this point of analysis at least. Several absolutely identical CE separation structures arranged on a large-area chip were supposed to bring about an increase in capacity and further reduction of the analysis time [2]. Recently, a total of 96 identically shaped microfluidic CE separation structures were realized for the first time on the format of a standard microtiter plate [3].

2. Design and Functioning Principle of a Microtiter Plate with 96 Integrated CE Structures

To realize this novel analytical approach in terms of manufacture, 96 miniaturized CE units have to be arranged on a standard microtiter plate. In the very simple case, typical CE structures consist of two crossing microchannels and four reservoirs (cf. Fig. 1). A reservoir is located at the beginning and end of each microchannel. The four reservoir points are switched as injection cathode (sample material), injection anode (waste), buffer cathode, and buffer anode. The separation channel is located directly downstream of the crossing area. At its end, fluorescence spectroscopy takes place. The volume of the crossing area of both microchannels defines the volume of the substance to be separated. After feeding in the sample, it is first transported into the sample channel (injection channel) by application of a voltage. The sample located in the crossing volume is then injected into the separation channel as soon as voltage is reversed. There, the substance is separated electrophoretically into its constituents. The separated substances are analyzed successively in the detector system.

For future simultaneous and parallel capillary electrophoresis of the samples from the 96 wells of a standard microtiter plate, all 96 CE structures have to be arranged on the microtiter plate in an optimized way. The design selected for this purpose is shown schematically in Fig. 1 based on the example of a CE structure. The cross section of the microchannels is 100 μm x 50 μm. The reservoirs have a diameter of 1.4 mm. Because of space limitations, the separation channels have to be arranged in a meandering form, where each separation channel has a total length of 40 mm. Of crucial importance is a very precise geometry of the channel crossings (rectangular crossing areas), as these areas determine the sample volumes. Here, the sample volume amounts to about 600 pl.

534

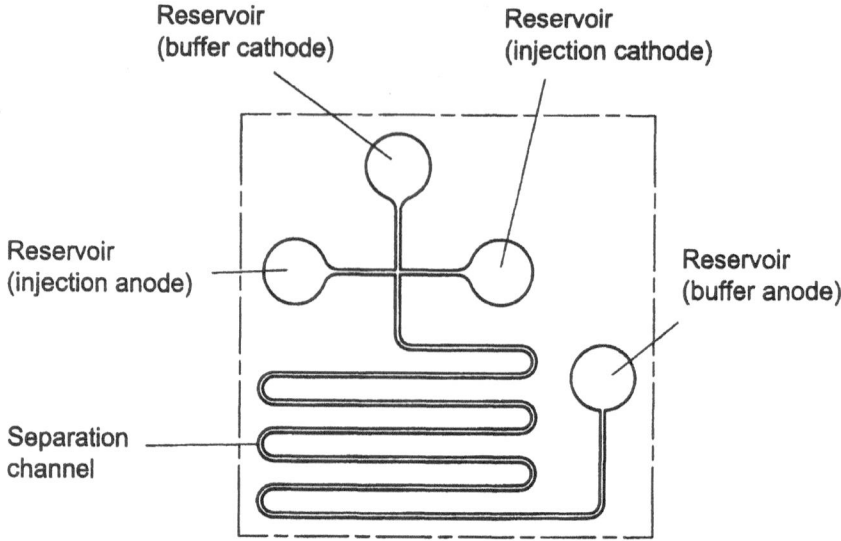

Reservoir
(buffer cathode)

Reservoir
(injection cathode)

Reservoir
(injection anode)

Reservoir
(buffer anode)

Separation
channel

Fig. 1:   Schematic design of a miniaturized CE structure. Due to space limitations, a meandering arrangement of the separation channel was selected.

On the surface area of a standard microtiter plate, the 96 miniaturized CE structures have a 9 mm lateral pitch and the reservoirs are positioned at a 2.25 mm
lateral distance, according to the 1536-microtiter plate format. This, in principle,
allows these novel microtiter plates with the 96 CE structures to be filled by commercially available pipetting and liquid handling robots and to be analyzed in
reader systems. Liquid management or separation of the samples within the
96 CE structures would have to be performed by several microelectrodes that are
lowered simultaneously. For this, 4 x 96 microelectrodes would have to be introduced into the 384 reservoirs of the CE structures. This seems to be feasible in
principle. Alternative solutions are presented in Sec. 5.

As the above microtiter plates with the 96 CE structures are usually disposable
products made of plastic, they have to be produced in large series at low costs. A
suitable manufacturing method from the microtechnical point of view is replication in plastics by means of vacuum hot embossing or injection molding using a
large-area metal molding tool that contains the inverse structures of the microchannels. To obtain real microfluidic CE structures, the channel structures molded
into the plastic material have to be closed subsequently with a correspondingly
large cover plate. In addition, the fluid connections to the respective CE structures
are incorporated in this cover plate.

3. Manufacture of Miocrotiter Plates with 96 Integrated CE Structures

3.1 Manufacturing Mold Inserts by Microcutting

Conventional microtiter plates are produced by means of injection molding using a large-area metal molding tool made of steel. Thus, series of 100000 pieces and more can be produced. The injection molding tools required for this purpose are frequently manufactured by EDM technology. Manufacture of the large-area microtiter plate with the integrated 96 CE structures described above requires an accordingly large microstructured metal molding tool. Presently, only a few microfabrication processes allow to properly machine such large areas of approx. 128 mm x 85 mm. The mold insert used here was cut into a metal base plate (128 mm x 85 mm) by mechanical micromachining [4] (cf. Fig. 2).

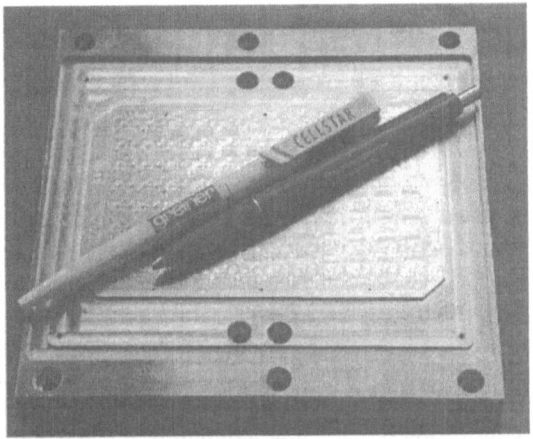

Fig. 2:   Photograph of a large-area metal mold insert. On an area of 128 mm x 85 mm, 96 inverse CE structures are milled into a brass substrate by mechanical micromachining.

Using diamond microtools and conventional high precision CNC machines equipped with high speed spindles, the inverse shapes of the 96 CE structures can be cut into the brass substrate rather rapidly. In accordance with the design represented in Fig. 1, the inverse CE structure consists of two crossing metal wall structures. The reservoirs have the shape of protruding cylinder structures (cf. Fig. 3). The areas between the metal walls and reservoirs are milled using various endmills with 100 and 400 µm diameter. The wall crossing (i.e. later channel crossing) represents a rather critical microstructure as the radii of curvature in the crossing section, which are determined by the end-mill diameter, have to be as small as possible. At present, they amount to about 50 µm. For adjusted bonding of the cover plate onto the structure in the subsequent back-end processes, the metal mold insert contains six pockets for to be replicated alignment pins.

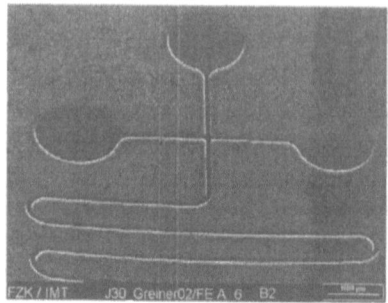 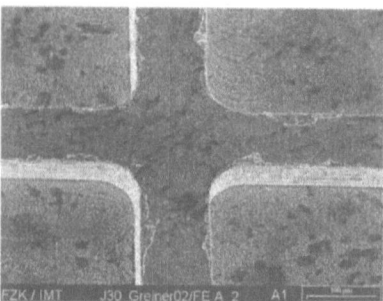

Fig. 3: SEMs of an inverse CE structure on a metal mold insert made of brass. The crossing metal wall structures or protruding cylinder structures forms the later channels or reservoirs on the plastic substrate. The curve radii in the crossing area of the metal walls have to be as small as possible.

3.2 Manufacturing Large-area Plastic Substrates by Vacuum Hot Embossing

All 96 microfluidic CE structures of the large-area mold insert can be molded into accordingly large plastic substrates by vacuum hot embossing. Presently, this is the only technology that allows for a very precise replication of filigree channel structures in polymers. A hot embossing facility mainly consists of two plates that are located opposite each other. During the embossing process, these plates are moved towards each other in a very precise manner. The lower, non-structured counterplate usually contains the semi-finished plastic product to be embossed. The upper plate is equipped with the microstructured mold insert [5]. In the first step of the hot deformation process, the molding tool and counterplate as well as the plastic substrate are heated to a temperature that exceeds the glass transition temperature of the plastic material to be embossed (heating phase). In the second step (embossing), the metal mold insert is lowered into the plastic material. This is done under vacuum in order to ensure complete filling of the microstructured embossing tool with the plastic. The plastic fills the embossing tool to the complete extent; as a result, the 96 microfluidic CE structures are molded precisely. Following a cooling phase, demolding takes place in a third step (demolding) and the mold insert is removed from the plastic material. Up to now, this novel microtiter plate has been molded into polymethylmetacrylate (PMMA) and cyclic olefin copolymers (COC). In principle, polycarbonate (PC), polysulfone (PSU), polyoxymethylene (POM), and polyvinylidene fluoride (PVDF) can also be applied [6].

Fig. 4 shows two molded plastic substrates of PMMA with 96 integrated microfluidic CE structures each. On the standard microtiter plate (128 mm x 85 mm), the

CE structures are arranged in the typical arrangement of eight rows and twelve columns. An alphanumerical grid allows rapid assignment of the individual CE structures. According to the design given, each of the 96 CE structures consists of two crossing microchannels and four reservoirs (cf. Fig. 1 and 5). The microchannels have a cross section of 100 μm x 50 μm (cf. also Fig. 7). The obtained geometry of the crossing area is already very close to the desired 90° corners. In the crossing area of both microchannels, defined substance volumes of about 600 pl can be generated. While the shorter channel represents the sample channel and has a total length of 3.1 mm, the separation channel downstream of the channel crossing area is arranged in a meandering way with a total length of 40 mm.

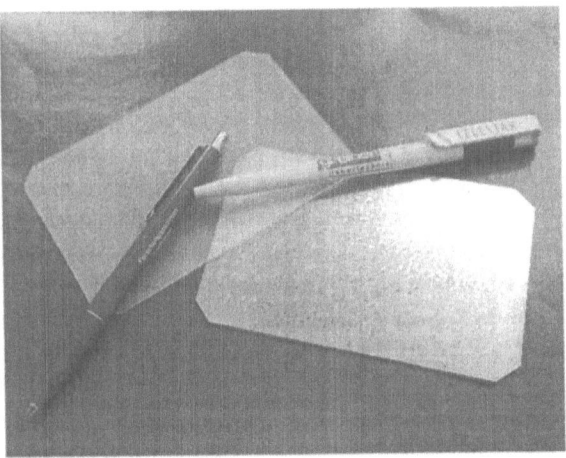

Fig. 4:   Photograph of two microtiter plates made of PMMA with 96 integrated microfluidic
            structures. To increase the optical contrast, the right plate was sputtered with gold.

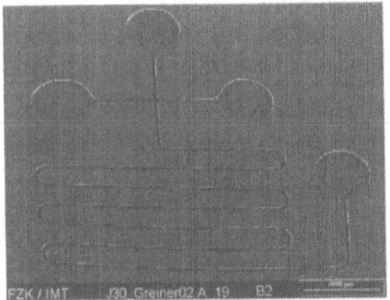 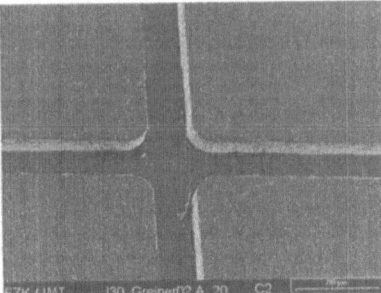

Fig. 5:   SEMs of a molded CE structure with two crossing microchannels and four reservoirs.
            The volume of the crossing area determines the substance volume that can be sepa-
            rated. Here, it amounts to about 600 pl.

3.3 Assembly and Bonding Technology (Back-end Processes)

To obtain functioning microfluidic CE structures, the microstructures embossed into the plastic material have to be covered by accordingly large, adapted, and precisely positioned cover plates. In case of the 96 CE structures, this cover plate has to contain 384 openings, such that all 384 reservoirs of the 96 CE structures can be reached and filled for the tests. Safe and adjusted bonding of the micro-structured base plate to the cover plate is ensured by six positioning aids. The large-area cover plate is approx. 128 mm x 85 mm in outer dimension and 1 mm thick. All 384 bores are produced in advance by a separate precision drilling process. The holes are 1.4 mm in diameter. Depending on the thickness of the cover plate, the reservoir volumes of the CE structures can be adjusted individually between 0.1 and 1 µl.

Various welding or bonding technologies allow to permanently connecting the microstructured base plate with the cover plate. The microfluidic CE structures have to be kept absolutely free from adhesives and impurities for the microchannels maintaining their given geometry and no deformation of the channel cross sections taking place. Therefore, largely gap-free bonding between the microstructured base plate (microtiter plate) and the cover plate is envisaged. In principle, various techniques are available, for example, laser welding [7], chamber bonding [8], capillary bonding [9], and solvent bonding [10]. Fig. 6 shows a full area covered microtiter plate. The sample material is fed in via one of the four bores of each microfluidic structure. The three other bores are used for buffer supply and waste.

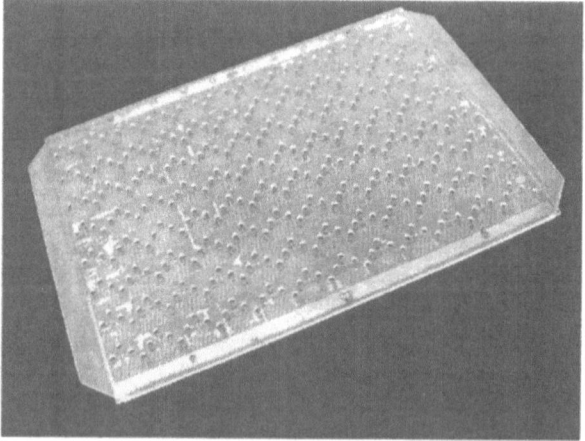

Fig. 6:   Photograph of a full area covered microtiter plate. Each of the CE structures can be reached via four bores such that sample supply and outflow of the spent substances are possible.

The detailed view in Fig. 7 shows the longitudinal section of the crossing area with the outgoing microchannels (sample channel and separation channel) and a reservoir. At the beginning of the meandering separation section, channel width is 100 μm and channel height amounts to 50 μm.

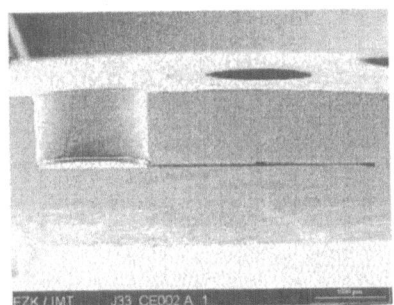

Fig. 7:   SEMs of a covered CE structure. The sectional view on the left hand side shows the cross section of a reservoir, the outgoing microchannel, the crossing area, and the entry area of the meandering separation channel. On the right, the cross section of the meandering separation channel of 100 μm x 50 μm in dimension is shown.

4.   Fluidic Tests of CE Structures

The covered microfluidic CE structures were subjected to first fluidic tests. These tests are necessary to verify the efficiency of coverage of each CE structure and to test the ability of filling the microchannels by capillary forces, respectively.

The six successively taken photographs in Fig. 8 illustrate the filling behavior of a CE structure. Via a pipette, a colored aqueous solution is applied to the waste reservoir (cf. Fig. 1 and 8). Due to the capillary forces, the separation channel is filled and after about 10 seconds, the liquid reaches the channel crossing area. Detailed video evaluation revealed a homogeneous flow behavior in the linear part of the microchannel. In the curve areas of the meandering separation channel, however, this flow behavior is changed significantly [11].

Test measurements on various CE structures indicate a favorable filling and flow behavior of the microfluidic CE structures and show that the geometry of the separation channel is clean and homogeneous to a large extent. In the areas directly outside of the microchannels, no coloring is observed such that perfect sealing of the microchannels by the bonded cover plate can be assumed. No gaps and dead spaces were found.

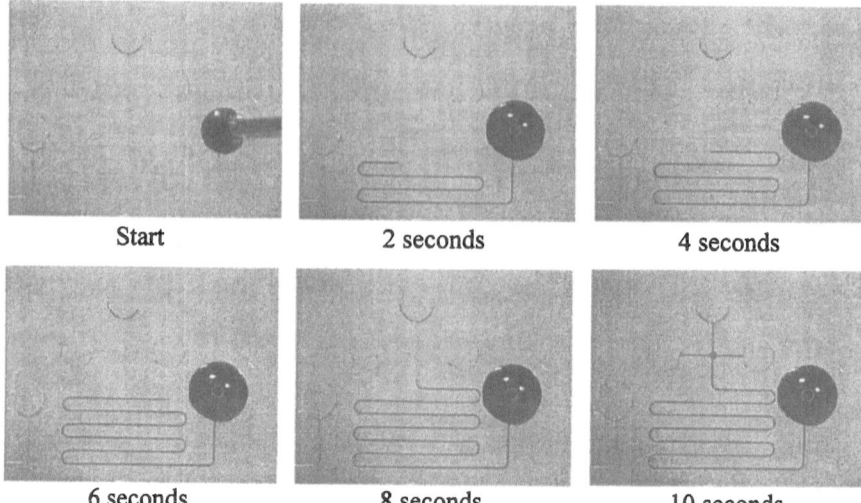

Start	2 seconds	4 seconds
6 seconds	8 seconds	10 seconds

Fig. 8:   Filling behavior of a microfluidic CE structure. Within 10 seconds, a colored liquid reaches the channel crossing area. In the range of the meandering separation channel, no crosstalk is observed.

5.   Summary and Outlook

For the first time, large-area microtiter plates with integrated microcapillary structures have been generated by plastic molding. On a surface of approx. 128 mm x 85 mm 96 microfluidic CE structures are arranged. The molding tool required for this purpose was milled into an accordingly large brass substrate by mechanical micromachining. Meanwhile, this tool has been applied for more than 100 molding processes into PMMA and COC. The microstructured plastic substrates can be sealed permanently and tightly with accordingly large cover plates, as was demonstrated by first fluidic tests.

Using pipetting and liquid handling robots, all reservoirs of all CE structures can be filled with sample material, buffer solutions, and fluorescence solutions. First test measurements by fluorescence spectroscopy resulted in a clear detection of the coloring substance in the separation channel.

If necessary, the design of the microfluidic CE structures still can be varied over a wide range. In the curve areas, for instance, tapered channel geometry appears feasible. An approach towards the machining of more delicate structure details, namely, sharp inner edges, is being made by development of a microplaning/microslotting process. First test mold inserts made of brass and steel are available for replication and will be tested soon [12].

This novel type of microtiter plates with integrated CE structures can be used in the future for the capillary electrophoresis of substance mixtures in HTS and UHTS screening. Handling and analysis of these novel microtiter plates may be further simplified by integrating the 384 microelectrode structures in all 96 CE structures. As microtiter plates practically always represent disposable products for cross contamination reasons, low-cost production by injection molding is envisaged.

6.   References

[1]   R. Heller: Technical notes and applications for laboratory work, Greiner Labortechnik, Forum No. 1 (1999)

[2]   I. Gibbons: Microfluidic arrays for high-throughput submicroliter assays using capillary electrophoresis, Drug Discovery Today Vol. 1, pp. 33-37 (2000)

[3]   A. Gerlach, G. Knebel, A. E. Guber, M. Heckele, D. Herrmann, A. Muslija, T. Schaller: Microfabrication of single-use plastic microfluidic devices for high-throughput screening and DNA analysis, Proceedings Micro System Technologies 2001, Düsseldorf (2001)

[4]   Th. Schaller, M. Heckele, R. Ruprecht: Mechanical micromachining for mold insert fabrication and replication, ASPE Proceedings Vol. 19, pp. 3-8 (1999)

[5]   M. Heckele, W. Bacher, H. Ulrich, T. Hanemann: Hot embossing and injection molding for microoptical components, SPIE Vol 3135 (1997)

[6]   Microfabrication Technologies, Leaflet Forschungszentrum Karlsruhe Microsystem Technologies Program (PMT) (1999)

[7]   J. W. Chen, T. Hessler: Transmission laser welding of plastics for microsystem packaging, Proceedings Micro Engineering 99, Stuttgart, pp. 81-88 (1999)

[8]   D. Maas, B. Büstgens, J. Fahrenberg, W. Keller, P. Ruther, W.K. Schomburg, D. Seidel: Fabrication of microcomponents using adhesive bonding techniques, Proceedings of International Workshop on Micro Electro Mechanical Systems, MEMS '96, 11th – 15th Feb. 1996 in San Diego, USA, pp. 331 – 336 (1996)

[9]   A. Gerlach, H. Lambach and D. Seidel: Propagation of adhesives in joints during capillary adhesive bonding of microcomponents, Microsystem Technologies 6, pp. 19 – 22 (1999)

[10]   P. Volk, R. Bader, P. Jacob, H. Moritz: DE19851644A1 (1998)

[11]   J. I. Molho, A. E. Herr, B. P. Mosier, J. G. Santiago, T. W. Kenny, R. A. Brennen, G. B. Gordon: Designing corner compensation for electrophoresis in compact geometries, Proceedings of the µTAS 2000 Symposium, 141th – 18th May 2000 in Enschede, The Netherlands, pp. 287- 290 (2000)

[12]   Th. Schaller: Microstructures machined by planing and slotting, to be published in: Book of Abstracts of HARMST 2001, June 17-19, 2001 Baden-Baden, Germany

Diagnostic Analyses by Biochemical Reactions and Separations on a Chip

Minoru SEKI[1]*, Jong Wook HONG[1&2], Ryusuke AOYAMA[1],
Ryutaro EZAKI[1], Yasuhiro KAKIGI[1], Masumi YAMADA[1],
Teruo FUJII[1] and Isao ENDO[2]

[1] The University of Tokyo, Japan
[2] The Institute of Physical & Chemical Research, Japan
*The University of Tokyo, Graduate School of Engineering,
7-3-1, Hongo, Bunkyo-ku, Tokyo 113-8656, Japan
mseki@chembio.t.u-tokyo.ac.jp

Abstract

Firstly we report the development of the hybrid polydimethylsiloxane (PDMS)-glass microchip for genetic analysis by functional integration of PCR and capillary gel electrophoresis (CGE) and related temperature control systems for PCR on a PDMS-glass hybrid microchip. The microchip was produced by molding PDMS against a microfabricated master with comparatively simple and inexpensive methods. PCR was successfully carried out on the PDMS-glass hybrid microchip with 500 bp target of λDNA and the amplified gene was analyzed by CGE on the same PDMS-glass microchip consequently. The chip could be considered as an inexpensive single use apparatus unlike the glass or silicon-made microchips for the same purpose.

Secondly, a PDMS microchip for a multiple diagnostic analysis system that can detect glucose and albumin concentrations by enzymatic and chemical reactions is demonstrated. This microchip is suitable for single-use to avoid a cross-contamination in clinical diagnostic tests. Because a PDMS microchip can be produced inexpensively by replica molding with the micro-fabricated master and used with simple optical detection device due to its transparency.

The sample solution containing glucose and serum albumin was introduced into the micro-channel and analytical reagents were also introduced through the other inlet-port. Glucose determination was conducted according to Oxidase/POD method. The pigment formation in the micro-channels depended linearly on the glucose and albumin concentrations within the tested ranges. We achieved to develop a multiple diagnostic analyzing system that can detect two major substrates for blood test by both enzymatic and chemical reactions on a PDMS microchip. A novel method for microfluidic introduction and mixing of samples was developed for small volume of liquid.

Thirdly, a novel aqueous two-phase system on a microchip is proposed for biological substance separations. An aqueous two-phase system was introduced into a microfabricated channel with a width of 50-400 micrometer, in which the two-phase flow with stable interface was formed. This micro-system successfully enabled the continuous separation of macromolecules or various kinds of cells from their mixtures. The proposed system will contribute the development of a microchip for diagnosis, gene analysis, or cell sorting.

Finally, the chromatographic separation of proteins on a PDMS microchip is shown. Here, we report the development of microchip systems for pressure-driven ion-exchange chromatography on an inexpensive material. We have devised the fabrication step and have successfully separated a mixture of albumin and immunoglobulin by ion-exchange chromatography on a PDMS microchip. These components of microfluidic sample manipulation will be integrated into more complicated procedures of diagnostic analyses on a microchip.

1. Introduction

Microfabrication technology based on photolithography mainly used for semiconductor related processes has surprisingly developed with electronics industry in the last fifty years. In recent ten years, this technology has been applied to non-electrical research areas, such as chemical and/or biological analysis, drug delivery, high throughput screening *etc.* [1-2]. Various types of miniaturized bio/chemical analysis systems have been developed by using microfabrication chip technology [3–12]. Among them, diagnostic analyses are one of the most promising application suitable for a miniaturized and integrated analytic system on a chip. The potential merits of miniaturized analysis systems are : (1) reduced amounts of sample, reagent consumption and waste discharge, (2) equivalent or increased analytical performance, (3) automation, (4) portability, (5) highly integrated & parallel processes, (6) inexpensiveness *etc.* These characteristics give benefits on a POC clinical diagnostics or an emergency health care system. However, the microchip diagnostic system requires many kinds of functional components and needs sophisticated microfluidic manipulation and operation strategies. In this paper, some examples as diagnostic analysis components using biochemical reactions and separations on a chip are presented along with the advantages of micro-scale systems.

2. Microchip for Gene Amplification [13-15]
2.1 CE and PCR on a Chip

A capillary electrophoresis (CE) on a chip have been studied extensively by many researchers for the electrophoretic sizing of biomolecules such as DNAs, RNAs or proteins [3, 7-9]. The attempts to integrate PCR with a CE system on a chip have also been tried by some research groups [16-19]. In these reports, however, the main substrate for the chip is silicon, glass or quarts. Therefore, the time-consuming fabrication process is inevitable for chip manufacturing. Consequently the chip is expensive and inappropriate for single-use to prevent cross-contamination, otherwise labor-intensive careful rinse processes are required for multiple use. Hence, inexpensive microchip systems are required for their wide application. We have firstly reported DNA amplification reaction (PCR) and consecutive electrophoretic sizing by capillary gel electrophoresis (CGE) for genetic analysis on an inexpensive PDMS based microchip [13-15].

2.2 Materials and Methods

2.2.1 Fabrication

As shown in Fig.1, the chip is composed of a reaction chamber for PCR, a separation channel for CGE, a connection channel between the reaction chamber and the separation channel, and two ports for the introduction of gel and electrical access. The hybrid microchip is constructed by covering a PDMS microchip with a glass plate. In the fabrication process, ultrathick photoresist SU-8 was spin-coated onto a silicon wafer to create a mold master with up to 100 μm thickness. The pattern on the mask was photolithographically transferred to the SU-8 coated silicon wafer by using a mask aligner. Before pouring prepolymerofPDMS,themasterwastreatedwithfluorocarbon plasma using a reactive ion etching system.

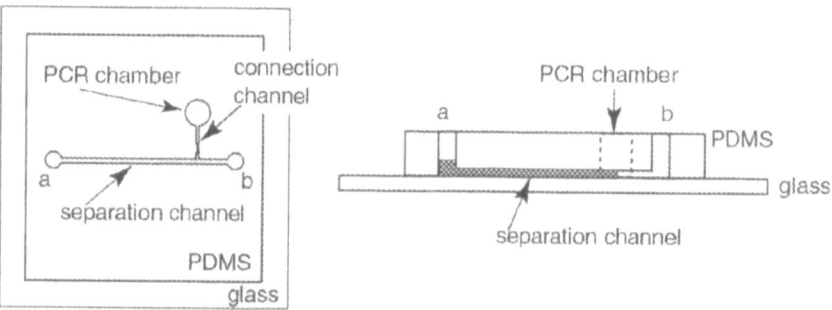

Fig.1 Layout of the PDMS-glass hybrid microchip for PCR and CGE (not to scale)

2.2.2 Curing of PDMS and the surface treatment

The mixture of PDMS prepolymer and curing agent was poured onto the master, cured at 65°C for 1h, and then placed at 135°C for 15 min. After curing, the PDMS replica was peeled off from the master. Reaction chamber and access ports were drilled on the chip. This PDMS replica was attached to a glass plate. Surface hydrophilization was carried out only in the separation channel by filling concentrated HCl solution into the channel.

2.2.3 Temperature control system

A Peltier element (TE) was used for heating the PCR chamber, and a thermocouple was attached for temperature feedback. The TE was connected to a computer via a programmable DC-power supply. The thermal cycle was controlled by a commercial computer program, LabVIEW 5.1.

2.2.4 PCR on the PDMS-glass microchip

PCR reaction on the chip was carried out using a PCR amplification kit (R011, Takara Shuzo Co., Ltd., Shiga, Japan). Template is λDNA and the length of target DNA is 500 bp. The three step PCR method was adopted, i.e., 94°C for denaturation, 54°C and 72°C for annealing and elongation. The reaction mixture in the chamber on the chip was covered by silicone oil to prevent evaporation. Reaction volume of the PCR was 30 to 50 µL. BSA treatment of reaction chamber was carried out prior to PCR reaction. PCR products were analyzed by conventional slab gel electrophoresis and a commercial capillary electrophoresis apparatus.

2.2.5 CGE on the PDMS-glass hybrid microchip

DNA fragments amplified in the PCR chamber was analyzed in a separation channel on the same microchip after finishing the PCR reaction. 2.0% of agarose gel solution was introduced into the separation channel through the hole 'a' (Fig. 1) and the edge of the gel was controlled to stop just before the junction of separation channel and the connection channel. The PCR product was stained on the microchip by adding SYBR Green I before separation. The PCR product in reaction chamber was introduced into the separation channel by pressure and sample plug was formed in the edge of the gel by applying 100 V, for 1 sec through the separation channel. Then, the buffer was refilled into the separation channel. Finally, electrical voltage was applied for the separation. Electrophoretic separation of the PCR product was visualized by a fluorescence microscope and a CCD camera.

2.3 Results and Discussion

2.3.1 PCR reaction on a PDMS-glass hybrid microchip

Conditions for PCR reaction on the PDMS-glass hybrid microchip were evaluated by varying Taq polymerase concentration, length of reaction time, *etc.* Fig. 2 shows electropherograms of the PCR product. 500 bp peak of the PCR product was confirmed by thermal cycling with 94°C, 30 sec, 54°C, 30 sec, and 72°C 30 sec for 25 cycles. The concentration of the amplified DNA under these conditions was 33 ng/µl. But further improvement and optimization of the temperature control system would lead to faster gene amplification.

2.3.2 CGE on a PDMS-glass hybrid microchip

The results of genetic analysis of the 500 bp target in λDNA demonstrate that the amplified 500 bp peak was detected at about 12 sec on the functionally integrated PDMS-glass hybrid microchip. Required time for separation is about 5 times shorter than the commercial system which is used for checking the PCR performance in Fig. 2. The time for completing CGE after the gene amplification, including gel introducing, gel maturing, sample plug formation and electrophoretic separation, was less than 10 min. Because of some intrinsic characteristics of PDMS such as high gas permeability and low thermal conductivity, we chose a hybrid microchip having glass plate as the lower part of the chip and heat is transferred through this glass plate to the reaction mixture.

Also, optical observation and detection are possible through glass plate. It is expected, furthermore, that fully integrated PDMS-glass hybrid microchip system could be realized by fabricating a heater and electrodes onto a glass substrate.

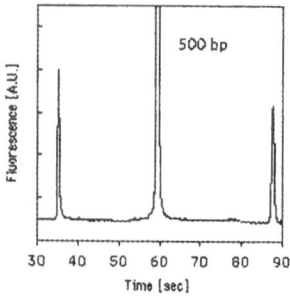

Fig.2 PCR product analysis of the 500 bp target gene in lDNA. The peaks appearing at around 35 s and 87 s indicate low and high molecular reference of 50 bp and 10380 bp, respectively.

3. Blood Analysis on a Chip for Health Check [20]

3.1 Clinical Blood Test Using a Microchip

A large amount of research has been applied to the development of new diagnostic apparatus. For example, novel analysis systems and assays have been introduced for monitoring blood glucose concentration of diabetic patients. The blood test systems, which have more than $50 billion cost worldwide, could be eventually be replaced by low-cost microbiochemical systems in a PC card size [21]. Most of diabetics monitor their blood glucose by themselves using disposable amperometric biosensor connected hand-held electrochemical analyzer. Such biosensors are generally based on screen-printed electrode strips with immobilized glucose oxidase and an adequate electron mediator [22]. Although their responses are considerably rapid, such devices have limits in versatility. For instance, oxidizable components in sample solutions commonly interfere with the amperometric assay. Electrode must be cleaned for avoiding cross-contamination. For a single-use chip, the electrode fabrication is too elaborate and high cost-consuming process.

In conventional microchip systems, the devices were fabricated on a silicon wafer or a planner glass. However, for the ongoing commercialization of this technology, these substrates represent certain disadvantages, such as costs of materials, the fabrication process including many steps and harmful wet chemistry and limitation in geometrical design due to the isotropicity of the etching process. If neither the detection nor material requirement were compromised, and if additional clinically relevant species could be measured simultaneously, an approach should greatly enhance the ability of diagnostic devices for blood test, and of other clinical analyzing systems, in general [23].

Therefore, we have developed a PDMS microchip for a multiple diagnostic analyzing system with optical detection [20]. The blood glucose assay is one of the indispensable biochemical tests for diagnosis and remediation of diabetes and other disease on glucose metabolism. Almost all of serum albumin is produced at liver, so serum albumin assay has great significance for diagnosis.

3.2 PDMS Microchip System for Multiple Diagnostic Analyses

3.2.1 PDMS Microchip

The fabrication process of the PDMS microchip used in this study is almost similar to those described above. A design of the PDMS microchip used in this study is shown in Fig. 3. The chip has three inlets, one for the sample solution and others for reagents, and two outlets for wastes. The chip also has two microchannels, one for glucose assay and another for albumin assay, whose width is 400 μm and the depth is 150 μm. After curing, the PDMS replica was peeled off from the master and placed onto a PMMA plate with stainless pipes for inlets and outlets to form closed microchannels.

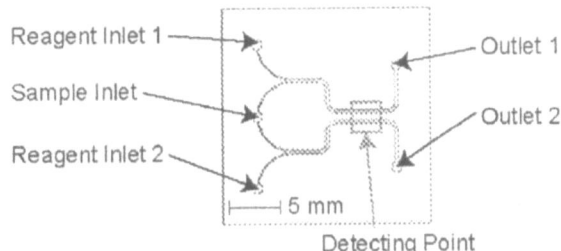

Fig.3 Design of the PDMS microchip for diagnostic analyses

3.2.2 Analysis System

The analysis system consists of a microscopic system with halogen lamp, syringe pumps, a thermo-controller, a charge-coupled device (CCD) camera connected to a PC.   The flow rates of the reagents and samples were controlled by three syringes and micro-syringe pumps.   Each syringe needle was connected to a stainless inlet connector through a Teflon tube. Reagents and samples loaded by pressure driven flow were mixed near the Y-shape junctions by diffusion and reacts to form dyes by enzymatic and chemical reactions. The microscopic images of detecting point were captured by CCD camera and stored in a PC for image processing.   The outlets were also connected to stainless outlet connectors and silicone tubes. The temperature of the PDMS microchip was controlled at 37 °C by the thermo-controller.

3.2.3 Image processing

The substrate concentrations in samples were calculated by image processing. The CCD images are containing 3 color data, red, green and blue, per single pixel. Absorbance of the specific wavelength can be determined from contributions of each data.

3.3 Glucose and Albumin Assay on a PDMS Microchip

3.3.1 Preparation for multiple diagnostic analyses

Glucose concentration was determined by Oxidase/POD method, which contains three enzymatic reactions. Three enzymes, hog kidney mutarotase, glucose oxidase and horseradish peroxidase, and two substrates, 4-amino-antipyrine and phenol, were dissolved in 60 mM phosphate buffer and adjusted pH 7.1. Albumin concentration was determined by BCG method. Bromocresol green (BCG) was dissolved in 75 mM succinate buffer and adjusted pH 4.2. The sample solutions were containing glucose at the concentration ranging from 0 to 180 mg/dl and bovine serum albumin (BSA) at the concentration ranging from 0 to 8.0 g/dl. The normal concentration of human blood glucose is ranging from 70 to 110 mg/dl and that of human serum albumin is ranging from 3.7 to 5.2 g/dl.

3.3.2 Glucose assay by oxidase/POD method

The sample solution loaded into the center inlet port was mixed in a micro-channel with the reagent solution containing three enzymes and two substrates. After mixing, α-D-glucose in sample solution is converted to β-D-glucose by mutarotase immediately. β-D-glucose is oxidized to D-gluconic acid by glucose oxidase and H_2O_2 is produced simultaneously. H_2O_2 reacts quantitatively with 4-aminoantipyrine and phenol to form a red pigment which is catalyzed by peroxidase. The digital images of detecting point were captured by CCD camera and processed by PC to calculate the concentration of glucose. This analysis was performed by a stopped flow method, and the temperature of microchip was controlled at 37 °C for enzymatic reactions. Fig. 4 shows the time-course of the formation of the quinone dye. The enzymatic reactions finished within 80 sec.

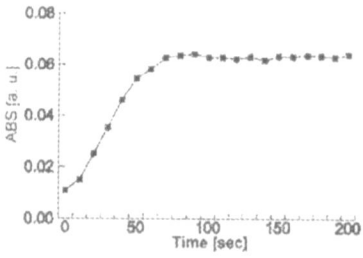

Fig.4 Time course of red dye formation in a microchannel for glucose analysis.

3.3.3 Albumin assay by BCG method

The sample solution containing BSA at the concentration ranging from 0 to 8.0 g/dl was mixed with reagent solution containing BCG. After diffusion mixing, BSA and BCG reacted immediately to form blue dye. This analysis was also performed by stopped flow, but the reaction occurred very fast.

3.3.4 Multiple diagnostic analyses on a chip

A multiple diagnostic analyzing system was developed and two major tests, glucose and albumin assay, were demonstrated on the system. These tests on the PDMS microchip system finished within two minutes with an inexpensive single-use PDMS-microchip. On going research is to develop a multiple diagnostic analyzing system that can test sufficient numbers of contents in human blood. For this purpose, we have devised a novel liquid injection method with wedge-shaped microchannel on a PDMS microchip for diagnostic analyses.

4. Continuous Separation of Particulates and Macromolecules in a Microchannel Separator [24]

Aqueous two-phase systems, generated by mixing aqueous solutions of two polymers, have been widely used in the separation of biological materials such as proteins, nucleic acids, cells or organelles by partitioning. However, this separation process generally requires both mixing and time-consuming settling procedures. As the results, the process is inevitably limited in a single or successive batch-wise operation. As in a micro-scale structure, the interfacial tension is more dominant and a surface-area/volume ratio is increased compared to macro-scale systems, the stable interface of two phases and the enhanced mass-transfer are expected. Therefore, an aqueous two-phase system was introduced into a micro-fabricated channel with a width of 50-400 micrometer, in which the two-phase flow with stable interface was formed. This micro-system successfully enabled the continuous separation of various kinds of cells and macromolecules such as proteins from their mixtures. The proposed system will contribute the development of a microchip for diagnosis, gene analysis, or cell sorting.

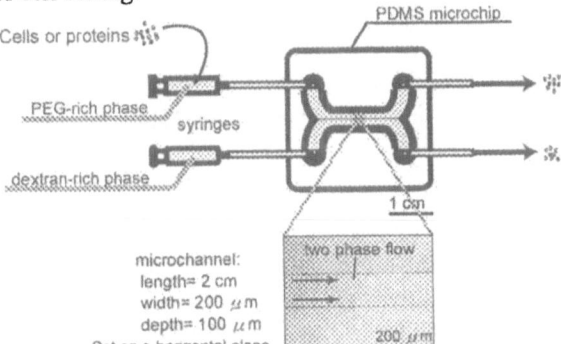

Fig.5 Schematic diagram of an aqueous two-phase flow system for cell separation from the mixture using polyethylene glycol (PEG) and dextran.

5. Chromatographic Separation of Proteins on a Microchip [25]

Although capillary chromatography on a microchip has been studied by many researchers, the chip substrates were limited to quartz or glass, which needs complex and expensive fabrication processes. Moreover, those studies mostly used electro-osmotic flow, so that the applications were only for analyses of neutral or less polar compounds.

We reported the development of microchip systems for pressure-driven ion-exchange chromatography on inexpensive material such as PDMS [25]. PDMS is useful material for its optical transparency and its simple sealing method but difficult to resist relatively higher pressure during LC operation. Therefore, we have devised the fabrication step and have successfully separated a mixture of proteins by ion-exchange chromatography on a PDMS microchip.

We have developed an ion-exchange chromatography system on a PDMS microchip. The system was composed of a three-dimensional microfluidic channel packed with mono-sized beads for anion-exchange, that was made by molding PDMS against two masters. With this system, we have separated a mixture of proteins using pressure-driven flow. We have also designed an on-chip static mixer for a gradient separation mode. The system enables the efficient separation of various charged compounds by ion-exchange without the effects of electrokinetic interaction, so that it is suitable for most analytes, even polar compounds.

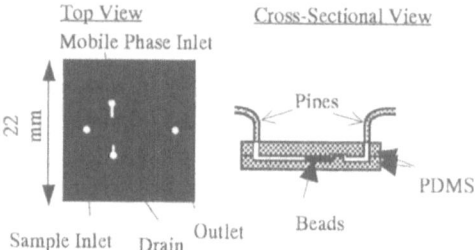

Fig.6   PDMS microchip for capillary chromatography (not to scale)

6. Conclusions

We have developed several microreaction and microseparation systems fabricated mainly on a PDMS microchip for diagnostic applications. These components of microfluidic devices will be integrated into more complicated processes of diagnostic analyses on a microchip.

7. Acknowledgements

This research was supported in part by grants for Scientific Research on Priority Areas (A) (No. 10145104 & 13025216) and Millennium Project (No.12319) from the Ministry of Education, Culture, Sports, Science and Technology in Japan.

8. References

[1] Santini Jr., J. T., Cima, M. J., Langer, R. A., Science 1999, 397, 335-337.

[2] Nashat, A. H., Moronne, M., Ferrai M., Biotech. & Bioeng. 1998, 60, 137-146.

[3] Manz, A., Harrison, D. J., Verpoorte, E. M. J., Fettinger, J. C., Paulus, A., Luedi, H., Widmer, M., J. Chromatogr. 1992, 593, 253-258.

[4] Harrison, D. J., Fluri, K., Seiler, K., Fan, Z., Effenhauser, C. S., Manz, A., Science 1993, 261, 895-897.

[5] Mastrangelo, C.H., Burns, M. A., Burke, D. T., Porc. IEEE. 1998, 86, 1769-1787.

[6] Ramsey, J. M., Jacobson, S. C., Knapp, M. R., Nature Med. 1995, 1, 1093-1096.

[7] Effenhauser, C. S., Manz, A., Widmer, H. M., Anal. Chem. 1993, 65, 2637-2642.

[8] Jacobson, S. C., Hergenroeder, R., Koutny, L. B., Warmack, R. J., Ramsey, M., Anal. Chem. 1994, 6, 1107-1113.

[9] Wooley, A. T., Mathies, R., Proc. Natl. Acad. Sci. USA. 1994, 91, 11348-11352.

[10] Effenhauser, C. S., Paulus, A., Manz, A., Widmer, M., Anal. Chem. 1994, 66, 2949-2953.

[11] Jacobson, S. C., Moore, W. A., Ramsey, J. M., Anal. Chem. 1995, 67, 2059-2063.

[12] Simpson, P. C., Wooley, A. T., Mathies, R. A., J. Biomedical Microdevices 1999, 1, 7-26.

[13] Hong, J. W., Hosokawa, K., Fujii, T., Seki, M., Endo, I., Proc. of IEEE Transducers'99 1999, 2, 760-763.

[14] Hong, J. W., Hosokawa, K., Fujii, T., Seki, M., Endo, I., Progress in Biotechnology, 2000, 16, 69-74.

[15] Hong, J. W., Hosokawa, K., Fujii, T., Seki, M., Endo, I., Electrophoresis, 2001, 22, 328-333.

[16] Wooley, A.T., Hadley, D., Landre, P., deMello, A.J., Mathies, R., Northrup, A., Anal. Chem. 1996, 68, 4081-4086.

[17] Ibrahim, M.S., Lofts, R.S., Jahrling, P.B., Henchal, E.A., Weedn, V.W., Northrup, M.A., Belgrader, P., Anal. Chem. 1998, 70, 2013-2017.

[18] Northrup, M.A.; Benett, B., Hadley, D., Landre, P., Lehew, S., Richards, J., Straton, P., Anal. Chem. 1998, 90, 918-922.

[19] Waters, L.C., Jacobson, S.C., Kroutchinina, N., Khandurina, J., Foote, R.S., Ramsey, M., Anal. Chem. 1998, 70, 5172-5176.

[20] Seki, M., Aoyama, R., Hong, J.W., Fujii, T. and Endo, I., Proceedings of First International IEEE EMBS Special Topic Conference on Microtechnology in Medicine and Biology, 2000, pp. 21-24.

[21] Lauks, I. R., Acc. Chem. Res., 1998, 31, 317- 324.

[22] Wang, J., Chatrathi, M.P., Tian, B. and Polsky, R., Anal. Chem., 2000, 72, 2514-2518.

[23] McDonald, J.C., Duffy, D.C., Anderson, J.R., Chiu, D.T., Wu, H., Schueller, O.J.A. and Whitesides, G.M., Electrophoresis, 2000, 21, 27-40.

[24] Seki, M., Kakigi, Y., Hong, J.W., Fujii, T. and Endo, I., Book of Abstract, PacifiChem (2000).

[25] Seki, M., Ezaki, R., Hong, J.W., Fujii, T. and Endo, I., "Ion-Exchange Chromatography on a PDMS Microchip," HPCE 2001, p.120 (2001)

IN SITU AND EX SITU FABRICATION OF DNA CHIPS BY MICRODEPOSITION

M. CABRERA, M. JABER, J.P. CLOAREC, J.R. MARTIN, E. SOUTERYAND, V. DUGAS, J. BROUTIN, F. BESSUEILLE, M. BRAS, J.P. CHAUVET

Laboratoire d'Ingénirie et Fonctionnalisation des Surfaces "IFOS" – UMR 5621 du CNRS – Ecole Centrale de Lyon – 36 avenue Guy de Collongue – 69130 Ecully Cedex – Tél : (33) 04 72 18 62 56 - Fax : (33) 04 78 33 15 77 - email : michel.cabrera@ec-lyon.fr

Key words : DNA chips, Fabrication, Microdeposition   Speaker : Dr. Cabrera

Introduction

DNA chips are devices associating the specific recognition properties of two DNA single strands through hybridisation process with the performance of microtechnology. It is expected that the use of microtechnology allows to obtain reproducible massively parallel analysis at a low cost [1].

DNA chips are constituted of solid supports on which well-defined areas are designed. Each area contains a type of known nucleic acid probes, for example oligonucleotides. When DNA chips are put in contact with the solution containing different unknown targets, hybridisation occurs only on the areas where the complementary strands are present. The detection of these areas on the solid support allows to identify the targets. Currently the detection of hybridisation events on the support is performed by using fluorescent or radioactive markers previously fixed on the targets. Other methods are being developed. For example, in our laboratory, we are working on the direct detection of hybridisation by measuring the surface potential of a Si/SiO_2 structure excited locally by a laser beam[2]. This method free of labelling, which description is out of the scope of this article, led us to study the functionnalization of DNA chips substrate and finally to suggest improvements for the overall manufacturing process. We shall discuss below this point specifically.

Improvement of the chemical stability of DNA chips

Insufficient attention has been paid to the quality and reproducibility of DNA chips. For example, chips are often used once. It seems that they do not stand without damage the chemical and thermal treatments needed to perform several hybridisation / denaturation cycles[3]. It is expected that any improvement in the chemical stability of DNA chips leads to a widening of applications (see figure 1).

Therefore, our laboratory IFOS, together with our partners, the CNRS laboratories LMOPS and LCOO, developed a new silanization process of glass and Si/SiO_2 substrates for the preparation of DNA chips. This process is based on the preparation of a self assembled silane monolayer which is functionnalized either with a carboxylic acid for the immobilisation of presynthetised DNA strands or with an alcohol for the DNA direct synthesis onto the surface.

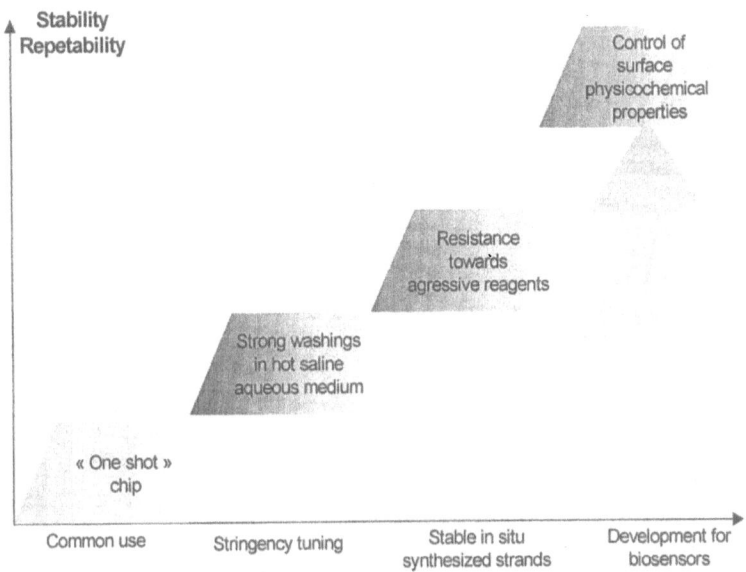

Figure 1 : Potential use of DNA chips / chemical stability

The main result is an improvement of the chemical stability of DNA chips, excellent sensitivity and high signal to noise ratio as illustrated in figure 2 which shows the reuse of a DNA chip during 25 hybridisation / denaturation cycles without loss of signal.

In situ manufacturing of DNA chips

The improvement of the surface chemistry of the substrates leads to very strong structures which can stand the chemical treatments needed not only to use, but also to manufacture DNA chips.

In situ manufacturing by microprojection of the 4 bases A, T, C, and G with independent piezoelectric devices has been proposed by different teams. Although several groups or companies like Agilent[4], Protogene[5], or Rosetta Impharmatics[6] in the USA use this process, no detailed work has been yet published. In theory this process is very flexible and uses only reactants available commercially, so it is surprising to note that it is not as extensively used as in situ synthesis by photochemistry[1].

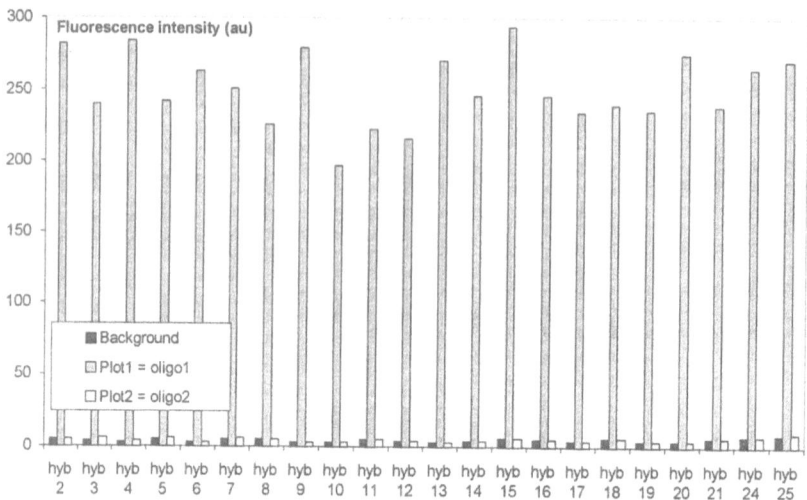

Figure 2 : Reuse of a DNA chip chemical stability after 25 cycles of hybridisation / denaturation

As our aim was to improve this process, we studied in situ manufacturing of DNA chips at different scales of microreaction, from 30 picolitre to 50 nanolitre drops, depending on the system used for the projection. Unlike the other groups, in order to obtain the best chemical yield of synthesis, we decided to use the phosphoramidite chemistry and to stay as close as possible to the process used in commercial oligonucleotide synthesisers, including acetonitrile as major solvent. In fact, oligonucleotides are currently built in porous glass column owing to a synthesiser. The idea was to combine such a synthesiser with a projection device with multiple heads and to replace the porous glass column with a glass or silicon substrate.

Starting with a solid support previously treated with an alcool to react, the first step to manufacture a DNA chip with this device consist to fix the 3'-OH from terminal nucleoside on the substrate. Then the fabrication process is a cycle with 4 repetitive steps :

- the deprotection step aims to remove protecting group to active nucleotide for next coupling ; classically, this deprotection is obtained under chemical treatment: for instance, DMT group largely used as protection can be taken off by TCA (trichloroacetic acid) treatment ;
- the coupling step concerns the addition of 3' activated phosphoramidite for reacting on 5' end fixed nucleotide ; as a function of cycle and the sequence to be built, nucleosides differ from one another in carried base and the introduction of one type of base is performed by a specific microprojection device ; the effective coupling is achieved by the microprojection of tetrazole coupling agent with another specific device ;

- the optional capping step is useful to prevent defect in the oligonucleotide building ; if some nucleosides have not reacted during coupling step, growth of these sequences must be stopped, and an acetylation process hinders them to react in next cycles, these sequences are thus aborted ;
- the oxydization step aims to stabilise the internucleotide phosphodiester bond by transforming unstable trivalent phosphor into stable pentavalent phophor.

At the end of manufacturing, a total deprotection step with NH4OH removes all protecting groups on heterocyclic bases and yields free the 5' and 3' OH terminals. Oligonucleotides probes on the chip surface are then ready for eventual hybridisation process.

Solvent, acids, oxidizers and bases are involved in this manufacturing process, most of them repetitively. Therefore we believe that the use of chemically stable substrates is a real improvement for the future use of this technology.

We developed a prototype of a machine which principle is shown in figure 3. It is made with a home made oligonucleotide synthesiser, 5 independent microprojection systems for the projection of A, C, T, G and tetrazole diluted in acetonitrile with a capacity to eject 50 nl drops, a XY positioning system, and a chemical reactor. We used a reactor to prevent the evaporation of the acetonitrile drops during the coupling step (acetonitrile evaporates in a few seconds in the air). For this, a flow of argon saturated with acetonitrile in a bubbling system is directed in the reactor during the coupling step so as to make sure that the reaction is effective. This gas flow is replaced by a flow of dry argon during the other steps. For the deprotection, capping, oxydization, and washing, it is not necessary to deposit the chemicals by microprojection, therefore we used an automatic 6 ways / 1 way valve to send the flow of reagents on the entire surface of the chip. A pump is part of the device to send all products to waste.

The algorithm used to manufacture a DNA chips in the machine is the following :

- send dry argon to the reactor ;
- send deprotection agent to the entire substrate ;
- wash the entire substrate with acetonitrile ;
- wash the entire substrate with anhydrous acetonitrile ;
- dry the substrate ;
- send argon saturated with acetonitrile to the reactor ;
- meanwhile, project the bases on the substrate together with tetrazole ;
- allow the coupling reaction for at least 1 minute ;
- send dry argon to substrate ;
- wash the entire substrate with acetonitrile ;
- send capping agent to the entire substrate ;
- wash the entire substrate with acetonitrile ;
- send oxidization agent to the entire substrate ;
- wash the entire substrate with acetonitrile ;
- return to the second step until achievement.

556

Figure 4 shows the first experimental set up which was used to prove the feasibility of this process. It was found that the use of acetonitrile is indeed a very strong requirement.

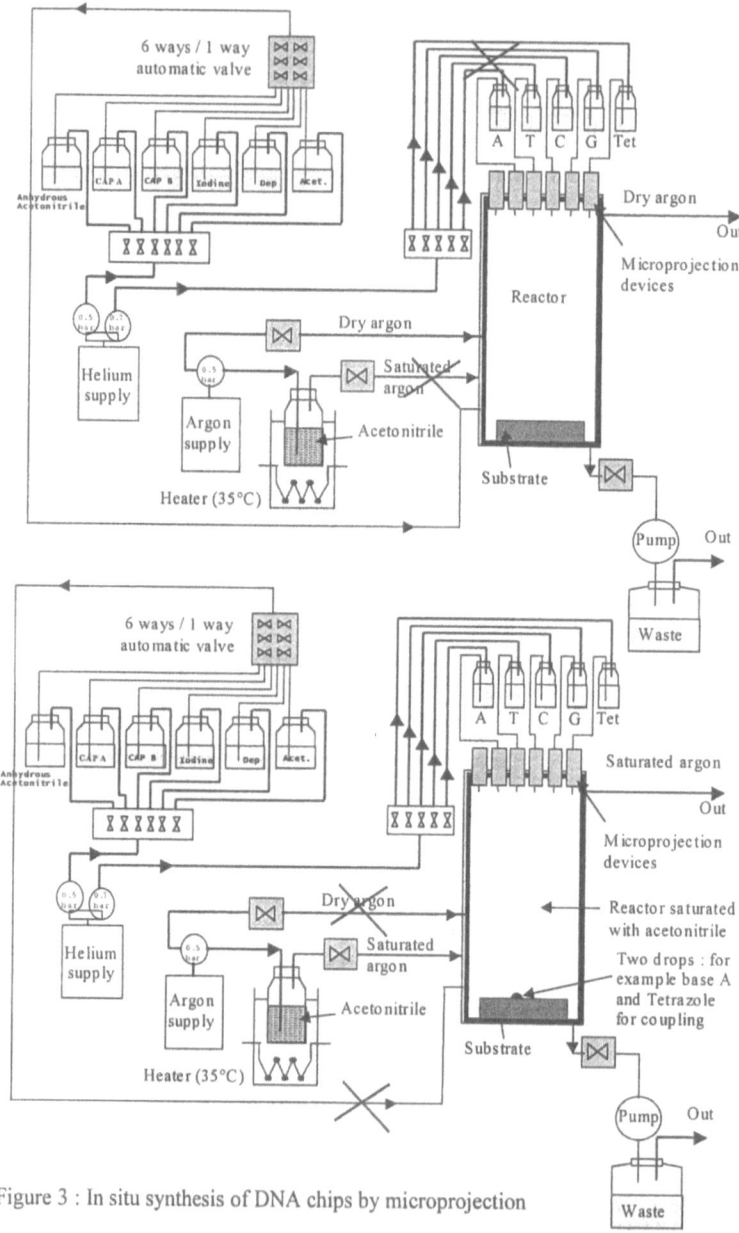

Figure 3 : In situ synthesis of DNA chips by microprojection

Figure 4 : The in situ synthetiser of DNA chips by microprojection

To avoid the corrosion of the machine by the reagents and specially by acetonitrile, we selected very carefully all components so as to use PTFE and stainless steel as much as possible. It was decided to put the XY positioning system outside the chemical reactor. It is necessary to keep the projection devices close to the substrate (1 mm) and to translate them, so as to project all the reactants on the chip. Therefore the chemical reactor is specially designed as a close chamber made in PTFE with a variable volume obtained with a bellows (see figure 4).

Another problem is that acetonitrile drops tend to spread onto the substrate so that it is impossible to get directly a good "printing" resolution. This problem was fixed by using a substrate with a polymer superstructure so that the spots of the DNA chip are physically delimited by 3D wells.

One of a first example of DNA chip manufactured with this device is shown in figure 5. It was used to study the concept from the chemical point of view. We are actually preparing a second generation prototype based on the same concept with a better resolution and a capacity to manufacture low and medium density DNA chips till 1000 wells.

558

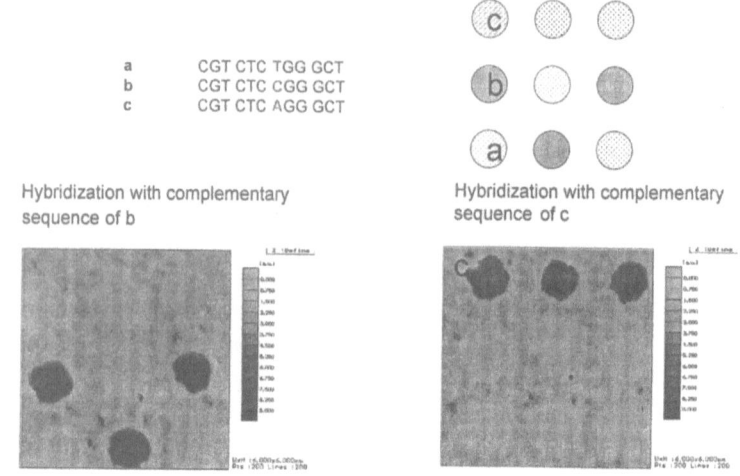

a   CGT CTC TGG GCT
b   CGT CTC CGG GCT
c   CGT CTC AGG GCT

Hybridization with complementary
sequence of b

Hybridization with complementary
sequence of c

Figure 5 : 9 dots DNA chip with 12 mer oligonucleotide probes manufactured by in situ microprojection ; hybridisation with a single strand fluorescent 84 mer ; 1 base mismatch sensitivity

Ex situ manufacturing for mass production

We shall also introduce a new process for the production of low density chips by microdeposition of presynthetized probes (ex situ manufacturing). The process is based on an optimised chemistry for the covalent bonding of the probes on the substrate and a machine made of 256 independent – free of contamination – microdeposition systems (5 nanoliters drops). The design of the reservoirs and the principle of the microdeposition allow a very stable and cost effective manufacturing process with no evaporation of solvent and no "touching" of the substrates, unlike spotters[7]. Figure 6 shows an example of low density DNA chip obtained by ex situ manufacturing.

Ex situ versus in situ manufacturing of DNA chips

Ex situ manufacturing is very cost effective for long run production and has a potential application in mass production of high quality DNA chips, for example for diagnostics. It makes possible to use highly purified probes. One drawback is its lack of versatility. For short run production, particularly for the R&D tests required to design DNA chips, we believe that in situ manufacturing by microprojection is more flexible and less expensive. If we compare this process specifically to spotting techniques, it has the potential to be less expensive at least by one order of magnitude. For low density chips, a good strategy is to combine these two techniques : in situ manufacturing to validate the chip designs and ex

situ manufacturing to perforn
of this strategy. Note that the
composition although chip co

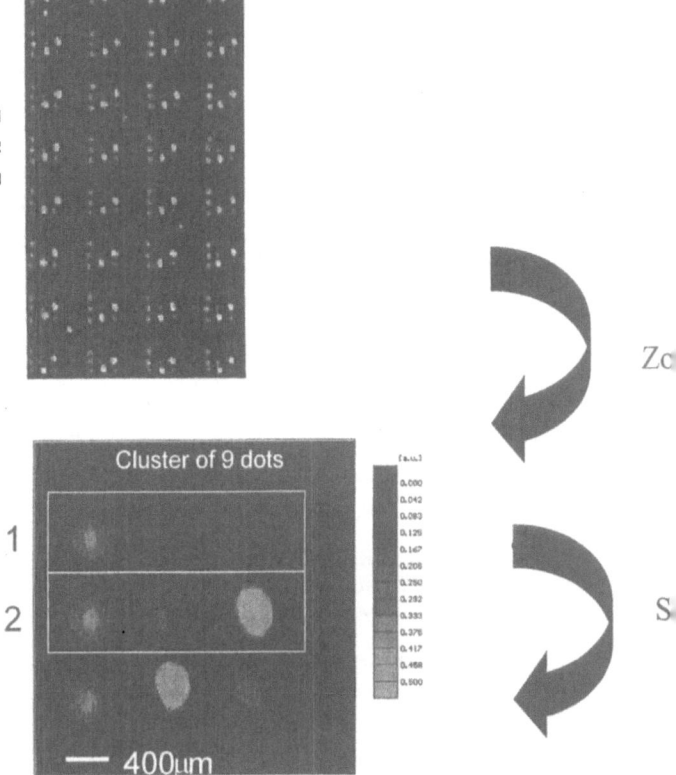

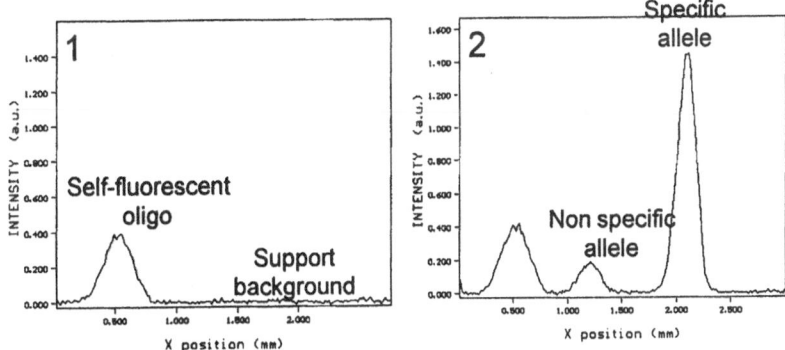

Figure 6 : Example of ex situ low density DNA chip manufactured by microdeposition for the
detection of SNP ; hybridization figure with a single strand 84 mer .

Conclusion

As a conclusion, DNA chips are considered as new tools allowing important advances in genomics. These tools are perfectible in terms of surface chemistry of chips for a better quantification, reproducibility, and robustness and in terms of process. We show two different manufacturing processes : in situ manufacturing by microprojection which is a low cost, flexible technique useful for R&D and a new process more effective for mass production.

Bibliography
[1] S.P. Fodor, J.L. Read, M.C. Pirrung et al., *Science*, **1991**, 251, 763-773
[2] E. Souteyrand, C. Chen, J.P. Cloarec, X. Nesme, P. Simonet, I. Navarro, J.R. Martin, *Applied Biotechnology and Biochemistry* , **2000**, 89, 195-207
[3] D.E. Gray et al., *Langmuir*, **1997**, 13, 2833-2842
[4] www.agilent.com
[5] A. Blanchard, R. Kaiser, L. Hood, *Biosensors and Bioelectronics* **1996**, 11, 687-690
[6] www.protogen.com
[7] P. Brown et al., *Science*, **1995**, 270 (5235), 467-470.

This work was financed by the french CNRS, "programme GENOME" (ROSA and ROSA2 projects) and "programme MICROSYSTEMES", as well as the ANVAR

Chemical and biological sensors based on modified electrodes with electropolymerized diamines

Boris LAKARD[a], Guillaume HERLEM[a], Bernard FAHYS[a],
Michel de la BACHELERIE[b], William DANIAU[b], Gilles MARTIN[b],
Web : http://electrochimie.univ-fcomte.fr
E-mail : guillaume.herlem@univ-fcomte.fr
Phone : (33) 3 81 66 62 94

[a] Laboratoire de Chimie des Matériaux et Interfaces – pôle Electrochimie
Université de Franche-Comté
UFR Sciences & Techniques
16 Route de Gray – La Bouloie
25 030 BESANCON Cedex, France

[b] Laboratoire de Physique et Métrologie des Oscillateurs
CNRS, UPR 3203
32 Avenue de l'Observatoire
25 044 BESANCON Cedex
France

Abstract

The electrochemical anodic oxidation of conductive liquid alkyldiamine-based electrolytes leads to the coating of the electrode surface by a thin polyalkyleneimine polymer film[1]. This later modifies durably the electrochemical behavior at the electrode surface and is suitable for miniaturized electrode applications. This film is of prime interest for both theoretical and technical areas such as new reference electrodes, sensors and biosensors, adhesion or heavy metal trace detection[2]. This is the reason why we have tested some electrodes modified with resultant polyalkyleneimine films as new pH sensors, as new NH_3 sensors and as based biosensors thanks to the enzyme trapped into the polymer matrix.

Introduction

The purpose of this work was to find the simplest new way for modifying an electrode (or a micro-electrode) surface with a high impedance polymer (containing or not an enzyme), reducing drastically the interferents from the signal response, and to design a chemical or biochemical sensor with it. We have shown[1] that the anodic oxidation of some pure alkylenediamines at platinum, gold, silicon or glassy carbon surfaces yields their electropolymerization into polyalkyleneimine films. Then we have chosen the polyalkyleimine type polymers as transductors firstly because this is a new way of synthesis that simplifies the electrode modification, secondly because polyalkyleneimine polymers are good candidates as transductors due to the fact they are strongly bonded to the electrode surfaces during the electropolymerization step and thirdly the enzymes trapped into the polymer matrix are known to be stabilized by polyalkyleneimine[3]. So these polymers can be used in miniaturized analytical sensors. These films can grow thick enough because of their non-conducting properties and they give high impedance to the modified electrode providing a better protection to interferents. Due to the presence of the amino groups onto the modified surfaces it was possible to imagine proton or ammonia sensitive sensors. This was successfully tested in aqueous and non aqueous buffered solutions leading to a linear relationship between the potential electrode and the pH. Linear polyethyleneimine (L-PEI) coatings being insoluble in cold water and in main organic solvents we proposed L-PEI modified electrodes as a new reliable pH sensor in aqueous and in some non aqueous electrolytes at a lower cost than a glassy standard pH electrode. Because L-PEI contains a high concentration of amino groups it is possible to use it as a bio-component for immobilizing enzymes. We easily modified this polymer by incorporating the glucose oxidase (GOX) into the polymer matrix by anodic oxidation of GOX dissolved in ethylenediamine (EDA). We obtained good sensitive amperometric responses as indicated by the linear response of current density versus glucose concentration between 0 and 7 mmol; this range includes the value 5-6 mmol, which is the maximum concentration that can be reached in blood patients with diabetes[4].

Methods

We have used our new way for synthesizing quickly linear (or crystalline) polyethyleneimine (L-PEI) from anodic oxidation of pure solvent ethylenediamine (EDA) monomer electrolytes. We will see further that a modified electrode with electropolymerized alkylenediamine at the electrode surface gives far a long-term electrochemical response compared to polyalkyleneimine synthesized by chemical route and spread on the electrode. It is possible to extend this electropolymerization to 1,3-diaminopropane (1,3-DAP) and diethylenetriamine (DETA) that yields polypropyleneimine (PPI) and polyethyleneimine respectively to modify, if needed, the chemical properties of the polymer. Several methods were employed to characterize these polymers such as electrochemical quartz crystal microbalance (EQCM) coupled to cyclic voltammetry, IR-ATR, Raman and XPS spectroscopies, *ab initio* calculations and SEM[2,5].

Results

1. Elaboration of pH sensors
♦ Electrodeposition of polyethyleneimine films on platinum electrodes:
The L-PEI films were electrodeposited onto a smooth platinum electrode by means of cyclic voltammetry in solutions composed of 10^{-2} M lithium trifluoromethanesulfonimide (LiTFSI) in ethylenediamine. Five cycles between 0 and 3 V vs. Ag^+/Ag (used as reference electrode) were performed. During the first scan the electrode was biased at 3 V for 30 minutes (Figure 1). While the L-PEI film growths at the electrode surface there is a current drop from 6 µA to 130 nA that we can link to the electrical insulating property of this polymer. Following four scans were performed to check the stability and the insulating property of the

coating. Then the modified platinum electrodes were rinsed by water and acetone, dried in an oven (at 40°C), and tested in several buffered solutions at different pH values.

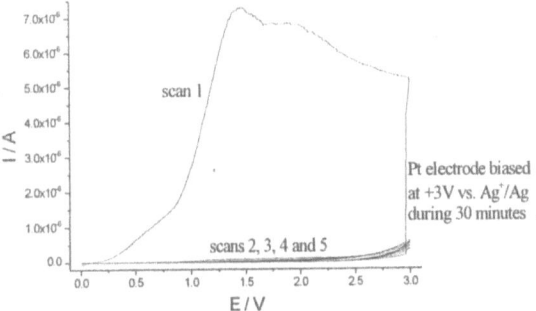

Figure 1: cyclic voltammogram of 10^{-2} M LiTFSI in EDA at a smooth platinum electrode surface, scan rate : 50 mV.s^{-1}.

♦ Potentiometric response to pH changes of a platinum electrode coated with a polyethylenimine film:

▪ Effect of the polyethyleneimine coating on the potentiometric response of a platinum electrode:

We have examined the potentiometric response of the modified platinum electrodes as a function of the changes in pH. In all cases the electrodes were immersed into different buffered solutions and the potential equilibrium response time was less than 15 seconds. The potentiometric response to changing [H$_3$O$^+$] of the solution is linear in the range [2; 10] pH units (Figure 2), because of the affinity of the coating amino groups to the protons in solutions. Table 1 and Figure 2 show the potentiometric responses of the same modified electrode 1 day (noted D+1), 15 days (D+15) and 30 days (D+30) after the deposition of the PEI film onto the platinum electrodes. The slope of the linear potentiometric response varies, between the 1st day and the 30th day, in the range [-32;-37] mV per pH unit. Notice that between (D+1) and D+15 or (D+30) the passivated electrodes were stored at the air (+20°C). It has been found that the potential range [-32; -37 mV per pH unit] can be reduced tighter when the coated electrodes are stored in pure water (at +20°C) between (D+1), (D+15) or (D+30) measurements. So the coating is quite stable and does not evolve or decompose with time. Moreover the increasing of the slope with time is interesting because it means that the sensitivity of the sensor becomes more and more important. Thus a good linearity and a good stability in time of the potentiometric response was observed and consequently our electrodes seem to be good pH sensors.

PH	E (mV) for a platinum electrode coated with PEI (D+1)	E (mV) for a platinum electrode coated with PEI (D+15)	E (mV) for a platinum electrode coated with PEI (D+30)
2.03	862	933	965
2.97	845	914	925
4.02	827	892	913
5.03	775	852	854
6.99	703	770	777
8	681	735	741
9.23	654	699	702
10	614	664	671

Table 1 : evolution of the potentiometric response (mV/SRE, SRE : silver reference electrode) as a function of the changes in pH for the same platinum electrode coated with PEI (see text). Influence of aging time. Three different aging times (in days) were checked, (D+1), (D+15), and (D+30).

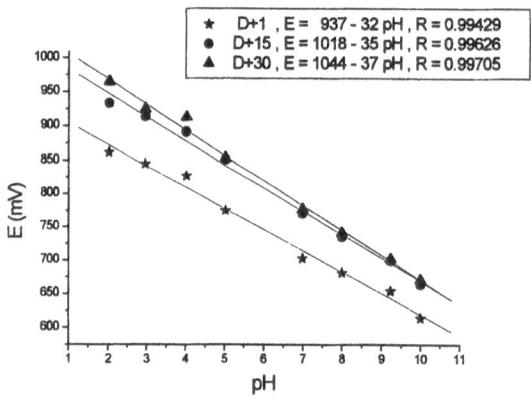

Figure 2: potentiometric response versus the pH of an electrode modified during 30 minutes (see text). Influence of aging: 1, 15 or 30 days. R is the linear correlation factor

- Influence of the electrodeposition time or influence of the thickness:

L-PEI was deposited on the first electrode called E_0 by means of five cycles between 0 and 3V vs. SRE. More, as the electropolymerization of ethylenediamine occurs for potentials superior to 2 V vs. Ag^+/Ag (Figure 1) several platinum electrodes were biased for different duration (5, 30 and 60 minutes) at +3 V vs. Ag^+/Ag during the first scan of the five cycles) in order to determine the bias time for which the potentiometric response versus the pH values in solution is the most important and the most reliable. These electrodes were called respectively E_5, E_{30} and E_{60} (Table 2). The potentiometric responses of the different biased electrodes to different pH values were gathered Figure 3. All the platinum modified electrodes show a linear behavior in the pH range 3-10. The more the polymer thickness is high the more the modified electrode is sensitive. Indeed the slope increases from -17 mV/pH for the thinnest electrode to −33 and -38 mV/pH for the two thickest ones. So the two thickest electrodes seem to be the two most efficient sensors because they show the best sensitivities to changes in pH.

PH	E_0 (mV)	E_5 (mV)	E_{30} (mV)	E_{60} (mV)
2.03	686	709	862	903
2.97	664	704	845	887
4.02	658	702	827	862
5.03	626	674	775	815
6.99	592	635	703	738
8	575	618	681	709
9.23	560	594	654	669
10	551	562	614	632
10.97	555	579	634	654

Table 2: evolution of the potentiometric response (vs. SRE) as a function of pH changes for different PEI thicknesses from different bias times.

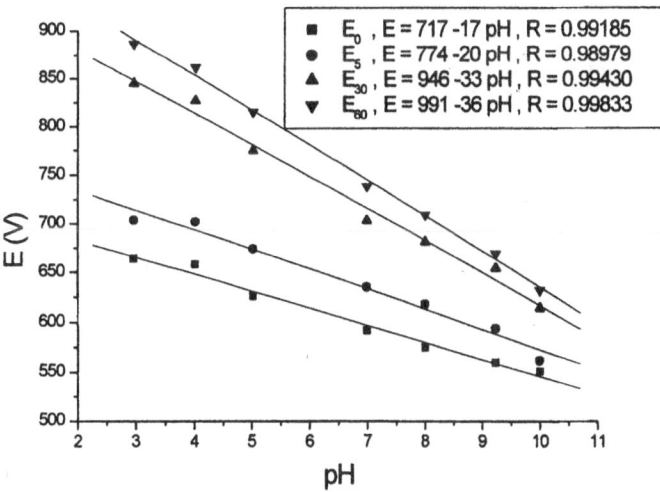

Figure 3: Influence of the PEI thickness: potentiometric response versus the pH of electrodes modified during different duration.

♦ Elaboration of pH transductors : deposition of platinum by lift-off process :
As we have tested successfully platinum electrodes modified with L-PEI, we can now try to make miniaturized pH transductors using the same chemical principles. The aim is to control the microsensors technology by making these pH transductors so as to make later biosensors allowing to detect biocomponents of medical interest.

■ Photolithography :
This method allowed us to draw geometrical patterns in a thin film of photoresist deposited onto a silicon oxide wafer surface. The first step consists in drawing the required pattern with the software Cadence. Then a mask, on which the negative of the pattern has been drawn, is made with a mask generator of Optical Pattern Generator Electromask type. Next, when the deposition of the resist has been done on a 4'' oxidized silicon wafer by centrifugation, we can insulate the pattern under UV radiation through the previous mask (Figure 4). Thus the pattern has been transferred on the resist and it must be developed so as to dissolve the resist where the metal deposition will appear.

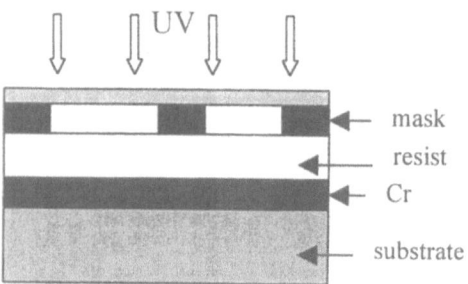

Figure 4 : principles of photolithography by UV lamp.

■ Deposition of the platinum by ion etching :
We used a plasma etching method, called ion etching, to deposit platinum onto our wafer. This method was chosen because it gives a better adhesion to the layers deposited than the methods of evaporation. We have deposited a layer of 150 nm of platinum above a layer of 30 nm of titanium. This later allows a good adhesion of the platinum on the silicon wafer. Thus the thin film of platinum is now deposited onto the wafer and we just have to eliminate the resist by dissolution in acetone and water to obtain a platinum pattern drawn on our wafer.

■ Description of the pattern:
We decided to make interdigitated microarray electrodes so as to increase the surface of contact between the electrode coated with the polymer and the solution to analyze (Figure 6).

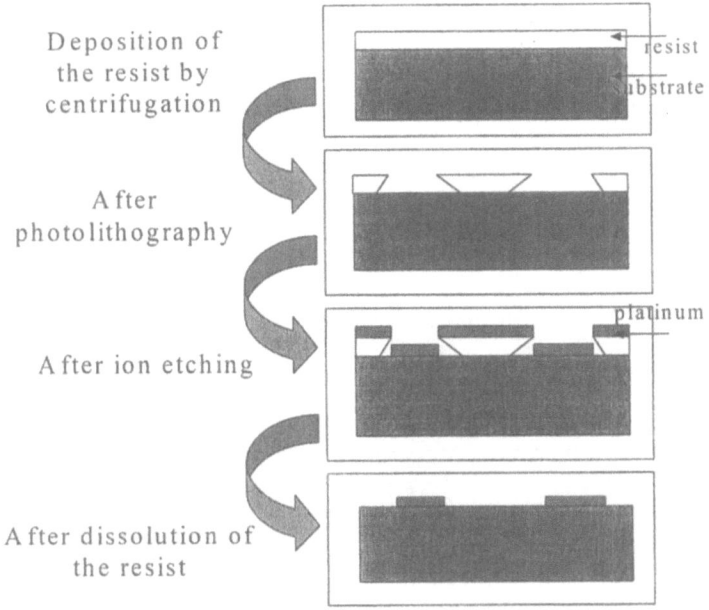

Figure 5 : Successive steps of the lift-off process.

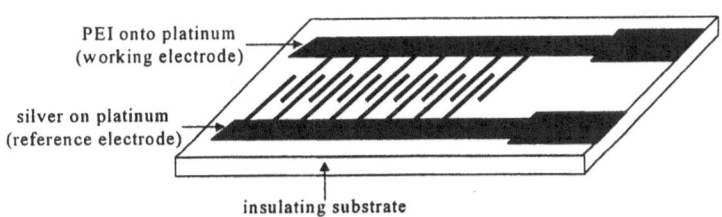

Figure 6: Schematic drawing of interdigitated microarray electrode

- Electrodes of the chemical transductor:

We have described the lift-off process which allows us to deposit a thin film of platinum on an isolating substrate so as to have two platinum electrodes. Now we have first to electrodeposit silver on one of the two platinum electrodes to obtain the reference electrode and second to electrodeposit polyethyleneimine on the second platinum electrode to obtain the working electrode. The silver is deposited thanks to the electrolysis of a solution composed of KCN, K_2CO_3 and $KAg(CN)_2$[6]. The polyethyleneimine is deposited as described before.

◆ Potentiometric response of the chemical transductors to pH changes:
■ Effect of the polyethyleneimine coating on the potentiometric response of the transductor:

It appears that the potentiometric response of the transductor with PEI is linear but that the potentiometric response of the transductor without PEI gives a random potentiometric response (Table 3 and Figure 7).

pH	E (mV) of the bare platinum electrode	E (mV) of the L-PEI coated Pt electrode
2.03	1026	1031
2.97	892	989
4.02	689	897
5.03	707	835
6.99	218	828
8	327	761
9.23	460	679
10	701	738
10.97	765	716

Table 3: evolution of the potentiometric response as a function of the pH changes for a transductor with PEI deposited and for a transductor without PEI deposited onto one of the platinum electrode.

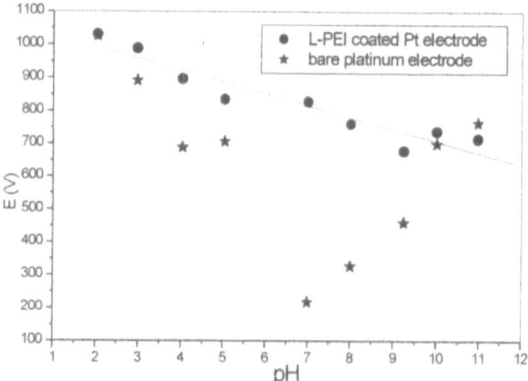

Figure 7: potentiometric response of transductors as a function of the pH, with and without PEI deposited onto one of the two platinum electrodes

2. Elaboration of a NH₃ sensor:

As the detection of urea is of medical interest, we tried to make chemical sensors capable of determine the changes in NH_3 concentration since NH_3 is one of the products of the decomposition of urea (reaction 1) in the presence of an enzyme called urease which catalyses the reaction (1).

We have tested the potentiometric response of platinum electrodes coated with polyethylenimine (the electrochemical way for modifying the electrode was the

$$\text{urea} \xrightarrow{\quad \text{urease} \quad} 2\,NH_3 + CO \quad \text{(reaction 1)}$$

same as above-mentioned) as a function of the ammonia concentration changes. The potentiometric response was not linear as in the case of the pH but it was an exponential decay (E (mV) = -57 +159.exp (- [NH₃] / 0.52) +36 exp.(- [NH₃] / 9.1), the exponential correlation factor R = 0.98683). So it is possible to make a NH_3 transductor similar to the pH sensor above-mentioned.

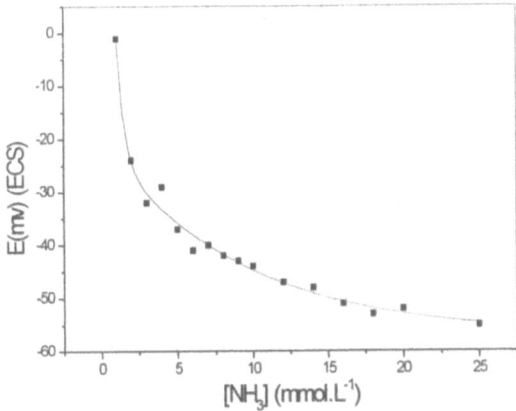

Figure 8: potentiometric response to [NH₃] changes of the platinum electrode.

3. Elaboration of biological sensors

♦ Elaboration of a gluconic acid sensor:

As we want to detect the presence of glucose from the reaction 2, we have tested the potentiometric response of a platinum electrode coated with polyethylenimine as a function of the D-gluconic acid concentration changes.

The day after the deposition of L-PEI on the electrode (D+1), the potentiometric response was not linear in the whole range [1;25] mmol of gluconic acid (Figure 8) but was linear in the range [3;9] mmol of gluconic acid (Figure 9). Then, we

$$\text{D-glucose} + O_2 \xrightarrow{\quad \text{GOX} \quad} \text{D-gluconic acid} + H_2O_2$$

immersed the modified electrode in water and we have tested it after 10 days : the potentiometric response was still linear So it is possible to make a gluconic acid transductor similar to the pH sensor above-mentioned.

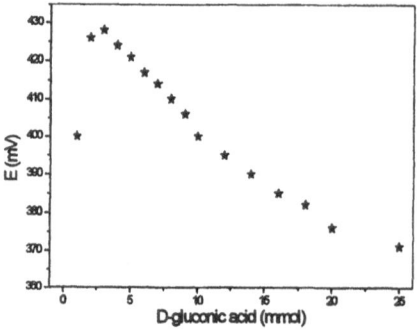
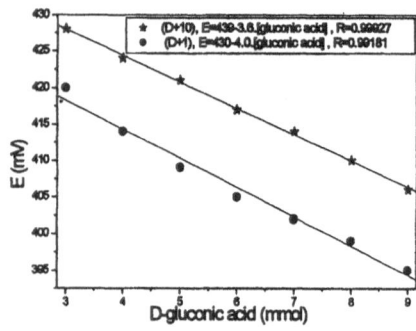

Figure 8: potentiometric response of the modified platinum electrode in the range [1; 25] mmol to [gluconic acid] changes.

Figure 9: potentiometric response of the modified platinum electrode in the range [3; 9] mmol to [gluconic acid] changes.

♦ Elaboration of a glucose sensor:

We modified platinum electrodes by a thin film composed of polyethylenimine and GOX trapped into the polymer by anodic oxidation of GOX (2.5 mg/mL) dissolved in EDA containing LiTFSI 10^{-2} M. We obtained good sensitive amperometric responses as indicated by the linear response of current density versus D-glucose concentration between 0 and 7 mmol of glucose (Figure 11). This concentration range includes the value 5-6 mmol, which is the maximum concentration that can be reached in blood patients with diabetes. In the range [0; 25] mmol, the amperometric response to [glucose]· variation stops its linear behavior and has an exponential decay (Figure 10). So we have made glucose transductors on the same process as the pH transductors above-mentioned.

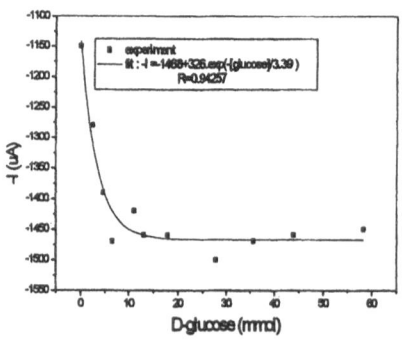
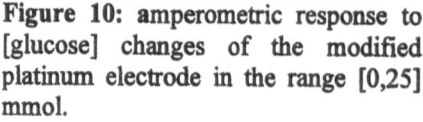
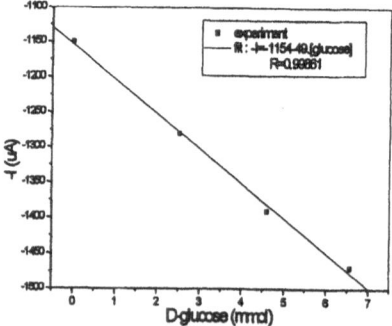

Figure 10: amperometric response to [glucose] changes of the modified platinum electrode in the range [0,25] mmol.

Figure 11: amperometric response to [glucose] changes of the modified platinum electrode in the range [0,7] mmol.

Conclusion

We compiled here our results obtained from the amperometric and potentiometric measurements made with electropolymerized alkylenediamine modified electrodes. So far, in the literature, these systems were studied using electrodes spread with commercially available branched or linear PEI and the coating was limited in time because of its solubility in common organic solvents and poor adhesion[7,8,9,10].So we validated new pH transductors in acidic and basic media such as NH_3 and gluconic acid, and a D-glucose biosensor suitable for a miniaturized analytical device because of the good adhesion of the polymer onto the electrode surface.

References

[1] G. Herlem, C. Goux, B. Fahys, F. Dominati, A.-M. Gonçalves, C. Mathieu, E. Sutter, A. Trokourey and J.-F. Penneau, *J. Electroanal. Chem.*, **435**, pp. 259-265 (1997).
[2] G. Herlem, B. Lakard and B. Fahys, in Recent Research Developments in Electroanalytical Chemistry, Vol.2 (2001), Transworld Research Network Ed.
[3] M. M. Andersson, R. Hatti-Kaul, *J. of Biotechnology*, **72** (1999) 21-31.
[4] G. Herlem, French Patent pending number 00.04690 (04-12-2000).
[5] G. Herlem, K. Reybier, A. Trokourey and B. Fahys, *J. Electrochem. Soc.*, Vol. 147, N°2, p. 597 (Feb. 2000).
[6] D.Pletcher and F.C.Walsh, in Industrial Electrochemistry, 2nd.ed., Chapman and Hall, New York, 1990.
[7] X.-J. Tang, B. Xie, P.-O. Larsson, B. Danielsson, M. Khayyami and G. Johansson, *Anal. Chimi. Acta*, **374** (1998) 185-190.
[8] C. L. Chuang, Y. J. Wang and H. L. Lan, *Anal. Chim. Acta*, **353** (1997) 37-44.
[9] D. Mandler,A. Kaminsky and I. Willner, *Electrochim. Acta*, 37,2765 (1992).
[10] A. Kaminsky, I. Willner and D. Mandler, *J. Electrochem. Soc.*, Vol. 40, N° 3, L215-27 (March 1993) and references therein.

A FLUIDIC MEMS FOR PROTEIN IDENTIFICATION BY ELECTROSPRAY IONIZATION/MASS SPECTROMETRY

Christian Druon(1), Xavier Mélique(1), Christian Rolando(2) and Pierre Tabourier(1),

(1) IEMN UMR 8520, Cité Scientifique, BP.69, Av. Poincaré
59652 Villeneuve d'Ascq, France
Tel :+33 (0)3 20 19 79 62 , Fax :+33 (0)3 20 19 78 80
Email : Christian.Druon@iemn.univ-lille1.fr
(2) UPRESA CNRS 8009, UFR de Chimie Organique et Macromoléculaire,
Université des Sciences et Technologies de Lille

Abstract

We describe here our proposal for a microfluidic lab-on-chip devoted to protein identification by ESI/MS/MS. As a starting point we have realized silicon etched microchannels (10 µm depth) connected to a standard silicon needle. This elementary prototype ready for the sample mass spectrometer (MS) screening can be viewed as a first step before implementing basic functions as fluid actuation, Solid Phase Extraction (SPE) column, nanoemitter, necessary to cope with the nanospray problems.

Introduction

A proteome has been defined as the protein complement expressed by the genome of an organism or in multicellular organisms as the protein complement expressed by a tissue or differentiated cell. A schematic diagram of a typical procedure for identification of proteins is shown in Fig.1

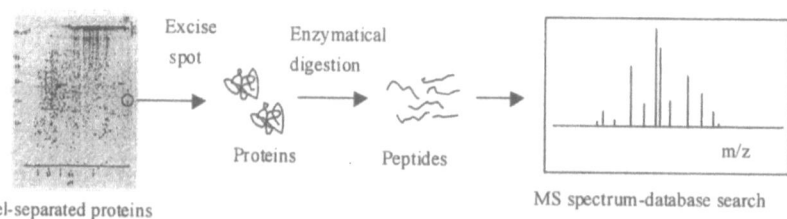

Fig.1 Schematic diagram of a typical procedure for identification of proteins

For the routine analysis of proteins at high sensitivity, the most significant challenge is the handling of minute amounts of sample [1]. Therefore the device miniaturization already brings many advantages while its on-chip total integration is expected to still greatly improve the experimental conditions [2]. We briefly recall some recent results of the literature before describing the basic elements of our proposal.

Microdevices for ESI/MS

Electrospray ionization (ESI) is widely used for peptide ionization [3]. In conventional devices, the emitter consists of a tapered capillary of fused silica. In the best reported case [4] the tip of the needle has an aperture of 0.4µm and the flow is reduced to a few nanoliters per minute. Besides glued or connected silica tips [5], different prototypes of emitters have been proposed in view of their on-chip integration. In a first MS coupling trial, the sample infusion was achieved from the planar edge of the chip right at the microchannel end [6]. Then overhanging silicon nitride or parylene [7-8] needles have been fabricated. Recently, pillar nozzles perpendicular to the silicon wafer with 15 µm of inner diameter have been etched by RIE-ICP and successfully tested [9]. Moreover integrated key functions are desirable: fluid actuation and switching, mixing, sample injection, separation and preconcentration of analytes via SPE microcolumns, filtration…

So far electro-switching and electroosmotic pumping have been extensively used [10] although they necessitate high voltages and salts addition. Sample filtration is thus necessary prior to MS delivery.

Recently however valveless flow rectifiers actuated by a piezoelectric element have been proposed as microfluidic actuators [11]. Interesting bubbles generating systems have also been published [12]. Moreover fluidic manifolds with different types of valves or hydrophobic events [13] are also efficient. Lately, the dielectrophoretic actuation of nanoliter to microliter water volumes has been demonstrated [14]. All these new elements could contribute to set up an attractive solution for the actuation of neutral/charged liquids in an integrated microfluidic device.

But despite outstanding results, to our knowledge no true µTAS for MS is available in the required standard: reliable, cheap, disposable and designed for high throughput screening.

A fluidic MEMS for ESI/MS

On-chip demonstrator proposal

Fig.2 gives a schematic representation of the on-chip demonstrator including three basic elements: novel thermal micropumps, a SPE column and a nanoemitter [15].

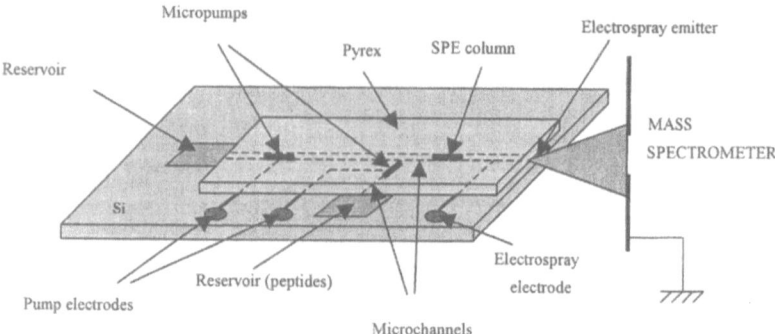

Fig. 2 Schematic representation of the demonstrator

The pump sketched Fig.3 has no moving parts. Three sharp rises corresponding to three narrowed cross sections of the channel, are fitted with a resistive implanted track. The central track is the thermal actuator while the two external are valves like steady parts with a rectifying function. These effects are obtained by controlling the local temperature with an appropriate monitoring of the applied voltages. A first assess assuming the following conditions: microchannel ($10\mu m$ depth, $20\mu m$ width) with a $100\mu m$ length electrode, 10% rectification efficiency; gives a net flow rate of 1.5nl/min and an electrical consumption of some μW at a working frequency of 50Hz.

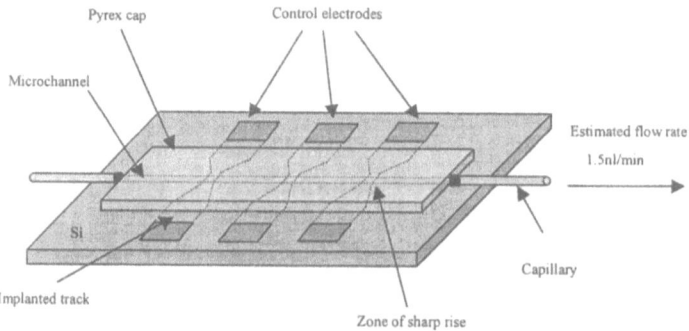

Fig.3 Sketch of the micropump: the three sharp rises in the microchannel are thermally monitored to generate a rectified flow.

The SPE column corresponds to a functionalized zone of the microchannel. It can be fabricated by using a polymer or a coated structure designed from beads or etched pillars. In a first step we plan to fill the microchannel with a monomer and flash it with U.V. through a mask in order to obtain a polymerized column. The nature of the polymer or the nature of the grafted biomaterial determines the required functionality of this column (filtering, concentrating, enzymatic digestion).

The nanoemitter can be designed either as an overhanging hollow cantilever lying in extension of the microchannel (see Fig.4a) or as a pillar nozzle [9] perpendicular to the wafer (see Fig.4b). This later structure can be etched by RIE-ICP.

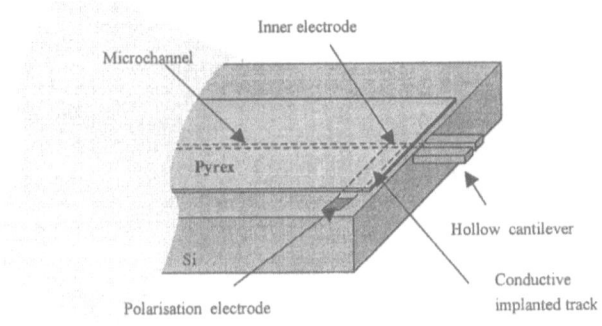

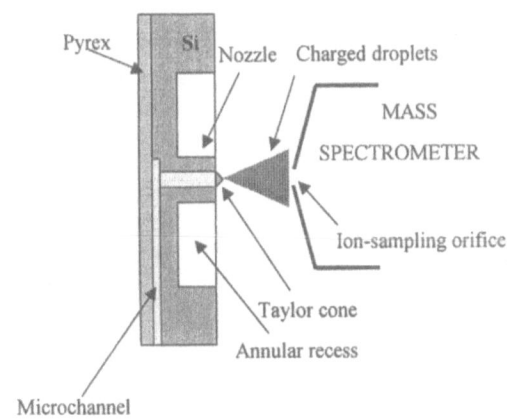

Fig.4 -a (top): schematic representation of a cantilever in-plan nanoemitter
-b (bottom): cross-sectional view of the pillar nozzle ($\varnothing_{IN}$ = 10µm,
$\varnothing_{OUT}$ = 20µm, length = 50 µm, annular recess $\varnothing$ = 1mm)

First microchannel realization

A basic operation for designing the silicon demonstrator is the etching of connected channels of different sizes. These channels constitute the manifold for the fluid circulation but also the sheath of the SPE column.

We have thus realized microchannels in a 2'' silicon (100) wafer using conventional technological methods to etch V grooves which are tightly closed by the anodic bonding of a pyrex lid [16]. Fig. 5 shows the fabricated chip including nine identical patterns with a central microchannel (10 μm depth and 10 mm length) and housing recesses at both ends. An electron beam top view of a recess – microchannel junction is shown in Fig. 6.

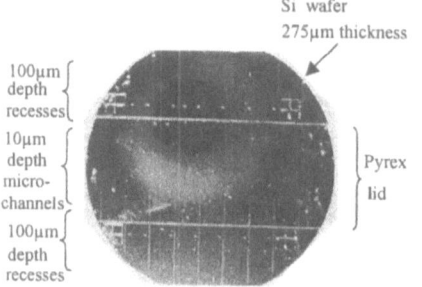

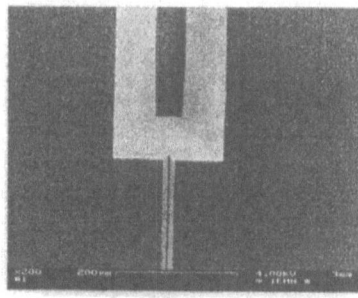

Fig.5. Photo of the first chip fabricated on a 2'' Si wafer. A pyrex lid is sealed to the Si wafer above the 10μm depth channels. On both sides recesses of 100μm depth are etched for the capillary connections.

Fig.6. Electron beam top view of a recess-microchannel junction (micro-channel depth: 10μm, recess depth: 100μm).

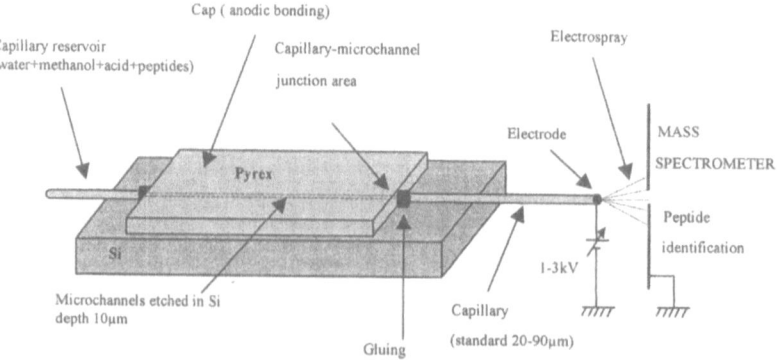

Fig.7. Representation of the first realization of a microchannel implementation

A commercial capillary is housed in each of the two recesses at the microchannel ends and glued on the chip. Fig.7 gives a representation of the first realization.

Photo 8 shows the wafer connected to the inlet capillary and photo 9 shows the connection to an electrospray needle using a T connector. The electrode contacting the fluid consists of a Platinum wire plugged in one branch of the T connector.

Fig. 8 Photo of the fluidic micro-system with inlet and outlet capillaries

Fig. 9 Photo of the fluidic micro-system connected to an electrospray needle

Prior to MS coupling, the prototype is tested on a bench shown in Fig. 10 in view to check the spray efficiency from the ionization current recording.

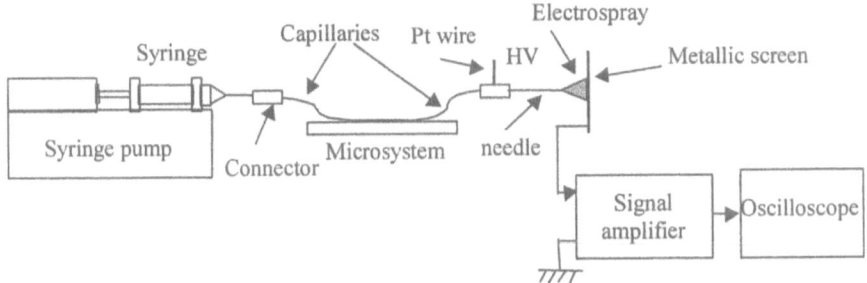

Fig. 10 Setup of the system used to test the first prototype

Next, the channel walls will be specifically coated to avoid the adsorption of the targeted peptides. The so called biocompatibility (absence of peptide adsorption) will be studied by the MS screening of the channel flushing solvent used for the peptide elution.

References

[1] P.A.Haynes, S.P.Gygi, D.Figeys, R.Aebersold, *Electrophoresis*, 19, pp.1862-1871, 1998

[2] R.D.Oleschuk, D.J.Harrison, *trends in Anal. Chem.*, vol.19, n°6, pp.379-388, 2000

[3] M.S. Wilm and M. Mann, *Intl. J. of Mass Spectr.&Ion Proc.*, vol.136, pp.167-180, 1994

[4] S. Geromanos, G. Freckleton and P. Tempst, *Anal. Chem.*, vol. 72, n°4, pp.777-790, 2000

[5] M. Lazar, R.S. Ramsey, S. Sundberg and J.M. Ramsey, *Anal. Chem.*, vol. 71, n°17, pp.3627-3631, 1999

[6] R.S. Ramsey and J.M. Ramsey, *Anal. Chem.*, vol. 69, n°6, pp.1174-1178, 1997

[7] A. Desai, Y.C. Tai, M.T. Davis and T.D. Lee, *Transducers'97, Proc. 9th intern. Conf. On Solid State sensors and Actuators*, pp.927-930, 1997

[8] L. Licklider, X.Q. Wang, A. Desai, Y.C. Tai and T.D. Lee, *Anal. Chem.*, vol. 72, n°2, pp.367-375, 2000

[9] G.A.Schultz, T.N.Corso, S.J.Prosser and S.Zhang, *Anal. Chem.*, vol. 72, n°17, pp.4058-4063, 2000

[10] Q. Xue, F. Foret, Y.M. Dunayevskiy, P.M. Zavracky, N.E. McGruer and B.L. Karger, *Anal. Chem.*, vol. 69, n°3, pp.426-430, 1997

[11] S. Matsumoto, A. Klein and R. Maeda, *IEEE MEMS'99*, 1999

[12] F.G. Tseng, C.J. Jin, C.I. Kim and C.M. Ho, *IEEE MEMS'98*, pp.57-62, 1998

[13] K.Hosokawa, T.Fujii and I.Endo, *Anal. Chem.*,71, n°20, pp.4781-4785, 1999

[14] M.Gunji, T.B.Jones, M.Washizu, *IEEE MEMS'01*,pp.385-388, 2001

[15] Projet Réseau Micro et Nano Technologies (RMNT) labeled on June 16th 2000 and financially supported by the Ministère de l'Education Nationale de la Recherche et de la Technologie.

[16] P.Tabourier, C.Druon, C.Rolando, P.Lefebvre, *1st Annual International IEEE-EMBS Special Topic Conference on Microtechnologies in Medecine & Biology*, Lyon, 12-14 Octobre 2000, Proceedings pp. 439-444, 2000